Bio-management of Postharvest Diseases and Mycotoxigenic Fungi

World Food Preservation Center Book Series

Series Editor

Charles L. Wilson

Postharvest Extension and Capacity Building for the Developing World
Majeed Mohammed and Vijay Yadav Tokala

Animal Sourced Foods for Developing Economies
Preservation, Nutrition, and Safety
Edited by Muhammad Issa Khan and Aysha Sameen

Bio-management of Postharvest Diseases and Mycotoxigenic Fungi
Edited by Neeta Sharma and Avantina S. Bhandari

For more information about this series please visit: http://worldfoodpreservationcenter.com/crc-press.html

Bio-management of Postharvest Diseases and Mycotoxigenic Fungi

Edited by

Neeta Sharma and Avantina S. Bhandari

CRC Press is an imprint of the
Taylor & Francis Group, an **informa** business

First edition published 2021
by CRC Press
6000 Broken Sound Parkway NW, Suite 300, Boca Raton, FL 33487-2742

and by CRC Press
2 Park Square, Milton Park, Abingdon, Oxon, OX14 4RN

CRC Press is an imprint of Taylor & Francis Group, LLC

Library of Congress Cataloging-in-Publication Data

Names: Sharma, Neeta, Dr. editor. | Bhandari, Avantina S., editor.
Title: Bio-management of postharvest diseases and mycotoxigenic fungi / Neeta Sharma, Avantina S. Bhandari.
Other titles: World Preservation Center book series.
Description: First edition. | Boca Raton, FL : CRC Press, 2021. | Series: World Preservation Center book series | Includes bibliographical references and index.
Identifiers: LCCN 2020020521 (print) | LCCN 2020020522 (ebook) | ISBN 9781498797313 (hardback) | ISBN 9781003089223 (ebook)
Subjects: LCSH: Food crops--Diseases and pests--Biological control. | Mycotoxins. | Toxigenic fungi. | Biological decontamination.
Classification: LCC SB608.F62 B563 2021 (print) | LCC SB608.F62 (ebook) | DDC 632/.9--dc23
LC record available at https://lccn.loc.gov/2020020521
LC ebook record available at https://lccn.loc.gov/2020020522

ISBN: [978-1-4987-9731-3] (hbk)
ISBN: [978-1-003-08922-3] (ebk)

Typeset in Times
by Deanta Global Publishing Services, Chennai, India

Dedication

To

ISHITA,

our lifeline

Contents

Preface

The origins of civilization can be traced back to humanity's discovery and assurance of an accessible, available and affordable continuous food supply. By the time man finally came out of the trees and placed his feet firmly on terra firma, he must have sought out food in the wild as a significant part of his hunting and gathering diet.

Over the past five decades, the combined effects of the green revolution and improved farming techniques and technologies have allowed the world's food production to increase many times. To ensure an abundant supply of sufficient food for every inhabitant of Earth, both in quality and in quantity, native ecosystems are rapidly being converted for human use by destroying forests, soils and native plants and animals. Although there are claims that food production has increased, they are accompanied by counterclaims that emphasize the depletion of our natural resources. Such a projection presents mankind with wide-ranging social, economic, environmental and political issues that need to be addressed today in order to ensure a sustainable future tomorrow. However, there is constant pressure to ensure sufficient food and feed is supplied for living beings.

One key issue is the production of sufficient food for everyone in this world of finite resources. Although we are in the twenty-first century, the astonishing advances in agricultural productivity and human ingenuity have not been translated into a world free of hunger and malnutrition. Loss of food and feed arguably poses the greatest vulnerability to food security worldwide. Approximately half of the population in the developing world do not have access to adequate food supplies.

The global population over the years callously used up the world's resources; substituting a blatant use of chemicals to sustain the system and provide for and feed the increasing population. This has resulted in dark gray clouds looming large and threatening future generations with climate change, new strains of human and plant pests, food losses, food scarcity and chemical-laden soils. As we approach the upper limits of the Earth's human-carrying capacity, it is questionable whether advances in crop production can address these problems in a timely manner and keep up with food demand and supply.

It is projected that in the current scenario of losses and constraints due to more challenging climatic conditions such as degradation of soil fertility, dwindling availability of water and arable land, limited low cost energy supplies, production short of demand, greater geographical inequity in production and demand, and above all, food losses due to various pathogenic diseases, the world will struggle to produce more food in the face of increasing population. Subsequent speculation in food markets could leave us in an even worse condition than seen in the current crisis if appropriate options for increasing food supply and security are not considered and implemented. The main reasons why the food supply is tightening are the population explosion, accelerated urbanization and motorization, changes in dietary habits, extreme and unpredictable environmental changes and to top the list, the presence of potentially life-threatening pathogens in our environment and the toxins produced by them that contaminate our food.

Although considerable attention has been drawn to the enormity of food spoilage and waste due to microbial deterioration and mycotoxin contamination, a limited number of resources have been directed toward its solution. Promoting food security through loss reduction is the most feasible and sustainable method of increasing food production.

Ever since humans started the practice of crop cultivation to meet food requirements, the problem of food safety has always troubled mankind. Humanity has been constantly searching for effective strategies to protect crops from deterioration and to safely store them, as well as ensuring an abundant and continuous supply. Food shortage is threatening the future of civilization, therefore, our top priority should be to draw a comprehensive plan in order to tackle this impending disaster.

There is an ever-increasing demand for more food but one of the stumbling blocks to achieving this goal seems to be yield loss due to plant diseases caused by an invasion of pests and pathogens.

The problem of plant diseases, particularly in developing countries, is exacerbated by the paucity of resources devoted to their study. Beyond good agronomic and horticultural practices, the strategies employed to manage these postharvest diseases and mycotoxins decontamination so far include established physical, cultural and chemical methods. Growers often heavily rely on chemical fertilizers and pesticides. Conventional insecticides possess inherent toxicities that endanger the health of farm operators and consumers and pollute the environment. In the recent past, the application of chemicals to reduce decay and deterioration caused by various pathogens has been increasingly impeded because these hazardous chemicals contaminate the environment, enter the food chain, destroy beneficial microorganisms and insect pests by targeting non-target microorganisms, increase the incidence of pest resurgence due to the development of resistance in pests and pathogens toward these chemicals, build residual toxicity in plants and animal tissues and cause health hazards in humans such as cancer, hormonal imbalances and respiratory troubles, coupled with legislative and executive actions by state and federal governments and negative public perception. In view of environmental and health concerns, a determined effort is being made to reduce the use of chemicals. The distinct propensity toward "Trek back to nature" is gaining popularity as evident in the recent past in the field of pesticides. The current trend in modern intensive, sustainable agriculture is characterized by three major interventions: consumer demand for healthy and pesticide-free products, environmentally safe measures and effective pest control. Understandably, safe, eco-friendly and non-polluting alternatives to chemical pesticides are the call of the hour.

The present volume is a timely attempt to link scientists from different and complementary disciplines to achieve a unified synthesis. This book deals with the current state and future prospects of using various bio-management techniques which are natural, eco-friendly and environmentally safe. It aims to increase awareness of their potential as well as sensitizing the readers to the various aspects of biologicals in pest control.

This book comprises 14 chapters contributed by eminent scientists and addresses different topics related to the bio-management of postharvest diseases and mycotoxigenic fungi. Chapter 1 deals with the arduous journey from the first generation of antagonists to the present concept of microbiomes. Chapter 2 highlights the classical versus new techniques adopted to manage postharvest diseases. Chapters 3, 4, 5, 6, 7 and 8 focus on the use of eco-friendly, alternative antimicrobial approaches to combat the deterioration of food. Chapters 9, 10, 11 and 12 deal with types of mycotoxin infestation of food and feed stuff and how these products can be decontaminated by using natural biopesticides. In Chapters 13 and 14, two novel approaches, viz. nanotechnology and CAP, are discussed in view of managing fungal spoilage and mycotoxin decontamination. It is an attempt to disseminate notable and diversified scientific work carried out by leading scientists in their own fields.

This book aims to provide the reader with a 360-degree perspective of quality pre- and postharvest and mycotoxin decontamination research conducted at present and details future ideas proposed to ensure a world which is both food secure and pesticide free.

Our goal is that this book will serve as the current and comprehensive treatise on the emerging field of bio-management of postharvest diseases and mycotoxin decontamination by "generally regarded as safe" products.

Neeta Sharma

Avantina S. Bhandari

Acknowledgments

We are grateful to all the contributors, who despite their busy academic schedules, so willingly offered their cooperation in enabling this book to come to fruition. Without their encouragement, enthusiasm and timely submission of the manuscript, this work would not have been possible.

We express our deep sense of gratitude to our family members for their constant support and encouragement in helping us shape this project in the present form.

Our special thanks to Stephen Zollo, Senior Editor, and Laura Piedrahita, Editorial Assistant, for their constant support. I am thankful to Iris Fahrer, Project Editor, and Lillian Woodall, Project Manager, for sparing no pains to ensure a high standard of publication.

Editors

Neeta Sharma, PhD, is an internationally recognized expert in Postharvest Plant Pathology. She has participated in academic and research activities since she joined Lucknow University's Botany Department in 1988.

Dr. Sharma has published more than 250 research papers and articles in leading international and national journals during her career span of more than 40 years. She has mentored 32 students for PhDs and ten students for MPhil degrees. She has also authored five books with well-known national and international publishers such as Wiley and Springer.

An outstanding teacher and research guide, Dr. Sharma is an active member of various professional societies. She is recipient of the prestigious fellowships of the Indian Pathological Society, the Society of Ethnobotanists, and the Society for Conservation of Natural Resources, to name a few.

Dr. Sharma is an academically acclaimed and widely traveled expert in her field. She has been conferred with many awards during her career, to name a few, Young Scientist Award, Scientist of the Year Award, and Distinguished Service Award in the field of Biological Sciences. Besides these, she has earned the prestigious Leadership Award in Plant Pathology conferred by the Indian Phytopathological Society and the Sharda Lele Award in recognition of her pioneering work in postharvest diseases of fruits and vegetables.

Dr. Sharma established the Bio Control Lab in 2013, a major contribution to the university. This is a unique initiative which connects the lab to the land, empowering farmers to efficiently use biocontrol agents in their farming methods.

Presently, she is associated with the World Food Preservation Center, Shepherdstown, West Virginia, USA, an organization working to impart postharvest knowledge relative to developing countries.

Avantina Sharma Bhandari, PhD (Nutrition), UGC-NET is a Nutrition Sustainable Agriculture consultant and has worked in the Department of Nutrition at City University of New York, USA and Isabella Thoburn College, India. She has also worked with international humanitarian agencies across the globe.

She remains current with the latest research as an active member attending various international and national seminars and conferences. She has contributed book chapters, research papers, review articles and popular articles in various national and international journals as well as newspapers. She is also the author of *Textbook of Food Science and Technology* (3rd edition), *Principles of Therapeutic Nutrition and Dietetics*, *Food Product Development* and *Basic Plant Pathology* by CBSPD.

Contributors

Afroz Alam
Department of Bioscience and Biotechnology
Banasthali Vidyapith University
Jaipur, India

Ambreen Bano
Department of Biochemistry
Dr. RML Awadh University
Ayodhya, India

Rachna Chaturvedi
Amity Institute of Biotechnology
Amity University
Lucknow, India

M. H. Tharika Dilhari
Food Technology Section
Industrial Technology Institute
Colombo, Sri Lanka

Tahseen Fatima
Division of Crop Protection
ICAR-Central Institute for Subtropical Horticulture
Rehmankhera, India

Gundappa
Division of Crop Protection
ICAR-Central Institute for Subtropical Horticulture
Rehmankhera, India

Anmol Gupta
Department of Biosciences
Integral University
Lucknow, India

Ilmi G. N. Hewajulige
Food Technology Section
Industrial Technology Institute
Colombo, Sri Lanka

Nidhi Kumari
Division of Crop Protection
ICAR-Central Institute for Subtropical Horticulture
Rehmankhera, India

Gustav Komla Mahunu
School of Food and Biological Engineering
Jiangsu University
Zhenjiang, People's Republic of China

and

Department of Food Science and Technology
Faculty of Agriculture
University for Development Studies
Tamale, Ghana

Narendra Shankar Pandey
Botany Department
NREC (P.G.) College
Khurja, India

Pramila Pandey
Department of Plant Molecular Biology and Genetic Engineering, NDUAT
Kumarganj, Ayodhya, India

Neelam Pathak
Department of Biochemistry
Dr. RML Awadh University
Ayodhya, India

Smita Rai
Department of Biosciences
Integral University
Lucknow, India

Swati Sharma
Department of Biosciences
Integral University
Lucknow, India

Amritesh Shukla
Department of Botany
University of Lucknow
Lucknow, India

P. K. Shukla
Division of Crop Protection
ICAR-Central Institute for Subtropical Horticulture
Rehmankhera, India

Madhu Prakash Srivastava
Department of Botany
Maharishi University of Information Technology
Lucknow, India

Abhishek Tripathi
Uttarakhand Civil Services
Government of Uttarakhand
India

Charles L. Wilson
World Food Preservation Center® LLC
Shepherdstown, WV

Qiya Yang
School of Food and Biological Engineering
Jiangsu University
Zhenjiang, People's Republic of China

Hongyin Zhang
School of Food and Biological Engineering
Jiangsu University
Zhenjiang, People's Republic of China

1 Postharvest Biological Control
Antagonists, Microbiomes, and Beyond

Charles L. Wilson

CONTENTS

1.1 INTRODUCTION AND BACKGROUND

In 1984, at the ARS USDA Appalachian Fruit Research Station in Kearneysville, West Virginia, a postdoc Larry Pusey and I had just returned from the orchard to our laboratory. On the laboratory bench was a water suspension of an antagonistic bacterium *Bacillus subtilis* we were testing in the orchard to control peach brown rot (*Monilinia fructicola*) blossom blight infections. Alongside the bacterial suspension were some peaches harvested to eat later.

With *Bacillus subtilis* not working so well in the field, Dr. Pusey and I decided to wound the peaches on the table and inoculate the wounds first with *Bacillus subtilis* and later with the brown rot fungal pathogen *Monilinia fructicola* (Pusey and Wilson, 1984). Little did we know that this simple experiment would launch a major field of science for the biological control of postharvest diseases. The wounds treated with *Bacillus subtilis* remained uninfected by *Monilinia fructicola* while the untreated fruit rotted (Wilson and Pusey, 1985).

This simple experiment caused us to realize that the biological control of postharvest diseases had many advantages over biological control in the field. The advantages were that: (1) harvested commodities are concentrated before they are treated with antagonistic organisms, making antagonists easier to target toward the pathogens than in the field; (2) the postharvest or packinghouse environment (temperature and humidity) is much more controllable and constant than the field environment, thus allowing more controllable and predictable results; and (3) because of previous investments in the culture, harvesting, and processing of commodities in their production, more expensive forms of biological control are affordable for postharvest biological control technologies.

Based on our success in using an antagonistic bacterium to control a postharvest disease, the postharvest biological control program at the USDA ARS Appalachian Fruit Research Station grew rapidly and was joined by Michael Wisniewski, Ahmed El Ghaouth, Wojchiech Janisiewicz, and multiple visiting scientists from other countries. We reached out to Israel and its BARD program and were introduced to Drs. Edo Chalutz and Samir Droby. This highly productive relationship between Israel and the ARS USDA Appalachian Fruit Research Station, funded by the Binational Agricultural Research and Development Fund (BARD), continues today. It has resulted in multiple seminal publications and workshops over the past 35 years (Droby et al., 2016; Wisniewski et al., 2016a).

The cooperative BARD-supported postharvest biocontrol research between the United States and Israel focused initially on the development of antagonistic yeasts as biological control agents rather than bacteria (Droby et al., 2009). The argument was made, correctly or not, that antagonistic bacteria would be more likely to produce antibiotics and cause human diseases. Therefore, yeast would be safer than bacteria to use as antagonists on harvested commodities. It was recognized early on that antagonistic organisms including yeast existed on the surfaces of harvested fruits and vegetables. This was because, when commodities were washed, they decayed more rapidly than when they were not, suggesting that washing removed putative antagonist organisms against postharvest pathogens. A method was sought to isolate putative antagonistic yeasts from the surfaces of fruits and vegetables that might be used for biological control products. Antagonistic yeasts do not produce inhibition zones as bacteria do in washings from plant surfaces plated out in Petri dishes seeded with the pathogen. This classical method of identification and isolation of antagonists was, therefore, not applicable. We explored other methods to isolate antagonistic yeasts from the surfaces of fruits and vegetables.

A simple, highly effective procedure was developed to "fish out" antagonistic yeasts against postharvest pathogens on the surfaces of fruits and vegetables (Wilson et al., 1993). It involved treating wounds on fruits and vegetables with washings from the surfaces of fruits and vegetables and subsequently inoculating the wounds with postharvest pathogens. Most of the inoculated wounds resulted in infections and decay, but on occasion, an inoculated wound would not develop an infection. It was suspected that this uninfected wound had been colonized by an antagonist from the surface washings that prevented the infection. Isolations were then made from the non-infected wound surfaces. If monocultures of yeast were isolated, it was suspected that this organism was an antagonist. To prove this, pure cultures of the isolated putative yeast antagonist were placed in wounds before they were inoculated with the pathogen to see whether they could prevent infections. This procedure has been used worldwide to identify multiple antagonists against postharvest pathogens.

The use of single antagonistic yeasts and bacteria to control postharvest diseases expanded globally during the 1990s into the 2000s with multiple antagonists being commercialized as postharvest biocontrol agents (Wisniewski et al., 2016b). During this period, potential antagonistic organisms were identified taxonomically based on their morphology and physiology. This form of identification greatly limited the number and diversity of organisms that were considered as potential antagonists for the biological control of postharvest diseases.

1.2 ADVENT OF MICROBIOMES IN POSTHARVEST BIOLOGICAL CONTROL

The advent of DNA sequencing and molecular identification of microorganisms completely changed our perspective on potential postharvest biological control antagonists and their microecology. These advanced identification techniques revealed that we had been looking at only a minuscule number of the microorganisms on the surfaces of fruits and vegetables. Also, these newly discovered populations of microorganisms existed in communities (microbiomes) with highly sophisticated relationships and systems for communication (Droby and Wisniewski, 2018). This discovery led to a major shift in the field of postharvest biocontrol to an emphasis on microbiomes rather than individual antagonists (Orozco-Mosquedo et al., 2018). Research is underway to identify, understand, and manipulate microbiomes associated with harvested commodities to affect postharvest disease biocontrol (Wiesniewski and Droby, 2019). It is hoped that new postharvest biological control products will result from the understanding, application, establishment, and enhancement of biological control microbiomes on and in fruits and vegetables.

1.3 POSTHARVEST BIOLOGICAL CONTROL BEYOND ANTAGONISTS AND MICROBIOMES

Early on, plant pathologists were "trapped" in their thinking of biocontrol by following the early definition of biological control set forth by entomologists. In entomology, biocontrol is defined as the control of one "organism" (an insect) with another "organism" (parasite, predator, or pathogen). But a plant disease is not an "organism." It is a "process." Therefore, biocontrol of a plant disease requires interruption of a biological disease "process" rather than just the control of an "organism." A different definition of biological control for plant diseases is therefore in order.

I propose the following definition for the biological control of plant diseases: "*A biologically based control agent or procedure that interrupts a plant disease process.*" This broader definition for biological control of plant diseases opens up additional new possibilities for control of plant diseases other than just the use of antagonistic organisms. It brings into play manipulation of the host resistance response to plant pathogens either through induced resistance or constitutive resistance. It also involves the use of non-synthetic chemical biological antimicrobial compounds such as essential oils (Wilson et al., 2007) and chitosan. Utilizing this broader definition of postharvest biocontrol, a number of control methods have been developed combining natural plant and animal antimicrobials with antagonistic organisms (El Ghaouth et al., 2000).

1.3.1 Induced Resistance to Postharvest Diseases

Under this broader definition of the biological control of plant diseases, we explored the induction of resistance in harvested commodities as a means of controlling postharvest diseases (Wilson et al., 1994). The early pioneering work of the late Dr. Clauzell Stevens at Tuskegee University on the use of low-dose UV-C light to induce resistance to postharvest diseases opened up a new prospective on the biological control of postharvest diseases. We reached out to Dr. Stevens at the USDA ARS Appalachian Fruit Research Station and developed a highly productive research relationship over a number of years. We were jointly able to demonstrate the control of multiple postharvest diseases through UV-C irradiation (Stevens et al., 2004, 2005). It was also discovered that not only UV-C light but also antagonists (El Ghaouth et al., 2003a) and natural compounds (Wang et al., 2009b) could induce resistance to postharvest diseases in harvested commodities. Studies were undertaken on the mode-of-action of UV-C light in inducing resistance to postharvest diseases. It was demonstrated that UV-C light induced a number of resistance compounds in irradiated tissue (El Ghaouth et al., 2003b).

1.3.2 Breeding for Resistance to Postharvest Diseases

Early on, plant breeders inadvertently bred plants more susceptible to postharvest diseases.

This was because in selecting plants with more digestible and less bitter edible qualities they were inadvertently selecting plants more susceptible to infection by postharvest pathogens. Plants with more digestible cell walls are easier for postharvest plant pathogens to penetrate and infect because they have less lignin (Miedes et al., 2014). Breeding fewer bitter plants involved the development of plants with a reduced content of bitter compounds such as alkaloids and phenolics that are resistant compounds against postharvest plant pathogens (Othanan et al., 2019). Only recently has the breeding of plants with resistance to postharvest diseases been pursued (Bally et al., 2013; Norelli et al., 2014).

1.3.3 Postharvest "Biocontrol Genes"?

Are there postharvest biocontrol genes that can be manipulated to make harvested commodities more resistant to postharvest diseases? There is growing evidence that epiphytic microorganisms on the surfaces of plants can contribute to their resistance to disease and that some of these organisms are under the genetic control of the host plant. If we can identify possible genes that control antagonistic organisms to postharvest pathogens on the surfaces of fruits and vegetables, then these can be used in a conventional breeding program to breed commodities resistant to postharvest diseases. Lindau and his associates at the University of California (Lindau et al., 2003) have been able to demonstrate that epiphytic microorganisms on the surfaces of plants can be controlled by specific genes that regulate leaf exudates. They found that a *Pseudomonas florescence* strain engineered to catabolize the opine mannityl had a competitive advantage on tobacco leaves engineered to produce and secrete the opine mannityl. Balint-Kurti et al. (2010) have shown that bacteria on the surface of corn that impart resistance to a leaf disease are under the control of host genes. If postharvest "biocontrol genes" are identified, they can be manipulated and incorporated into breeding programs to impart postharvest resistance in fruits and vegetables to postharvest pests.

1.3.4 Conservation of Biological Control of Postharvest Diseases

It was suspected early on that there were "natural" postharvest biocontrol systems that exist on the surfaces of fruits and vegetables. This can be inferred when washed commodities decay faster than unwashed commodities, indicating that "natural" antagonists had been removed in the washing. There needs to be more research exploring this possibility. Wilson (1998) has discussed how we can sustain and promote existing natural biological control systems to control postharvest diseases of fruits and vegetables.

With more sophisticated identification technologies it has been discovered that there is a complex microbiome on the surfaces and within plant tissues. Knowledge of this "second genome" opens up unlimited possibilities for the manipulation of these microbiomes to develop both production and postharvest biocontrol systems. Zuanovic and Rogers (2019) explore the role that plant endophytes may play in host resistance.

1.3.5 Making Postharvest Commodities "Better than Fresh"

One of the more exciting developments in the biological control of postharvest diseases has been the discovery that while controlling postharvest diseases we can also make the harvested commodity "Better than Fresh." Studies on the mode-of-action of UV-C light induced resistance to postharvest diseases have demonstrated the induction of a number of resistance compounds in plants by UV-C irradiation. Among the UV-C induced resistance compounds are antioxidants, phenolics, and phytoalexins (Erkan et al., 2008; Wang et al., 2009b). Some of these UV-C induced compounds are also excellent nutrients and nutraceuticals for humans and other animals.

UV-C irradiation of strawberries (Erkan et al., 2008) and blueberries (Wang et al., 2009a) has been shown to effectively control Botrytis rot. It has also been shown that UV-C irradiation of these commodities induces an array of compounds that are effective resistance compounds against postharvest pathogens. Some of these same compounds are excellent nutrients and nutraceuticals for humans and other animals such as resveratrol (Cantos et al., 2000). Therefore, we have the opportunity of controlling postharvest diseases while at the same time making the treated commodity "Better than Fresh." This has been suggested as a breakthrough technology to enhance the world's food supply (Wilson, 2016). Besides producing more food, we can increase the nutrient content of the food we already produced.

1.4 CONCLUSION

Synthetic fungicides are predominantly used to control postharvest diseases of fruits and vegetables. Even after over 35 years of research on the biological control of postharvest diseases, a very small number of products are used as alternatives to synthetic fungicides (Wisniewski et al., 2016a).

This is the scenario in spite of an increased awareness of the potential carcinogenicity of many of our commonly used postharvest fungicides such as imazalil. A study conducted by the National Academy of Sciences concluded that synthetic fungicides pose more of a health risk to humans, especially children (National Academy of Science, 1993), than synthetic insecticides and herbicides combined. Synthetic fungicides as postharvest treatments are of particular concern as they are applied close to the time the commodity is consumed and constitute a major component of the synthetic pesticide residue on the consumed product.

A more educated consumer in both the developed and developing world is demanding healthier food that is more nutritious and synthetic pesticide and genetically modified organism (GMO) free. A greater demand for organically produced food creates a greater demand for biological control products for postharvest diseases. A wide range of promising approaches for the biological control of postharvest diseases have been established. Our challenge is to take this intellectual foundation and create biological postharvest disease control products as effective alternatives to synthetic pesticides. The demand is there. Let's provide a supply of these biological products sufficient to meet the demand!

REFERENCES

Balint-Kurti, P., S. J. Simmons, J. E. Blum et al. 2010. Maize leaf epiphytic bacteria diversity patterns are genetically correlated resistance in fungal infections. *MPMI* 23: 473–484.

Bally, S. E., C. Grise, N. Dillon et al. 2013. Screening and breeding for genetic resistance to anthracnose in mango. *Acta Hortic.* 992: 239–241.

Cantos, E., C. Garcia-Viguera, T. S. de Pascual et al. 2000. Effect of postharvest ultraviolet irradiation on resveratrol and other phenolics of cv. Napoleon table grape. *J. Agric. Food Chem.* 480: 4406–4672.

Droby, S., and M. Wisniewski. 2018. The fruit microbiome: A new frontier for postharvest biocontrol and postharvest biology. *Postharvest Biol. Technol.* 140: 107–112.

Droby, S., M. Wisniewski, D. Macarisin et al. 2009. Twenty years of postharvest biocontrol research: Is it time for a new paradigm? *Postharvest Biol. Technol.* 52: 137–145.

Droby, S., M. Wisniewski, N. Teixid6 et al. 2016. The science, development, and commercialization of postharvest biocontrol products. *Postharvest Biol. Technol.* 122, 22–29.

El Ghaouth, A., J. L. Smilanick, M. Wisniewski et al. 2000. Improved control of apple and citrus fruit deca with a combination of *Candida saitoana* and 2-deoxy-d-glucose. *Plant Dis.* 84: 249–253.

El Ghaouth, A., C. L. Wilson, and A. M. Callahan. 2003b. Induction of chitinase, beta-1-3 glucanase, and phenylalanine ammonia lyase in peach fruit by UV-C treatment. *Phytopathology* 93: 349–355.

El Ghaouth, A., C. L. Wilson, and M. Wisniewski. 2003a. Control of postharvest decay of apple fruit with *Candida saitoana* and induction of defense responses. *Phytopathology* 93: 344–348.

Erkan, M., S. Y. Wang, and C. Y. Wang. 2008. Effect of UV treatment on antioxidant capacity, antioxidant enzyme activity and decay in strawberry fruit. *Postharvest Biol. Technol.* 48: 163–171.

Lindau, S. E., and M. T. Brandl. 2003. Microbiology of the phyllosphere. *Appl. Environ. Microbiol.* 69: 1075–1883.

Miedes, E., R. Vanholme, W. Boejan et al. 2014. The role of the secondary cell wall in plant resistance to pathogens. *Front. Plant Sci.* 5: 358–361.

National Research Council, and National Academy of Sciences. 1993. *Pesticides in the Diets of Infants and Children.* National Academy Press, Washington, DC.

Norelli, J., M. Wisniewski, and S. Droby. 2014. Identification of a QTL for postharvest disease resistance to *Penicillium expansum* in *Malus sieversii. Acta Hortic.* 105: 199–203.

Orozco-Mosqueda, M., M. del Carmen Rocha-Granados, B. R. Glick et al. 2018. Microbiome engineering to improve biocontrol and plant growth-promoting mechanisms. *Microbiol. Res.* 208: 25–31.

Othanan, L., A. Sieiman, and R. M. Aabdel-Massih. 2019. Antimicrobial activity of polyphenols and alkaloids in middle eastern plants. *Front. Microbiol.* 10: 911–914.

Pusey, P. L., and C. L. Wilson. 1984. Postharvest biological control of stone fruit brown rot by *Bacillus subtilis. Plant Dis.* 68: 753–756.

Stevens, C., V. A. Khan, C. L. Wilson et al. 2005. The effect of fruit orientation of postharvest commodities following low dose ultraviolet light-C treatment on host induced resistance to decay. *Crop Prot.* 24: 756–759.

Stevens, C., J. Liu, V. A. Khan et al. 2004. The effects of low dose ultraviolet light- C treatment on polygalacturonase activity, delay ripening and Rhizopus rot development of tomato. *Crop Prot.* 23: 551–554.

Wang, C. Y., C. Chen, and S. Y. Wang. 2009a. Change of flavonoid content and antioxidant capacity in blueberries after illumination with UV-C. *Food Chem.* 117: 426–431.

Wang, K., P. I. Jin, S. Cao et al. 2009b. Methyl jasmonate reduces decay and enhances antioxidant capacity in Chinese bayberries. *J. Agric. Food Chem.* 57: 5809–5815.

Wilson, C. L. 1998. Conserving epiphytic microorganisms on fruits and vegetables for biological control. In: *Conservation Biological Control* (pp. 335–344), Pedro Barbosa, ed. Academic Press, New York.

Wilson, C. L. 2016. Breakthrough technology to preserve and enhance food (Chapter 10). In: *Postharvest Management* of *Horticultural Crops*. Taylor and Francis Group, Boca Raton, FL.

Wilson C. L., A. El-Ghaouth, E. Chalutz et al. 1994. Potential of induced resistance to control postharvest diseases of fruits and vegetables. *Plant Dis.* 78: 837–844.

Wilson, C. L., and P. L. Pusey. 1985. Potential for biological control of postharvest plant diseases. *Plant. Dis.* 69: 375–378.

Wilson, C. L., J. M. Solar, A. El Ghaouth et al. 2007. Rapid evaluation of plant extracts and essential oils for antifungal activity against Botrytis cinerea. *Plant Dis.* 81: 204–210.

Wilson, C. L., M. Wisniewski, and S. Droby. 1993. A selection strategy for microbial antagonist to control postharvest diseases of fruits and vegetables. *Sci. Hortic.* 53: 183–189.

Wisniewski, M., S. Droby, J. Norelli et al. 2016a. Alternative management technologies for postharvest disease control: The journey from simplicity to complexity. *Postharvest Bio. Tech.* 122: 3–10.

Wisniewski, M., J. Norelli, S. Droby et al. 2016b. Genomic tools for developing markers for postharvest disease resistance in Rosaceae fruit crops. *Acta Hortic.* 1144: 7–16. DOI: 10.17660/ActaHortic.2016.1144.2.

Zuanovic, A., and L. Rogers. 2019. The role of fungal endophytes in plant pathogen resistance. *BIOS* V89: 192–197.

2 Managing Postharvest Diseases
Classical versus New Technologies

Neeta Sharma

CONTENTS

2.1 INTRODUCTION

We are approaching the upper limits of the Earth's human-carrying capacity, and with this, the impending disaster of food shortage is becoming a threat to all living beings. The global population has, over the years, callously used up the world's resources; substituting a blatant use of chemicals to sustain the system and to provide for and feed the growing population (Sharma, 2014). This has resulted in the dark clouds of climate change, plant pests, food losses, food scarcity and chemical-laden soils, all looming large, discombobulating and disheartening future generations.

The origins of human civilization can be traced back to the discovery and reassurance of an available, accessible and affordable food supply. The total dependence of humans on cereals, legumes, oil crops, fruits and vegetables for sustenance and survival has always been of paramount importance. Even the historical journey from the "Garden of Eden" to Earth is of great significance, because of the "Fruit of Wisdom." By the time man had finally come out of trees and placed his feet firmly on terra firma, he must have sought out plants and their edible products in the wild as a significant part of his hunting and gathering diet. Although it is not known when cultivation of the first plants took place, people of diversified lands selected the bulk of our present-day plants, which provided innumerable plant products required to satisfy the needs of living beings. The earliest written records discovered to date are in the form of cave paintings from early modern humans, dating from the upper Paleolithic period (c. 40,000–c. 10,000 years ago).

2.2 POSTHARVEST/MARKET PATHOLOGY

Plant products are essential sources of carbohydrates, proteins, oils, antioxidants and micronutrients needed for healthier diets. The potential of these products is to generate positive economic and nutritional impacts. It is likely that climatic variability due to temperature patterns and precipitation are some of the problems faced by the producers, which affect different aspects of agricultural production.

Cereals and grains not only provide the carbohydrates and proteins required for healthy living, but the aggregation of cereals and grains is essential as a buffer stock throughout the year as an insurance against crop failure in times of drought, excessive rainfall or any other natural calamity, on the one hand, while on the other hand, their importance as seeds cannot be denied (Sharma and Bhandari, 2014).

Consumer awareness that diet and health are linked has resulted in a rapid increase in the demand for fresh and healthy products. These dietary changes have increased the importance of fruits and vegetables, which are rich sources of vitamins and essential mineral nutrients.

Although technological advances have focused their efforts on the development of new varieties, crop management techniques and innovations in postharvest handling and processing, the production of these crops is not increasing as rapidly as would be expected. There is enough evidence that public funding for research in agriculture is less than expected. As plant species were selected for higher yield potential and better quality, the levels of resistance to biotic as well as abiotic stresses were depleted over time. Consequently, these factors inflicted considerable losses on standing crops as well on/in harvested yield, and reduced the aesthetic value and storage life of agricultural crops. Postharvest diseases develop due to various abiotic and biotic factors (Figure 2.1) on the harvested parts of the plants, such as seeds, fruits and vegetables.

The importance of postharvest diseases was emphasized for the first time in 1906 by G.H. Powell who named this facet of plant pathology "Market Pathology." Diseases occurring on fresh fruits and vegetables during transit, in storage and in market and the means of controlling these diseases constitute the special field of *market pathology* or *postharvest pathology*. The plant parts may get infected in the field, but the expression of disease symptoms may take place later, at any stage before final consumption. This special and independent field of postharvest pathology is distinguished from other aspects of science on the basis of following characteristics:

- The products affected by diseases are the detached dormant storage organs and their tissues are physiologically different from those of actively growing plant parts.
- The conditions under which the problems develop are usually different from those that pertain in the field.
- The people who handle the harvested products are not the same as those who cared for it during the growing season.
- The locale of the market pathologist has traditionally been urban centered and not rural centered (exception being latent infection).

Food losses refer to the decrease in edible food mass (dry matter) or nutritional value (quality) of food that was originally intended for human consumption (FAO, 2011). Food losses take place at the production, postharvest and processing stages in the food supply chain (Parfitt et al., 2010). Food losses are mainly due to poor infrastructure and logistics, lack of technology, insufficient skills, knowledge and management capacity of supply chain actors and lack of markets.

Food waste refers to food appropriate for human consumption being discarded, whether or not it is kept beyond its expiry date or left to spoil. Food waste occurs in the food chain at retail as well as final consumption and relates to "retailer and consumer" behavior. Food waste or loss is measured only for products that are intended for human consumption, excluding feed and parts of products, which are not edible.

2.3 TYPES OF FOOD SPOILAGE

The quantity of the plant product becomes reduced due to various abiotic and biotic factors and also the quality of the product is rendered unfit for human consumption. The term *contact spoilage* is used when microbial spoilage is the result of direct contact or proximity between the food and

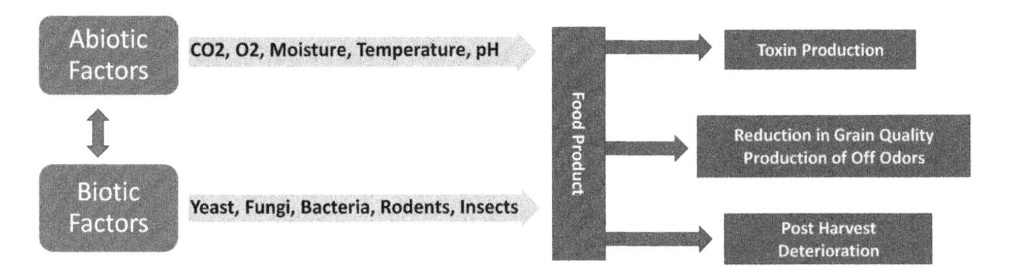

FIGURE 2.1 Abiotic and biotic factors affecting food quality.

any contaminated or unclean surface such as shelves, food preparation boards or unwashed hands. It also includes food-to-food contact, for example, between rotting fruit and healthy fruit. *Physical spoilage* is due to physical damage to food during harvesting, processing or distribution. The damage increases the chance of chemical or microbial spoilage and contamination because the protective outer layer of the food is bruised or broken, and microorganisms can enter the foodstuff more easily. For example, when an apple skin is damaged, the apple rots more quickly. In *chemical spoilage*, chemical reactions in food are responsible for changes in the color and flavor of foods during processing and storage. Foods are of best quality when they are fresh, but as they are harvested, chemical changes automatically begin within the foods that can lead to a deterioration in quality. The process of *enzymic spoilage* involves use of specialized *enzymes* to drive chemical reactions in the cells. After death, enzymes play a role in the decomposition of once-living tissue, through a process called *autolysis* (self-destruction) or enzymic spoilage. Once enzymic spoilage is under way, it damages the product, thus facilitating entry of various pathogens inside the food and speeding the process of decay. In food spoilage, the changes in appearance or texture of the food, such as rottenness, softness and change in color, taste or odor are usually obvious, whereas in contaminated foods such characteristics may not be noticed.

The process of ripening basically signals the end of development and the beginning of senescence. It can involve a number of changes in the fruit, such as conversion of starch to sugars, increase in pH, increase in fruit softness, development of aromas, reduction in chlorophyll content and a corresponding increase in yellow and orange pigment levels.

2.4 TYPES OF INFECTION

The main reasons for deterioration are due to microorganisms—mainly fungi followed by bacterial and viral contamination. These pathogens are further categorized as *field pathogens*, which cause infection during the development of plants or their products before harvest, developing inconspicuous symptoms to be noted at the time of harvest, and *storage pathogens*, which cause infection during transit and storage. These pathogens invade the host through different ways. The main types of penetration by postharvest pathogens have been observed as follows.

2.4.1 Latent Infections

Fungi and bacteria responsible for in-storage deterioration often originate in the field or orchard. Pathogen penetration takes place in the field and reaches the storehouse within the host tissue without eliciting any decay symptoms. However, during one of the phases, between reaching the host and the development of progressive disease, further development is arrested until after harvest, when physiological and biochemical changes occurring within the host will enable their renewed growth. Such arrested infections are described as "latent infections" in which pathogen growth is temporarily inhibited. The latent stage of the pathogen is linked to dynamic balance between the host, the pathogen and the environment. The physiological and biochemical changes occurring in the host tissues after and during storage might affect the host–pathogen interrelationship and lead to the activation of the latent pathogen, for example, the blight pathogen *Phytophthora* infecting potato, tomato and citrus; soft watery rot caused by *Sclerotinia* on carrot, cabbage, cauliflower, cucumber and lettuce; and Anthracnose caused by *Colletotrichum gloeosporioides.*

2.4.2 Contact infections

Fruits and vegetables that are free from any pathogen invasion by any means of penetration might be infected through contact with infected produce during storage. The development of *Botrytis* infection in stored strawberries, which turn them into "mummies" covered with a gray layer of spore bearing mycelium, causes contact infection and makes other fruits susceptible to disease. Similarly, any product infected with *Rhizopus* can spread the infection within the container. In citrus fruits, infection of green and blue mold is very common due to contact.

2.4.3 Invasion through Natural Openings

Penetration through stomata and lenticels can be observed in *Colletotrichum gloeosporioides* and *Monilinia fructicola*. The penetration of germ tubes of the spores of pathogens into young fruits of papaya and stone fruits takes place through stomata. Penetration through lenticels has also been described for *Alternaria alternata* conidia in mango and *Gloeosporium perrenans* in apple. The bacterial soft rot of potato tubers is caused by *Erwinia carotovora*. The bacteria remain inactive in the lenticels until the conducive conditions such as mechanical pressure, low oxygen pressure and existence of free water enhance the tuber sensitivity to invasion by pathogens.

2.4.4 Penetration through Wounds

Most of the storage pathogens are incapable of penetrating the cuticle and epidermis of the host but do require an injury or wound to facilitate their invasion. Therefore, these pathogens are also referred to as wound pathogens. Careless separation from the parent plant might result in injury liable to attack by the pathogen. A possible penetration point is the stem-end separation area where damage often occurs during picking. Each incision, scratch or mechanical injury inflicted on the harvested product during handling, transporting, sorting, packing and storing might present adequate penetration points for the pathogen to enter. Growth cracks present on the harvested commodity are natural avenues of infection. The extent of injury caused by mechanical harvesting is far greater than caused by manual operation. Large amounts of bacterial cells and fungal spores accumulate at the injured site and invade the host. Wound infection is caused by *Rhizopus stolonifer*, *Alternaria*, *Geotrichum candidum*, *Aspergillus*, *Cladosporium* and *Trichothecium*.

2.4.5 Penetration Owing to Senescence

Tissue senescence during continuing storage reduces disease resistance. The rate of decay increases with prolonged duration of storage as the tissue gradually dies. A senescence onion that has commenced sprouting is more vulnerable to decay caused by *Fusarium*. This similar to melons, as they become more sensitive to *Trichothecium* and various species of *Penicillium* if stored for long durations.

2.4.6 Physical Damage

Injuries caused by low temperature, extreme temperature, heat, oxygen shortage or any other environmental stress increase the sensitivity to storage fungi. The physiological damage can be extremely expressed through tissue browning and splitting, thus forming sites vulnerable to invasion of wound pathogens.

Extreme environmental conditions enhance sensitivity to an attack without any visible external sign of damage. Sun scald lesions in apple lead to attacks by *Alternaria alternata* and *Stemphylium botryosum*.

2.5 FACTORS AFFECTING FOOD SPOILAGE

Several factors, both intrinsic as well as extrinsic, are responsible for the occurrence and spread of postharvest diseases. The growth of microorganisms in food products can be affected by *extrinsic factors* and *intrinsic factors*. By understanding the factors that affect the growth of microorganisms in food, one can know how to keep food safe to eat. This knowledge can also help us to work out how to preserve food for longer periods.

2.5.1 Extrinsic Factors

Extrinsic factors are factors in the environment *external* to the food, which affect both the microorganisms and the food itself during processing and storage. Extrinsic factors include temperature, humidity and oxygen.

2.5.1.1 Water Content/Moisture Content (Water Activity)

Microorganisms need a moist environment to grow in. The water activity (a_w) of pure water is 1.00, while the water activity of most fresh foods is 0.99. This means that they have a high water content and can support a lot of microbial growth. Most foodborne pathogenic bacteria require a_W to be greater than 0.9 for growth and multiplication; however, *Staphylococcus aureus* may grow with a_W as low as 0.86. But even *Staphylococcus aureus* cannot grow and multiply in drier foods like bread, which has $a_w = 0.7$.

2.5.1.2 Temperature

The availability of a suitable temperature is one of the most important factors required for the establishment and progress of pathogens. Different microorganisms grow over a wide range of temperatures. Some microorganisms like to grow in the cold, some like to grow at room temperature and others like to grow at high temperatures. It has been observed that the temperature range 10 °C to 35 °C favors the growth and sporulation of most postharvest pathogens. The optimum temperature favoring the progress of rot causing organisms in/on the produce is slightly higher than the optimum temperature for growth in vitro. This is because higher temperatures enhance the process of ripening which results in the increase of sugar content and the decrease in acidity as well as the firmness of the fruit tissue.

2.5.1.3 Humidity

The humidity of the storage environment is an important factor for the growth of microorganisms. The spores and cells of pathogens require high humidity for germination and to cause infection. Also, the condensate from the transpiration of the commodity and exudates resulting from minor bruises and insect punctures can provide sufficient humidity needed for germination and subsequent penetration. Once the disease is initiated, moisture never becomes a limiting factor inside the tissue.

2.5.1.4 Oxygen

Many microorganisms, which need oxygen in order to develop and reproduce, are known as *aerobic* microorganisms. A good example is *Escherichia coli*, a fecal bacterium that grows readily on many foods. If you keep food in a low oxygen environment, aerobic bacteria cannot grow and multiply. Conversely, there are some microorganisms that grow without oxygen, called *anaerobic* microorganisms. An example of this is *Clostridium botulinum*, the bacterium causing botulism, which can survive in very low oxygen environments such as tinned foods.

2.5.2 Intrinsic Factors

These factors exist as part of the food product itself. The following common intrinsic factors affect the growth and multiplication of microorganisms in food. Most microorganisms grow best at close to neutral pH value (pH 5.6 to 7.5). Only a few microorganisms grow in very acidic conditions, below a pH of 4.0. Bacteria grow at a fairly specific pH for each species, but fungi grow over a wider range of pH values. In order to grow, multiply and function normally, microorganisms require a range of nutrients such as nitrogen, vitamins and minerals. Microorganisms therefore grow well on nutrient-rich foods. The natural covering of some foods provides excellent protection against the entry and subsequent damage by spoilage organisms.

Condition of produce at harvest determines how long the crop can be safely stored. For example, apples are picked slightly immature to ensure that they can be stored safely for several months. The onset of ripening and senescence in various fruits renders them more susceptible to infection by pathogens. On the other hand, fruits can be made less prone to decay by management of crop nutrition. It is important to maintain sanitary conditions in all areas where produce is packed. Organic matter (culls, extraneous plant parts, soil) can act as substrates for decay-causing pathogens. For example, in apple and pear packing houses, the flumes and dump tank accumulate spores and may act as sources of contamination if steps are not taken to destroy or remove them.

2.6 ENORMITY OF POSTHARVEST LOSSES

Consumer awareness that diet and health are linked has resulted in a rapid increase in the demand for fresh and healthy products. Consumption of whole and lightly processed foods is also increasing due to greater choice and year-round availability. At harvest, once the products are detached from the parent plant, they get deprived of essential nutrients, water and other requirements, which enhances their ability to overcome adverse effects and recover, thus making them fall victim to various abiotic and biotic adversities. Apart from rendering the product unfit for human consumption, they reduce the aesthetic appeal and also bring about a substantial decrease in their food value and organoleptic quality.

Postharvest loss is a complex phenomenon and the extent of losses reported may differ depending upon the region selected and the consumer acceptance of the product in that region. The very wide range of the extent of reported losses from 0 to 80% in cereals, fruits, vegetables, root and tuber depends on many factors such as the nature of the crop (whether it is highly perishable, moderately perishable or non-perishable), the physiology of different crops and disease incidence in the field, as well as other contributing factors such as time of harvest and time from harvest, temperature during handling, weather conditions, type of packages used, duration of handling, transport and storage (Kitinoja, 2010; Kitinoja et al., 2018)

The Food and Agricultural Organization of the United Nations estimates that globally about one-third of all food produced is either lost or wasted before consumption.

In some African, Caribbean and Pacific ACP countries, where tropical weather and poorly developed infrastructure contribute to the problem, wastage can regularly be as high as 40–50%. Approximately half of the population in the developing world is facing starvation. It is estimated that nearly 30–40% (1.2–2 billion tons) of all food produced never reaches a human. In developed countries, the losses and wastage of food have been estimated to be between 10 and 60%. In developing countries, these losses run to more than 50%. A recent FAO report indicates that at the global level, volumes of lost and wasted food in high income regions are higher in downstream phases of the food chain, but are the opposite in low-income regions where more food is lost and wasted in upstream phases (FAO, 2011). In a recent FAO report, the director general Qu Dongyu has questioned, “How can we allow food to be thrown away when more than 820 million people in the world continue to go hungry every day” (FAO, 2019).

Loss of quantity is more common in developing countries. In regions such as Sub-Saharan Africa (SSA), where food-insecurity is highly prevalent, approximately 20% of all the grains, 44% of roots and tubers and 52% of fruits and vegetables are lost between harvest and consumption (AGRA, 2013, 2014). For grains and legumes/pulses, the range of losses varied between 1 and 40%, based on the climate and season (wet or dry season), volume of precipitation during the period following harvesting, incidence of pests/pathogens and the utilization (or not) of improved processing and storage methods. Nanda et al. (2012) measured postharvest losses (PHLs) in India for 37 grains, legumes/pulses, root/tuber, fruits and vegetable crops. Losses of highly perishable crops when there are gluts can be enormous, since the offered price for the produce may be so low that it does not make sense to harvest or transport the crop to the market. This was the case in 2017 in India for tomatoes when the price fell to 30 INR/20 kg, the equivalent of US$0.025/kg.

The national level food loss estimates for India in 2013–2014 were higher for perishable crops than for staple crops: cereals 4.65 to 5.99%; pulses 6.36 to 8.41%; oilseeds 3.08 to 9.96%; fruits 6.7 to 15.88%; and vegetables 4.58 to 12.44%. (CIPHET, 2015).

In Bangladesh, postharvest loss of fruits and vegetables ranged from 23.6 to 43.5% (Hassan et al., 2010). Kitinoja and AlHasan (2012) reported that the percent mechanical damage for individual samples of cabbage varied from farm to wholesale and the retail market. Losses for highly perishable leafy green vegetables have been measured to be as high as 70 to 80% in West Africa, and losses in fruits to be 50 to 70%, especially during the rainy season.

Qualitative losses can be very high and result in economic losses (FAO, 2011, 2015), data were collected at the farm, wholesale and the retail market for a large number of different fruits and vegetable crops in India and SSA (WFLO, 2010) using the methodology of modified CSAM (LaGra et al., 2016). Qualitative loss data were reported for many types of vegetables and fruits in Bangladesh (Ahmed, 2013) and for vegetable crops in Ghana (Appiah, 2013a, 2013b), Kenya (Owino, 2013) and Rwanda (Musenase and Kitinoja, 2017).

Measured damage to leafy greens transported in bunches tied in cloth reached 89% in Benin; green cooking bananas in Rwanda experienced 98% damage by the time they reached the wholesale market and suffered 100% damage at the retail market level, mainly due to rough handling and transport by bicycle (Rwubatse and Kitinoja, 2017).

Fewer recent PHL studies have included reports on measurements of economic data. WFLO (2010) reported on economic losses at the retail level for amaranth (30%) and pineapple (33%) in Benin due to mechanical damage and weight losses. Sharma and Rathi (2013) provided data on economic losses for wheat and soybeans in India, reported losses in kg per acre and then calculated the change in market value.

An FAO (2015) study in the Caribbean calculated total estimated economic losses in cassava as they were measured as "unfit for sale" in Trinidad and Tobago (US$500,000) and Guyana (US$839,000). Gautam and Buntong (2015) reported on economic losses/hectare and estimated at approximately US$4213 per ha for tomato and US$2208 per ha for leafy mustard in Cambodia. Massive quantities of food are lost due to spoilage and infestations on the journey from field to consumers. Only a few studies reported PHLs in terms of nutrition or calories (Kitinoja and Dandago, 2017). Five PHL studies for COMCEC reported calorie losses up to 10 to 45% in maize in Uganda. Maize losses at the farm level equals 280,000 to 420,000 tons/year taking a conservative estimate on farm losses of 10 to 15%. Maize has a food value of 3700 kilocalories/kg, which means that PHLs in food value is a minimum of 1.04 trillion kilocalories (Kitinoja et al., 2016). This amount could have fed 1.14 million people for a full year at 2500 Kcal/day or 3.4% of Uganda's population of nearly 34 million. Worldwide, approximately 9000 species of insects and mites, 5000 species of plant pathogens and 8000 species of weeds cause damage to crops. An insect pest causes an estimated 14% of loss, plant pathogens cause 13% and weeds a 13% loss. It has been estimated that a minimum of 47,000,000 metric tons of durable and 60,000,000 metric tons of perishable crops fall prey to various pathogens (Sharma, 2014). This indicates an urgent need for greater attention toward reducing PHLs in order to address the world's food security challenges and to determine what may be happening to these lost foods. However, considerable attention has been drawn to the enormity of losses and wastage due to food spoilage, but only a limited number of resources has been directed toward finding an appropriate solution

2.7 FOOD SECURITY: AN URGENT NEED

However, in spite of large claims of excessive and ample food production, there are counter claims that focus on the depletion of our natural resources; further trends like increasing urban populations and the shift of lifestyle and diet patterns of the rising middle class in emerging economies, along with climate change, puts considerable pressure strain on the planet's resources. The world will struggle to produce food in the face of rising population, limited energy supplies and the degradation of soil and fresh water. At the close of the 20th century, astonishing advances in agricultural productivity and human ingenuity have not been translated into a world free of hunger and malnutrition (Sharma, 2014).

According to the FAO, food production will need to grow by 70% to feed the world population that will reach 9 billion by 2050. Demand for cereals, for both food and animal feed uses, is projected to reach some 3 billion tons by 2050, up from today's nearly 2.1 billion tons. The estimates show that feeding the world population of 9.1 billion people in 2050 would require raising overall food production by 70% between 2005/2007 and 2050. The International Food Policy Research Institute estimated that global demand for cereals will increase by 41% to 2490 metric tons by 2020 and for roots and tubers will be raised by 40% to 855 million tons.

International agencies, monitoring world food resources, have acknowledged that promoting food security through loss reduction is the most feasible and sustainable method of increasing sufficient food for everyone in a world of finite resources.

Consequently, there is a need for an integrated and innovative approach to the global effort of ensuring sustainable food production and consumption. Obviously, one of the major ways of strengthening food security is by reducing these losses.

2.8 CHEMICALS IN USE

As with many inventions, development of the first fungicide was the result of good observations. The first use of brining of grain with salt water took place in the middle of the 17th century to control bunt of wheat. The observations made by Millardet in France provided sufficient evidence that grape vines that had been sprayed with a bluish-white mixture of copper sulfate and lime to deter pilferers, could also effectively control downy mildew. Up until the 1940s, chemical disease control relied upon inorganic chemical preparations. The decade from 1960 to 1970 saw a rapid expansion of research and development along with a fast growth of these chemicals in markets. In this decade, the most widely used protectant fungicides, mancozeb and chlorothalonil, were introduced. The dithiocarbamates and later the phthalimides represented a major improvement over the previously used inorganic fungicides in that they were more active, less phytotoxic and easier to prepare by the user.

Broadly, these pesticides are classified as non-systemic and systemic, and act as "protectants" or "eradicants." Several inorganic and organic sulfur compounds, copper pesticides, organomercurial compounds, phthalimides and quinones are designated as non-selective chemicals. However, benzimidazoles, carboxins, several sterol inhibiting chemicals such as azole, morpholine, piperidine, piperazine and pyridine, phenylamides and site-specific pesticides such as strobilurins (QoI) and methyl bromide and halogenated hydrocarbons are the commonly used chemicals for control of various diseases. Fungicides, commonly applied as dips or sprays, include benomyl, thiabendazole, prochloraz and imazalil. Fumigants, such as sulfur dioxide, carbon dioxide, ozone and ammonia are sometimes used for disease control. Fruit wraps or box liners impregnated with the fungicide biphenyl are used in some countries for the control of *Penicillium* in citrus. All uses of methyl bromide are being phased out to avoid any further damage to the protective layer of ozone surrounding the earth. These pesticides can control plant diseases by altering the host's cell wall constitution or the host metabolism. In pathogens, these chemicals may damage the cell membrane, inhibit synthesis of cell wall substances, inactivate enzymes and co-enzymes or may precipitate pathogenic proteins.

2.9 CHEMICALS: A HAZARDOUS THREAT

Pragmatic approaches to control diseases by chemicals have been practiced since ancient times. Application of chemicals became a norm as it was observed that crop loss from pests declined to 35% from 42%, otherwise the loss of cereals, fruits and vegetables from pathogenic diseases reached 32%, 54% and 78%, respectively. About 23 million kg of fungicides are applied to fruits and vegetables annually and it is generally accepted that production and marketing of these products could not be possible without their use. In India, above maximum pesticide residues were found on vegetables (Srivastava and Sharma, 2015). The largest pesticide producers and consumers of pesticides in the world are the United States, China, France, Brazil and Japan. Projected growth of the agrochemical market is 3.4% over the next five years.

The history of pesticide development has been instructive in terms of benefits derived as well as the hazards, which accompany the indiscriminate use of these chemicals. The application of any chemical to a crop or food raises the question of risks and benefits. These hazardous chemicals contaminate the environment, enter the food chain, build residual toxicity in animal and plant tissues and introduce health hazards to humans such as cancer, hormonal imbalances and respiratory troubles. They also aim at non-target microorganisms, destroy native insects and increase the incidence of pest resurgence due to the development of resistance in pests and pathogens toward these chemicals.

The first case of benzimidazole resistance was reported in powdery mildew in greenhouses in 1969, one year after its introduction. By 1984, resistance had been reported on many of the pathogens against which benzimidazoles are active (Brent and Hollman, 2007). The reason for the rapid development of resistance was that these fungicides were single site inhibitors of fungal microtubule assembly during mitosis, via tubulin-benzimidazole-interactions.

The world became so obsessed with the idea of increasing crop productivity that instead of the era of "Green Revolution," it was dominated by shades of "Gray" (Sharma and Bhandari, 2014). It completely forgot that produce should be safe for human consumption and the environment. According to a report by the WHO and UNEP, there are more than 26 million human pesticide poisonings with about 220,000 deaths/year worldwide. Despite recent technological advances in the development of resistant varieties of crop plants using genetic engineering approaches, and with discoveries of novel, site-specific fungicides and continually evolving crop protection practices, plant pathogens continue to find opportunities to destroy crop plants. The aim is not only to prevent losses but "*Go Green*" in order to recover and reap the benefits of the "*Stolen Harvest*." Hence, the dire need to prevent biodeterioration of food through safe and effective measures is brought to the forefront (Figure 2.2).

2.10 THE PATH BEYOND 2020: TOWARD FOOD SECURITY

Whether due to a unidirectional approach, shortsightedness or lack of adequate planning and mismanagement in the pre- and postharvest environment, the inputs directed toward increasing productivity could not generate the desired results and instill a sense of food security. The challenges for producers in managing these diseases are ever increasing as consumer demands for year-round production of fresh products with reduced or no pesticides residues continue to grow. Concerns over the potential impact of disease management practices including the use of fungicides on the environment or on consumer health have promoted producers to examine alternatives to combat these postharvest diseases. The world's arsenal is not enough to treat all the parasitic organisms infecting our food crops and thus we need to explore new strategies for the management of pathogens that are on increase.

2.11 STRATEGIES FOR POSTHARVEST DISEASE MANAGEMENT

Postharvest disease management (Figure2.3) may be achieved by preventing infection, eradicating infection or delaying symptom development so that the product can be safely marketed and consumed before diseases appear (Fallik, 2014).

2.11.1 Prevention of Injury

As many postharvest pathogens gain entry through wounds or infect physiologically damaged tissue, prevention of injury at all stages during production, harvest and postharvest

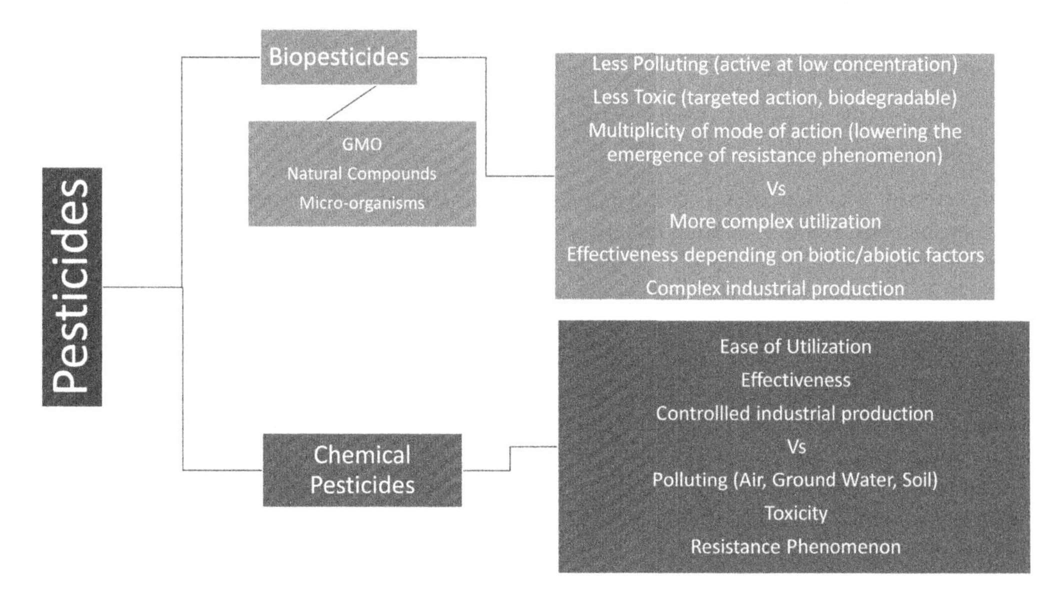

FIGURE 2.2 Advantages and limitations of biopesticides and chemical pesticides.

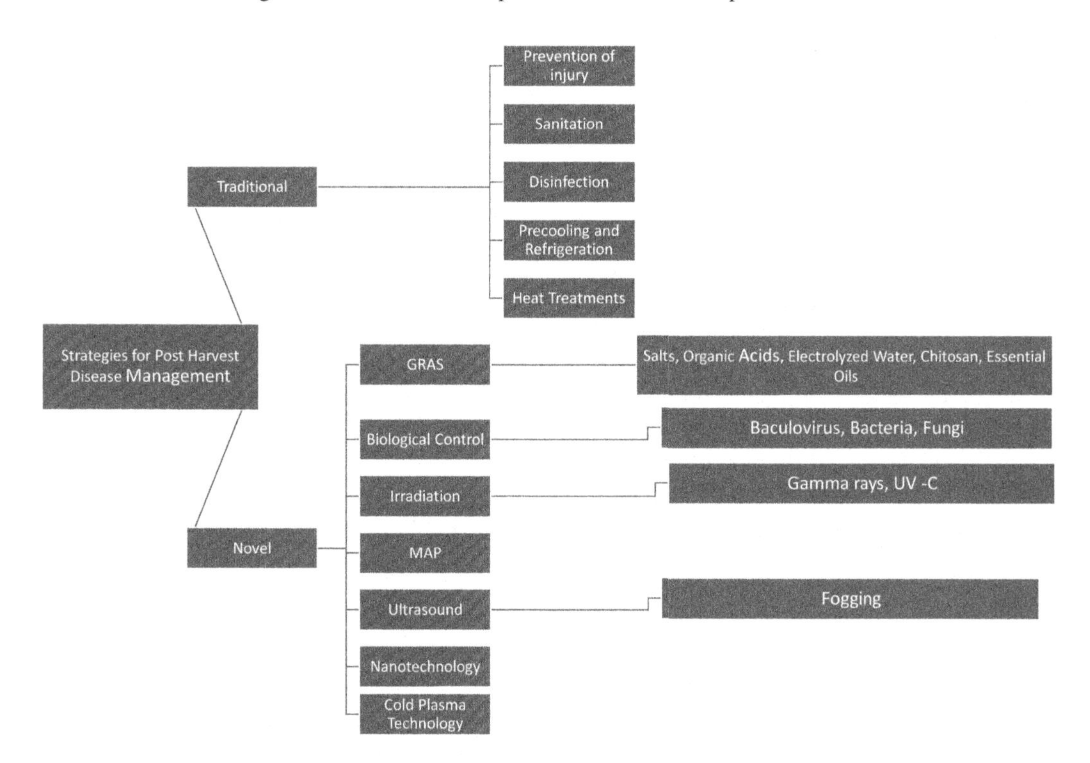

FIGURE 2.3 Strategies for postharvest disease management.

handling is critical. Injuries can be mechanical (e.g. cuts, bruises and abrasions), chemical (e.g. burns), biological (e.g. insect, bird and rodent damage) or physiological (e.g. chilling injury, heat injury). Injuries can be minimized by careful harvesting and handling of produce, appropriate packaging of produce, controlling insect pests in the field, storing produce at the recommended temperature and applying postharvest treatments correctly. Where injuries are present, the process of wound healing can be accelerated in some instances through manipulation of the postharvest environment (e.g. temperature and humidity) or by application of certain chemical treatments.

2.11.2 Sanitation

Sanitation during production, harvesting and handling is critical in minimizing sources of inoculum for postharvest diseases. A good understanding of the life cycle of the pathogen is essential to effectively reduce sources of inoculum and initial inoculum potential. Practices, which make the crop environment less favorable to pathogens, will help reduce the amount of infection that occurs during the growing season. For example, in tree crops, pruning and skirting can increase ventilation within the tree canopy, making conditions less favorable for the development of pathogens; removing dead branches and leaves entangled in the tree canopy is also helpful in minimizing inoculum build-up. In many soil-borne pathogens, minimizing contact of leaves and fruit with the soil will reduce contamination and infection.

Water used for washing or cooling produce can become contaminated with pathogen propagules if not changed on a regular basis and if a disinfectant is not incorporated. In many diseases, overhead irrigation can encourage pathogen spread and infection, so trickle or micro-sprinkler irrigation systems may be more appropriate.

Packing and grading equipment that is not cleaned and disinfected on a regular basis can also be a major source of inoculums. Containers used for storing and transporting the goods can harbor pathogen propagules, particularly if recycled a number of times without proper cleaning. Reject produce that has not been discarded from the packing shed or storage environment provides an ideal substrate for postharvest pathogens. Inoculum for infections occurring after harvest commonly originates from the packing shed and storage environments.

2.11.3 Disinfection

Decontamination of fresh produce and storage facilities is generally a prerequisite for postharvest decay control as the pathogens survive if they reside within the wounds on the host or if they are present as incipient, latent or quiescent infections within host tissue. Sanitizers cause rapid mortality of pathogens they come in contact with and do not deposit a persistent antimicrobial residue in treated products. These disinfectants are characterized by their low impact on the environment, leaving no or non-toxic residues on the food matrix, having multiple mechanisms of action, it is quite unlikely that resistance in target microorganisms could develop. The possibility to use disinfectants as fumigants is interesting for postharvest decay control since their use requires minimal product handling and wetting and drying operations are not required.

Commonly used disinfectants are chlorine, chlorine oxide, hydrogen per oxide, ozone and ethanol. All of these compounds are approved for use in some food contact roles. A level of 50 to 100 ppm of active chlorine readily kills microorganisms suspended in dump tanks and flumes. Although chlorine effectively kills spores in water, it does not protect wounded tissue against subsequent infection from spores lodged in wounds. Chlorine as a hypochlorous acid can be obtained by adding chlorine gas, sodium hypochlorite or dry calcium hypochlorite. Levels of chlorine must be constantly monitored as organic matter in water inactivates chlorine.

There is no disinfectant that is absolutely better than the other. Their applicability to control postharvest deterioration depends on many aspects, i.e. on fresh produce, the length of the produce storage, the characteristics of the postharvest facilities, the orientation toward organic or conventional agriculture and the possibility to integrate the disinfection operation with other technologies. Their mode of action is through inhibiting germination and oxygen consumption of sporulating structures and ultrastructural disorganization after lethal sequential oxidative treatment, showing severe cellular damage.

In the USA, the United States Environmental Protection Agency (EPA), the Food and Drug Administration (FDA, 2018) and state agencies regulate sanitizers. In Europe, the Food Safety Authority (EFSA) is the independent advisory agency concerning risks associated with the food chain and the safety of all food additives and plant protection products that are authorized in Europe. European member states are responsible for controlling the usage of sanitizers according to recommendations by EFSA.

A possible limitation to the large-scale use of some disinfectant, such as ozone, is the initial cost associated with the equipment, while hydrogen peroxide and organic acids could be considered more economic alternatives. On the other hand, to be fully effective, high doses of disinfectants are sometimes needed, therefore, potential damage to facilities, threats to operators and phytotoxicity signs on fresh produce should be considered and monitored. However, in some cases, disinfection is a precondition to successful implementation of major postharvest technologies and, in particular cases, it can become the major technology (Feliziani et al., 2016).

2.11.4 Pre-cooling and Refrigeration

In the field, the heat generated by the sun and the respiration of the produce combine to heat up the produce, this accumulation of "field heat" reduces the postharvest life of the product and it has to be transported rapidly to the packing shed without delay. Pre-cooling requires a greater refrigeration capacity than cool storage and is often best done as a separate step. Hydro-cooling with cold water that drenches and forces air cooling through stacks, which ensures proper air distribution, and packing with ice are the systems most commonly used, with the choice depending on the individual requirements of the commodity.

Refrigeration is the most important tool for extending the shelf life of the product. To maintain storage temperature of 0 °C the temperature of the coils will have to be appreciably below 0 °C. The lower the average temperature of the cooling coils, the more moisture will be removed. The drier and cooler air then circulates around the room where it warms and picks up moisture. The more moisture that freezes on the refrigerator coils, the greater the frequency of defrost cycles and these make temperature management control more difficult to attain. An even distribution of air produces a room with a consistent temperature, but if the flow of cooled air is "short circuited" back to the coiling coils, the areas starved of circulation will become warmer.

2.11.5 Heat Treatments

Heat treatments have been in use for decades to control diseases and pest infestation when the effectiveness of hot water in controlling smuts on/in cereal seeds was reported. The specific temperature at which the germination or growth of the pathogen is completely inhibited is known as the thermal inactivation point. Heat treatments control decay via three mechanisms: (1) direct germicidal effect on pathogens; (2) inducing defense mechanisms in the plant host; and (3) melting and spreading the distribution of cuticular waxes on the fruit surface, thus obstructing and sealing open stomata, wounds and microcracks, thereby limiting the sites of pathogen penetration. Because of the known deleterious effects of excessive heat on cell physiology, temperatures too high and/or times too long could easily cause damage to the commodity (Escribano and Mitchem, 2014). Luengwilai et al. (2012a) applied hot water from 30 to 50 ÅãC for 3–9 min to mature green tomatoes of different sizes, prior to 2.5 ÅãC storage. Temperatures higher than 40 ÅãC damaged the fruits, while 40 ÅãC for 7 min was the most effective combination. Treatments at 60 ÅãC for 60 s reduced the incidence of brown rot from 80% to less than 2% in plums, while in nectarines, decay incidence decreased from 100% to less than 5% on fruit stored at 20 ÅãC and from 73 to 28% on cold-stored fruit. Liu et al. (2012) treated peaches with hot water at 40 ÅãC for 10 min, and indicated that decay control was due to a direct effect of the heat on the host by enhancing the defense-related enzyme phenylalanine ammonia lyase, and on *Monilinia fructicola*, it was associated with an increase in intracellular reactive oxygen species, mitochondrial dysfunction and a decrease in ATP. Lafuente et al. (2011) studied the effect of hot air at 37 ÅãC for 1 2 days on tolerance of "Fortune" mandarins to a single (16 days) or double (32 days) quarantine treatment at 1.5 ÅãC followed by 4 days at 20 ÅãC. No off-flavors or any changes were found in the flavonoids, vitamin C or antioxidant capacity, weight loss, firmness, color or acidity. Li et al (2013), working with papaya, showed that hot water at 54 ÅãC for 4 min controlled *Colletotrichum gloeosporioides* in the fruit peel by inducing the local expression of defense-related proteins. In addition, heat melted the fruit wax, creating a mechanical barrier against pathogen penetration. Similar results were obtained by Yuan et al. (2013) who studied the effects of hot water dipping at 53 ÅãC for 3 min in muskmelon. The treatments reduced decay caused by *Trichothecium roseum, Alternaria alternata, Fusarium* spp. and *Rhizopus stolonifer*. The treatment cleaned the surface of the fruit, melted the epicuticular waxes, covered and sealed stomata and also enhanced the activities of the defense-related enzymes phenylalanine ammonia lyase, cinnamate-4-hydroxylase, 4-coumarate: CoA ligase, polyphenol oxidase and peroxidase. In addition, it increased the antifungal compounds, cinnamic, coumaric, caffeic and ferulic acids. The heat treatment also resulted in higher levels of phenolic compounds, flavonoids, lignin and hydroxyproline-rich glycoproteins, and maintained fruit firmness by suppressing the activities of cell wall degrading enzymes. Jing et al. (2010) obtained very encouraging results in strawberry with hot water rinsing and brushing at 55–60 ÅãC for 20 s; fruit was not damaged but heat clearly redistributed the epicuticular wax layer preventing pathogen penetration. After cold storage, treated fruits showed lower decay incidence (0–22.2%) as compared with 58.6% in control fruits.

Heat is generally applied as a short treatment preceding cold storage. There are five main types of application: (1) forced air; (2) vapor; (3) water baths; (4) water rinsing; and (5) water rinsing with brushing. The protocol for each type varies depending on the fruit species, cultivar, size, harvest maturity and growing conditions. Treatment effects are localized rather than systemic; uniform coverage is necessary to attain optimal benefits.

One of the main goals for effective future research should be to minimize the adverse effects of heat treatments on the flavor quality of fresh fruits and vegetables, which has been overlooked for many years, to understand the physiological processes occurring in internal and external tissues during and after heat treatment. The identification of biological markers to facilitate the selection of adequate heat protocols that are not injurious would also be highly beneficial. A combination of heat treatments with other postharvest technologies such as ethylene suppressors, plant growth regulators, edible coatings, biological control agents and adequate packaging could not only maintain, but also improve the sensory quality of the commodity.

2.12 NEW TECHNOLOGIES FOR POSTHARVEST DISEASE CONTROL

Efforts of researchers have led to the development of novel control tools that can be used as alternatives to synthetic fungicides. This kind of preharvest application can be grouped into the following categories: (1) decontaminating agents that are "generally recognized as safe" (GRAS); (2) biological control agents (BCAs); (3) irradiation; (4) ultrasound; (5) nanotechnology; and (6) cold plasma technology. All of these tools can be used alone or as combinations to benefit from additive or synergistic effects (Romanazzi et al., 2009).

2.12.1 GRAS Agents

Many compounds produced naturally by microorganisms and plants have fungicidal properties. Plants possess various biochemical and structural defense mechanisms, which protect them against infection (Palou, 2018). Some of these mechanisms are in place before the arrival of the pathogen (i.e. constitutive resistance), while others are only activated in response to infection (i.e. induced resistance). Although there is a growing interest in this area, relatively little is known about host defense responses in harvested commodities. Being naturally synthesized and biodegradable, these compounds may be more desirable than synthetic chemicals from a consumer viewpoint but their potential toxicity to humans needs to be evaluated before useable products are developed. Many compounds produced naturally by microorganisms and plants have fungicidal properties. Chitosan, for example, is not only an elicitor of host defense responses but also has direct fungicidal action against a range of postharvest pathogens. Antibiotics produced by various species of *Trichoderma* have potent antifungal activity against *Botrytis cinerea*, *Sclerotinia sclerotiorum*, *Corticium rolfsii* and other important plant pathogens.

Low-toxicity chemicals that are recognized as GRAS include some essential oils, plant extracts and other natural compounds of varying composition, but also synthetic inorganic or organic salts. Noteworthy research results as well as successful commercial applications emphasize the suitability of these antimicrobial agents as an alternative treatment for the control of postharvest diseases.

2.12.1.1 Salts

Salts such as acetates, benzoates, bicarbonates, carbonates, parabens, silicates and sorbates have demonstrated bioactivity against various postharvest pathogens (Cavalcante et al., 2014; Mshraky et al., 2016). When compared with other GRAS compounds, the advantages of these salts are their great availability, ease of handling and use and low cost (Cerioni et al., 2013). It has been shown that salts such as calcium propionate completely inhibit the mycelial growth of *B. cinerea* at a concentration of 5% (w/v). This may be due to the fact that it changes the plasma membrane, thus inhibiting essential metabolic functions. High concentrations of sodium benzoate inhibited the growth of *A. alternata* (Montesinos et al., 2016) and this is attributed to the fact that weak acids within the cell create a dissociation, causing protons and anions to accumulate, which cannot cross the plasma membrane again (Palao et al., 2016). Bicarbonate salts were also effective in inhibiting the growth of various pathogens, the efficiency of this salt is attributed to the fact that it creates cellular ionic imbalances affecting the synthesis of polyamines and DNA during cell division (Minocha et al., 1992: Vilaplana et al., 2018). In a study, it was found that when potassium sorbate was applied at low concentrations (1%) on infected tomato fruits, a significant reduction of disease incidence was obtained against *Rhizoctonia solani, Colletotrichum coccodes, Botrytis cinerea* and *Alternaria solani* (Jabnoun et al., 2016). The effectiveness of potassium sorbate is attributed to its dissociated form, which has the ability to penetrate cell membranes, causing an internal imbalance affecting enzymes related to the growth of microorganisms. In a study on citrus, potassium sorbate at 3% solution was applied in combination with heat treatment (62 °C, 60 s) to evaluate their effectiveness in reducing the disease incidence caused by *Penicillium* strains. The results showed that the treatments can reduce the disease incidence on "Clemenules" (20%) and "Nadorcott" mandarins (25%), "Fino" lemons (50%), tangerine "Ortanique" (80%) and "Valencia" oranges (95%) stored 20 °C for 7 days. Besides, when infected and treated, fruits were stored at 5 °C for 60 days, on "Valencia" oranges, the green mold (*Penicillium digitatum*) was reduced by up to 95% and blue mold (*Penicillium italicum*) up to 80% (Montesinos-Herrero et al., 2016). Green mold caused by *P. digitatum* was reduced by up to 80% on infected oranges by the application of sodium benzoate (3%) in combination with hot water (53 °C) for 60 s, and fruits were stored at 20 °C for 7 days (Palao et al., 2018). The effectiveness of sodium benzoate is related to its undissociated form of benzoic acid, which can enter the cell membrane, and its neutralization within the cell leads to acidification of the intracellular space, thus affecting the growth of fungus. It has been found that the application of calcium chloride salts delay ripening and senescence, reduces respiration, extends shelf life, maintains firmness and reduces physiological disorders of many fruits and vegetables. $CaCl_2$ has been found to delay fruit color and development in tomatoes (Arah et al., 2016).

These organic salts cause inhibition of mycelial growth and alteration of the morphology of the hyphae, and, in addition, the germination of conidia is inhibited and causes alterations in external morphology. The application of silicate as a postharvest treatment presents results similar to those reported with chitosan and tebuconazole. In addition, the application of silicates can improve some quality attributes of fruits like the maintenance of weight, while reducing the respiration rate due to the capacity of the silicates to deposit between the cell wall and the cell membrane, thus decreasing the permeability. The stomata are covered, maintaining the humidity of the fruit and reducing its respiration (Tesfay et al., 2011; Moscoso- Ramirez et al., 2016).

These salts have been applied on orange, melon, avocado (Coskun et al., 2018) and papaya (Bandara et al., 2015), with good results due to the treatments, which form a barrier against pathogens on the surface of the fruit. With the accessibility of these compounds and their effectiveness, it is possible that they can be adopted to reduce the use and application of traditional fungal agents.

2.12.1.2 Organic Acids

The applications of organic acids to control postharvest decay refers to the use of disinfecting antimicrobial agents that are allowed in food industries and are applied on the surface of eatables. They include short-chain organic acids, characterized by one or more carboxyl groups. However, the term "organic acids" generally refers to organic compounds that have acidic properties and includes both aliphatic and aromatic components that sometimes could even be strong acids and are not suitable for food matrices. Some other natural compounds from plants, such as jasmonic acid and salicylic acid, are organic acids, used in postharvest disease management not as disinfecting agents, but mainly for their ability in induction of plant defense mechanisms (Kumar, 2014; Saavedra et al., 2016). The FDA (2018) classifies acetic, citric, malic, tartaric and propionic acids as GRAS compounds. When applied as fumigants, they prove to be efficient candidates for postharvest decay control since their use requires minimal product handling and wetting and drying operations are not required. For best effectiveness, it is important to establish an optimal combination of volatile concentration and duration of fruit exposure. They can be added to food matrices as preservatives, since they can act as antimicrobials, preventing food spoilage from bacteria and fungi, or as antioxidants, slowing or preventing changes in color, flavor or texture and delaying rancidity.

2.12.1.2.1 Acetic Acid, Peracetic Acid, and Hydrogen Peroxide

The use of acetic acid ($C_2H_4O_2$), peracetic acid ($C_2H_4O_3$) and hydrogen peroxide (H_2O_2) on the processing and marketing industries of fruits and vegetables have been considered as a useful tool to control different kinds of pathogens (Feliziani et al., 2016). The FDA has approved the use of these compounds because their decomposition products are water, oxygen and acetic acid. These are not toxic compounds and are friendly with the environment (FDA, 2018). The inactivation capacity of these compounds on pathogens is based on their high oxidizing power, producing reactive oxygen species (ROS) that generate instability in biomolecules such as DNA, lipids and proteins, which are vital for the correct cellular functioning of pathogens. Usually, the application of these sanitizers at the postharvest stage in fruits and vegetables is by spraying, dipping and fumigation. Indeed, high concentrations of vapors could induce fruit injury or alteration of sensory characteristics, while insufficient concentrations are ineffective. Furthermore, most fumigants are unable to penetrate into fruit tissue, so they cannot control latent infections. The use of acetic acid (CH_3COOH) as a disinfecting agent is an ancient practice. During the last couple of decades, the effectiveness of antimicrobial activity of acetic acid vapors for the control of postharvest decay of fruit and vegetable has been investigated and encouraging results have been collected on produce, including apples, pears, tomato, kiwi, table grapes and others. Culture media amended with 3 g /L of acetic acid completely inhibited the growth of *Colletotrichum gloeosporioides* (Kang et al., 2003). The inhibition of this fungus by acetic acid accelerated along with a decrease in pH from 6.0 to 4.0, suggesting that inhibition might be enhanced by an undissociated form of acetic acid. It has been suggested that the antimicrobial activity of acetic acid relied on its ability to inhibit fungal respiration, as measured by the augmented dissolved oxygen in the culture media containing acetic acid. The hypothesis of structural damage of fungal cells by acetic acid was discarded as when acetic acid was eliminated from the culture media, the fungus was able to restart growing (Kang et al., 2003). In another study, when acetic acid was used as a fumigant, 4 mL/L for 6 min can be considered a limit required to completely inhibit the growth of *B. cinerea* (Lagopodi et al., 2009). Complete inhibition of the growth of *A. alternata* occurred when the fungus was exposed to fumes of 8 mL/L acid concentration or the culture medium was amended with 1.7 mL/L of acetic acid (Alawlaqi and Alharbi Asmaa, 2014). Acetic acid reduced decay caused by *M. fructicola and R. stolonifer* on peaches and sweet cherries at a concentration higher than 2 mg/L; acetic acid vapors gave phytotoxic signs as light brown streaks and pitting on fruit surfaces. On the contrary, acetic acid fumigation on apples before storage did not affect the quality or left vinegar aroma and it was found to be as effective as thiabendazole in reducing decay (Sholberg et al., 2001). Similarly, fumigation of table grapes controlled both *B. cinerea and P. expansum* decay, without producing any quality deficiencies and was as effective as sulfur dioxide fumigations applied at commercial rates. Application of acetic acid vapors to pears reduced the pathogenic microflora to lower levels and after the fruit had been stored for 4 months, reduced rots by 51% on fruit inoculated with *B. cinerea* (Sholberg et al., 2004). Triple fumigation of 30 min each with 2 mg/L acetic acid vapors resulted in a 56% reduction of decay caused by *B. cinerea* in strawberry (Hassenberg et al., 2010). Acetic acid completely prevented decay of tomato when the fruit were dipped in 4% acetic acid solution or exposed to its vapor at 40 mL/L (Alawlaqi and Alharbi Asmaa, 2014). In another study, even vinegar vapors, containing 4–6% acetic acid, effectively prevented conidia germination of *M. fructicola, B. cinerea* and *P. expansum* causing decay on stone fruit, strawberries and apples, respectively (Sholberg et al., 2000). In vivo studies on fresh and fresh-cut horticultural products confirm their antimicrobial capacity with the reduction of human pathogens such as *Staphylococcus aureus, Escherichia coli, Streptococcus mutant, Salmonella Thompson* and *Listeria monocytogenes.* The application of acetic acid, peracetic acid and hydrogen peroxide also decreases the microbial pollution of aerobic mesophiles, molds and yeast. In fruits like guavas, peaches and tomatoes, the antifungal effect of these sanitizers has been confirmed with the inhibition of phytopathogens such as *Rhizopus stolonifer, Monilinia fructicola, Alternaria alternata, Botrytis cinerea, Fusarium solani* and *Rhizoctonia solani.*

Apple fruit treated using hot acetic acid solutions (1, 2 and 3%) at 50 °C for 1, 2 and 3 min had lower decay during storage compared with untreated fruit, in particular, the most effective combination was 2% acetic acid solution for 3 min and 3% acetic acid solution for 2 min (Radi et al., 2010).

The inhibition of *A. alternata* and *B. cinerea* in tomato fruits treated with acetic acid by immersion (50 ml/L for 3 min) and fumigation (50 µl/L for 30 min) was evaluated (Alawalaqi and Alharbi, 2014). The growth inhibition of the pathogens tested ranged from 90 to 100% by immersion and fumigation, respectively. In strawberry fruits infected with *B. cinerea* and treated with peracetic acid, the lower incidence (66%) of gray mold disease was obtained with the fumigation compared with the immersion method (80%). In peppers, fruits treated with the solution of hydrogen peroxide and applied by dipping at 15 mM for 30 min increased their shelf life and the fruits maintained their appearance after 2 weeks stored at 20 °C compared with control fruits.

2.12.1.2.2 Oxalic Acid

There is evidence suggesting the beneficial effects of applying oxalic acid to delay quality deterioration and extend the shelf life of various fruits. Oxalic acid ($C_2H_2O_4$) is used as an antioxidant compound since it can inhibit enzymes responsible for browning such as polyphenol oxidase by chelating copper from the active site of the enzyme itself (Altunkaya and Gökmen, 2009). Oxalic acid has antimicrobial properties against phytopathogenic fungi as well (Kang et al., 2003). Immersion of artichokes into a solution of oxalic acid delayed postharvest deterioration by reducing respiration, delaying color change and chlorophyll degradation, and reduced microbial populations, but not influencing total phenol content or antioxidant activity (Ruíz-Jiménez et al., 2014). However, when spinach leaves were immersed in a solution of oxalic acid, it yellowed their leaves but preserved phenol content, and its effectiveness to reduce mesophilic aerobic bacterial populations was similar to hypochlorite washing (Cefola and Pace, 2015). In plums treated with oxalic acid, reduced ethylene production and delayed softening of the fruit were observed which in turn slowed down ripening and senescence (Wu et al., 2011). The inhibition of plum softening was associated with decreased enzymes activities and retarded pectin solubilization/degradation. Preharvest application of oxalic acid to kiwi plants led to fruit with higher ascorbic acid content at harvest, slowed the decreases in fruit firmness and decreased the natural disease incidence and lesion diameter in fruit inoculated with *P. expansum*, indicating that the organic acid treatments improved kiwifruit quality (Zhu et al., 2016). When mango fruits were immersed in an oxalic acid solution, incidence of fruit anthracnose caused by *C. gloeosporioides* was reduced and even the ripening of fruits was delayed (Razzaq et al., 2015). Similarly, immersion of litchi fruit in oxalic acid reduced pericarp browning due to an increase of membrane integrity and to the retention of relatively low peroxidase activity during storage (Zheng and Tian, 2006). A proteomic approach on jujube fruit revealed that when oxalic acid was applied it affected ethanol and ethylene metabolism, causing a delay in senescence and an increase resistance of jujube fruit against blue mold caused by *P. expansum* (Wang et al., 2009).

2.12.1.2.3 Jasmonic and Salicylic Acid

In general, plants have their own defense mechanisms acting against pathogen resistance processes such as pre- or post-existing antimicrobial compounds that induce defense mechanisms. Resistance induced to disease in plants by biotic and abiotic elicitors is a very effective method for restricting the spread of fungal infection (Soylu et al., 2003). Jasmonic acid (JA), methyl jasmonate (MeJA) and salicylic acid (SA) have been found to occur naturally in a wide range of higher plants. The signal molecules SA, JA and MeJA are endogenous plant growth substances that play key roles in the development and response to environmental stresses. These signal molecules induce specific enzymes which catalyze certain biosynthetic reactions to form defense compounds such as polyphenols, alkaloids or pathogenesis-related (PR) proteins resulting in induction of defense responses and provide protection for the host from pathogen attack. In the plant–microorganism interaction, salicylic acid activates the enzyme phenylalanine ammonia lyase (PAL), which is key in the biosynthesis of phenolic compounds (Potlakayala et al., 2007). Peroxide has an antibiotic activity against pathogens; it could intervene in the signaling cascade for the expression of defense genes. SA regulates activities of enzymes, peroxidase (POD) and polyphenol oxidase (PPO), which are related to the induced defense of plants and fruits against biotic and abiotic stresses (Idrees et al., 2011).

MeJA modulates many physiological processes including responses to environmental stresses. Studies indicate that acquired systemic resistance depends on signaling mediated by MeJA and is associated with some signal transduction systems, which induce particular enzymes that catalyze biosynthetic reactions to form defense compounds such as polyphenols, alkaloids, ROS or PR proteins. The exogenous application of MeJA induces and increases the activity of defense enzymes such as β-1,3-glucanase (β-Gluc), chitinase, PPO and PAL, which are enzymes providing resistance to diseases (Haggag et al., 2010). Application of MeJA effectively suppressed gray mold rot caused by *Botrytis cinerea* in strawberry (Moline et al., 1997) and decreased fruit decay on papaya fruit infected by *C. gloeosporioides* and *Alternaria alternata* (Gonzalez-Aguilar et al., 2003). For grapefruits inoculated with *Botrytis cinerea*, the application of MJ (0.01 mM) increased the enzymatic activity of PAL and PPO (Jiang et al., 2015). Similar behavior was observed on avocado fruits with an increase in the activity of the resistance enzymes, chitinase, β-1,3-glucanase and PAL. The application of MeJA (10 mM) in cranberry fruit inoculated with *Penicillium citrinum* increased the levels of POD and PAL activity (Wang et al., 2014). There are several reports enlarging on the application and effects of SA in fruits for the induction of defense mechanisms against pathogens. In tomato fruit, resistance against *Botrytis cinerea* using SA as a resistance inducer showed a significant increase in the expression level of the PR1 gene while a lower expression was observed in the PR2 and PR3 genes (Wang et al., 2011). The postharvest application of SA (2 mg/mL) showed a decrease in the severity of anthracnose in mango cv. Kensington Pride (Joyce et al., 2001). Thus, the use of inducers offers several advantages for postharvest disease control; further, they can be combined with other methods to enhance their efficacy (Jin et al., 2009).

2.12.1.2.4 *Cinnamic, Malic, Sorbic, Decanoic, Propionic, Lactic, Benzoic, Citric and Formic Acids*

Cinnamic acid ($C_9H_8O_2$), an organic acid isolated from cinnamon bark, is generally used as a common food additive but also has antimicrobial activity against postharvest pathogens such as *B. cinerea, P. expansum* and *A. niger* (Sadeghi et al., 2013). When cinnamic acid was tested on table grapes, it controlled gray mold by inhibiting the growth of pathogens and inducing resistance in the host (Zhang et al., 2015). Cinnamic acid can damage the integrity of the plasma membrane and induce the production of intracellular reactive oxygen species, which were responsible for the reduction of the growth rate of *B. cinerea.*

Malic acid ($C_4H_6O_5$) significantly inhibited the growth of *C. gloeosporioides* through the inhibition of fungal respiration (Kang et al., 2003) and when it was used in vivo it prolonged the storage of banana dipped in a solution of 10.7 g /L malic acid (Huang et al., 2016).

Similarly, addition of sorbic ($C_6H_8O_2$), propionic ($C_3H_6O_2$) and decanoic ($C_{10}H_{20}O_2$) acids to culture media inhibited the growth of *Penicillium commune*, and treatments with sorbic acid in the presence of hydrochloric acid made litchi fruit less prone to decay caused by *P. commune* (Zhang et al., 2005). Complete inhibition was observed in the linear growth of *Geotrichum candidum, P. digitatum* and *P. italicum* when exposed to benzoic ($C_7H_6O_2$), citric ($C_6H_8O_7$) and sorbic organic acids at concentrations of 4% (El-Mougy et al., 2008). Fumigations with vapors of formic (CH_2O_2) and propionic acids controlled decay caused by *M. fructicola, P. expansum* and *R. stolonifer* on sweet cherry, *P. digitatum* on citrus and *P. expansum* on apples that were inoculated with spores of these fungi. In these cases, formic acid increased fruit pitting and stem browning of fumigated sweet cherry compared with the control and browning of the fruit peel was observed on grapefruit and oranges fumigated with formic acid (Joas et al., 2005).

These results confirm that the capacity of inhibition of these GRAS substances against phytopathogens depends not only on the concentration used but also on the exposure time, the microorganism tested and the application method. The individual application of acetic acid, peracetic acid and hydrogen peroxide has controlled microbial contamination in an acceptable way, but different reports have shown that their combination with other compounds and technologies such as ultrasonic, organics salts, essential oils, ultraviolet light, hot water and steam increases microbial control.

2.13 ELECTROLYZED WATER

Electrolyzed water (EW) was developed in Russia for water decontamination and regeneration, and great interest in this technology followed in other countries (Hao and Wang, 2019) for the sterilization of utensils, meats, cutting boards and, more recently, in livestock management and the sanitation of the washing waters of fresh and minimally processed fruit and vegetables (Fallanaj et al., 2015, 2016). Their applicability to control postharvest decay depends on the fresh produce, the orientation toward organic or conventional agriculture, the length of produce storage, the characteristics of the postharvest facilities and the possibilities to integrate the disinfection operations with other technologies (Rahman et al., 2016).

EW is typically produced by electrolysis of dilute solutions of NaCl, KCl or $MgCl_2$ in an electrolysis cell with or without a diaphragm, which separates the anode and cathode. In an electrolysis cell divided by a membrane, two types of EW are produced: the acidic electrolyzed water (AEW) and the basic electrolyzed water (BEW).

EW is generally considered safer and less expensive than most traditional preservation methods. The on-site production of EW, whatever the use, represents a great advantage because there are no chemicals to purchase or store, except for an inexpensive salt (NaCl or other salts). EW has minimal impact on the environment (Koseki et al., 2002). It is safe for the environment and the operators since little chlorine is released to the air. If non-chlorine salts (e.g. $NaHCO_3$) are used as electrolytes, health concerns with regard to chlorine in the air and in water are avoided and, consequently, the formation of respiratory irritant chlorinated organic compounds including chloramines (NH_2Cl), dichloramines ($NHCl_2$) and trichloromethanes ($HCCl_3$) is also avoided (Fallanaj et al., 2013). EW reverts to normal water after use, and its effectiveness has been verified within a large pH range (Park et al., 2004a). Since EW has multiple mechanisms of action, it is quite unlikely that resistance in target microorganisms will develop (Al-Haq et al., 2005).

Controlling fungal pathogen populations in wash water to reduce contamination on fresh fruit and vegetables is the primary reason for its postharvest use (Issa-Zacharia et al., 2010), as well as controlling spoilage microorganisms in minimally processed vegetables (Pinto et al., 2015). Relatively thin-walled species such as *Botrytis* spp. and *Monilinia* spp. died within 30 s or less, whereas thick-walled, pigmented fungi such as *Curvularia* spp. and *Helminthosporium* spp. survived 2 min or longer. EW immediately reacted with *Botryosphaeria berengeriana*, a pathogen colonizing the first few layers of the pear surface, and could not control growth of bacteria that entered into the fruit deeper than 2 mm. Koseki et al. (2004) reported that cucumbers and strawberry washed with BEW (pH 11.3) for 5 min and then soaked in AEW (pH 2.6) for 5 min showed a strong reduction of bacterial and fungal populations, being more effective as compared with ozone and sodium hypochlorite treatments. Recently, Guentzel et al. (2010) obtained interesting results on grapes and peaches artificially inoculated with *B. cinerea* and *M. fructicola*, respectively. The use of EW did not prevent lesion formation on apple fruit previously inoculated with *P. expansum*, but cross-contamination of wounded apples from decayed fruit or by direct addition of spores to a simulated dump tank was substantially reduced (Okull and Laborde, 2004).

Whangchai et al. (2010) observed the effect of EW on the reduction of *P. digitatum* growth on tangerines. In commercial trials, a 93% reduction *of Penicillium* spp. population in citrus wash water was observed 1 h after beginning the electrolysis process when water was supplemented with 1.25% of sodium bicarbonate (SBC); whereas, in the electrolyzed tap water without salt, similar results were observed only after 7 h. In addition, no rot development was observed in fruit exposed to electrolyzed SBC solution, whereas in absence of the salt, it reached 70% (Fallanaj et al., 2013). The use of EW on various food commodities did not negatively affect the organoleptic properties, color, scent, flavor or texture (Huang et al., 2008).

AEW and free chlorine content may be corrosive to some metals and may induce synthetic resin degradation (Tanaka et al., 1999) and effectiveness may be hindered by the presence of organic substances. Depending on the electrolyte and pH (e.g. in AEW), pungent chlorine gas is formed that can cause discomfort to operators and can be potentially toxic for plant produce. AEW can induce phytotoxicity, for example, white spots and slight necrosis were observed on flowers and leaf edges of some ornamental bedding plants following an AEW foliar spray.

On the other hand, to be fully effective, high doses of disinfectants are sometimes needed, therefore, potential damage to facilities, threats to operators and phytotoxicity signs on fresh produce should be taken into consideration and monitored. In view of the potential benefits of prolonging fruit storage furnished by disinfectant agents, further studies could optimize their integration into current practices of postharvest manipulation.

2.14 CHITOSAN

Chitosan is another elicitor of host defense responses. It is a natural compound present in the cell wall of many fungi as well as being derived from crab or prawn shell after deacetylation of chitin. It has proven efficacy for controlling several postharvest diseases (Sharif et al., 2018). It is the second most naturally and abundantly available biodegradable biopolymer, next only to cellulose. The antimicrobial activity of chitosan depends on its concentration, pH of the solution, molecular weight, degree of deacetylation and the target organism. On the fruit surface, chitosan forms a mechanical barrier (coating) offering several advantages like maintaining fruit firmness and color, a reduction in respiration rate and avoiding water losses, thus, extending the shelf life of fruits (Betchem et al., 2019). Chitosan can stimulate a number of processes including production of chitinase, accumulation of phytoalexins and increase in lignification. Several mechanisms of action have been proposed for chitosan: the interaction of the biopolymer (Romanazzi et al., 2017, 2018) with the microorganism causes changes on cell permeability affecting biochemical processes like homeostasis, fungal respiration and nutrient uptake, and the synthesis of proteins cause severe damage on fungal cells (Gutierraz-Martinez et al., 2018b). On the other hand, the induction of defense systems is initiated in the affected commodities by the production of important enzymes (phenylalanine ammonium lyase, polyphenol oxidase, among others) and plant immunity, favoring the adaptation of plants to biotic and abiotic stresses (Katiyar et al., 2015).

The induction of defense systems has been reported by the application of chitosan at the postharvest stage, preventing the development and dispersion of important pathogens such as *Colletotrichum gloeosporioides*, *Alternaria alternata*, *Rhizopus stolonifer* and *Fusarium oxysporum* (Hewajulige, 2012; Gutierraz-Martinez et al., 2015). Enzymatic activity is also affected by the curative application of chitosan, and it increases the activity of PPO, POD and PAL that induces the expression genes of β-1,3-glucanase and chitinase, which are involved in the defense against pathogens (Gutierraz-Martinez et al., 2017; Khaliq et al., 2016). Some studies reported an enhanced content of total soluble solids, ascorbic acid, nutritional value and acceptability (Devi and Kumari, 2015; Xoca-Orozoco et al., 2017). Chitosan is compatible with other substances like organic salts, gums and essential oils, and this alternative can improve their efficacy against pathogens due to a synergistic effect (Chavez-Magdaleno et al., 2018; Zeray et al., 2017). Although adequate information is available on the use of chitosan in postharvest tropical and subtropical fruits, it is still necessary to generate information on the regulation (activation and suppression) of genes that participate in both systems of acquired resistance and those that control the processes of physiological, enzymatic and physicochemical factors of maturation at the postharvest stage.

2.15 ESSENTIAL OILS

Essential oils (Eos), also known as volatile oils or aromatic plant essence, are multi-component, volatile, aromatic liquids extracted from the whole plant or their parts; utilization of these EOs to control postharvest diseases is gaining popularity due to their antimicrobial, antiviral, antifungal, antiparasitic and insecticidal properties, safety features and biodegradable nature (Ding and Lee, 2019). Their bioactivity depends upon various factors such as chemical composition that is determined by agronomic practices, environmental conditions, geographical origin of the plant species and the applied dose concentration. The mode of action of essential oils depends on the cell wall structure and outer membrane arrangement of the target organism. Their action increases cell wall permeability and reduces various processes such as inhibition of electron transport, protein translocation, phosphorylation steps and other enzyme-dependent reactions. Their application retards fungal growth, sporulation and germ tube elongation in plant pathogens. The application of EOs in harvested products has advantages such as high effectiveness against several pathogens and low toxicity to non-target microorganism and humans (Pavela and Benelli, 2016).

The bioactivity of essential oils depends on the application method, as EOs with smaller phenolic components such as allyl isothiocyanate and citral are more efficient when added as volatiles, whereas thymol and eugenol with larger phenolic components showed more activities when applied directly (Cosic et al., 2014). The bioactivity results from the complex interactions between different component compounds of the essential oil. The degree of inhibition by the essential oil and sensitivity of pathogenic fungi to various oils also differ. Essential oils extracted from *Thymus*, *Cestrum*, *Zataria*, *Magnolia*, *Salvia*, *Eugenia*, *Rosmarinus*, *Pimpinella* and *Cinnamomum* exhibited antimicrobial activities against various plant pathogenic fungi. Essential oils of copaiba and eucalyptus were tested against *Alternaria alternata* and *Colletotrichum musae* in vitro, and the results showed good efficacy at low concentrations of the treatments (0.0–1.0%) (da Nobrega et al., 2019). In a recent study, anthracnose and stem-end rot in green-skinned avocado fruits was successfully controlled by the application of thyme oil in combination with a prochloraz solution;

furthermore, this treatment improved fruit quality (firmness) during storage time (Obianom and Sivakumar, 2018). The efficacy of black caraway (*Carum carvi*) and anise (*Pimpinella anisum*) essential oils was tested against *Penicillium digitatum* in vitro as well as in vivo (on oranges) evaluations. The results showed that treatments were capable of controlling fungi development; besides, the quality of oranges was preserved by the application of the treatment of 600 μL/L (Aminifard et al., 2018). In a study, lemongrass oil was tested against *Colletotrichum coccodes, Botrytis cinerea, Cladosporium herbarum, Rhizopus stolonifer* and *Aspergillus niger* in vitro. It was found that essential oil extracted from epicarp of *Citrus sinensis* (Sharma and Tripathi, 2006, 2008), *Ocimum gratissimum* and *Hyptis suaveolens* inhibited mycotoxigenic strains of storage mycoflora (Sharma, 2007). The efficacy of essential and vegetable oils in preventing infestation by bruchids, maize and rice weevils on stored products has been well documented. Thus, utilization of EOs for controlling diseases can be a viable alternative to chemical treatments.

2.16 PLANT-DERIVED PRODUCTS

Farmers in their traditional wisdom have identified and used locally available plants for control of pests and pathogens, especially in storage. As many as 2131 plant species are reported to possess pest/pathogen management properties; 1005 species of plants exhibit insecticidal character, 389 with antifeedant, 300 with repellant, 27 with attractant and 31 with growth inhibiting traits. Higher plants, sources of a wide spectrum of secondary metabolites such as alkaloids, flavonoids, phenolics, quinines, saponins, sterols and tannins offer resistance to pathogens and some of these can be utilized as biopesticides (Rodriguez-Casado, 2014). The most widely used botanical compound is neem oil, an insecticidal chemical extracted from seeds of *Azadirachta indica*. Pyrethrins, fast acting insecticidal compound produced by *Chrysanthemum cinerariaefolium*, have low mammalian toxicity but degrade rapidly after application promotes the development of synthetic pyrethroids.

In this sense, some extracts from plants have proven to be an attractive option for the extraction of substances with antimicrobial activity with high efficiency and low toxicity. Root extracts of *Asparagus* and *Tagetes* as nematicides and *Chenopodium* and *Bougainvillea* as antiviruses have also been observed. In a study, aqueous and ethanolic extracts from garlic (*Allium sativum*) and garlic creeper (*Mansoa alliacea*) were evaluated on citrus fruits (*Citrus sinensis, Jaffa* and *Valencia*) and it was suggested that the highest antifungal activity of the extract can be produced by the presence of allicin in the soluble fractions of extracts from garlic (Chen et al., 2019). In the same way, extracts of garlic, ginger and celery have been shown to have different effects on the control of the incidence of *Penicillium* sp. in fruits of the species *Citrus reticulata Blanco* (Gong et al., 2018). In a similar study, soybean extracts were evaluated as a protection method on oranges infected with *P. digitatum*, and green mold was significantly reduced (88–100%) due to the presence of β-conglycinin in the soy protein fraction (Osman et al., 2016). In recent years, extracts of some angiosperm species have been studied as natural fungicides, such as *Orobanche crenata* and *Sanguisorba minor,* which have shown high efficiency in the control of the diseases produced by *Monilinia laxa* in stone fruits such as apricot, cherry and nectarine (Gatto et al., 2013, 2016b). In the case of *O. crenata* extracts, the antifungal activity is attributed to the presence of the phenolic compound verbascoside, while in *S. minor* extracts, the efficacy is related to the presence of a combination of phenolic compounds like caffeic acid, quercetin, luteolin and kaempferol (Gatto et al., 2016a). In a recent investigation, the application of guava leaf extracts and lemon on banana fruits (*Musa sapientum* L.) at the postharvest stage considerably improved the shelf life of the fruits (up to 8 days) compared with untreated fruits (only 4 days), having a positive effect on the conservation of the physicochemical characteristics of fruit during storage time (Tabassum et al., 2018).

A wide range of microorganisms has been reported to synthesize secondary metabolites, which are useful as biopesticides. Soil actinomycetes are one of these and are utilized for this purpose. A mixture of two macrolide compounds, Spinosad, from *Saccharopolyspora spinosa,* has a very low mammalian toxicity and degrades rapidly in the field. Farmers and growers have used it extensively since its introduction in 1997, but due to the development of resistance in some pests such as western flower thrips, it was withdrawn. A macrocyclic lactone compound Abamectin synthesized by *Streptomyces avermitilis* was found to be active against several pests but resistance to tetranycids developed. A wide range of predatory animals use insect-specific toxins to kill their prey; these toxins can be employed as pesticides and incorporated in crops to provide resistance against insect attack. Microorganism-derived natural products currently in use as biopesticides have been summarized in Table 2.1.

Semiochemicals are organic compound markers or signals that mediate interactions between individuals of the same species, i.e. pheromones, or different species, i.e. allelochemicals. These can be volatile or non-volatile signals that operate at a long or short range to modify the behavior of the recipient. Plant- and insect-derived semiochemicals can be used for pest control to cause behavioral disruption of pests and/or their natural enemies (Mauchline et al., 2017). Pheromones are largely being used in the field for insects belonging to various orders. The sex pheromone of the silkworm moth was the first one to be chemically identified in 1959 and is still used in pest management. Plant-derived products from neem, pongamia and mahua have been found effective against rice cutworm, rice green hopper and tobacco caterpillar, and several species of aphids and mites. High repellency values were obtained for *Cymbopogon flexuosus, Lavandula angustifolia, Litsea cubeba* and *Mentha arvensis*. Trap cropping is another strategy that utilizes semiochemicals, released by plants to deter pests from the main crop. Members belong to the *Brassica* group such as *B. juncea, B. nigra, B. rapa* and *Sinapis alba*; other plants such as *Tagetes* show this quality to trap pests.

2.17 COATINGS AND EDIBLE FILM FROM NATURAL SOURCES

Edible coatings and films are thin layers of edible material applied directly by dipping, spraying or brushing to create a modified atmosphere to the product surface (Brychcy-Rajska, 2017). These coatings act as a barrier to moisture, oxygen and solute movement for food during processing, handling and storage by delaying the deterioration of food, improving its quality and extending the shelf life, in addition to or as a replacement for natural protective waxy coatings. The desirable qualities of edible film/coatings should be that it is easily synthesized and economically viable; non-toxic, non-allergic and consisting of non-digestible components; serve as a carrier for desirable additives such as flavor, fragrance, coloring, nutrients and vitamins; should be structurally stable to prevent mechanical damage during transportation and handling; have good adhesion to the surface of the food to be protected, providing uniform coverage; provide semi-permeability to maintain internal equilibrium of gases involved in aerobic and anaerobic respiration; should be able to control water migration both in and out of protected food to maintain desired moisture content; and be able to prevent loss or uptake of components that stabilize aroma, flavor, nutritional and organoleptic characteristics necessary for consumer acceptance while not inversely altering the taste or appearance. It should be able to provide biochemical and microbial surface stability while protecting against contamination, pest infestation, microbe proliferation and other types of decay (Martha et al., 2018).

The efficiency and functional properties of the coatings and films depend on their application and the characteristics of the fruit on which they are being used, as well as the nature of the compound that is used to produce the coating; thus both its physical and chemical properties, as well as its mechanical and permeable properties must be taken into consideration (Akhtara et al., 2015). An edible coating is a thin layer of edible material formed as a coating on the surface of a food product, while an edible film is a preformed standalone thin layer of edible material, which once formed can be placed on or between food components (Suput et al., 2015). The main difference between them is that edible coatings are applied in liquid form on the food, usually by immersing the product in a solution-generating substance formed by the structural matrix (carbohydrate, protein, lipid or multi-component mixture), and edible films are first molded as solid sheets, which are then applied as a wrapping on the food product. The use of wax coating of fruits by dipping is one of the age-old methods that were in vogue in the early 12th century. This was practiced in China, essentially to

TABLE 2.1
Microorganisms and Plant-derived Natural Products Used as Biopesticides

Study no.	Source	Product	Trade name
		Microorganism-derived natural products	
1	*Bacillus thuringiensis*	D endotoxin	Able (Thermo Trilogy), Bactospeine (Valent BioSciences), Dipel (Valent BioScience), Javelin (Thermo Trilogy), Thuricide (Thermo Trilogy) and Xentari (Violent BioScience)
2	*Saccharopolyspora spinosa*	Spinosad	Tracer, Conserve, Success and Spintor (Dow Agro Sciences)
3	*Streptomyces aureus*	Polynactins	Mitecidin (Plus Fenobucard) and Mitedown (Plus Fenbutatin Oxide) (Eikou Kasei)
4	*Streptomyces avermitilis*	Avermectins	Dynamic Avid, Zephyr and Agrimek (Novartis) and Abacide (Mauget)
5	*Streptomyces griseochromogenes*	Blasticidin	Bla-S (Kaken, Kumiai and Nihon Noyaku)
6	*Streptomyces griseus*	Streptomycin	Agrimycin 17 and AS-50 (Novartis), Plantomycin (Aries Agro-Vet Industries) and Paushamycin (Paushak)
7	*Streptomyces hygroscopicus*	Validamycin	Validacin, Valimun (Takeda) Solacol (Takeda and Aventis) Mycin (Sanonda) and Vivadamy (Vietnam Pesticide)
8	*Streptomyces kasugaensis*	Kasugamycin	Kasugamin and Kasumin (Hokko)
9	*Streptomyces natalensis, S. Chattanoogensis*	Natamycin	Delvolan (Gist-Brocades)
10	*Streptomyces rimosus*	Oxytetracycline	Mycoshield and Terramycin (Novartis) and Phytomycin (plus streptomycin sulfate, Ladda)
11	*Streptomyces avermitilis*	Emamectin	Proclaim and Affirm
12	*Streptomyces cacaoi var asoensis*	Polymyxin B	Polyoxin AL (Kaken, Kumini Nihon Nohyaku and Hokko)
13	*Streptomyces hygroscopicus*	Milbemectin	Milbe knock (Sankyo)
14	*Streptomyces hygroscopicus, S. Viridochromogenes*	Bilanafos	Meiji Herbiace (sodium salt) (Meiji Seika)
15	*Streptoverticillium rimofaciens* strain b 98891	Mildiomycin	Mildiomycin (Takeda)
		Plant-derived natural products	
1	*Azadirachta indica* A Juss	Azadirachtin	Neemix 90, Neemazid, Trilogy 90 (neem oil for disease control) Triact 90 (neem oil for disease control) Bioneem, Margosan-0 Azation, Align, Turplex and Bollwhip
2	*Chrysanthemum cinerariaefolium*	Pyrethrins	Alfadex (Novartis), Pyricide and evergreen (MGK), Pyronyl (Mixture), Excite R and Prentox Pyrethrum extract (all Prentiss), Milon (Delicia), Pycon (for concentrated mixture with Piperonyl Butoxide), Agropharm and Checkout (Consep)
3	*Derris, Lonchocarpus* and *Tephrosia* species	Rotenone	Chem. Sect, Cube Root and Rotenone Extract (all Tifa), Noxfire and Rotenone EK-11 (Agreva Environmental Health) and Prentish (mixture), Prentiss, Synpren fish (mixture) and Prentox (all Prentiss)
4	*Nicotiana rustica* L	Nicotina	Nico soap (United Phosphorus Ltd), No-Fid (Hortichem) XL-All Nicotine (Vitax) and Nicotine 40% shreds (Dow AgroSciences)
		Ryania extracts	Nature GRO R 50 and Nature GRO Triple Plus (Agrisystems International and Ryan 50 (Dunhill Chemical)
5	Plant-derived fatty acids		Thinex and Scythe (Mycogen) and Gantico (Japan Tobacco)

retard water transpiration loss in lemons and oranges. In the 1930s, hot-melt paraffin waxes became commercially available as edible coatings for fresh fruits such as apples and pears. Later, fat coating on food products, specifically called "larding" was quite prevalent in England. "Sausage casing" used very commonly nowadays is a film thickness of 2.5 mm derived from gelatin, a protein source (Suput et al., 2015). Since the 1930s, carnauba wax, beeswax and paraffin wax have been commercially used for coating fresh fruits and vegetables with the purpose of reducing moisture loss and surface abrasion during fruit handling and controlling internal gas composition of the fruits (Bodini et al., 2013). They are substantially more resistant to moisture transport than other lipid or non-lipid coatings.

Edible films and coatings can be divided into three categories based on their components: hydrocolloids include proteins (including corn zein, wheat gluten, soy protein, whey protein, casein, collagen/gelatin, pea protein, rice bran protein, cottonseed protein, peanut protein and keratin) and polysaccharides (including starch and starch derivatives, cellulose derivatives, alginate carrageenan, pectin, pullulan, chitosan and various gums). Proteins and polysaccharides generally have a good barrier to oxygen at low relative humidity due to their tightly packed hydrogen-bonded network structure but have a poor moisture barrier due to their hydrophilic nature. Therefore, hydrophobic compounds are generally added to make edible coatings. Carrageenan, extracted from several red seaweeds, is a complex mixture of several polysaccharides and is another potential coating material for fruits and vegetables (Lin et al., 2018). Carrageenan-based coatings have been applied to fresh apples for reducing moisture loss, oxidation and disintegration.

Lipids include waxes, resins, triglycerides (natural lipids) and fatty acids, while composites contain both hydrocolloid and lipids. Lipid-based edible films or coatings include neutral lipids, fatty acids, waxes (beeswax, candelilla wax, carnauba wax, rice bran wax) and resins (shellac, wood rosin). However, as lipids are not polymers, they form films or coatings with poor mechanical properties (Cagri et al., 2004).

Several other compounds such as plasticizers and emulsifiers can be added to edible coatings when lipids and hydrocolloids are combined. The purpose of adding plasticizers in film-forming material is to decrease the intermolecular forces between polymer chains, which result in greater film flexibility, elongation, toughness and permeability. Plasticizers used for edible coatings include glycerol, sorbitol, sucrose, propylene glycol, polyethylene glycol, fatty acids and monoglycerides. Emulsifiers or surfactants are surface-active agents that are used to improve the stability of the emulsion and ensure good surface wetting, spreading and adhesion of the coating to the food surface. Common emulsifiers used for coatings are ethylene glycol monostearate, fatty acids and their esters, glycerol monostearate, lecithin, sucrose ester, polysorbates and sorbitan monostearate (tweens).

In addition, food additives as antioxidants, colorants, flavoring agents and antimicrobial compounds can also be added to edible coatings (Cha and Chinnan, 2004; Han, 2002). Composite coating formulation using lipid and hydroxy propyl methyl cellulose (HPMC), MC and lipid, MC and fatty acid, corn zein, MC and fatty acid, whey isolate and lipids, casein and lipids, gelatin and soluble starch, hydroxypropyl starch and gelatin, corn zein and corn starch, gelatin and fatty acid, soy protein isolate and gelatin, soy protein isolate and polylactic acid (Hassan et al., 2017) can be tailor-made to suit to the needs of a specific commodity or farm produce. Edible coatings made of CMC, MC, HPC and HPMC have been applied to some fruits and vegetables for providing barriers to oxygen, oil or moisture transfer (Maftoonazad and Ramaswamy, 2005) and for improving better adhesion. McGuire and Hagenmaier (2001) applied shellac coating to grapefruit and oranges over a CMC layer and observed that a shellac formulation at pH 9.0 with 5.2% ethanol was more toxic to the coliform bacteria *Enterobacter aerogenes* and *Escherichia coli* than a formulation at pH 7.25 with 12% ethanol.

Recently, incorporation of minerals, vitamins and fatty acids into edible film and coating formulations was found to enhance the nutritional value of some fruits and vegetables.

Tapia et al. (2008) reported that the addition of ascorbic acid (1% w/v) to the alginate and gellan-based edible coatings preserved the natural ascorbic acid content in fresh cut papaya throughout its storage. The development of chitosan coatings containing high concentrations of calcium, zinc or vitamin E also provided alternative ways to fortify fresh fruits and vegetables. Han (2002) reported that chitosan-based coatings have the capability to hold high concentrations of calcium or vitamin E and therefore significantly increased their content in fresh and frozen strawberries and red raspberries. Similarly, Hernandez- Munoz et al. (2006) observed that chitosan-coated strawberries retained more calcium gluconate than strawberries dipped in calcium solutions. Han et al. (2004) improved the nutritional and physicochemical quality of strawberries and raspberries by means of chitosan-based coatings enriched with calcium and vitamin E. These coatings not only prevented moisture loss and caused surface whitening, but also significantly increased the calcium and vitamin E content of carrots (Mei et al., 2002).

There are several commercial products available in the market for postharvest use, such as Bio-Save (*Pseudomonas syringae*) registered in the USA and "Shemer" (*Metschnikowia fructicola*) registered in Israel (Droby et al., 2009). Pro-long, TAL-Prolong (Courtaulds Group, London, UK), Semperfresh (AgriCoat Industries Ltd., Berkshire, UK) and Natural Shine 9000 (Pace International, Seattle, WA) are commercially available composite coating formulations based on CMC (Nisperos-Carriedo et al., 1992).

Nature-Seal (Eco-science Product System Division, Orlando, FL) is another cellulose-based edible coating formulation used for delayed ripening of tomatoes and mangoes. Nature-Seal in combination with antimicrobials, plasticizers, antioxidants and so on has also been used to coat fresh-cut apples and potatoes. The coating significantly reduced weight loss of apples and potatoes more than those treated with water solutions and was not objectionable in taste during several weeks of storage (Baldwin et al., 1996). The coating has also been used to effectively reduce the discoloration of mini-peeled carrots without affecting microbial and chemical quality (Howard et al., 1995). A commercial fruit coating, Nutri-Save (Nova Chem, Halifax, Canada), was developed to serve as both film former and natural preservative and to create a modified atmosphere for whole apples and pears to reduce respiration rate and desiccation of these commodities.

Edible coatings and films on fruits made from compounds obtained from natural sources such as *Aloe vera* have shown promising results in the preservation of tropical fruits. It is capable of forming a uniform layer on the surface of the fruit and being easily applied (Suriati et al., 2020). *Aloe vera*-based coatings have been successfully tested in *Mangifera indica* L. cv. Kensington Pride fruit ripening. *Aloe vera* coating reduced aroma volatile biosynthesis in the fruit pulp. Likewise, it was found that coatings delayed ripening of the fruit compared with the control. They state that this effect was characterized by the suppression in respiration and/or delayed climacteric peak, late fruit color development and a greater firmness in the coated fruit in comparison with the uncoated ones. Starch-based edible coatings preserved the properties of fruits and vegetables in postharvest conditions (Guimaraes et al., 2016; Sapper and Chiralt, 2018). Coating with a concentration of 4% potato starch and 20% *Aloe vera* extended the shelf life of the guava (*Psidium guajava*) by reducing the weight loss and the respiratory rate of the fruit and increasing the firmness and retention of the vitamin C content for up to 10 days. Similarly, *Aloe vera* gel coating was found suitable on peach fruits (Hazrati et al., 2017). Bio-coatings containing CMV and guar gum also provided protection from various postharvest diseases (Shah et al., 2016).

Carnauba wax isolated from the leaves of the Brazilian palm *Copernicia cerifera* is being extensively used as an edible coating because it has been observed that it reduces water loss, improves appearance and prolongs shelf life in a wide variety of fruits. Fresh cut apples were found to be better preserved when cassava starch–carnauba wax edible coatings were applied (Chiumarelli and Hubinger, 2012). An edible coating based on cassava starch and carnauba wax adding organic acids and calcium chloride was evaluated in mangoes cv. Tommy Atkins (Dussan-Sarria et al., 2013). According to the results, the attributes of sensory, physical and chemical qualities were maintained, and the useful life of fruit was possible to prolong up to 24 days under refrigeration conditions (5 ± 1 °C and $90 \pm 2\%$ RH).

An edible coating based on candelilla wax improved the quality of avocado and pear fruits and extended shelf life compared with control fruits (Saucedo-Pompa et al., 2009; Cruz et al., 2015). Also, the addition of ellagic acid to the edible film showed an important effect, as it reduced the damage caused by the fungus *C. gloeosporioides* (the main phytopathogenic fungus for avocados) and significantly improved the quality and shelf life of avocado. Another coating based on mesquite gum–candelilla wax was evaluated in Persian limes (Wu et al., 2005) The results showed that coatings decreased the weight loss of the fruit. In addition, by adding mineral oil (33%) to the emulsion, they observed that water vapor permeability was significantly improved, as well as its appearance.

2.18 RADIATION

Radiation technology can complement and supplement existing technologies and be efficiently used for insect disinfestations, sprout inhibition, maintenance of nutritional or functional components during storage and finally for controlling postharvest diseases.

Irradiation is a process of controlled application of energy from ionizing and non-ionizing (gamma and UV-C) radiations to increase the safety and shelf life of food. Ionizing radiation is another physical treatment that can be used after harvest to reduce disease in some commodities. The effectiveness of irradiation on postharvest disease management depends upon the dosage of irradiation used and on the type of pathogen, its growth stage and the number of viable cells on or within the tissue (Kader et al., 1986).

Irradiation kills microbes primarily by destroying DNA, hence, preventing the proliferation of microorganisms such as bacteria, viruses, mold and yeast. The sensitivity of the organism increases with the complexity of the organism. Thus, viruses are most resistant to destruction by irradiation, while insects and parasites are more sensitive. Spores and cysts are quite resistant to the effect of irradiation because they contain little DNA and are in a highly stable resting state (Temur and Tiryaki, 2013). Bacterial spores are more resistant to ionizing radiation than those of the vegetative cells. Gram-positive bacteria are more resistant than Gram-negative bacteria. The resistance of yeasts and molds vary considerably, but some are more resistant than bacteria. The effect of irradiation depends on the type, application dose, oxygen, moisture content, composition of food and storage conditions (Aziz et al., 2006).

It was found that the penetrating power of gamma rays was greater, thereby allowing them to reach microorganisms inside the produce of different sizes and shapes that are not accessible to chemicals. Irradiation at doses below 1 kGy is an effective insect disinfestations treatment against various insect species of quarantine significance in fresh fruits and vegetables (Kader, 1986). Dosage of γ-irradiation (0.06–0.5 kGy) inhibited cap opening and browning, stalk elongation, reduction in the level of microbial contamination (surface molds) and extended shelf life (Deepshika et al., 2017).

Non-ionizing radiation, particularly *UV-C*, has potential for controlling postharvest diseases (Wilson et al., 1997). It has been proven to be effective in delaying ripening and senescence, diminishing decay and even in increasing the content of beneficial compounds owing to its germicidal properties. Low dosage of shortwave ultraviolet light (UV-C, 190–280 nm wavelengths) can control storage rots of many fruits and vegetables by targeting the DNA of microorganism in addition to this direct germicidal and mutagenic activity. UV-C irradiation at the appropriate wavelength and dose rate can modulate induced defense in plants. UV-C irradiation can stimulate accumulation of stress-induced phenylpropanoids and pathogenesis-related proteins (Porat et al., 2000).

These methods can be efficiently used to delay the ripening of fruits, inhibit germination, improve nutritional quality, minimize insect infestation and deactivate viruses. The short shelf life of mushrooms can be extended by inhibiting cap opening and browning, stalk elongation, reducing the level of microbial contamination and finally by increasing the concentration of vitamin D_2 significantly, without causing any adverse effect on its taste.

The use of irradiation technology has also been approved by various agencies as the Food and Agriculture Organization (FAO), the World Health Organization (WHO), the International Atomic

Energy Agency (IAEA) and the Codex Alimentarius Commission. Irradiated fresh commodities carry the "radura" symbol placed on the "country of origin" or product locator unit (PLU) stickers, and a statement about its treatment with irradiation. About 100 countries have approved this process in more than 100 food items.

Although this technology has received increased attention, parameters like irradiation dose, intensity, type of produce and pathogen should be considered, otherwise it might cause acute radiation injury (as dose increases) or other detrimental effects. Several issues regarding the wider use of irradiation for postharvest treatment include the high costs coupled with poor consumer acceptability due to the lack of knowledge in relation to the safety and benefits of irradiated foods. In order to fetch maximum benefit, irradiation technology has to be combined with other chemical and physical methods, which will provide synergistic effects for the control of postharvest diseases of fruit and vegetables (Deepshikha et al., 2017).

2.19 ULTRASOUND

Decontamination of fresh product by ultrasound is a relatively recent, economically and environmentally friendly alternative for quality control of fresh fruit and vegetables in pre- and postharvest processes (Mizrach et al., 2008). Ultrasound can be applied directly to the medium (water) or can be used in combination with some organic salts, organic acids, chitosan and so on to achieve better results. Low intensity ultrasound is used for the inactivation of microorganisms caused by the cavitation phenomenon. The efficiency of the ultrasound process is affected by several factors such as power level, treatment time and temperature (Pinheiro et al., 2015). Application of ultrasound at low frequencies (20 and 40 kHz) has demonstrated a decrease in the microbial load of mesophilic aerobes in lettuce (0.9 log CFU/g) and strawberry (1.49 log CFU/g) (Ajlouni et al., 2006). Furthermore, the combined treatment did not impair the quality parameters of peach fruit after 6 days of storage at 20 °C (Yang et al., 2011). A combination of ultrasound and sanitizers could increase pathogen reduction without affecting product quality, while concentration of sanitizers could be reduced as well as treatment time required, saving time and money, and avoiding significant risks to consumers.

2.20 FOGGING

Various technologies have been developed to prolong the shelf life of harvested fruits and vegetables that maintain their integrity as well as their nutritional properties. The use of ultrasonic nebulization (fogging) is one such technique little explored at present, which serves to prevent or control pathogenic diseases on harvested products. Fogging has been used successfully for the spraying and distribution of disinfectants such as acetic acid, chlorine dioxide, ethanol, hydrogen peroxide and sodium hypochlorite to control epiphytic microorganisms present on the surface of strawberry, thus reducing the decay index by up to 83.2%, demonstrating that nebulization is an effective method for the reduction of diseases at the postharvest stage (Vardar et al., 2012). The combination of fogging and refrigeration for white asparagus provides better results as compared with controls (Tirawat et al., 2017). When peracetic acid at low concentration was applied on strawberry by ultrasonic nebulization as a disinfectant, the results showed that the anthocyanin and phenolic compound contents were well preserved (Van De Velde et al., 2016). In a study on figs, Karabulut et al. (2009) observed that 80% inhibition of gray mold disease was achieved with the application of chlorine dioxide (1000 μL/L). Ultrasonic nebulization as a conservation method in the postharvest period gives a number of benefits in different ways such as reducing the quantity of substances applied and the exposure time as well as a better distribution of the treatments. Its application in fruits and vegetables has not been explored, thus the development of this technology can be an attractive option.

2.21 BIOLOGICAL CONTROL

Postharvest biological control offers several advantages over conventional biological control as exact environmental conditions can be established and maintained, the biocontrol agent can be targeted much more efficiently and are cost-effective on harvested food. The first biological control agent developed for postharvest use was a strain of *Bacillus subtilis*, which controlled peach brown rot (Wilson et al., 1997).

Such organisms can be isolated from a variety of sources including fermented food products and the surfaces of leaves, fruit and vegetables. Once isolated, these organisms (bacteria, yeasts/filamentous fungi) can be screened in various ways for inhibition of target pathogens. In most of the investigations, maximum pathogen inhibition was observed when the antagonist was applied prior to infection taking place. Unless an antagonist has eradicant properties or has some effect on host defense responses, control of quiescent field infections using postharvest applications of antagonists is often more difficult to achieve than the control of infections occurring after harvest. It has been observed that both field and postharvest applications of *Bacillus subtilis* and *B. licheniformis* suppress anthracnose (caused by *Colletotrichum gloeosporioides*) and stem-end rot (caused by *Dothiorella* spp. and other fungi) development in avocado.

Several naturally occurring or genetically manipulated microorganisms (Table 2.1) are being used as bioagents against plant pathogens, pestiferous insects and weeds. A number of direct and indirect modes of action are involved in these pathogen–antagonist interactions, including biofilm formation, nutrient and space competition, production of diffusible and volatile antimicrobial compounds, hyperparasitism and release of hydrolases by induced resistance and priming, production of oxidative stress and site exclusion (Kohl et al., 2019). To be effective against wound pathogens, an antagonist must be able to successfully colonize wound sites for the exclusion of the pathogen.

Baculoviruses, belonging to the family *Baculoviridae*, have a major advantage of being highly specific to a limited number of insects, leave no toxic residues in crops and have not demonstrated any toxicity against any other living forms, including humans (Sharma and Srivastava, 2013). Their target viruses include nuclear polyhedron viruses (NPVs), granuloviruses (GVs) and non-occluded viruses (NOVs) that can infect and destroy numerous important plant pests. Products are produced as concentrated powders and can be formulated as liquids, suspensions and wettable powders. Some of the baculoviruses used as bioagents are *Anagrapha falcifera* NPV (AfNPV), *Anticarsia gemmatalis* NPV (Polygen Multigen), *Helicoverpa zea* (Disparvirus), *Lymantria dispar* (Gemstar), *Spodoptera exigua* NPV (SPOD-X) and *S. littoralis* NPV (Spodoptrin) against pests of cotton, corn, soybeans and so on. There are certain shortfalls such as slow speed of kill, narrow spectrum of biological activity, photo stability and short residual effect. However, the role of natural baculoviruses might be enhanced if the time required to kill the target pathogen is shortened and their synergistic interaction with chemical pesticides could be evaluated, which can result in a reduction of the rate of application of the latter, thereby reducing the chemical load in the environment.

The prominent *bacterial antagonists* are the K84 strain of *Agrobacterium radiobacter, Bacillus thuringiensis, B. subtilis, B. licheniformis, B. megaterium, Pseudomonas aureofaciens* and *P. fluorescens.* The bacteria-based products contain endotoxins and live bacterial cells and when derived from plasmid conjugation, result in cry gene exchanges producing novel toxin arrays. They are registered and sold as Able, Bactophene, Condor, Crymax, Cutlass, Design DiPel, Foil, Javelin, Thuricide and Xentari, and are being used against a wide range of plant pathogens.

Yeasts such as *Aureobasidium pullulans, Candida oleophila, Cryptococcus laurentii, Debaryomyces hansenii (Pichia guilliermondii)* and *Metschnikowia fructicola* act against *Botrytis, Penicillium, Rhizopus and Aspergillus* causing deterioration of various plant products (Spadaro et al., 2016).

The commercial products developed for postharvest application are Aspire™ based on *C. oleophila* has been registered in the USA; Boni-Protect based on *A. pullulans* was developed in Germany; Candifruit based on *C. sake* was developed in Spain; Yield Plus™ was based on *Cryptococcus albidus* in South Africa; Shemer™ registered in Israel based on a heat tolerant strain of *M. fructicola*; and NEXY in Belgium based on *C. oleophila*.

In addition to bacteria and yeasts, the application of *Trichoderma*, alone or in combination with other alternative control systems in different fruits for the control of pathogenic fungi in fruits, has been found to be an effective measure with favorable results (Gonzalez-Estrada et al., 2018, 2019). Existing species of *Trichoderma* with high antagonistic capacity are *T. asperellum*, *T. viride* and *T. harzianum*. Different species of *Trichoderma* present variation in degrees of pathogen control at postharvest stage if applied individually, but if the strains are combined, the potential of biocontrol increases. The combinations of *T. viride*, *T. harzianum*, and *T. koningii* reduced the incidence of crown rot caused by *Colletotrichum musae*.

The combination of strains of *Pseudomonas-Trichoderma* and *Trichoderma viride-Bacillus subtilis* was effective in controlling *Penicillium digitatum*, *Penicillium expansum* and *Fusarium moniliforme*, respectively, in citrus and grapes. A synergistic effect was observed when the combination of *P. syringae* and *Trichoderma* antagonists was evaluated against *Botrytis cinerea* and *Fusarium oxysporum*. This biocontrol agent attacks and penetrates fungal cells, causing an alteration with the consequent degradation of the cell wall, causing retraction of the plasma membrane and disorganization of the cytoplasm (Woo et al., 2006)

Filamentous entomopathogenic fungi *Beauveria bassiana* and *Metarhizium anisopliae* have been developed and used against insects and acarine orders in glasshouse crops, fruits and field vegetables as well as broad acre crops. In Brazil alone, around 750,000 ha of sugarcane and 250,000 ha of grassland are annually using commercial biopesticides based on *M. anisopliae* against spittlebugs. The fungus has been recommended for locust and grasshopper pests in Africa and Australia, and even the FAO has recommended the use of the biofungicide for locust management.

Ideally, an antagonist should be unique and effective against a broad spectrum of pathogens on a wide range of harvested products and be able to be produced on inexpensive growth media. Formulations incorporating such antagonists should have a long shelf life and be able to be manufactured at a low cost. Most antagonists do not satisfy all of these criteria. Many are quite specific in their activity against pathogens and for this reason may not be particularly appealing to prospective investors.

While the potential for biological control of postharvest diseases clearly exists, future success relies on the ability to achieve consistent results in the field and after harvest. It will be necessary to enhance the efficacy of biological control agents against postharvest diseases and commercialize the technology involved.

2.22 MODIFIED OR CONTROLLED ATMOSPHERES

Alterations in O_2 and CO_2 concentrations are sometimes provided around fruit and vegetables because the pathogen respires as does produce; lowering the O_2 or raising the CO_2 above 5% can suppress pathogenic growth in the host. In crops such as stone fruits, a direct suppression occurs when fungal respiration and growth are reduced by the high CO_2 of the modified atmosphere. Low O_2 does not appreciably suppress fungal growth until the concentration is below 2%. Important growth reductions result if O_2 is lowered to 1% or lower, although there is a danger that the crop will start respiring anaerobically. Modified atmosphere packaging utilizes a predetermined composition of both of these gases. The packaging material in MAP allows the diffusion of gases through them until a stable equilibrium is reached between external and internal gases. The advantage of using MAP not only provides a modified atmosphere to control ripening but also helps in reducing shrinkage and other mechanical injuries, enhances hygiene and reduces water loss, which in turn checks the spread of diseases (Arah et al., 2016).

2.23 NANOTECHNOLOGY

The threatening challenges of sustainability, food security and climate change are engaging scientists in exploring nanotechnology as an exciting and novel frontier in the field of agriculture, including horticulture with a profound impact on human life and welfare (Shoala, 2018). The development of polymeric nanoparticles was first reported in the pharmaceutical field for drug delivery systems. Less than two decades ago, these systems captured the interest of the food sector as an emerging innovative application that could provide solutions to issues such as active or intelligent packaging, protecting nutraceuticals from degradation, the delivery and controlled release of nutraceuticals and agrichemicals across edible coatings and taste-masking (Zambrano-Zaragoza et al., 2018), as well as plant hormone delivery, seed germination, transfer of target genes, nanobarcoding and nanosensors (Elmer and White, 2018a). It is anticipated that due to the worldwide popularity and expansion of this technology, its market value will reach US$75.8 billion by 2020 (Research and Markets, 2015).

Nanotechnology deals with the fabrication, manipulation and study of matter at atomic level in the size range of 1–100 nm, which are designated as "nanoparticles." Nanoparticles can be designed with unique biological, chemical and physical properties to enhance all the properties of the parent metal at the nanoscale. These particles have entirely new properties which enable them to demonstrate hardiness, breaking strength and toughness at low temperature; elevated chemical reactivity and surface energy; extended plasticity at high temperature; and more mobility in the body of organism including cellular entry (Banik and Sharma, 2011).

Several conventional physical and chemical protocols such as laser pyrolysis or ablation, melt mixing, microemulsion and sol-gel, photo reduction, sputtering, thermolysis, ultrasonic fields, ultraviolet irradiation and vapor deposition have been employed for nanoparticle synthesis for decades (Singh et al., 2018a). Transformation of a bulk element to the nano level, not only reduces its size, but also leads to the formation of different shapes such as spherical, triangular, truncated triangular, octahedral, rod and flower shaped. The variation in geometric shapes is advantageous in the application of these particles as antimicrobials since the bioactivity of nanoparticles is directly proportional to the surface area available for interaction with biological components. These particles have the potential to inhibit diseases of seeds, roots, seedlings, foliage and fruits, providing protection against pathogens like bacteria, viruses, fungi and insects.

Current research in biosynthesis of nanometals using plant extracts is very cost effective, and therefore can be used as an economic and valuable alternative for large-scale production of metal nanoparticles (Kharissova et al., 2013). The biological synthesis of metal nanoparticles (especially gold and silver nanoparticles) using plants (inactivated plant tissue, plant extracts and living plants) has received more attention as a suitable alternative to chemical procedures and physical methods (Hussain et al., 2016).

Nanosized particles of different metals, pesticides and growth promoters have been explored for plant disease management either as nanoparticles alone, acting as protectants, or as nanocarriers for fungicides, herbicides, nematicides, insecticides and RNA-interference molecules (Abdellatif et al., 2016; Worrall et al., 2018). Nanoparticles are commonly used as carriers to absorb or attach, encapsulate and entrap the active molecules in developing effective agricultural formulations.

Nanoparticles (NPs) act against pathogens through several different methods such as direct attachment to the cell surface, disruption of cell membrane and membrane potential, damage of cellular components, generation of oxidative stress and inhibition of proteins/enzymes (Iravani et al., 2014).

Nanomaterials as unique carriers of agrochemicals facilitate the site-targeted controlled delivery of nutrients with increased crop protection. In addition; nanomaterials offer a wider specific surface area to fertilizers and pesticides. Nanobiosensors support the development of high-tech agricultural farms due to their direct and intended applications in the precise management and control of fertilizers, pesticides and herbicides. Nanostructured catalysts will be available on demand doses that will increase the efficiency of pesticides. Nanoparticles for the delivery of drugs or nutrients for the therapy of nearly all pathological problems of plants are underway. Smart sensors and smart delivery systems will help to combat viruses and other crop pathogens.

It is noteworthy that regulation of the use of nanosystems in foods is a controversial issue. In the United States, Regulation 258/97 establishes that if a nanomaterial is used as a primary ingredient, then it needs to be considered as a "Novel Food" (Gallocchio et al., 2015). When the term "food additive" is employed (1333/2008), even if its composition is authorized as a raw material, it is not necessarily applicable to the nano form, and the nanosystem requires consideration as a different additive; thus, it needs to be evaluated as a new, safe material before it can be placed on the market.

Use of nanotechnology could permit rapid advances in agricultural research, such as alleviating stress effects, conversion of agricultural and food wastes to energy and other useful byproducts through enzymatic nano-bioprocessing, disease prevention and treatment in plants and animals, early detection of stresses and in reproductive science and technology (Mishra et al., 2017).

2.24 COLD PLASMA TECHNOLOGY

Plasma technology is a newcomer to the field of agriculture and the food industry. Plasma treatment is one of the most promising and unique preservation techniques. It is effective at ambient temperatures, responsible for microbial destruction and surface modification of substrate and also deactivates mycotoxins in/on food; thereby, having minimum thermal effects on nutritional and sensory quality parameters of food with no chemical residues, as conventional preservative techniques have some detrimental effects on nutritional quality. It has been convincingly demonstrated at the experimental level that this technique can efficiently destroy fungi present on the surface of food and decontaminate the mycotoxins that these organisms secrete.

Plasma is a quasi-neutral ionized gas primarily composed of photons, ions and free electrons, as well as atoms in their ground or excited states with a roughly zero net electrical charge (Turner, 2016). The strong electric fields used in generating cold plasma accelerate electrons to energies capable of ionizing, electronically exciting, dissociating and heating the constituents of the background gas. This results in the ionic species of H^+, H^-, O^+, O^-, H_3O^+, OH^-, and N_2^+; the excited and ground states of N_2, O, NO, and OH; and longer-lived species such as O_3, NO_2 and N_2O; collectively, these are referred to as reactive oxygen and nitrogen species (RONS) (Sakiyama et al., 2012). Based on the properties of plasma, it is used in various fields like textiles, electronics, life sciences, packaging and so on (Misra et al., 2016b), and now it is extended to food industries as a novel technology.

Different mode of actions have highlighted a reduced growth of microorganism via etching phenomenon, cell disruption by electroporation and so on. The reaction mechanisms resulting in the formation of active plasma chemical species include electronic impact processes (vibration, excitation, dissociation, attachment and ionization), ion–ion neutralization, ion–molecule reactions, Penning ionization, quenching, three-body neutral recombination and neutral chemistry, as well as photoemission, photo-absorption and photo ionization (Sahu et al., 2017). Plasma treatment results in cell wall destruction and inhibition of the cell membrane function, making it permeable and thus allowing the leakage of intracellular components; intracellular nanostructural changes; at low doses, fungal cells undergoing apoptosis; ROS from plasma causing oxidation of intracellular organelles; and lipid phosphates oxidized by ROS to lipid peroxide through a chain reaction. In later stages, the oxidation of genomic DNA and cellular proteins may also occur. So far, these studies have emphasized the dominant role of ROS, while air plasma is a mixture of ROS and RNS. The roles and contributions of RNS and ultraviolet radiation in cold plasma remain less researched considering the difficulty in separating the effects of ROS and RNS. Therefore, further research to understand fungal interactions with UV and RNS from plasma sources is desirable.

The efficiency of microbial inactivation depends on the surface of the treating produce, plasma device, gas composition and mode of exposure. Produce like potatoes and strawberries took more time for complete destruction of microbes due to grooves and uneven surfaces (Misra et al., 2019). The efficiency of microbial reduction was improved with increases in humidity of air.

In favor over many of the traditional food decontamination methods, plasma-based decontamination methods are generally lower cost and ecologically benign. Compared with several conventional and non-thermal approaches, cold plasma treatments act rapidly against molds, require low energy input and have a relatively milder impact on quality. Plasma sterilization provides high efficacy, preservation and does not introduce toxicity to the medium. The most important point is to select particular gases that already possess germicidal properties so that the efficiency of plasma sterilization can be increased. Cold plasma is used efficiently for sterilization and modifications of packaging polymers. Plasma can be used for starch modification as an additive and as a filler component in packing materials.

One of the important challenges associated with plasma technology is ensuring a high microbial inactivation while maintaining sensory qualities that ensure their fresh appearance. It presents a research opportunity to further explore the effects of cold plasma on the physico-chemical and sensory properties of food products at the molecular level. Optimization studies are also required to avoid the negative impacts on quality, such as accelerated lipid oxidation, loss of vitamins and sensory characteristics. The precise understanding of the mechanisms and control over the quality attributes will be required for plasma technology to realize its full potential at a commercial scale (Pankaj et al., 2018).

Future research activities are likely to yield more information for the development of validation protocols for antimycotic or mycotoxin reduction action in food and feed, paving the way for potential upscaling of the technology to industrial adoption. Although plasma technology is not yet used commercially on a large scale, the equipment should be readily scalable. However, research efforts must be taken to evaluate the expenditure for the treatment for large quantities of food commodities at industry level and also the quality, safety and wholesomeness of food commodities.

Given the diverse nature of the food industry, it is continually seeking and willing to adopt new technologies for sustainability, safety, profitability, consumer trust and continued success (Misra et al., 2018a). Therefore, treatment with plasma could be one approach to reduce the amount of postharvest inoculum in different food products. One thing that has to be kept in mind is that besides managing decontamination of these infected commodities, the novel methods should also have to be environmentally benign and economically suitable and also preserve the quality of food products. Considering the above-mentioned requirements, cold plasma technology offers a promising non-thermal decontamination approach (Raviteja et al., 2019).

2.25 INTEGRATED DISEASE MANAGEMENT

The reduction of postharvest food loss is a critical component of ensuring future global food security, which is again a critical component of integrated resource management (Elangbam et al., 2015). Effective control will need to adopt integrated strategies in which, besides new non-polluting postharvest antifungal treatments, all factors affecting disease epidemiology and incidence will need to be taken into account, including preharvest factors. The main purpose of combination applications is to increase effectiveness and to decrease the negative effects of application by exposure to lower doses when compared with single applications.

2.25.1 Combining GRAS with Other Treatments

Although the acidic forms of some GRAS salts can also show substantial antimicrobial activity, salt compounds are preferred as potential postharvest treatments because of their superior solubility and ease of manipulation and application. Salts such as calcium chloride, calcium propionate, sodium carbonate, sodium bicarbonate, potassium meta bisulfate, ethanol and ammonium molybdate have been found to be successful when combined with microbial antagonists for the control of postharvest diseases. The use of GRAS substances like sodium bicarbonate with *Trichoderma harzianum* was effective to control *Colletotrichum musae* and *Fusarium verticillioides* (crown rot) in banana fruits (Alvinidia, 2013). The effectiveness of the antagonist and salt combination was dependent upon the concentration of both partners, their mutual compatibility and duration and time when applied.

Moreover, the additional antifungal activity against important postharvest pathogens of cations such as Na^+, K^+ and $NH4^+$ has been proven for many salts. The addition of hydrogen peroxide as a sanitizer into solutions of pyrimethanil, azoxystrobin or fludioxonil, which are common fungicides used by the citrus industry, prevented the build-up of microbial populations to avoid contamination of healthy fruit and to minimize selection of fungicide-resistant pathogen strains (Kanetis et al., 2008). The control of sour rot and green and blue molds on lemons was poor with salt solutions alone but significantly improved in treatments including hydrogen peroxide followed by potassium sorbate, sodium bicarbonate or potassium phosphate. Applications of either potassium sorbate or a sequence of hydrogen peroxide followed by potassium phosphite were the most promising treatments, primarily because they controlled most of the postharvest lemon diseases without the need to heat the solutions, which avoids considerable expense and an increased risk of fruit injury (Cerioni et al., 2013). A sequential oxidative treatment using sodium hypochlorite and hydrogen peroxide in the presence of a cupric salt inhibited growth, conidial germination and fungal infectivity of *P. digitatum* and *P. expansum* on lemons and apples (Cerioni et al., 2013), respectively. In numerous studies, acetic acid, in the form of a fumigant or solution, was used in combination with heat treatments. Mandarin fruit, inoculated with *P. digitatum*, were cured at 36 °C for 36 h and then fumigated with 0, 5, 15, 25, 50, 75 and 100 mL L/1 of acetic acid vapors for 15 min. Curing or fumigations performed alone reduced decay with respect to untreated fruit, but the best control was achieved with combined treatments (Venditti et al., 2009). Fumigation with acetic acid followed by use of modified atmosphere packaging and cold storage reduced the percentage of decayed table grapes from 94% in the control to 2% and of decayed strawberries from 89% in the control to zero.

Organic acids with a low carbon number, such as formic, acetic and lactic acids, were considered the ideal solvents for practical grade chitosan and more effectively controlled gray mold on table grapes than those with higher carbon numbers (maleic, malic, succinic, L-ascorbic and L-glutamic acids) or inorganic acids such as hydrochloric or phosphorous acids (Romanazzi et al., 2009). Citric acid or tartaric acid embedded into a chitosan coating increased the shelf life of litchi fruit by at least 3 weeks compared with untreated control fruit, which browned rapidly (Ducamp-Collin et al., 2008). Similarly, addition of ascorbic acid to chitosan caused an extension in shelf life and better retention of the quality of plums during storage (Liu et al., 2014). A commercial mixture of organic acids and calcium was effective in reducing the in vitro growth of *B. cinerea, Monilinia laxa, R. stolonifer* and *A. alternata* (Feliziani et al., 2013). Sweet cherry treated immediately after harvest with this mixture of organic acids and calcium showed reduced postharvest decay (Feliziani et al., 2013, 2014). Similar results were obtained with strawberry treated after harvest or during the season from flowering to maturity (Romanazzi et al., 2013; Feliziani et al., 2015). Fumigation with acetic acid followed by use of modified atmosphere packaging and cold storage reduced the percentage of decayed table grapes from 94% in the control to 2% and of decayed strawberries from 89% in the control to zero (Moyls et al., 1996).

Essential oils can be applied directly in the vapor phase or incorporated as active microbial agents in different matrices such as films and coatings, among others, at preharvest or postharvest stages. In a recent investigation, when cinnamon essential oil was added into biodegradable polyester nets, the development of fungus *A. alternata* was inhibited by 72% as mycelial growth and germination was totally checked (Black-Solis et al., 2019). The presence of essential oil did not alter the biodegradability of nets as well as their efficiency to maintain fruit quality and disease control on infected fruits by reducing disease incidence. Recently, incorporation of limonene into the protein matrix (Gonzales-Estrada et al., 2017) inhibited the mycelial growth and germination process of *P. italicum* isolated from infected limes. The applications of films based on chitosan, oleic acid/

beeswax and lemon essential oils were tested on tomatoes for preserving their quality. These films improved the fruit quality by reducing water losses and maintaining the appearance of treated strawberry fruit (Landi et al., 2014).

The addition of *Trichoderma harzianum* in polymeric matrices like chitosan has been evaluated against *Fusarium oxysporum* with good results, inducing defense mechanisms and the production of lytic enzymes as well as parasitism. In strawberry fruits, the incorporation of *Trichoderma* in chitosan and the application of a physical treatment (hot air) were effective in reducing microorganisms and maintaining fruit quality (Nitu et al., 2016; Aml et al., 2017).

2.25.2 Combination of Irradiation with Other Treatments

When integrated with other biologically based technologies, a complex defense response against pathogens arise in irradiated tissue, which includes the induction of a wide range of genes and the modification of enzyme activities, as well as de novo synthesis of PR proteins and the increase of secondary metabolites. The effect of irradiation is more promising when applied in combination with hot water treatment, chemicals such as SO_2 fumigation and cold storage treatment (Sripong et al., 2015). The effects of the integration of 3% aqueous sodium carbonate dips at 20 °C for 150 s in treatments and X-ray irradiation at doses of 510 and 875 Gy for *Penicillium* decay control on "Clemenules" clementine mandarins was observed and the resultS showed that combined treatments, especially at the X-ray highest dose of 875 Gy, significantly reduced disease incidence and the severity of both green and blue molds (*Penicillium digitatum* and *Penicillium italicum*) at 5 °C (Palou et al., 2007). In contrast to fungal growth, pathogen sporulation, especially that of *P. digitatum*, was clearly inhibited on inoculated clementine by the combined treatments. Since sodium carbonate does not exert anti-sporulant activity, this effect should be attributed to irradiation. The integration of curing treatments (35 °C for 72 h) or hot water dips (50–55 °C for 2 min) in association with UV-C illumination was found superior to either the treatment alone in reducing decay or in maintaining fruit quality. When UV-C treatment preceded heat treatment, the elicitation of phytoalexins in fruit rind was inhibited. Dipping fruit in water at 52 °C for 5 min followed by gamma irradiation at low dose (500 Gy) delayed the *P. digitatum* by up to 40 days. Combination of irradiation with cold storage is more promising. The degree of sensitivity of these storage pathogens on PDA to gamma rays at 3 to 4 °C was found (from resistant to sensitive): *B. cinerea* > *Alternaria tenuissima* > *P. expansum* > *R. stolonifer*. The combination of ultraviolet-C irradiation and biocontrol treatments to control postharvest decay showed promising results (Sharma and Tewari, 2014). The combination of yeast *Candida oleophila* with UV-C irradiation evidenced a synergistic effect in reducing *Penicillium digitatum* mold, whereas no synergistic effects were observed when UV-C was combined with bacteria *Bacillus subtilis*. In a similar study, it was found that infection on fruits of tomato, gooseberry and mango were effectively controlled by combined treatment with yeast and irradiation (Sharma and Srivastava, 2014).

2.25.3 Combination of Ultrasound with Other Treatments

It was found that the effectiveness of ultrasound (40 kHz, 5 min) alone and in combination with 0.3, 0.5, 0.7, 1.0 and 2.0% of malic acid, lactic acid and citric acid, respectively, reduced *Escherichia coli* O157:H7, *Salmonella Typhimurium* and *Listeria monocytogenes* in fresh lettuce. For all the three pathogens, the combined treatment of ultrasound and organic acids resulted in an additional 0.8–1.0 log reduction compared with individual treatments without causing significant quality change (color and texture) on lettuce during 7-day storage. The maximum reductions of *E. coli* O157:H7, *S. Typhimurium* and *L. monocytogenes* were 2.75, 3.18 and 2.87 log CFU/g observed after combined treatment with ultrasound and 2% organic acid for 5 min, respectively (Sagong et al., 2011).

Cao et al. (2010) observed that the addition of low weight chitosan (1000 ppm) enhanced the inactivation of *Saccharomyces cerevisiae* by ultrasound (20 kHz) at 45°C in Sabouraud broth (pH 5.6). After 30 min of exposure to chitosan, approximately one log cycle reduction of the yeast was obtained leading to a final reduction of more than three log cycles after 30 min of the ultrasonic treatment (Guerrero et al., 2005). The investigations on peach fruit infected with *Penicillium expansum* treated with ultrasound (40 kHz, 10 min) and salicylic acid (0.05 mM) either separately, or in combination, showed that the application of salicylic acid alone could reduce blue mold, while the use of ultrasound alone had no effect. Results also revealed that salicylic acid combined with ultrasound treatment was more effective in inhibiting fungal decay during storage than the salicylic acid treatment alone. The combined treatment increased the activities of defense enzymes such as chitinase, β-1,3-glucanase, phenylalanine ammonia lyase, polyphenol oxidase and peroxidase, which were associated with higher disease resistance induced by the combined treatment.

2.25.4 Combining Hot Water with Other Treatments

The combination of hot water treatments with other alternatives was found to be effective in managing development of postharvest diseases (Janisiewicz and Conway, 2010). Sodium bicarbonate in combination with hot water treatment was effective in reducing the disease caused by *P. digitatum*. Gutierrez-Martinez et al. (2018a) experimenting with mango achieved complete control of *Pestalotia mangiferae* and *Curvularia lunata* by treatment with 300 ml/L ethanol and hot water rinsing at 50 ÅãC for 60 s, while fruit ripened normally.

Application of hot water brushing and yeast antagonist *Candida oleophila* reduced decay development of blue mold in grapefruits. Standalone treatments, with hot water, *C. guilliermondii* or *Pichia membranaefaciens*, reduced disease incidence from 81.7% in the control to 61.7%, 40.0% or 45.0%, respectively. The synergistic effect of the combined treatment decreased the disease incidence to 21.7% for *C. guilliermondii* and 26.7% for *P. membranaefaciens*.

Hot water treatment and yeast antagonists *C. guilliermondii* and *P. membranaefaciens* were investigated separately and in combination for controlling *Botrytis cinerea* infecting tomato (Zhong et al., 2010). Hot air at 38 ÅãC for 36 h was combined with the antagonist *Pichia guilliermondii* to control anthracnose in loquat. After 10 days of storage, the decay index was 20, 6.67, 7.78 and 3.33 in control fruit, and fruit treated with heat, biological control and the combination, respectively.

Combining hot water treatment with the bacterial antagonist *Burkholderia cepacia* gave more effective control of anthracnose, blossom end and crown rot of banana. Hong et al. (2014) applied hot water at 42 ÅãC for 2 min to mandarins together with *Bacillus amyloliquefaciens* HF-01 and sodium bicarbonate, reducing decay associated with *Penicillium digitatum, P. italicum* and *Geotrichum citri-aurantii* by more than 80% compared with the control, showing a better performance than treatments with the fungicide imazalil.

The combination of HWT with an antagonist proved to be better in comparison with standalone treatments. Heat-shock has been reported to enhance the efficacy of some biocontrol agents, although the physiological and biochemical mechanisms are not so well understood (Liu et al., 2013). The mechanism by which hot water treatment enhanced the biocontrol activity of yeasts may be attributed to the elicitation of a biochemical defense response such as a significant increase in the activities of PAL, chitinase and β-1,3-glucanase in fruits (Zhong et al., 2010).

MeJa plays an important role in responses to environmental stresses. Jin et al. (2009) combined treatment of peaches with hot air at 38 ÅãC for 12 h with 1 mol/L MeJa prior to storage at 0 ÅãC for 3–5 weeks. The treatments were shown to induce phenylalanine ammonia lyase, superoxide dismutase and polygalacturonase.

The hot air treatment caused severe mealiness in peaches without MeJa, but chilling injury was counteracted by MeJa treatments.

Combination of heat treatments with other technologies could not only maintain but also improve the sensory quality of the commodity. Special attention should be given to the order of applications of combined treatments, as commodity response can be very different. Sharma and Tripathi (2008) observed that combining hot water, UV-C and essential oil of *Hyptis suaveolens* gave better results in comparison with standalone treatments in the postharvest management of *Fusarium* rot.

2.25.5 Nanotechnology and Edible Coatings

There is an increasing tendency to use natural solid and liquid lipids that can function to encapsulate active substances and, at the same time, provide some benefit to either food products or directly to consumer health. The submicron size of nanoparticles represents an area of opportunity that provides a vehicle to transport certain EOs, vitamins and other plant extracts, such as polyphenols, with antimicrobial and antioxidant properties incorporated in hydrophilic and lipophilic substances, which can be released during storage periods to increase the shelf life of diverse products (Zambrano-Zaragoza et al., 2018). It offers a new way to modify gas transport properties and the release of natural products, while improving mechanical resistance, transparency, functionality and antioxidant and antimicrobial activity. Edible coatings have included various polymeric nanoparticles, nanoemulsions (Severino et al., 2014) and nanocomposites, among others, in an effort to control the release of EOs, polyphenols and fat-soluble vitamins, compared with their function in microcapsules and other forms of incorporation into polymeric matrices (Cortez- Vega et al., 2014)

The biolipid phase of nanoemulsions consists of oily solutions in which lipophilic components such as essential oils, oily flavors, oily colors, vitamins and so on are first dissolved in corn, soybean, sunflower or olive oil. The lipophilic compounds in nanoemulsions are bioactive, flavoring, antimicrobial, antioxidant and nutraceutical lipids (Silva et al., 2012, Junqueira-Goncalves et al., 2017)). A potential advantage of reducing the size of the oil drops to nanometer range is that it increases the solubility of the bioactive lipids in the surrounding water phase, which enhances bioactivity, desirability and palatability (McClements, 2015). The lipophilic materials that can be incorporated into nanoemulsions include essential oils from plants (e.g. oregano, sage, clove, mint, limonene), fatty acids (ω-3 fatty acids, conjugated linoleic acid, butyric acid), carotenoids (β-carotene, lycopene, lutein, zeaxanthin), antioxidants (tocopherols, flavonoids, polyphenols), phytosterols (stigmasterol, β-sitosterol, campesterol) and quinones (coenzyme Q_{10}) (Salvia-Trujillo et al., 2017).

Curcumin, a polyphenol extracted from the rhizome of turmeric, is an antioxidant with antimicrobial properties. Liu et al. (2013) also reported elaborating curcumin nanospheres using chitosan as the polymer, while Lv et al. (2014) reported the incorporation of two biocompatible polymer gelatins and arabic gum by complex coacervation of jasmine essential oil nanocapsules. Coradini et al. (2014) reported the co-encapsulation of resveratrol and curcumin in PCL/grape seed oil (polymer/oil) nanocapsules.

Plant extracts such as peppermint oil were employed due to their antimicrobial effect (Ghayempour et al., 2015), and lutein was encapsulated into PCL nanocapsules by the interfacial deposition technique (Brum et al., 2017). Polyphenols, such as epigallocatechin gallate (Liang et al., 2017), quercetin (Kumar et al., 2015; Kumari et al., 2014; Hao et al., 2017) and hydroxycinnamic acids (Granata et al., 2018) have been incorporated into polymeric nanoparticles with potential for use in edible coatings. Carotenoids, an antioxidant molecule, have been successfully included in nanocapsules. PCL nanocapsules incorporated with β-carotene by the emulsification-diffusion method also showed potential for use in food technology (Galindo-Pérez et al., 2015, ; Galindo et al., 2017). Assis et al. (2017) incorporated lycopene-PCL nanocapsules into cassava starch films to observe and evaluate them as biodegradable coatings.

Nanostructured lipid carriers (NLCs) are a second generation of lipid-based nanoparticles, developed by Müller in 2000 to overcome problems with solid lipid nanoparticles (SLNs) due to the process of the full crystallization of fat that reduces drug solubility and causes expulsion of the active substances from lipid particles (Tamjidi et al., 2013). Huang et al. (2017) developed NLCs to encapsulate quercetin (because of its therapeutic effects, which include antioxidant, anticancer, antibacterial and anti-inflammatory actions, and its potential ability to prevent neurodegenerative diseases) using glyceryl monostearate as the solid lipid, linseed oil as the liquid lipid, and a mixture of Tween® 80 and polyglycerol monostearate as the surfactant. Linseed oil is known to be a good replacement for fish oil because of its high α-linoleic acid content, which is used to prevent cardiovascular diseases and hypertension.

Nanocomposite coatings that incorporate nanosystems mixed with organic and inorganic substances represent a recent option for improving the properties of edible coatings, since they permit better mechanical resistance, transparency, controlled release and more effective gas barrier properties (Dang et al., 2017). The most widely used inorganic components for modifying the properties of edible coatings include montmorillonite (MMT), nano-SiOx, nano-TiO_2, and nano-ZnO, as well as silver nanoparticles (Shah et al., 2016), though it is important to note that the latter can only be used to coat whole fruits and vegetables (Shi et al., 2018). Nano-SiO_2 is widely used in food products and is registered in Codex Alimentarius as a food additive (E551). It is mainly used to thicken pastes, as an anti-caking agent to maintain flow properties in powdered products and as a carrier for fragrances or flavors in food and non-food products.

Natrajan et al. (2015) prepared alginate–chitosan nanocapsules of turmeric oil and lemongrass oil by the pre-gelation of oil in a water-based nanoemulsion of alginate by adding a calcium chloride solution. Finally, chitosan was added to crosslink the preformed alginate nanocapsules. Lemon grass oil has also been incorporated into cellulose–acetate nanocapsules by the nanoprecipitation method (Liakos et al., 2016). In another case, a natural material recognized as safe, tragacanth gum, was used to prepare peppermint oil nanocapsules by the microemulsion method (Ghayempour et al., 2015). The use of EO nanocapsules in food technology has been described by Mohammadi et al. (2016) recently as they encapsulated *Zataria multiflora* essential oil in chitosan nanoparticles to prepare a coating that improved antioxidant activity and extended the shelf life of cucumbers.

Although edible coatings based on a solution of chitosan have been reported for decades (Petriccione et al., 2015; Duan et al., 2011), few studies have reported using chitosan nanoparticles (Deng et al., 2017), although Mustafa et al. (2013) included them in an edible coating to preserve postharvest tomatoes. In their study, they found that the coating had low adhesion and durability, as evident by the breakdown of its moisture barrier properties, which meant that this edible coating required a support matrix, however, the delay in color evolution in the coated fruits and the maintenance of quality during storage were the main benefits of the nanoparticles. Pilon et al. (2015) revealed the differences between the conventional chitosan coating process (dipping into chitosan gel) and nanoparticle coating. They found that when the nanoparticles were sprayed on freshly cut apples, a non-continuous coating was formed that compromised the moisture barrier, though they provided a well-dispersed coating that had a greater antimicrobial effect on microorganisms due to greater surface interaction. Eshghi et al. (2014) in an experiment applied chitosan nanoparticles prepared by ionotropic gelation on fresh strawberries. In this case, the nanoparticles provided an effective control in reducing weight loss and maintaining firmness, while also delaying changes in the respiration rate for 3 weeks. Recently, Martínez-Hernández et al. (2017) reported the encapsulation of carvacrol, a major component of the essential oil of oregano, thyme, marjoram and summer savory into polymeric chitosan nanoparticles, prepared by ionotropic gelation, which was used to protect freshly cut carrots.

Recently, silver nanoparticles with *Trichoderma* have been synthesized and their antifungal capacity against postharvest pathogenic fungi such as *Alternaria, Penicillium,* and *Fusarium* has been evaluated; good control on mycelial growth and development of the pathogenic fungi was reported at low concentrations of *Trichoderma* spp. (Elamawi et al., 2018)

2.26 NON-POLLUTING INTEGRATED DISEASE MANAGEMENT

Considering the new tendencies in the fruit industry and marketing, the use of alternative methods represents a suitable approach not only for controlling postharvest diseases but also for maintaining fruit quality of postharvest products. The establishment of non-polluting integrated disease management (NPIDM) strategies against target postharvest diseases should be based upon the in-depth knowledge of pathogen biology and epidemiology and should consider all preharvest, harvest and postharvest factors that can influence disease appearance and incidence in order to minimize decay losses. They should have cost-effective action plans in the field or after harvest that should not adversely affect fruit quality. The basis of successful NPIDM strategies is the commercial adoption of suitable physical, chemical or biological non-polluting alternative postharvest treatments to replace the use of conventional postharvest fungicides. Chemical alternatives should be compounds with known and minimal toxicological effects on mammals and impact on the environment and they should be affirmed as GRAS by the United States Food and Drug Administration (US FDA), as food additives by the EFSA or as an equivalent status by national legislations of other countries because as substances they will be in contact with fresh produce. Such a "non-polluting integrated disease management" concept should not be confused with traditional "integrated disease management" (IDM) in the context of agricultural "integrated production," which often implies fruit production in compliance with particular national or regional regulations and programs that will still include the use of postharvest fungicides. In Valencia (Spain), for instance, the regional administrative rule for IDM of citrus fruit allows the use of postharvest conventional fungicides when necessary, and the only restriction is that they have to be applied under technical supervision (Palou, 2016, 2018).

REFERENCES

Abdellatif, K.F., R.H. Abdelfattah and M.S.M. El-Ansary. 2016. Green nanoparticles engineering on root knot nematode infecting eggplants and their effect on plant DNA modification. *Iran. J. Biotechnol.* 14(4): 250–59.

AGRA. 2013. *Establishing the status of post-harvest losses and storage for major staple crops in eleven African countries (Phase I).* Nairobi, Kenya: Alliance for a Green Revolution in Africa (AGRA).

AGRA. 2014. *Establishing the status of post-harvest losses and storage for major staple crops in eleven African countries (Phase 2).* Nairobi, Kenya: Alliance for a Green Revolution in Africa (AGRA).

Ahmed, M.D.S. 2013. Needs Assessment of Postharvest Handling and Storage for Selected Vegetable Crops in Bangladesh. Prepared for the AVRDC - The World Vegetable Center.

Ajlouni, S., H. Sibrani, R. Premier et al. 2006. Food microbiology and safety ultrasonication and fresh produce (Cos lettuce) preservation. *J. Food Sci.* 71(2): M62–M68.

Akhtara, J., P.K. Omreb and Z.R.A. Ahmad Azadc. 2015. Edible coating for preservation of perishable foods: A review. *J Ready Eat Food.* 2: 81–88.

Al-Haq, M.I., J. Sugiyama and S. Isobe. 2005. Applications of electrolyzed water in agriculture and food industries. *Food Sci. Technol. Res.* 11: 135–150.

Alawlaqi, M.M. and A. A. Asmaa. 2014. Impact of acetic acid on controlling tomato fruit decay. *Life Sci. J.* 11: 114–119.

Aminifard, M.H. and H. Bayat. 2018. Antifungal activity of black caraway and anise essential oils against *Penicillium digitatum* on blood orange fruits. *Int. J. Fruit Sci.* 18(3): 307–319.

Altunkaya, A. and V. Gökmen. 2009. Effect of various anti-browning agents on phenolic compounds profile of fresh lettuce (*L. sativa*). *Food Chem.* 117: 122–126.

Alvinidia, D. 2013. Sodium bicarbonate enhances efficacy of *Trichoderma harzianum* DGA01 in controlling crown rot of banana. *J. Gen. Plant Pathol.* 79: 136–144. doi: 10.1007/s10327-013-0432-z

Aml, A., A. Ezzat, M. Rageh et al. 2017. Effect of chitosan, biocontrol agents and hot air to reduce postharvest decay and microbial loads of strawberries. *Development* 6: 8.

Appiah, F. 2013a. *Report on Commodity System Assessment of Amaranth SPP in the Ashanti Region of Ghana.* Arusha, Tanzania: AVRDC.

Appiah, F. 2013b. *Report on Commodity System Assessment of Tomato in the Ashanti Region of Ghana.* Arusha, Tanzania: AVRDC.

Arah, I.K., G.K. Ahorbo, E.K. Anku et al. 2016. Postharvest handling practices and treatment methods for tomato handlers in developing countries: A mini review. *Adv. Agric.* 8. doi: 10.1155/2016/6436945

Assis, R.Q., S.M. Lopes, T.M.H. Costa et al. 2017. Active biodegradable cassava starch films incorporated lycopene nanocapsules. *Ind. Crops Prod.* 109: 818–827. doi: 10.1016/j.indcrop.2017.09.043

Aziz, N.H., R.M. Souzan and A.S.Azza. 2006. Effect of gamma irradiation on the occurrence of pathogenic microorganism and nutritive value off our principal cereal grains. *Appl. Radiat. Isot.* 64: 1555–1562.

Baldwin, E.A., M.O. Nisperos-Carriedo, X. Chen et al. 1996. Improving storage life of cut apple and potato with edible coating. *Postharvest Biol. Technol.* 9: 151–163.

Bandara, W.M.K.I., O.D.A.N. Perera and H.L.D. Weerahewa. 2015. Enhancing disease resistance and improving quality of papaya (*Carica papaya* L.) by postharvest application of silicon. *Int. J. Agric. For. Plant.* 1: 24–27.

Banik, S. and P. Sharma. 2011. Plant pathology in the era of Nanotechnology. *Indian Phytopathol.* 64(2): 120–127.

Black-Solis, J., R.I. Ventura-Aguilar, Z. Correa-Pacheco et al. 2019. Preharvest use of biodegradable polyester nets added with cinnamon essential oil and the effect on the storage life of tomatoes and the development of *Alternaria alternata*. *Sci. Hortic.* 245: 65–73.

Betchem, G., N.A.N. Johnson and Y. Wang. 2019. The application of chitosan in the control of post-harvest diseases: A review. *J. Plant Dis. Prot.* 126(6): 495–507.

Bodini, R., P. Sobral, C. Favaro-Trindade et al. 2013. Properties of gelatin-based films with added ethanol-propolis extract. *LWT Food Sci. Technol.* 51: 104–110.

Brent, K.J. and D.W. Hollomon. 2007. *Fungicide Resistance in Crop Pathogens: How Can It Be Managed?* (2nd Rev. ed.). Brussels, Belgium: Fungicide Resistance Action Committee (FRAC). CropLife Int'l. Available at: http://www.frac.info/frac/publication/anhang/FRAC_Monol_2007_100dpi.pdf

Brum, A.A.S., P.P. dos Santos, M.M. da Silva et al. 2017. Lutein-loaded lipid-core nanocapsules: Physicochemical characterization and stability evaluation. *Colloids Surf. A Physicochem. Eng. Asp.* 522: 477–484. doi: 10.1016/j.colsurfa.2017.03.041

Brychcy-Rajska, E. 2017. Edible protective films and coatings in food industry. *J Food Microbiol.* 1(1): 1–2.

Cao, S., Z. Hu, B. Pang et al. 2010. Effect of ultrasound treatment on fruit decay and quality maintenance in strawberry after harvest. *Food Control* 21(4): 529–532.

Cagri, A, Z. Ustunol and E.T. Ryser. 2004. Antimicrobial edible films and coatings. *J Food Prot.* 67: 833–848.

Cavalcante, R.D., W.G. Lima and R.B. Martins. 2014. Thiophanate-methyl sensitivity and fitness in *Lasiodiplodia theobromae* populations from papaya in Brazil. *Eur. J. Plant Pathol.* 140: 521–259.

Cerioni, L., M. Sepulveda, Z. Rubio-Ames et al. 2013. Control of lemon postharvest diseases by low-toxicity salts combined with hydrogen peroxide and heat. *Postharvest Biol. Tech.* 83: 17–21.

Cha, D.S. and M.S. Chinnan. 2004. Bipolymer-based antimicrobial packaging–A review. *Crit. Rev. Food Sci. Nutr.* 44: 223–237.

Chavez-Magdaleno, M.E., A.G. Luque-Alcaraz, P. Gutierrez-Martinez et al. 2018. Effect of chitosan-pepper tree (*Schinus molle*) essential oil biocomposites on the growth kinetics, viability and membrane integrity of *Colletotrichum gloesporioides*. *Biotechnology* 17(1): 29–45.

Chen, J., Y. Shen, C. Chen et al. 2019. Inhibition of key citrus postharvest fungal strains by plant extracts in vitro and in vivo: A review. *Plants* 8(2): E26. doi: 10.3390/plants80200026

Chiumarelli, M. and M. Hubinger. 2012. Stability, solubility, mechanical and barrier properties of Cassava starch- Carnauba wax edible coating to preserve fresh-cut apples. *Food Hydrocollids* 28: 59–67. doi: 10.1016/jfoodhyd.2011.12.006

CIPHET. 2015. *Report on Assessment of Quantitative Harvest and Postharvest Losses of Major Crops and Commodities in India*. Ludhiana: All India Coordinated Research Project on Postharvest Technology (ICAR).

Cortez-Vega, W.R., S. Pizato, J.T.A. de Souza et al. 2014. Using edible coatings from Whitemouth croaker (*Micropogonias furnieri*) protein isolate and organo-clay nanocomposite for improve the conservation properties of fresh-cut "Formosa" papaya. *Innov. Food Sci. Emerg. Technol.* 22: 197–202.

Cosic, J., K. Vrandecic and D. Jurkovic. 2014. The effect of essential oils on the development of phytopathogenic fungi. In Neeta Sharma (Ed.), *Biological controls for Preventing Food Deterioration: Strategies for Pre-and Postharvest Management* (pp. 273–291). Hoboken, NY: John Wiley & Sons, Ltd.

Coskun, D., R. Deshmukh, H. Sonah et al. 2018. The controversies of silicon's role in plant biology. *New Phytol.* 221: 67–85.

Coradini, K., F.O. Lima, C.M. Oliveira et al. 2014. Co-encapsulation of resveratrol and curcumin in lipid-core nanocapsules improves their in vitro antioxidant effects. *Eur. J. Pharm. Biopharm.* 88: 178–185. doi: 10.1016/j.ejpb.2014.04.009

Cruz, V., R. Rojas, S. Saucedo-Pompa et al. 2015. Improvement of shelf life and sensory quality of Pears using specialized edible coating. *J. Chem.* 2015: 138707. doi: 10.1155/2015/138707

da Nobrega, L.P., F.K.R. da Silva, T.S. Lima et al. 2019. In vitro fungitoxic potential of copaiba and eucalyptus essential oils on phytopathogens. *J. Exp. Agric. Inter.* 29(3): 1–10.

Dang, X., X. Cao, L. Ke et al. 2017. Combination of cellulose nanofibers and chain-end-functionalized polyethylene and their applications in nanocomposites. *J. Appl. Polym. Sci.* 134: 45387–45393. doi: 10.1002/app.45387

Deepshikha, B. Kumari, E.P. Devi et al. 2017. Irradiation as an alternative method for post–harvest disease management: An overview. *IJAEB* 10(5): 625–663.

Deng, Z., J. Jung, J. Simonsen et al. 2017. Cellulose nanocrystal reinforced chitosan coatings for improving the storability of Postharvest pears under both ambient and cold storages. *J. Food Sci.* 82: 453–462. doi: 10.1111/1750-3841.13601

Devi, E.P. and B. Kumari. 2015. A review on prospects of pre-harvest application of bioagents in managing post-harvest diseases of horticultural crops. *Int. J. Agric. Environ. Biotechnol.* 8: 933–941.

Ding, P. and Y.L. Lee. 2019. Essential oil for prolonging postharvest life of fresh fruits and vegetables. *Int. Food Res. J.* 26(2): 363–366.

Droby, S., M. Wisniewski, D. Macarisin et al. 2009. Twenty years of postharvest biocontrol research, is it time for a new paradigm? *Postharvest Biol. Technol.* 52: 137–145.

Duan, J., R. Wu, B.C. Strik et al. 2011. Effect of edible coatings on the quality of fresh blueberries (Duke and Elliott) under commercial storage conditions. *Postharvest Biol. Technol.* 59: 71–79. doi: 10.1016/j.postharvbio.2010.08.006

Ducamp-Collin, M.N., H. Ramarson, M. Lebrun et al. 2008. Effect of citric acid and chitosan on maintaining red coloration of litchi fruit pericarp. *Postharvest Biol. Technol.* 49: 241–246.

Dussan-Sarria, S., C.T. Leon and J.I.H. Leap. 2013. Effect of edible coating and different packaging during cold storage of fresh cut Tommy Atkins mango. *Inf. Technol.* 25(4): 123–140.

Elamawi, R., R. Al-Harbi and A. Hendi. 2018. Biosynthesis and characterization of silver nanoparticles using *Trichoderma longibrachiatum* and their effect on phytopathogenic fungi. *Egypt. J. Biol. Pest Control* 28: 28. doi: 10.1186/s41938-018-0028-1

Elmer, W. and J.C. White. 2018a. The future of nanotechnology in plant pathology. *Ann. Rev. Phytopath.* 56: 111–133. doi: 10.1146/annurevev-phyto-080417-050108

El-Mougy, N.S., N.G. El-Gamal and F. Abd-El-Karrem. 2008. Use of organic acids and salts to control post-harvest diseases of lemon fruits. *Egypt. Arch. Phytopathol. Plant Protect.* 41: 467–476.

Escribano, S. and E.J. Mitchem. 2014. Progress in heat treatments. *Stewart Postharvest Rev.* 3: 2.

Eshghi, S., M. Hashemi, A. Mohammadi et al. 2014. Effect of nanochitosan-based coating with and without copper loaded on physicochemical and bioactive components of fresh strawberry fruit (*Fragaria* x *ananassa* Duchesne) during storage. *Food Bioprocess Technol.* 72: 397–409. doi: 10.1007/s11947-014-1281-2

Fallanaj, F., S.M. Sanzani, C. Zavanella et al. 2013. Salt addition improves the control of citrus postharvest diseases using electrolysis with conductive diamond electrodes. *J. Plant Pathol.* 95: 373–383.

Fallanaj, F., A. Ippolito, A. Ligorio et al. 2016. Electrolyzed sodium bicarbonate inhibits *Penicillium digitatum* and induces defence responses against green mold in citrus fruit. *Postharvest Biol. Technol.* 115: 18–29.

Fallanaj, F., S.M. Sanzani, K. Youssef et al. 2015. A new perspective in controlling postharvest citrus rots: The use of electrolyzed water. *Acta Hortic.* 1065: 1599–1605.

FAO. 2019. *The state of Food and Agriculture 2019. Moving forward on food loss and waste reduction.* Rome Licence: CC BY-NC-SA 3.0IGO.

FAO. 2015. *Food Losses and Waste in Latin America and the Caribbean*, Bulletin No. 2. FAO: Rome

FAO. 2011. *Global Food Losses and Food Waste – Extent, Causes and Prevention.* FAO: Rome.

Fallik, E. 2014. Microbial quality and safety of fresh produce. In W. Florkwski, R. Shewfelt, S. Prussia, N. Banks, *Post Harvest Handling. A System Approach*, Cambride, MA: Academic Press. (pp. 313–339).

FDA. 2018. Code of Federal Regulations. 21 CFR 173, Section 173.370: Peroxyacids. Available at: https://www.accessdata.fda.gov/scrip ts/cdrh/cfdocs/cfcfr/cfrsearch.cfm?fr =173.370. [Review date: January 2019]

FDA. 2018. Code of Federal Regulations. 21CFR 173.356. Section 173.356 Hydrogen peroxide. Available at: https://www.accessdata.fda.gov/scripts/cdrh/cfdocs/cfcfr/CFRSearch.cfm?fr=173.356. [Review date: January 2019]

Feliziani, E., L. Landi and G. Romanazzi. 2015. Preharvest treatments with chitosan and other alternatives to conventional fungicides to control postharvest decay of strawberry. *Carbohydr. Polym.* 132: 111–117.

Feliziani, E., A. Lichter, J.L. Smilanick et al. 2016. Disinfecting agents for controlling fruit and vegetable diseases after harvest. *Postharvest Biol. Technol.* 122: 53–69. doi: 10.1016/j.postharvbio.2016.04.016

Feliziani, E., G. Romanazzi and J.L. Smilanick. 2014. Application of low concentrations of ozone during the cold storage of table grapes. *Postharvest Biol. Technol.* 93: 38–48.

Feliziani, E., M. Santini, L. Landi et al. 2013. Pre- and postharvest treatment with alternatives to synthetic fungicides to control postharvest decay of sweet cherry. *Postharvest Biol. Technol.* 78: 133–138.

Galindo, M., D. Quintanar, M.A. Cornejo-Villeges et al. 2017. Optimization of the emulsification-diffusion method using ultrasound to prepare nanocapsules of different food-core oils. *LWT- Food Sci.Tech.* 87: 333–341. doi: 10.1016/j.lwt.2017.09.008

Galindo-Pérez, M.J., D. Quintanar-Guerrero, E. Mercado-Silva et al. 2015. The effects of tocopherol nanocapsules/xanthan gum coatings on the preservation of fresh-cut apples: Evaluation of phenol metabolism. *Food Bioprocess Technol.* 8: 1791–1799. doi: 10.1007/s11947-015-1523-y

Gallocchio, F., S. Belluco and A. Ricci. 2015. Nanotechnology and food: Brief overview of the current scenario. *Procedia Food Sci.* 5: 85–88. doi: 10.1016/j.profoo.2015.09.022

Gatto, M.A., A. Ippolito, L. Sergio et al. 2016a. Extracts from wild edible herbs for controlling postharvest rots of fruit and vegetables. *Acta Horticult.* 1144: 349–354.

Gatto, M.A., S.M. Sanzani, P. Tardia et al. 2013. Antifungal activity of total and fractionated phenolic extracts from two wild edible herbs. *Nat. Sci.* 05(08): 895–902.

Gatto, M.A., L. Sergio, A. Ippolito et al. 2016b. Phenolic extracts from wild edible plants to control postharvest diseases of sweet cherry fruit. *Postharvest Biol. Technol.* 120(July): 180–187.

Gautam, S. and B. Buntong. 2015. *Assessment of Postharvest Loss in Vegetable Value Chains in Cambodia. Tomato and Leafy Mustard in Siem Riep and Battambang Provinces.* AVRDC and USAID.

Ghayempour, S., M. Montazer and M. Rad Mahmoudi. 2015. Tragacanth gum as a natural polymeric wall for producing antimicrobial nanocapsules loaded with plant extract. *Int. J. Biol. Macromol.* 81: 514–520. doi: 10.1016/j.ijbiomac.2015.08.041

Gong, M., Q. Guan and S. Xu. 2018. Inhibitory effects of crude extracts from several plants on postharvest pathogens of citrus. *AIP Conf. Proc.* 1956: 200431–200434. doi: 10.1063/1.5034295

Gonzalez-Aguilar, G.A., J.G. Buta and C.Y. Wang. 2003. Methyl jasmonate and modified atmosphere packaging (MAP) reduce decay and maintain postharvest quality of papaya 'Sunrise.'. *Postharvest Biol. Technol.* 28(3): 361–370.

Gonzalez-Estrada, R., F. Blancas-Benítez, B. Montaño-Leyva et al. 2018. *A Review Study on the Postharvest Decay Control of Fruit by Trichoderma.* Rijeka: IntechOpen.

Gonzalez-Estrada, R., F. Blancas-Benítez, R.M. Velázquez-Estrada et al. 2019. *Alternative Eco-Friendly Methods in the Control of Post-Harvest Decay of Tropical and Subtropical Fruits.* Rijeka: IntechOpen.

Gonzalez-Estrada, R.R., P. Chalier, J.A. Ragazzo-Sanchez et al. 2017. Antimicrobial soy protein based coatings: Application to Persian lime (*Citrus latifolia* Tanaka) for protection and preservation. *Postharvest Biol. Technol.* 132(June): 138–144.

Granata, G., G.M.L. Consoli, R. Lo Nigro et al. 2018. Hydroxycinnamic acids loaded in lipid-core nanocapsules. *Food Chem.* 245: 551–556. doi: 10.1016/j.foodchem.2017.10.106

Guerrero, S., M. Tognon and S.M. Alzamora. 2005. Response of Saccharomyces cerevisiae to the combined action of ultrasound and low weight chitosan. *Food Control* 16(2): 131–139.

Guimaraes, I.C., K.C. dos Reis, E.G.T. Menezes et al. 2016. Cellulose microfibrillated suspension of carrots obtained by mechanical defibrillation and their application in edible starch films. *Ind. Crops Prod.* 89: 285–294. doi: 10.1016/j.indcrop.2016.05.024

Gutierraz-Martinez, P., A. Ledezma-Morales, L. del Carmen Romero-Islas et al. 2018a. Antifungal activity of chitosan against postharvest fungi of tropical and subtropical fruits. In R. Dongre, *Chitin-Chitosan-Myriad Functionalities in Science and Technology.* Rijeka: IntechOpen.

Gutierraz-Martinez, P., A. Ramos-Guerrero, C. Rodriguez-Pereida et al. 2018b. Chitosan for postharvest disinfection of fruits and vegetables. In M.W. Siddiqui, *Postharvest Disinfection of Fruits and Vegetables* (1st ed., pp. 231–241). Cambridge, MA: Academic Press, Elsevier.

Gutierraz-Martinez, P., A. Ramos- Guerrero, H. Cabanillas-Beltran et al. 2015. Chitosan as alternative treatment to control postharvest losses of tropical and subtropical fruits. In A.M. Ndez-Vilas (Ed.), *Science within Food: Up-to-date Advances on Research and Educational Ideas* (pp. 42–47). Badajoz, Spain: Formatex Research Center.

Gutierraz-Martinez, P., S. Bautista-Baos, G. Bermen-Varela et al. 2017. In vitro response of *Colletotrichum* to chitosan. Effect on incidence and quality on tropical fruit. Enzymatic expression in mango. *Acta Agronomica J.* 66(2): 282–289.

Guentzel, J.L., K.L. Lam, M.A. Callan et al. 2010. Postharvest management of gray mold and brown rot on surfaces of peaches and grapes using electrolyzed oxidizing water. *Int. J. Food Microbiol.* 143: 54–60.

Haggag, W.M., Y.S. Mahmoud and E.M. Farag. 2010. Signaling necessities and function of polyamines/jasmonatedependent induced resistance in sugar beet against beet mosaic virus (BtMV) infection. *N. Y. Sci. J.* 3(8): 95–103.

Han, J.H. 2002. Protein-based edible films and coatings carrying antimicrobial agents. In Gennadios A (Ed.), *Protein-Based Films and Coatings* (pp. 485–499). Boca Raton, FL: CRC Press.

Han, C., Y. Zhao, S.W. Leonard et al. 2004. Edible coatings to improve storability and enhance nutritional value of fresh and frozen strawberries (*Fragaria ananassa*) and raspberries (*Rubus ideaus*). *Postharvest Biol. Technol.* 33: 67–78.

Hassan, B., S.A.S. Chatha, A. Hussain et al. 2017. Recent advances on polysaccharides, lipids and protein based edible films and coatings: A review. *Int. J. Biol. Macromol.* 109: 1095–1107. doi: 10.1016/j.ijbiomac.2017.11.097.

Hassan Kamrul, M., B.L.D. Chowdhury and N. Akhter. 2010. Post Harvest Loss Assessment: A Study to Formulate Policy for Loss Reduction of Fruits and Vegetables and Socioeconomic Uplift of the Stakeholders. NFPCSP, Final Report (Bangladesh), PR #8.08.

Hassenberg, K., M. Geyer and W.B. Herppich. 2010. Effect of acetic acid vapour on the natural microflora and *Botrytis cinerea* of strawberries. *Eur. J. Hortic. Sci.* 75: 141–146.

Hazrati, S., A.B. Kashkooli, F. Habibzadeh et al. 2017. Evaluation of Aloe vera gel as an alternativr edible coating for peach fruits during cold storage period. *Gesunde Pflanzen* 69(3): 131–137.

Hao, J., B. Guo, S. Yu et al. 2017. Encapsulation of the flavonoid quercetin with chitosan-coated nano-liposomes. *LWT Food Sci. Technol.* 85: 37–44. doi: 10.1016/j.lwt.2017.06.048

Hao, J. and Q. Wang. 2019. Application of Electrolysed water in fruits and vegetables industry. In T. Ding et al. (Eds.), *Electrolysed Water in Food: Fundamentals and Applications* (pp 67–111). Hangzhou: Zhejiang University Press.

Hernandez-Munoz, P., E. Almenar, M.J. Ocio et al. 2006. Effect of calcium dips and chitosan coatings on postharvest life of strawberries (*Fragaria ananassa*). *Postharvest Biol. Technol.* 39: 247–253. doi: 10.1016/j.jfoodeng.2017.07.002

Hewajulige, I. 2012. *Anthracnose and Storage life extension of Papaya using Chitosan* (PhD thesis publication). Saarbrücken, Germany: Lap Lambert Academic Publishing.

Hong, P., W. Hao, J. Luo et al. 2014. Combination of hot water, *Bacillus amyloliquefaciens* HF-01 and sodium bicarbonate treatments to control postharvest decay of mandarin fruit. *Postharvest Biol. Technol.* 88: 96–102.

Howard, L.R. and T. Dewi. 1995. Sensory, microbiological and chemical quality of mini-peeled carrots as affected by edible coating treatment. *J. Food Sci.* 60: 142–144.

Huang, Y.R., Y.C. Hung, S.Y. Hsu et al. 2008. Application of electrolyzed water in the food industry. *Food Control* 19: 329–345.

Huang, H., Q. Jian, Y. Jiang et al. 2016. Enhanced chilling tolerance of banana fruit treated with malic acid prior to low-temperature storage. *Postharvest Biol. Technol.* 111: 209–213.

Huang J., Q. Wang, T. Li et al. 2017. Nanostructured lipid carrier (NLC) as a strategy for encapsulation of quercetin and linseed oil: Preparation and in vitro characterization studies. *J. Food Eng.* 215: 1–12.

Hussain, I., N.B. Singh, A. Singh et al. 2016. Green synthesis of nanoparticles and its potential application. *Biotechol. Lett.* 38: 545–560. doi: 10.1007/s10529-015-2026-7

Idrees, M., M. Naeem, T. Aftab et al. 2011. Salicylic acid mitigates salinity stress by improving antioxidant defence system and enhances vincristine and vinblastine alkaloids production in periwinkle [*Catharanthus roseus* (L.) G. Don]. *Acta Physiol. Plant.* 33(3): 987–999.

Iravani, S., H. Korbekandi, S. Mirmohammadi. 2014. Synthesis of silver nanoparticles: Chemical, physical and biological methods. *Res. Pharm. Sci.* 9: 385–406.

Issa-Zacharia, A., Y. Kamitani, H.S. Muhimbula et al. 2010. A review of microbiological safety of fruits and vegetables and the introduction of electrolyzed water as an alternative to sodium hypochlorite solution. *Afr. J. Food Sci.* 4: 778–789.

Jabnoun-Khiareddine, H., R. Abdallah, R. El-Mohamedy et al. 2016. Comparative efficacy of potassium salts against soil borne and air-borne fungi and their ability to suppress tomato wilt and fruit rots. *J. Microbiol. Biochem. Technol.* 8(2): 45–55.

Jiang, L., P. Jin, L. Wang et al. 2015. Methyl jasmonate primes defense responses against *Botrytis cinerea* and reduces disease development in harvested table grapes. *Sci. Hortic.* 192: 218–223.

Jin, P., Y. Zheng, S. Tang et al. 2009. A combination of hot air and methyl jasmonate vapor treatment alleviates chilling injury of peach fruit. *Postharvest Biol. Technol.* 52: 24–29.

Jing, W., K. Tu, X.F. Shao. 2010. Effect of postharvest short hot-water rinsing and brushing treatment on decay and quality of strawberry fruit. *Jour. Food Quality* 33(1): 262–272.

Joas, J., Y. Caro, M.N. Ducamp et al. 2005. Postharvest control of pericarp browning of litchi fruit (Litchi chinensis Sonn cv Kwaï Mi) by treatment with chitosan and organic acids. *Postharvest Biol. Technol.* 38: 128–136.

Joyce, D.C., H. Wearing, L. Coates et al. 2001. Effects of phosphonate and salicylic acid treatments on anthracnose disease development and ripening of 'Kensington Pride' mango fruit. *Aust. J. Exp. Agric.* 41(6): 805–813.

Junqueira-Goncalves, M.P., G.E. Salinas, J.E. Bruna et al. 2017. An assessment of lactobiopolymer-montmorillonite composites for dip coating applications on fresh strawberries. *J. Sci. Food Agric.* 97: 1846–1853. doi: 10.1002/jsfa.7985

Kader, A.A. 1986. Potential applications of ionizing radiation in post-harvest handling of fresh fruits and vegetables. *Food Technol.* 40: 117–121.

Kang, H.C., Y.H. Park and S.J. Go. 2003. Growth inhibition of a phytopathogenic fungus: *Colletotrichum* species by acetic acid. *Microbiol. Res.* 158: 321–326.

Karabulut, O.A., K. Ilhan, U. Arslan et al. 2009. Evaluation of the use of chlorine dioxide by fogging for decreasing postharvest decay of fig. *Postharvest Biol. Technol.* 52: 313–315.

Kanetis, L., H. Foster and J.E. Adaskaveg. 2008. Optimising efficacy of new postharvest fungicides and evaluation of sensitizing agents for managing citrus green mold. *Plant Dis.* 92: 261–269.

Katiyar, D., A. Hemantaranjan and B. Singh. 2015. Chitosan as a promising natural compound to enhance potential physiological responses in plant: A review. *Ind. J. Plant Physiol.* 20(1): 1–9.

Khaliq, G., M.T.M. Mohamed, P. Ding et al. 2016. Storage behaviour and quality responses of mango (*Mangifera indica* L.) fruit treated with chitosan and gum Arabic coatings during cold storage conditions. *Int. Food Res. J.* 23(Dec): S141–S148.

Kharissova, O.V., H.V.R. Dias, B.I. Kharisov et al. 2013. The greener synthesis of nanoparticles. *Trends Biotechnol.* 31: 240–248. doi: 10.1016/j tibtech.2013.01.003

Kitinoja, L. 2010. Identification of Appropriate Postharvest Technologies for Improving Market Access and Incomes for Small Horticultural Farmers in Sub-Saharan Africa and South Asia. Slide Deck WFLO Appropriate Postharvest Technology Planning Project, BMGF Grant No. 52198.

Kitinoja, L. and H.Y. AlHassan. 2012. Identification of Appropriate Postharvest Technologies for Small Scale Horticultural Farmers and Marketers in Sub-Saharan Africa and South Asia – Part 1. Postharvest Losses and Quality Assessments. In *Proc. XXVIIIth IHC – IS on Postharvest Technology in the Global Market. Acta Hort. 934*, ISHS *2012.*

Kitinoja, L. and M. A. Dandago. 2017. *Maize CSAM Studies Report for the ABA World Bank Postharvest Project Rwanda, Nigeria and India Combined Report.* Davis, CA: WBG/Agribusiness Associates Inc.

Kitinoja, L., K. Hell, H. Chahine-Tsouvalakis et al. 2016. *Reducing On-Farm Food Losses in the OIC Member Countries.* Ankara: Standing Committee for Economic and Social Cooperation (COMCEC) for the Organization of Islamic Cooperation.

Kitinoja, L., V.Y. Tokala and A. Brondy. 2018. A review of global postharvest loss assessments in plant-based food crops: Recent findings and measurement gaps. *J. Postharvest Technol.* 6(4): 1–15.

Kohl, J., R. Kolnaar and W.J. Ravensberg. 2019. Mode of action of microbial biological control agents against plant diseases: Relevance beyond Efficacy. *Front. Plant Sci.* 10: 845. doi: 10.3389/fpls.2019.00845

Koseki, S., K. Fujiwara and K. Itoh. 2002. Decontaminative effect of frozen acidic electrolyzed water on lettuce. *J. Food Protect.* 65: 411–414.

Koseki, S., K. Yoshida, S. Isobe et al. 2004. Efficacy of acidic electrolyzed water for microbial decontamination of cucumbers and strawberries. *J. Food Protect.* 67: 1247–1251.

Kumar, D. 2014. Salicylic acid signaling in disease resistance. *Plant Sci.* 228: 127–134. doi: 10.1021/jf4009923

Kumar, D.V., P.R.P. Verma and S.K. Singh. 2015. Development and evaluation of biodegradable polymeric nanoparticles for the effective delivery of quercetin using a quality by design approach. *LWT Food Sci. Technol.* 61: 330–338. doi: 10.1016/j.lwt.2014.12.020

Kumari, A., S.K. Yadav, Y.B. Pakade et al. 2014. Development of biodegradable nanoparticles for delivery of quercetin. *Colloids Surf. B Biointer.* 108: 182–187. doi: 10.1016/j.colsurfb.2010.06.002

Lafuente, M.T., A.R. Ballester, J. Calejero et al. 2011. Effect of high temperature-conditioning treatments on quality, flavonoid composition andvitamin C of cold stored 'Fortune' mandarins. *Food Chem.* 128: 1080–1086.

Lagopodi, A.L., K. Cetiz, A. Koukounaras et al. 2009. Acetic acid, ethanol and steam effects on the growth of *Botrytis cinerea* in vitro and combination of steam and modified atmosphere packaging to control decay in kiwifruit. *J. Phytopathol.* 157: 79–84.

LaGra, J., L. Kitinoja and K. Alpizar. 2016. *Commodity Systems Assessment Methodology for Value Chain Problem and Project Identification.* San Jose, Costa Rica: IICA-San Jose.

Landi, L., E. Feliziani and G. Romanazzi. 2014. Expression of defense genes in strawberry fruits treated with different resistance inducers. *J. Agric. Food Chem.* 27: 235–242.

Li, X., X. Zhu, N. Zhao et al. 2013. Effects of hot water treatment on anthracnose disease in papaya fruit and its possible mechanism. *Postharvest Biol. Technol.* 86: 437–446.

Liakos, I.L., F. D'autilia, A. Garzoni et al. 2016. All natural cellulose acetate—Lemongrass essential oil antimicrobial nanocapsules. *Int. J. Pharm.* 51: 508–515. doi: 10.1016/j.ijpharm.2016.01.060

Liang, J., H. Yan, X. Wang et al. 2017. Encapsulation of epigallocatechin gallate in zein/chitosan nanoparticles for controlled applications in food systems. *Food Chem.* 231: 19–24. doi: 10.1016/j.foodchem.2017.02.106

Liu F., Y. Jiang and B. Du. 2013. Design and characterization of controlled-release edible packaging films prepared with synergistic whey-protein polysaccharide complexes. J. Agric. Food Chem. 61: 5824–5833.

Liu, J., Y. Sui, M. Wisniewski et al. 2012. Effect of heat treatment on inhibition of *Monilinia fructicola* and induction of disease resistance in peach fruit. *Postharvest Biol. Technol.* 65: 61–68.

Liu, K., C. Yuan, Y. Chen et al. 2014. Combined effects of ascorbic acid and chitosan on the quality maintenance and shelf life of plums. *Sci. Hortic.* 176: 45–53.

Luengwilai, K., D.M. Beckles and M.E. Saltveit. 2012a. Chilling-injury of harvested tomato (*Solanum lycopersicum* L.) cv. Micro-Tom fruit is reduced by temperature pretreatments. *Postharvest Biol. Technol.* 63: 123–128.

Lv, Y., F. Yang, X. Li et al. 2014. Formation of heat-resistant nanocapsules of jasmine essential oil via gelatin/gum arabic based complex coacervation. *Food Hydrocolloids* 35: 305–314. doi: 10.1016/j.foodhyd.2013.06.003

Maftoonazad, N. and H.S. Ramaswamy. 2005. Postharvest shelf-life extension of avocados using methylcellulose edible coating. *LWT-Food Sci.Technol.* 38: 617–624.

Martha, L.A., D.A. Ricardo and G.S. Jairo. 2018. Effect of edible coatings based on oxidised Cassava starch on color and textural properties of minimally processed Yam. *Adv. J Food Sci. Technol.* 15(SPL): 42–50.

Martínez-Hernández, G.B., M.L. Amodio and G. Colelli. 2017. Carvacrol-loaded chitosan nanoparticles maintain quality of fresh-cut carrots. *Innov. Food Sci. Emerg. Technol.* 41: 56–63. doi: 10.1016/j.ifset.2017.02.005

Mauchline, A.L., M.R. Herv and S.M. Cook. 2017. Semiochemical-based alternatives to synthetic toxicant insecticides for pollen beetle management. *Arthopod-Plant Interact.* 12(6): 835–847. doi: 10.1007/s11829-017-9569-6

McClements, D.J. 2015. *Food Emulsions: Principles, Practices, and Techniques.* Boca Raton, FL: CRC Press.

McGuire, R. and R. Hagenmaier. 2001. Shellac formulation to reduce epiphytic survival of coliform bacteria on citrus fruit postharvest. *J Food Prot.* 64: 1756–1760.

Mei, Y., Y. Zhao, J. Yang et al. 2002. Using edible coating to enhance nutritional and sensory qualities of baby carrots. *J. Food Sci.* 67: 1964–1968.

Mishra, S., C. Kesarvani, P.C. Abhilash et al. 2017. Integrated approach of agri-nanotechnology: Challenges and future trends. *Front. Plant Sci.* 8: 1896. doi: 10.3389/fpls.2017.00471

Misra, N. N., O. Schlüter and P.J. Cullen. 2016b. *Cold Plasma in Food and Agriculture.* Cambridge, MA: Academic Press.

Misra, N.N., B. Yadav, M.S. Roopesh et al. 2019. Cold plasma for effective fungal and mycotoxin control in foods: Mechanisms, inactivation effects, and applications. *Comp. Rev. Food Sci. Food Saf.* 18: 106–120.

Misra N.N., B. Yadav, M.S. Roopesh et al. 2018a. Cold plasma for effective fungal and mycotoxin control in foods: Mechanisms, inactivation effects, and applications. *Comp. Rev. Food Sci. Food Saf.* 18: 106–120. doi: 10.1111/1541-4337-12398

Mizrach, A. 2008. Ultrasonic technology for quality evaluation of fresh fruit and vegetables in pre- and postharvest processes. *Postharvest Biol. Technol.* 48(3): 315–330.

Mohammadi, A., M. Hashemi and S.M. Hosseini. 2016. Postharvest treatment of nanochitosan-based coating loaded with *Zataria multiflora* essential oil improves antioxidant activity and extends shelf-life of cucumber. *Innov. Food Sci. Emerg. Technol.* 33: 580–588. doi: 10.1016/j.ifset.2015.10.015

Moline, H.E., J.G. Buta, R.A. Saftner et al. 1997. Comparison of three volatile natural products for the reduction of postharvest decay in strawberries. *Adv. Strawberry Res.*

Montesinos-Herrero, C., P.A. Moscoso Ramerez and L. Palou. 2016. Evaluation of sodium benzoate and other food additives for the control of citrus postharvest green and blue molds. *Postharvest Biol. Technol.* 115: 72–80.

Minocha, R., S.C. Minocha, S.L. Long et al. 1992. Effects of aluminum on DNA synthesis, cellular polyamines, polyamine biosynthetic enzymes and inorganic ions in cell suspension cultures of a woody plant, *Catharanthus roseus. Physiol. Plant.* 85: 417–424.

Moscoso-Ramrez, P.A. and L. Palou. 2016. Potassium silicate: A new organic tool for the control of citrus postharvest green mold. *ISHS Acta Horticult.* 1144: 287–292.

Mshraky, A., F.K. Ahmed and G.A.M. El-hadidy. 2016. Influence of pre and post applications of potassium silicate on resistance of chilling injury of olinda valencia orange fruits during cold storage at low temperatures. *Middle East J. Agric. Res.* 5(04): 442–453.

Musenase, S. and L. Kitinoja. 2017. *Commodity Systems Assessment Study: Report on Green Chilies in Rwanda. USAID Reducing Postharvest Losses in Rwanda Project.* ABA Inc.

Mustafa, M.A., A. Ali and S. Manickam. 2013. Application of a Chitosan Based Nanoparticle Formulation as an Edible Coating for Tomatoes (*Solanum lycoperiscum* L.). *Acta Horticult.* 1012: 445–452. doi: 10.17660/ActaHortic.2013.1012.57

Nanda, S.K., R.K. Vishwakarma, H.V.L. Bathla et al. 2012. *Harvest and Post-Harvest Losses of Major Crops and Livestock Produce in India*. Ludhiana: ICAR.
Natrajan D., S. Srinivasan, K. Sundar et al. 2015. Formulation of essential oil-loaded chitosan–alginate nanocapsules. *J. Food Drug Anal.* 23: 560–568. doi: 10.1016/j.jfda.2015.01.001
Nisperos-Carriedo, M.O., E.A. Baldwin and P.E. Shaw. 1992. Development of an edible coating for extending postharvest life of selected fruits and vegetables. *Fla. State Horticult. Soc.* 104: 122–125.
Nitu, N., M. Masum, R. Jannat et al. 2016. Application of chitosan and *Trichoderma* against soil-borne pathogens and their effect on yield of tomato (*Solanum lycopersicum* L.). *Int. J. Biosci.* 9: 10–24.
Okull, D.O. and L.F. Laborde. 2004. Activity of electrolyzed oxidizing water against *Penicilium expansum* in suspension and on wounded apples. *J. Food Sci.* 69: 23–27.
Owino, W. 2013. *Final Report: Postharvest Loss Assessment Survey of Vegetables in Kenya*. Tainan, Taiwan: USAID AVRDC.
Obianom, C. and D. Sivakumar. 2018. Differential response to combined prochloraz and thyme oil drench treatment in avocados against the control of anthracnose and stem-end rot. *Phytoparasitica* 46(3): 273–281.
Osman, A., E. Abbas, S. Mahgoub et al. 2016. Inhibition of *Penicillium digitatum* in vitro and in postharvest orange fruit by a soy protein fraction containing mainly β conglycinin. *J. Gen. Plant Pathol.* 82(6): 293–301.
Palou, L. 2018. Postharvest treatments with GRAS salts to control fresh fruit decay. *Horticulturae* 4(4): 46. doi: 10.3390/horticulturae4040046
Palou, L., A. Ali, E. Fallik et al. 2016. GRAS, plant- and animal derived compounds as alternatives to conventional fungicides for the control of postharvest diseases of fresh horticultural produce. *Postharvest Biol. Technol.* 122: 41–52.
Palou, L., P.A. Moscoso-Ramrez and C. Montesinos-Herrero. 2018. Assessment of optimal postharvest treatment conditions to control green mold of oranges with sodium benzoate. *Acta Horticult.* 1194: 221–225.
Palou, L., A. Marcillia, C. Rojas-Argudo et al. 2007. Effect of X-ray irradiation and sodium bicarbonate treatments on post harvest *Penicillium* decay and quality attributes of *Clementine mandarins*. *Postharvest Biol. Technol.* 46(3): 252–261.
Pankaj, S.K., Z. Wan and K.M. Keener. 2018. Effect of cold Plasma on food quality: A review. *Foods* 7: 4. doi: 103390/foods7010004
Park, H., Y.C. Hung and D. Chung. 2004a. Effects of chlorine and pH on efficacy of electrolyzed water for inactivating *Escherichia coli* 157:H7 and *Listeria monocytogenes*. *Int. J. Food Microbiol.* 91: 13–18.
Pavela, R. and G. Benelli. 2016. Essential oils as ecofriendly biopesticides? Challenges and constraints. *Trends Plant Sci* 21(12): 1000–1007.
Petriccione M., F. Mastrobuoni, M. Pasquariello et al. 2015. Effect of chitosan coating on the postharvest quality and antioxidant enzyme system response of strawberry fruit during cold storage. *Foods* 4: 501–523. doi: 10.3390/foods4040501
Pilon, L., P.C. Spricigo, M. Miranda et al. 2015. Chitosan nanoparticle coatings reduce microbial growth on fresh-cut apples while not affecting quality attributes. *Int. J. Food Sci. Technol.* 50: 440–448. doi: 10.1111/ijfs.12616
Pinheiro, J., C. Alegria, M. Abreu et al. 2015. Influence of postharvest ultrasounds treatments on tomato (*Solanum lycopersicum*, cv. Zinac) quality and microbial load during storage. *Ultrasonics Sonochem.* 27: 552–559.
Pinto, L., A. Ippolito and F. Baruzzi. 2015. Control of spoiler *Pseudomonas* spp. on fresh cut vegetables by neutral electrolyzed water. *Food Microbiol.* 50: 102–108.
Porat, R., A. Lers, S. Dori et al. 2000. Induction of resistance against *Penicillium digitatum* and chilling injury in star ruby grapefruit by a short hot water-brushing treatment. *J. Hort. Sci. Biotechnol.* 75: 428–432.
Potlakayala, S.D., D.W. Reed, P.S. Covello et al. 2007. Systemic acquired resistance in canola is linked with pathogenesis-related gene expression and requires salicylic acid. *Phytopathology* 97(7): 794–802.
Powell, G.H. 1906. The decay of oranges while in transit from California. *Bulletin No. 123*. USDA Bureau of Plant Industry.
Radi, M., H.A. Jouybari, G. Mesbahi et al. 2010. Effect of hot acetic acid solutions on postharvest decay caused by *Penicillium expansum* on Red Delicious apples. *Sci. Horticult.* 126: 421–425.
Rahman, S.M.E., I. Khan and D.H. Oh. 2016. Electrolyzed water as a novel sanitizer in the food industry: Current trends and future perspectives. *Compr. Rev. Food. Sci.* 15: 471–490.
Raviteja, T., S.K. Dayam and J. Yashwanth. 2019. A study on cold plasma for food preservation. *J. Sci. Res. Rep.* 23(4): 1–14. doi: 10.9743/JSRR/2019/v23i430126

Razzaq, K., A.S. Khan, A.U. Malik et al. 2015. Effect of oxalic acid application on Samar Bahisht Chaunsa mango during ripening and postharvest. *LWT Food Sci. Technol.* 63: 152–160.

Research and Markets. 2015. Global nanotechnology Market Outlook 2015–2020. Available at: http://www.prnewswire.com/news-release/global-nanotechnology-market-outlook-2015-2020---industry-will-grow-to-reach-us-758-billion-507155671.html

Rodriguez-Casado, A. 2014. The health potential of fruits and vegetables. Phytochemicals: Notable examples. *Critical Rev. Food Sci. Nutr.* 56. doi: 10.1080/10408398.2012.755149

Romanazzi, G., E. Feliziani, S. Bautista-Baoos et al. 2017. Shelflife extension of fresh fruit and vegetables by chitosan treatment. *Crit. Rev. Food Sci. Nutr.* 57: 579–601.

Romanazzi, G., E. Feliziani, M. Santini et al. 2013. Effectiveness of postharvest treatment with chitosan and other resistance inducers in the control of storage decay of strawberry. *Postharvest Biol. Technol.* 75: 24–27.

Romanazzi, G., E. Feliziani and D. Shivkumar. 2018. Chitosan, a biopolymer with triple action on postharvest decay of fruit and vegetables: Eliciting, antimicrobial and film-forming properties. *Food Microbiol.* doi: 10.3389/fmicb.2018.02745

Romanazzi, G., F. Mlikota Gabler, D. Margosan et al. 2009. Effect of chitosan dissolved in different acids on its ability to control postharvest gray mold of table grape. *Phytopathology* 99: 1028–1036.

Ruíz-Jiménez, J.M., P.J. Zapata, M. Serrano et al. 2014. Effect of oxalic acid on quality attributes of artichokes stored at ambient temperature. *Postharvest Biol. Technol.* 95: 60–63.

Rwubatse, B. and L. Kitinoja. 2017. *Commodity Systems Assessment Study: Report on Cooking Bananas in Rwanda.* Davis, CA: USAID Horticulture Innovation Lab. ABA Inc.

Saavedra, G.M., N.E. Figueroa, L.A. Poblete et al. 2016. Effects of preharvest applications of methyl jasmonate and chitosan on postharvest decay, quality and chemical attributes of *Fragaria chiloensis* fruit. *Food Chem.* 190: 448–453.

Sadeghi, M., B. Zolfaghari, M. Senatore et al. 2013. Antifungal cinnamic derivatives from Persian leek (*Allium ampeloprasum* subsp. *persicum*). *Phytochem. Lett.* 6: 360–363.

Sagong, H., P. San-Hyun, C. Young-Jin et al. 2011. Inactivation of *Escherichia coli* 0157:H7.*Salmonella Typhimurium*, and *Listeria monocytogenes* in orange and tomato juice using ohmic heating. *J. Food Prot.* 74: 899–904.

Sahu, B. B., J.G. Han and H. Kersten. 2017. Shaping thin film growth and microstructure pathways via plasma and deposition energy: A detailed theoretical, computational and experimental analysis. *Phys. Chem. Chem. Phys.* 19(7): 5591–5610.

Sakiyama, Y., D.B. Graves, H.W. Chang et al. 2012. Plasma chemistry model of surface microdischarge in humid air and dynamics of reactive neutral species. *J. Phys. D Appl. Phys.* 45: 425201.

Salvia-Trujillo L., R. Soliva-Fortuny, M.A. Rojas-Graü et al. 2017. Edible nanoemulsions as carriers of active ingredients: A review. *Annu. Rev. Food Sci. Technol.* 8: 439–466. doi: 10.1146/annurev-food-030216-025908.

Sapper, M. and A. Chiralt. 2018. Starch based coating for preservation of tomato and vegetables. *Coatings* 8: 152. doi: 10.3390/coatings8050152

Saucedo-Pompa, S., R. Tojas-Molina, A.F. Aguilera-Carbo et al. 2009. Edible film based on candelilla wax to improve the shelf life and quality of avocado. *Food Res. Int.* 42: 511–515.

Severino, R., K.D. Vu, F. Donsì et al. 2014. Antibacterial and physical effects of modified chitosan based-coating containing nanoemulsion of mandarin essential oil and three non-thermal treatments against *Listeria innocua* in green beans. *Int. J. Food Microbiol.* 191: 82–88. doi: 10.1016/j.ijfoodmicro.2014.09.007

Shah, R.W., M. Aisar, M. Jahangir et al. 2016. Influence of CMC- and guar gum-based silver nanoparticle coatings combined with low temperature on major aroma volatile components and the sensory quality of kinnow (*Citrus reticulata*). *Int. J. Food Sci. Technol.* 512: 345–352.

Sharif, R., M. Mujtaba, M. Ur Rahman et al. 2018. The multifunctional role of chitosan in horticultural crops: A review. *Molecules* 23(4): 872. doi: 10.1111/ijfs.13213

Sharma, N. 2007. Complexity of stored grain ecosystem. In A. Arya and C. Monaco (Eds.), *Seed Borne Diseases: Ecofriendly Management* (pp. 57–69). Jodhpur: Scientific Publishers.

Sharma, N. 2014. Biologicals- green alternatives for plant disease management. In N. Sharma (Ed.), *Biological Controls for Preventing Food Deterioration: Strategies for Pre-and Postharvest Management* (pp. 1–24). Chichester, UK: Wiley & Sons Ltd.

Sharma, N. and A. S. Bhandari. 2014. Management of Pathogen of stored cereal grains. In E. Lichtfouse (Ed.), *Sustainable Agriculture Reviews* (pp 87–108). Basel, Switzerland: Springer International Publishing.

Sharma, H.O. and D. Rathi. 2013. *Assessment of Pre and Post Harvest Losses of Wheat and Soybean in Madhya Pradesh*. Study No. 109. Jabalpur, India: Agro-Economic Research Centre for Madhya Pradesh and Chhattisgar.

Sharma, N. and R. Srivastava. 2013. *In vitro* production of nuclear polyhedrosis virus of *Helicoverpa armigera*. *Persian Gulf Crop Protect.* 2(3): 18–31.

Sharma, N. and S. Srivastava. 2014. Hot water and UV-C as methods of physical control in postharvest losses of *Emblica officinalis* Gaertn. *Int. J. Curr. Microbiol. Appl. Sci.* 3(1): 487–493.

Sharma, N. and R. Tewari. 2014. Yeasts–Bio-bullets for post harvest diseases of horticultural perishables. In N. Sharma (Ed.), *Biological Controls for Preventing Food Deterioration- Strategies for Pre- and Postharvest Management* (pp. 41–60). Chichester, UK: Wiley & Sons Ltd.

Sharma, N. and A. Tripathi. 2006. Fungitoxicity of the essential oil of Citrus sinensis on post harvest pathogens. *World J Microbiol. Biotechnol.* 22: 587–593.

Sharma, N. and A. Tripathi. 2008. Integrated management of post harvest rot of Gladiolus corms using hot water, UV-C and *Hyptis suaveolens* (L.) Poit essential oil. *Postharvest Biol. Technol.* 47(4): 246–254.

Shi, C., Y. Wu, D. Fang et al. 2018. Effect of nanocomposite packaging on postharvest senescence of *Flammulina velutipes*. *Food Chem.* 246: 414–421. doi: 10.1016/j.foodchem.2017.10.103

Shoala, T. 2018. Positive impacts of nanoparticles in plant resistance against different stimuli. In K. Abd Elsalam and R. Prasad (Eds.), *Nanobiotechnology Applications in Plant Protection* (pp. 267–279). Cham: Springer Sciences.

Sholberg, P.L., M. Cliff and A.L. Moyls. 2001. Fumigation with acetic acid vapor to control decay of stored apples. *Fruits* 56: 355–366.

Sholberg, P.L. and A.P. Gaunce. 2000. Fumigation of stone fruit with acetic acid to control postharvest decay. *Crop Protect.* 15: 681–686.

Sholberg, P.L., T. Shephard, P. Randall et al. 2004. Use of measured concentrations of acetic acid vapour to control postharvest decay in d'Anjou pears. *Postharvest Biol. Technol.* 32: 89–98.

Silva, H.D., M.Â. Cerqueira and A.A. Vicente. 2012. Nanoemulsions for food applications: Development and characterization. *Food Bioprocess Technol.* 5: 854–867. doi: 10.1007/s11947-011-0683-7

Singh, P., S. Pandit, J. Garnaes et al. 2018a. Green synthesis of gold and silver nanoparticles from *Cannabis sativa* (industrial hemp) and their capacity for biofilm inhibition. *Int. J. Nanomed.* 13: 3571–3591.

Soylu, S., Ö. Baysal and E.M. Soylu. 2003. Induction of disease resistance by the plant activator, acibenzolar-S-methyl (ASM), against bacterial canker (*Clavibacter michiganensis* subsp. *michiganensis*) in tomato seedlings. *Plant Sci.* 165(5): 1069–1075.

Spadaro, D. and S. Droby. 2016. Development of biocontrol products for postharvest diseases of fruit: The importance of elucidating the mechanisms of action of yeast antagonists. *Trends Food Sci. Technol.* 47: 39–49

Srivastava, R. and N. Sharma. 2015. Indiscriminate pesticide use in vegetable crops at Bakshi ka Talab tehsil of district Lucknow: Frequency and intensity. *J. Biochem. Res.* 32(2): 920–926.

Suput, D.Z., V.L. Lazić, S.Z. Popović et al. 2015. Edible films and coatings-sources, properties and application. *Food Feed Res.* 42: 11–22.

Suriati, L., I. Utama, B. Harjosuwono et al. 2020. Stability Aloe vera gel as edible coatings. In *IOP Conference Series: Earth and Environmental Science*.

Tabassum, P., S. Ahmed, K. Uddin et al. 2018. Effect of guava leaf and lemon extracts on postharvest quality and shelf life of banana cv. Sabri (*Musa sapientum* L.). *J. Bangladesh Agric. Univ.* 16(3): 337–342.

Tamjidi, F., M. Shahedi, J. Varshosaz et al. 2013. Nanostructured lipid carriers (NLC): A potential delivery system for bioactive food molecules. *Innov. Food Sci. Emerg. Technol.* 19: 29–43. doi: 10.1016/j.ifset.2013.03.002

Tapia, M.S., M.A. Rojas-Grau, A. Carmona et al. 2008. Use of alginate and gellan-based coatings for improving barrier, texture and nutritional properties of fresh-cut papaya. *Food Hydrocolloids* 22: 1493–1503.

Tanaka, N., T. Fujisawa, T. Daimon et al. 1999. The effect of electrolyzed strong acid aqueous solution on hemodialysis equipment. *Artif. Organs* 23: 1055–1062.

Temur, C. and O. Tiryaki. 2013. Irradiation alone or combined with other alternative treatments to control post-harvest diseases. *African J. Agric. Res.* 8(5): 421–434.

Tesfay, S.Z., I. Bertling and J.P. Bower. 2011. Postharvest biology and technology effects of postharvest potassium silicate application on phenolics and other anti-oxidant systems aligned to avocado fruit quality. *Postharvest Biol. Technol.* 60(2): 92–99.

Tirawat, D., D. Flick, V. Merendet et al. 2017. Combination of fogging and refrigeration for white Asparagus preservation on vegetable stalks. *Postharvest Biol. Technol.* 124: 8–17.

Turner, M. 2016. Physics of cold plasma. In N. N. Misra, O. Schlüter and P. J. Cullen (Eds.), *Cold Plasma in Food and Agriculture* (pp. 17–51). San Diego: Academic Press.
Vardar, C., K. Ilhan and O.A. Karabulut. 2012. The application of various disinfectants by fogging for decreasing postharvest diseases of strawberry. *Postharvest Biol. Technol.* 66: 30–34.
Van De Velde, F., M.H. Grace, M. Lida et al. 2016. Impact of a new postharvest disinfection method based on peracetic acid fogging on the phenolic profile of strawberries. *Postharvest Biol. Technol.* 117: 197–205.
Venditti, T., A. Dore, M.G. Molinu et al. 2009. Combined effect of curing followed by acetic acid vapour treatments improves postharvest control of *Penicillium digitatum* on mandarins. *Postharvest Biol. Technol.* 54: 111–114.
Vilaplana, R., P. Alba and S. Valenciachamorro. 2018. Sodium bicarbonate salts for the control of postharvest black rot disease in yellow pitahaya (*Selenicereus megalanthus*). *Crop Protect.* 114(March): 90–96.
Wang, K., P. Jin, L. Han et al. 2014. Methyl jasmonate induces resistance against *Penicillium citrinum* in Chinese bayberry by priming of defense responses. *Postharvest Biol. Technol.* 98: 90–97.
Wang, Q., T. Lai, G. Qin et al. 2009. Response of jujube fruits to exogenous oxalic acid treatment based on proteomic analysis. *Plant Cell Physiol.* 50: 230–242.
Wang, A.Y., B.G. Lou, T. Xu et al. 2011. Defense responses in tomato fruit induced by oligandrin against *Botrytis cinerea. African J. Biotechnol.* 10(22): 4596–4601
WFLO. 2010. Identification of Appropriate Postharvest Technologies for Improving Market Access and Incomes for Small Horticultural Farmers in Sub-Saharan Africa and South Asia. Bill and Melinda Gates Foundation, World Food Logistics Organization Grant No. 52198.
Whangchai, K., K. Saengnil, C. Singkamanee et al. 2010. Effect of electrolyzed oxidizing water and continuous ozone exposure on the control of *Penicillium digitatum* on tangerine cv 'Sai Nam Pung' during storage. *Crop Protect.* 29: 386–389.
Wilson, C.L., A. El Ghaouth, B. Upchurch et al. 1997. Using an on-line apparatus to treat harvest fruit for controlling post-harvest decay. *Hort Technol.* 7: 278–282.
Woo, S., F. Scala, M. Ruocco et al. 2006. The molecular biology of the interactions between *Trichoderma* spp., phytopathogenic fungi, and plants. *Phytopathology* 96: 181–185. doi: 10.1094/PHYTO-96-0181
Worrall, E.A., A. Hamid, K.T. Mody et al. 2018. Nanotechnology for plant disease management. *Agronomy* 8: 285. doi: 10.3390/agronomy8120285.
Wu, F., D. Zhang, H. Zhang et al. 2011. Physiological and biochemical response of harvested plum fruit to oxalic acid during ripening or shelf life. *Food Res. Int.* 44: 1299–1305.
Wu, T., S. Zivanovic, F.A. Draughon et al. 2005. Physicochemical properties and bioactivity of fungal chitin and chitosan. *J. Agric. Food Chem.* 53: 3888–3894.
Xoca-Orozco, L., E.A. Cuellar-Torres, S. Gonzlez-Morales et al. 2017. Transcriptomic analysis of avocado hass (*Persea americana* Mill) in the interaction system fruit-chitosan colletotrichum. *Front. Plant Sci.* 8: 1–13.
Yang, Z., S. Cao, Y. Cai et al. 2011. Combination of salicylic acid and ultrasound to control postharvest blue mold caused by *Penicillium expansum* in peach fruit. *Innov. Food Sci. Emerg. Technol.* 12(3): 310–314
Yuan, L., Y. Bi, Y. Ge et al. 2013. Postharvest hot water dipping reduces decay by inducing disease resistance and maintaining firmness in muskmelon (*Cucumis melo* L.) fruit. *Scientia Horticult.* 161: 101–110.
Zambrano-Zaragoza, M. L., R. González-Reza, N. Mendoza-Muñoz et al. 2018. Nanosystems in Edible Coatings: A Novel Strategy for Food Preservation *Int. J. Mol. Sci.*19 (3): 705. doi: 10.3390/ijms19030705
Zeray, S. and L. Samukelo. 2017. Evaluating the efficacy of moringa leaf extract, chitosan and carboxymethyl cellulose as edible coatings for enhancing quality and extending postharvest life of avocado (*Persea americana* Mill.) fruit. *Food Pack. Shelf Life* 11: 40–48.
Zhang, Z., O. Dvir, E. Pesis et al. 2005. Weak organic acids and inhibitors of pH homeostasis suppress growth of *Penicillium* infesting litchi fruits. *J. Phytopathol.* 153: 667–673.
Zhang, Z., G. Qin, B. Li et al. 2015. Effect of cinnamic acid for controlling gray mold on table grape and its possible mechanisms of action. *Curr. Microbiol.* 71: 396–402.
Zhong, Y., J. Liu, B. Li et al. 2010. Effects of yeast antagonists in combination with hot water treatment on post harvest diseases in tomato fruit. *Biol. Control* 54(3): 316–321.
Zhu, Y., J. Yu, J.K. Brecht et al. 2016. Pre-harvest application of oxalic acid increases quality and resistance to *Penicillium expansum* in kiwifruit during postharvest storage. *Food Chem.* 190: 537–543.

3 Chitosan

An Antimicrobial Coating to Reduce Pre- and Postharvest Diseases of Fruits and Vegetables

M. H. Tharika Dilhari and Ilmi G. N. Hewajulige

CONTENTS

3.1 INTRODUCTION

Good quality and freshness of fruits and vegetables are imperative factors in food selection as consumer preferences depend on the appealing nature as well as the nutritional composition of a product. However, pre- and postharvest diseases caused by fungi, bacteria and viruses affect the internal and external quality of fruits and vegetables causing high economic losses of these commodities. Other than the economic losses, some fungi such as *Penicillium*, *Aspergillus*, *Fusarium* and *Alternaria* are known to produce mycotoxins, under certain conditions, causing health hazards to humans. The greatest loss of fruits and vegetables is a result of decay caused by pre- and postharvest fungi such as *Colletotrichum*, *Aspergillus*, *Fusarium*, *Botrytis*, *Geotrichum*, *Lasiodiplodia* and *Rhizoctonia*. Some postharvest disease-causing fungi show latent or quiescent infection where the pathogen initiates the infection to the host before harvest and remains dormant until the physiological status of the host tissue is appropriate to continue the infection process.

Fungicides are extensively used to control pre- and postharvest diseases of fruits and vegetables and are commonly applied as a dip or spray treatment either in the pre- or postharvest period. In some instances, fungicides are included in wax formulations and applied as a postharvest coating to control pathogens separately from the main functions of a coating—delaying the respiration and ripening process. However, with the increase in consumer concerns over pesticide residues remaining on fruit surfaces and related health consequences, non-chemical disease control measures are prompted.

Chitosan is a natural coating derived from crab or prawn shell after the deacetylation of chitin. After cellulose, it is the second most abundant naturally available and easily degradable biopolymer. Chitosan is known to form a semi-permeable film, which modifies the internal atmosphere of the fruit and decreases transpiration losses. Therefore, it delays the ripening process and respiration. Chitosan shows antifungal activity by inducing the defense barrier in the host tissue. It has also been demonstrated that chitosan is an exogenous elicitor of many inherent host defense responses, including accumulation of chitinase, β-1,3-glucanase (Arlorio et al., 1992). Hirano and Nagao reported that low molecular weight chitosan has more antifungal activity against several plant pathogenic fungi than high molecular weight chitosan (Hirano and Nagao, 1989). Further findings in a storage life extension study of papaya with different molecular weight chitosan coatings supported the above findings and confirmed that low molecular weight chitosan has more antifungal activity (Hewajulige et al., 2011). Chitosan was the first compound in the list of basic substances approved in the European Union for plant protection purposes (Reg. EU 66 2014/563) for both organic agriculture and integrated pest management (Romanazzi et al., 2018). This proves that chitosan is non-toxic and safe for humans. This chapter describes the potential use of chitosan as an antimicrobial coating to reduce pre- and postharvest diseases of fruits and vegetables.

3.2 STATUS OF POSTHARVEST LOSSES OF FRUITS AND VEGETABLES DUE TO DISEASES

In horticultural crops, postharvest losses occur from the time of harvesting the crop until it reaches the consumer. Among the various causes of such postharvest losses, losses due to pre- and postharvest diseases are significant. A wide range of microorganisms such as fungi, bacteria, viruses and mycoplasma affect horticultural crops and cause considerable damage to those crops, contributing to heavy economic losses. The perishable nature, availability of a high number of nutrients and moisture makes fruits and vegetables more susceptible to invasion by certain pathogenic fungi and bacteria. Most postharvest diseases are initiated through injuries created during and after harvest as some of these organisms need an entry point to enter into plant cells. In latent infections, infection occurs prior to harvest but the disease does not develop until the postharvest period (Coates and Johnson, 1997). However, fruits and vegetables may also be infected by direct penetration of certain fungi through the intact cuticle or through wounds and/or natural openings on their surfaces. The development of fungal infection during the postharvest phase can depend upon the physiological age of the host, mechanical injuries and storage conditions (Salunkhe et al., 1991).

Pre- and postharvest disease control involves the prevention of infection, eradication of infection or delaying of symptoms until the fruit would normally have been consumed. Other than the chemical methods used to control diseases, non-chemical means such as the use of GRAS (generally regarded as safe) compounds, inducing the plant resistance by the activity of PR proteins (pathogenesis-related) and some postharvest remedies, such as hot water treatment, use of polymeric films and wax coatings, are effectively being used.

3.3 EFFECT OF FRUIT COATINGS TO EXTEND THE STORAGE LIFE OF PERISHABLES

The application of various dipping solutions to provide a membrane over the fruit, limiting gaseous exchange, has been found to be a satisfactory method of achieving modified atmospheres. A number of technologies have been developed and are currently in use to prolong shelf life and enhance the quality of fruits and vegetables. Wax coatings are applied to many commodities to delay fruit ripening, control moisture loss and modify gas exchange, and thereby reducing the rate of respiration and extending the shelf life (Dhall, 2013). These coatings often create a modified atmospheric condition in fruits by modifying their internal gas composition. Wax formulations derived from biological compounds or GRAS compounds could be considered as biowaxes and edible coatings. It has been reported that edible coatings are able to carry active ingredients such as antibrowning agents, colorants, flavors, nutrients, spices and antimicrobial compounds that could extend product shelf life and reduce the risk of pathogen growth on food surfaces (Pranoto et al., 2005).

Chitosan has been reported to be a strong antifungal compound that can be used as an antimicrobial coating or as an ingredient in food coatings. Chitosan forms a semi-permeable film around the fruit that regulates gas exchange, reduces transpiration loss and slows down the ripening process (Hewajulige et al., 2009a). Several studies have further demonstrated that chitosan is an exogenous elicitor of many inherent host defense responses, including accumulation of chitinase, β-1,3-glucanase and phenolic compounds, induction of lignification and synthesis of phytoalexins by the infected tissue (Arlorio et al., 1992; Zhang and Quantick, 1998).

3.4 STRUCTURE AND PROPERTIES OF CHITOSAN

Chitosan (poly (1-4) β, D-glucosamine), the deacetylated form of chitin, which is soluble in acidic solutions, is a high molecular polymer. Chitosan, a non-toxic, bioactive agent has become a useful compound due to its fungicidal effects and elicitation of defense mechanisms in plant tissue (Wilson et al., 2010; Terry and Joyce, 2004). After cellulose, chitin is the second most abundant, natural polymer polysaccharide and it is found in the outer skeletons of insects, crabs, shrimps and other marine animals (Jeuniaux and Voss-Foucart, 1991). The limited solubility of chitin in common solvents limits its commercial applications. Therefore, conversion of chitin into chitosan, a more soluble form by the deacetylation process, is imperative.

Chitosan has many applications due to the presence of a reactive $-NH_2$ group at position 2 and two hydroxyl groups at positions 3 and 6, respectively, of the 2-deoxy-D-glucose residue. The positive charge of chitosan confers to this polymer numerous and unique physiological and biological properties with great potential in a wide range of industrial applications. These industries include bioconversion for the production of value-added food products (Revah-Moiseev and Carroad, 1981; Shahidi and Synowiecki, 1991), preservation of food from microbial deterioration (Chen et al., 2016), formation of biodegradable films (Kittur, Kumar and Tharanathan, 1998), recovery of waste material from food processing discards (Bough, 1975; Pinotti et al., 1997), purification of water (Deans and Dixon, 1992) and clarification and deacidification of fruit juices (Imeri and Knorr, 1988).

3.5 MODE OF ACTION OF CHITOSAN

The fungicidal activity of chitosan has been well documented for both *in vitro* and *in vivo* studies. The level of inhibition of fungi is highly correlated with chitosan concentration and it indicates that chitosan performance is related to the application of an appropriate concentration. It is believed that the polycationic nature of this compound is the key to its antifungal properties and that the length of the polymer chain enhances its antifungal activity (Hirano and Nagao, 1989). El-Ghaouth et al. (1992) reported that chitosan might have a possible effect on the synthesis of certain fungal enzymes. It acts as a water-binding agent and inhibits various enzymes (Young et al., 1982). Chitosan also acts as a chelating agent that selectively binds trace metals and thereby inhibits the production of toxins and microbial growth (Cuero et al., 1991). Chitosan also activates several defense processes in the host tissue (Ghaouth et al., 1992). Recent studies have shown that chitosan is not only effective in halting the growth of pathogens, but also induces marked morphological changes, structural alterations and molecular disorganization of the fungal cells (Benhamou, 1996; El-Ghaouth et al., 2007; Ait Barka et al., 2004).

The antifungal activity of chitosan depends on its molecular weight, degree of deacetylation (DDA), concentration, solution pH and the target organism (Devlieghere et al., 2004; Li, Feng and Yang, 2008). Hirano and Nagao reported that low molecular weight chitosan has more antifungal activity against several plant pathogenic fungi than high molecular weight chitosan (Hirano and Nagao, 1989). This finding was supported by Hewajulige et al. (2011) in a storage life extension study of papaya with different molecular weight chitosan coatings (Figures 3.1 and 3.2). Other studies reported no difference in the antifungal activity of low, medium and high molecular weight chitosan (Bautista-Baños et al., 2006). However, enhanced antifungal activity has been reported with increased chitosan concentration (0.5–2.0%) (Bautista-Baños et al., 2005).

The mechanism by which chitosan affects the growth of several phytopathogenic fungi has not been fully elucidated and several hypotheses have been postulated. Because of its polycationic nature, it is believed that chitosan interferes with negatively charged residues of macromolecules exposed on the fungal cell surface. This interaction leads to the leakage of intracellular electrolytes and proteinaceous constituents (Leuba and Stossel, 1986). Other mechanisms mentioned in the literature are the interaction of diffused hydrolysis products with microbial DNA, which leads to the inhibition of mRNA and protein synthesis (Hadwiger et al., 1986) and the chelation of metals, spore elements and essential nutrients (Cuero, Osuji and Washington, 1991).

3.6 APPLICATION OF CHITOSAN-BASED COATING FOR FRUITS AND VEGETABLES

3.6.1 Safety of Chitosan for Application in Fruits and Vegetables

Chitosan is an interesting material in food application due to its low toxicity (Bautista-Baños et al., 2006). The single dose toxicity of chitosan in rodents is low with reported LD_{50} values of >1500 mg/kg in rats (Minami et al., 1996) and no deaths or clinical signs of toxicity in a dose range of 1000–10,000 mg/kg of chito-oligomers in mice (Qin et al., 2006). The repeat dose toxicity studies

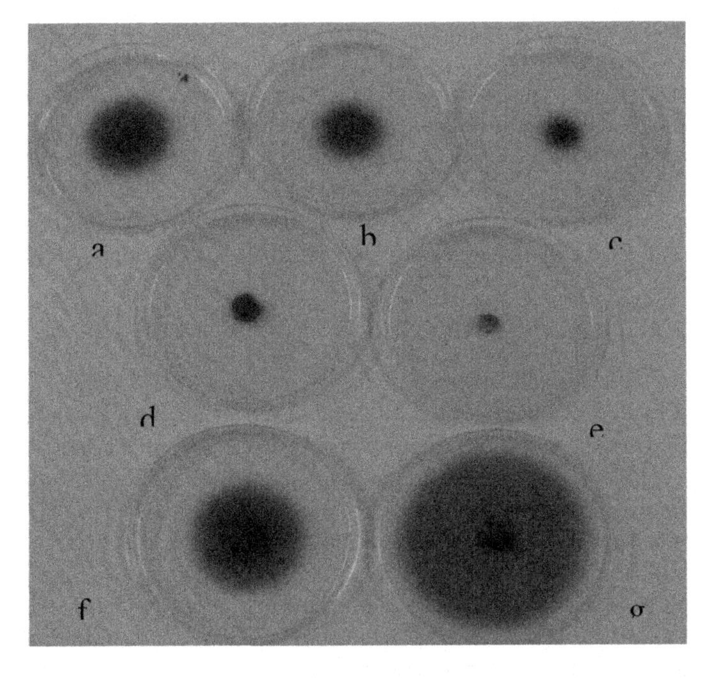

FIGURE 3.1 Effect of different concentrations of low molecular chitosan on radial mycelial growth of *Colletotrichum gloeosporioides* (a) 0.01%; (b) 0.05%; (c) 0.1%; (d) 0.5%; (e) 1%; (f) control with acetic acid; (g) control with distilled water.

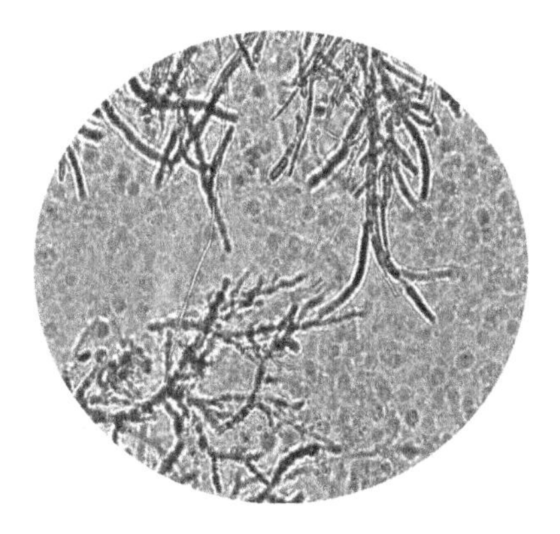
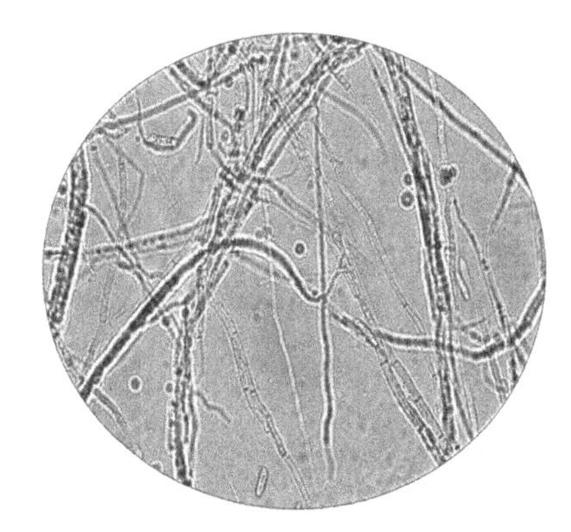

FIGURE 3.2 Microscopic observations of mycelia growth of low molecular weight chitosan treated (a) with abnormal swelling and growth retardation and control (b) with normal mycelia growth.

that have been performed with chitosan or related materials (including chitosan's copolymers glucosamine and *N*-acetylglucosamine) in rodents (Tago et al., 2007), rabbits (Illum, 1998) and dogs (Minami et al., 1996) using various parenteral and oral (gavage and dietary) routes showed no obvious toxicity or abnormal changes at low dose ranges. Human exposure of gram levels of chitosan has occurred from oral dietary supplementation intake and no adverse findings appear to have occurred. An observed safe level of oral intake for the chitosan constituent glucosamine, which is used in health supplements, of up to 2000 mg/day has been established for human use (Hathcock and Shao, 2007).

Chitosan was the first compound in the list of basic substances approved in the European Union for plant protection purposes (Reg. EU 66 2014/563), for both organic agriculture and integrated pest management (Romanazzi et al., 2018). Chitosan is approved as safe by the US Food and Drug Administration (FDA) (2005) for dietary use and wound dressing applications (Allwin et al., 2015). However, the toxicity of chitosan increases by increasing charge density and DDA (Remunan-Lopez and Bodmeier, 1997). Therefore, chitosan is considered as a biocompatible, non-toxic and safe raw material to use in fruits and vegetables (Uriarte-Montoya et al., 2010).

3.6.2 Film-forming Property and Biocompatibility of Chitosan

Film-forming property and biocompatibility are major characteristics to determine the potential for the preparation of edible coatings and films (Nadarajah et al., 2006). Chitosan is an excellent edible coating/film component due to its film-forming capacity, good mechanical properties and the formation of transparent films, which can fulfill various needs (Srinivasa et al., 2003). Film formation as a coating depends on both cohesion and adhesion (Sothornvit and Krochta, 2014), where the food is immersed or sprayed with chitosan solution (Casariego et al., 2008). Protonated amino groups, which are essential in forming ionic attractions between oppositely charged surfaces, and the wettability of the coating, which affects the constancy and thickness, are paramount factors in an effective coating (Hong et al., 2004).

Filmogenic property and the polycationic nature of chitosan increases the firmness of fruits and vegetables, which acts as a barrier to the outward nutrient flux, affecting the host–pathogen nutritional relationship (Ghaouth et al., 1994). The properties of the edible coating, such as thickness, gas and water permeability and viscosity, depend on the acid in which the biopolymer is dissolved in providing a hurdle for gas exchange, reducing respiration and ripening, which eventually decreases the sensitivity to postharvest decay (Romanazzi et al., 2018).

3.6.3 Chitosan-based Composite Coating

Single chitosan coating sometimes manifests blemishes and imperfections on preserved postharvest fruits and vegetables including limited inhibition of microorganisms leading to decay, senescence and poor coating structure (Kumar, 2000). Therefore, in order to enhance the properties of the chitosan coating, it can be combined with other substances such as essential oils (Grande-Tovar et al., 2018), organic acids (Romanazzi et al., 2009), collagen (Kaczmarek and Sionkowska, 2018), wax (Velickova et al., 2013), gelatin (Poverenov et al., 2014), inorganic/organic nano composites (Zambrano-Zaragoza et al., 2018) and metals/metal oxides (Kumar et al., 2017). In addition, chitosan can be integrated with non-coating treatments such as heat treatment (Shao et al., 2012), hypobaric treatment (Romanazzi et al., 2003), gas fumigation and modified atmosphere packaging (Olawuyi et al., 2019).

3.6.4 Chitosan as an Elicitor of Response Mechanisms in Plants

Plants are exposed to several biotic stresses all the time and disease control is largely based on chemical compounds toxic to plant invaders, causative agents or vectors. Environmental and health concerns on these hazards and their acute toxicity necessitated the search for harmless means of disease control, driving the attention toward inducible defense systems. Elicitation is the process of triggering cellular biochemical events leading to the activation of signaling cascades from outside the cell to specific gene activation, thereby inducing expression of enzymes responsible for enhanced synthesis of secondary metabolites.

3.6.4.1 Biochemical Defense Response in Preharvest Studies

Elicitors are compounds/macromolecules adept at inducing structural and/or biochemical defense responses, emerging either from the host plant (endogenous elicitors) or from the plant pathogen (exogenous elicitors) (Dixon et al., 1994) enhancing the resistance. Chitosan is one of the most important plant elicitors in promoting the quality, antioxidant activity and shelf life of fruits and vegetables (Gol et al., 2013), suggesting chitosan influences various secondary plant metabolites while triggering immunity in order to improve quality. Chitosan is an exogenous elicitor of host defense responses, including lignification (Barber, Bertram and Ride, 1989), phytoalexin biosynthesis (Hadwiger and Beckman, 1980), proteinase inhibition (Walker-Simmons and Ryan, 1984), callose apposition (Faoro and Iriti, 2007), generation of reactive oxygen and nitrogen species (Malerba and Cerana, 2015), raising cytosolic Ca^{2+} (Zuppini et al., 2003), activation of mitogen-activated protein (MAP) kinase (Yin et al., 2009), synthesis of abscisic acid (ABA), jasmonate and PR proteins (Iriti and Faoro, 2009). The elicitor activity of chitosan seems to depend on its physicochemical properties (deacetylation degree, molecular weight and viscosity) (Malerba and Cerana, 2015) and threshold concentration which can switch the induction of a cell death program into cytotoxicity, other than the plant model (Iriti and Faoro, 2009).

Preharvest chitosan treatment has shown positive effects on fruit firmness, viscosity, lignin, vitamin C, total phenol, antioxidant activity (Petriccione et al., 2015) and suppression of ethylene signaling pathway genes (He et al., 2018), as well as inducing the resistance in the host by increasing phenylalanine ammonia-lyase (PAL) activity, which is a key enzyme in the phenylpropanoid pathway, hindering the pathogen progression (Romanazzi et al., 2002) through inducing the fungal cell wall degrading enzymes such as chitinase and β-1,3-glucanase (Aziz et al., 2007; Hewajulige et al., 2011).

3.6.4.2 Biochemical Defense Response in Postharvest Studies

Plants, upon recognition of a potential pathogen or its byproducts, usually mount a highly coordinated cascade of defensive responses to ward off the infection. Application of chitosan coatings in postharvest tropical fruits such as mango (Abd-Alla and Haggag, 2010), banana (Jinasena et al., 2011), soursop, avocado (Gutierrez-Martinez et al., 2018) and papaya (Hewajulige et al., 2009b) exhibits an increase in early genetic expression of the *Colletotrichum* sp. (causal agent of anthracnose) infection, through a process of signal transduction, attaining the induction and increase of the gene expression (Berumen-Varela et al., 2015) of polyphenol oxidase (PPO) and peroxidase (POD) enzymes, activating the defense mechanisms (Meng et al., 2010). Chitosan impedes the production of fungal virulence factors such as cell wall degrading enzymes (polygalacturonase, pectate lyase and cellulose), organic acids (oxalic and fumaric acids) and host specific toxins (alternariol and alternariol monomethyl ether) and induced production of rishitin against tomato black mold (Reddy et al., 1998).

3.6.5 Maintaining Storage Quality and Shelf-life Extension

Fresh fruits and vegetables undergo putrefaction and decay during postharvest transportation and storage due to several physiological reactions such as respiration, ripening, ethylene production and senescence (Kumar et al., 2019). Environmental factors such as humidity, temperature, oxygen and light trigger nutrient and water loss as well as deterioration of texture and color (Nunes, 2008).

Low molecular weight chitosan coatings have demonstrated extended shelf life in fresh produce through slowing down the sensory property reduction, maintaining the total soluble solids, titratable acidity and ascorbic acid content of fruits (Chien et al., 2007). Hewajulige et al. reported that low molecular weight chitosan with 150 kDa molecular weight and 86% DDA significantly reduces the *in vitro* mycelia growth of the anthracnose causing organism of papaya, *Colletotrichum gloeosporioides* (Hewajulige et al., 2011). The study further confirmed mycelia growth retardation and abnormal swelling of the pathogen with low molecular weight chitosan treatment (Figures 3.1 and 3.2).

Coatings containing chitosan have demonstrated a beneficial role in slowing down fruit withering in rose or pitaya fruit (*Hylocercus undatus*), maintaining a fresh appearance along with prolonging the maximal storability at ambient temperature (Chutichude and Chutichude, 2011), reducing the respiration rate as well as transpiration rate of litchi fruits during storage (Lin et al., 2011), enhancing the shelf life of red kiwi fruits (Kaya et al., 2016), tomato (Parvin et al., 2018) and banana (Tattiyakul and Siripatrawan, 2017), delaying the climacteric peak, water loss, firmness and ultimately prolonging quality attributes (Silva et al., 2017) and delaying the ripening of guava (Silva et al., 2018) and table grapes (Melo et al., 2018).

3.6.6 Control Microbial Growth

Chitosan is a promising versatile biostimulant, eliciting the plant defense system in the pre- and postharvest sector. Chitosan is identified to be responsible for the hydrolysis of peptidoglycans (cell wall component), increasing electrolyte leakage and potentially causing the death of the pathogen (Goy et al., 2016).

3.6.6.1 Antifungal/Fungicidal Activity of Chitosan in Reducing Pre- and Postharvest Diseases

Chitosan shows fungicidal activity against several fungal species (Muzzarelli et al., 1990) and affects germination and hyphal morphology of economically important pre- and postharvest fungal pathogens (Verlee et al., 2017). The minimum inhibitory concentrations (MICs) reported for specific target organisms range from 10–5000 ppm (Rabea et al., 2003) and are influenced by a multitude of factors such as the pH of the growth medium, the degree of polymerization of chitosan and the presence or absence of interfering substances such as lipids and proteins (Ahmed, Ahmad and Ikram, 2014). *Botrytis cinerea* (gray mold in wine grapes) and *Micronectriella nivalis* demonstrated MIC at 10 ppm, *Fusarium oxysporum* at 100 ppm and *Rhizoctonia solani* (Rhizoctonia fruit rot in tomato) at 1000 ppm (Rabea et al., 2003). The maximum antifungal activity of chitosan is often observed around pH 5.0–6.0 (El-Hadrami et al., 2010).

Antifungal activity of chitosan is explained using several mechanisms. First, there is the suggestion that the polycationic nature of the biopolymer is the major factor that affects the antifungal properties which are enhanced by the polymer chain length (Hirano and Nagao, 1989). Another elucidation is that certain fungal enzymes and nucleic acid syntheses are negatively affected by the charges of chitosan (Bautista-Baños et al., 2006). According to recent studies, chitosan can induce severe morphological changes, molecular disorganization and structural alterations in the fungal mycelia other than fungal pathogen inhibition (Moussa et al., 2013).

Several recent studies have investigated the role of the plasma membrane in the sensitivity of fungi to chitosan (Palma-Guerrero et al., 2009). Chitosan-sensitive fungal membranes (e.g., *Neurospora crassa*) are highly fluidic and rich in polyunsaturated free fatty acid (FFA) such as linolenic acid (Aranda-Martinez, Lopez-Moya and Lopez-Llorca, 2016). On the contrary, chitosan-resistant fungi (*Pochonia chlamydosporia*) have low-fluidity membranes enriched with saturated FFA such as palmitic or stearic acid (Palma-Guerrero et al., 2010).

Chitosan has the potential to delay symptoms and to suppress root rots and other preharvest plant diseases. Seed-borne diseases caused by *Fusarium oxysporum* f.sp. *radicis lycopersici* (FORL) (Benhamou and Thériault, 1992) and *Pythium aphanidermatum* were significantly reduced when tomato seedlings and seeds were dipped in chitosan solutions (Khiareddine et al., 2016). Chitosan concentration is directly proportional to the number of root lesions on tomato seedlings inoculated with FORL (Benhamou and Thériault, 1992). Lower gray mold incidences were recorded on cucumber treated with 1% chitosan before inoculation with *Botrytis cinerea* (Ben-Shalom et al., 2003) and *Sphaerotheca fuliginea*, causing powdery mildew in cucumber cotyledons (Moret et al., 2009).

Postharvest disease development in fruits and vegetables during storage is a significant issue and chitosan reports effective control, delaying the onset of infection. Antifungal/fungicidal activity of chitosan was observed on the growth of several pathogens including *Sclerotinia sclerotiorum* causing sclerotinia rot in carrots (Wang et al., 2015), *Alternaria alternata* causing black mold in tomatoes (Reddy et al., 1998), *Alternaria kikuchiana* and *Physalospora pyricola* causing black spots and ring rot in pear (Meng et al., 2010), *Colletotrichum gloeosporioides* causing anthracnose in papaya (Bautista-Baños et al., 2003) and mango (Abd-Alla and Haggag, 2010) and so on.

3.6.6.2 Bactericidal Activity of Chitosan in Reducing Pre- and Postharvest Diseases

Chitosan inhibits the growth of a wide range of bacteria (Rabea et al., 2003). Chitosan binds non-covalently with teichoic acids integrated in the peptidoglycan layer of gram-positive bacteria (Raafat et al., 2008), while it correlates to the effect of different cations with chitosan chelation when the pH is above pKa (Wang et al., 2005), which disrupts the cell wall integrity, perturbing the uptake of important nutrients such as Ca^{2+}, Mg^{2+} and so on in gram-negative bacteria (Kong et al., 2010). Anionic parts of bacteria's outer membrane lipopolysaccharides interact with chitosan and can generate an electrostatic effect in gram-negative bacteria inducing membrane permeability (Helander et al., 2001). The minimal growth-inhibiting concentrations vary among species from 10–1000 ppm (Rabea et al., 2003).

Chitosan coating on postharvest crops has significantly affected bacterial growth and increased the quality and product stability against bacteria such as *Erwinia* spp. and *Xanthomonas* spp., causing bacterial soft rot in tomato (Oh et al., 2019), *Pseudomonas syringae* causing dry rot in avocado (Devi and Kumari, 2015) and *Escherichia coli* and *Staphylococcus aureus* preventing fruit degradation of red grapes (Zhang et al., 2017).

3.6.6.3 Antiviral Activity of Chitosan in Reducing Pre- and Postharvest Diseases

Chitosan was shown to inhibit the systemic propagation of viruses and viroids throughout the plant (Faoro, 2013) and to enhance the host's hypersensitive response to infection (Chirkov, 2002). Chitosan emulates a phytopathogen contact in the plant, stimulating a wide array of defensive reactions, which limits the systemic proliferation of the viruses and viroids over the plant and steering the development of systemic acquired resistance (Chirkov, 2002). The level of suppression of viral infections depends on several factors. The extent of viral suppression is directly associated with chitosan concentration, for example, 1 mg/mL chitosan treated potato and tomato plants acquired resistance to systemic infection with potato X virus (PVX), tobacco mosaic virus (TMV) and alfalfa mosaic virus of beans over 0.1 mg/mL (Chirkov et al., 2001). Antiviral activity also depends on the molecular structure of chitosan and its molecular weight (Kulikov et al., 2006). High molecular weight chitosans are more effective than oligomers in suppressing infections in plants, for example, chitosan preparations with molecular weight 50 kDa on beans and 120 kDa on potato exhibited greater activity over low molecular weight preparations (Chirkov et al., 2001). Antiviral effectiveness is determined by average degree of polymerization, the degree of N-deacetylation, the positive charge value and the character of the chemical modifications of the molecule (Rabea et al., 2003).

REFERENCES

Abd-Alla, M. A. and W. M. Haggag. 2010. New safe methods for controlling anthracnose disease of mango (*Mangifera indica* L.) fruits caused by *Colletotrichum gloeosporioides* (Penz.). *Fruits* 8(8): 361–367.

Ahmed, S., M. Ahmad and S. Ikram. 2014. Chitosan: A natural antimicrobial agent – A review. *J. Appl. Chem.* 3(2): 493–503.

Ait Barka, E., P. Eullaffroy, C. Clément et al. 2004. Chitosan improves development, and protects *Vitis vinifera* L. against *Botrytis cinerea*. *Plant Cell Rep.* 22(8): 608–614. doi: 10.1007/s00299-003-0733-3.

Allwin, S. I. J., K. I. Jeyasanta and J. Patterson. 2015. Extraction of chitosan from white shrimp (*Litopenaeus vannamei*) processing waste and examination of its bioactive potentials. *Adv. Biol. Res.* 9(6): 389–396. doi: 10.5829/idosi.abr.2015.9.6.96183.

Aranda-Martinez, A., F. Lopez-Moya and L. V. Lopez-Llorca. 2016. Cell wall composition plays a key role on sensitivity of filamentous fungi to chitosan. *J. Basic Microbiol.* 56(10): 1059–1070. doi: 10.1002/jobm.201500775.

Arlorio, M., A. Ludwig and T. Boller. 1992. Inhibition of fungal growth by plant chitinases and β-1,3-glucanases. *Protoplasma* 171(1–2): 34–43. doi: 10.1007/bf01379278.

Aziz, A., P. Trotel-aziz, A. Conreux et al. 2007. Chitosan induces phytoalexin synthesis, chitinase and β -1, 3-glucanase activities and resistance of grapevine to fungal pathogens. In *Macromolecules and Secondary Metabolites of Grapevine and Wines*. Jeandet, P., Clément, C. and Conreux, A. (eds.) Secaucus, NJ: Lavoisier. Intercept Publisher, pp. 83–88.

Barber, M. S., R. E. Bertram and J. P. Ride. 1989. Chitin oligosaccharides elicit lignification in wounded wheat leaves. *Physiol. Mol. Plant Pathol.* 34(1): 3–12. doi: 10.1016/0885-5765(89)90012-X.

Bautista-Baños, S., A. N. Hernández-Lauzardo and M. G. Velázquez-Del Valle. 2006. Chitosan as a potential natural compound to control pre and postharvest diseases of horticultural commodities. *Crop Prot.* 25(2): 108–118. doi: 10.1016/j.cropro.2005.03.010.

Bautista-Baños, S., M. Hernández-López, E. Bosquez-Molina et al. 2003. Effects of chitosan and plant extracts on growth of *Colletotrichum gloeosporioides*, anthracnose levels and quality of papaya fruit. *Crop Prot.* 22(9): 1087–1092. doi: 10.1016/S0261-2194(03)00117-0.

Bautista-Baños, S., H. López and H. Lauzardo. 2005. Effect of chitosan on in vitro development and morphology of two isolates of *Colletotrichum gloeosporioides* (Penz.) Penz. and Sacc. *Revista Mexicana de Fitopatologí* 23(1): 62–67.

Benhamou, N. 1996. Elicitor-induced plant defence pathways. *Trends Plant Sci.* 1(7): 233–240. doi: 10.1016/1360-1385(96)86901-9.

Benhamou, N. and G. Thériault. 1992. Treatment with chitosan enhances resistance of tomato plants to the crown and root rot pathogen *Fusarium oxysporum* f. sp. radicis-lycopersici. *Physiol. Molecul. Plant Pathol.* 41(1): 33–52. doi: 10.1016/0885-5765(92)90047-Y.

Ben-Shalom, N., R. Ardi, R. Pinto et al. 2003. Controlling gray mould caused by *Botrytis cinerea* in cucumber plants by means of chitosan. *Crop Prot.* 22(2): 285–290. doi: 10.1016/s0261-2194(02)00149-7.

Berumen-Varela, G., L. D. C. Partida, V. A. O. Jiménez et al. 2015. Effect of chitosan on the induction of disease resistance against Colletotrichum sp. in mango (*Mangifera indica* L.) cv. Tommy Atkins'. *Investigación y ciencia* 66: 16–21.

Bough, W. A. 1975. Reduction of suspended solids in vegetable canning waste effluents by coagulation with chitosan. *J. Food Sci.* 40(2): 297–301. doi: 10.1111/j.1365-2621.1975.tb02187.x.

Casariego, A., B. W. S. Souza and A. A. Vicente. 2008. Chitosan coating surface properties as affected by plasticizer, surfactant and polymer concentrations in relation to the surface properties of tomato and carrot. *Food Hydro* 22(8): 1452–1459. doi: 10.1016/j.foodhyd.2007.09.010.

Chen, C., W. Liau and G. Tsai. 2016. Antibacterial effects of N-sulfonated and N-sulfobenzoyl chitosan and application to Oyster preservation. *J. Food Prot.* 61(9): 1124–1128. doi: 10.4315/0362-028x-61.9.1124.

Chien, P., F. Sheu and H. Lin. 2007. Quality assessment of low molecular weight chitosan coating on sliced red pitayas. *J. Food Eng.* 79(2): 736–740. doi: 10.1016/j.jfoodeng.2006.02.047.

Chirkov, S. N. 2002. The antiviral activity of Chitosan. *Appl. Biochem. Micro.* 38(1): 5–13. doi: 10.1023/A.

Chirkov, S. N., A. V. Il'ina, N. A. Surgucheva, E. V. Letunova et al. 2001. Effect of chitosan on systemic viral infection and some defense responses in potato plants. *Russ. J. Plant Physiol.* 48(6): 774–779. doi: 10.1023/A:1012508625017.

Chutichude, B. and P. Chutichude. 2011. Effects of Chitosan coating to some postharvest characteristics of *Hylocercus undatus* (Haw) Brit. and Rose Fruit. *Int. J. Agric. Res.* 6(1): 82–92. doi: 10.3923/ijar.2011.82.92.

Coates, L. and G. Johnson. 1997. Postharvest diseases of fruit and vegetables, In *Plant Pathogens and Plant Diseases*. Brown, J. and Ogle, H. (eds.), 1st ed. Armidale: Rockvale Publications, pp. 533–547.

Cuero, R. G., G. Osuji and A. Washington. 1991. N-carboxymethylchitosan inhibition of aflatoxin production: Role of zinc. *Biotechnol. Lett.* 13(6): 441–444. doi: 10.1007/BF01030998.

Deans, J. R. and B. G. Dixon. 1992. Bioabsorbants for waste water treatment. In *Advances in Chitin and Chitosan*. Brine, C. J., Sandford, P. A. and J. P. Zikakis (eds.). London: Elsevier Applied Science, pp. 648–656.

Devi, E. P. and B. Kumari. 2015. A review on prospects of pre-harvest application of bioagents in managing post-harvest diseases of horticultural crops. *Int. J. Agric. Environ. Biotech.* 8(4): 933–941. doi: 10.5958/2230-732X.2015.00106.0.

Devlieghere, F., A. Vermeulen and J. Debevere. 2004. Chitosan: Antimicrobial activity, interactions with food components and applicability as a coating on fruit and vegetables. *Food Microbiol.* 21(6): 703–714. doi: 10.1016/j.fm.2004.02.008.

Dhall, R. K. 2013. Advances in edible coatings for fresh fruits and vegetables: A review. *Crit. Rev. Food Sci. Nutr.* 53(5): 435–450. doi: 10.1080/10408398.2010.541568.

Dixon, R. A., M. J. Harrison and C. J. Lamb. 1994. Early events in the activation of Plant defense responses. *Annu. Rev. Phytopathol.* 32(1): 479–501. doi: 10.1146/annurev.py.32.090194.002403.

El-Ghaouth, A., J. Arul and J. Grenier. 1992. Effect of chitosan and other polyions on chitin deacetylase in *Rhizopus stolonifer. Exper. Myco.* 16(3): 173–177. doi: 10.1016/0147-5975(92)90025-M.

El-Ghaouth, A., J. Arul, C. Wilson et al. 1994. Ultrastructural and cytochemical aspects of the effect of chitosan on decay of bell pepper fruit. *Physiol. Molecul. Plant Pathol.* 44(6): 417–432. doi: 10.1016/S0885-5765(05)80098-0.

El-Ghaouth, A., J. L. Smilanick and G. E. Brown. 2007. Application of *Candida saitoana* and Glycolchitosan for the control of postharvest diseases of apple and citrus fruit under semi-commercial conditions. *Plant Dis.* 84(3): 243–248. doi: 10.1094/pdis.2000.84.3.243.

Faoro, F. 2013. Induced systemic resistance against systemic viruses: A feasible approach? In *Induced Resistance in Plants Against Insects and Diseases*, Vol. 89. IOBC-WPRS Bulletin: Zurich, Switzerland, pp. 199–203.

Faoro, F. and M. Iriti. 2007. Callose synthesis as a tool to screen chitosan efficacy in inducing plant resistance to pathogens. *Caryologia -Firenze* 60(2): 121–124. doi: 10.1080/00087114.2007.10589558.

Gol, N. B., P. R. Patel and T. V. R. Rao. 2013. Improvement of quality and shelf-life of strawberries with edible coatings enriched with chitosan. *Postharvest Biol. Technol.* 85: 185–195. doi: 10.1016/j.postharvbio.2013.06.008.

Goy, R. C., S. T. B. Morais and O. B. G. Assis. 2016. Evaluation of the antimicrobial activity of chitosan and its quaternized derivative on *E. Coli* and *S. aureus* growth. *Brazilian Journal of Pharmacognosy* (Sociedade Brasileira de Farmacognosia) 26(1): 122–127. doi: 10.1016/j.bjp.2015.09.010.

Grande-Tovar, C. D., C. Chaves-Lopez, A. Serio et al. 2018. Chitosan coatings enriched with essential oils: Effects on fungi involve in fruit decay and mechanisms of action. *Trends Food Sci. Technol.* 78: 61–71. doi: 10.1016/j.tifs.2018.05.019.

Gutierrez-Martinez, P., A. Ledezma-Morales, L. C. Romero-Islas et al. 2018. Antifungal activity of chitosan against postharvest fungi of tropical and subtropical fruits, In *Chitin-Chitosan - Myriad Functionalities in Science and Technology*. Dongre, R. (ed.). Rijeka: Intech Open, pp. 311–327.

El-Hadrami, A., L. R. Adam, I. El-Hadrami et al. 2010. Chitosan in plant protection. *Mar. Drugs* 8(4): 968–987. doi: 10.3390/md8040968.

Hadwiger, A., D. F. Kendra and B. W. Fristensky. 1986. Chitosan both activates genes in plants and inhibits RNA synthesis in fungi. In *Chitin in Nature and Technology*. Boston Muzzarelli, R., Jeuniaux, C. and Gooday, G. W. (eds.). Berlin: Springer, pp. 209–214.

Hadwiger, L. A. and J. M. Beckman. 1980. Chitosan as a component of Pea-Fusarium solani interactions. *Plant Physiol.* 66(5440): 205–211.

Hathcock, J. N. and A. Shao. 2007. Risk assessment for glucosamine and chondroitin sulfate. *Regul. Toxicol. Pharmacol.* 47(1): 78–83. doi: 10.1016/j.yrtph.2006.07.004.
He, Y., S. K. Bose, W. Wang et al. 2018. Pre-harvest treatment of chitosan oligosaccharides improved strawberry fruit quality. *Int. J. Mol. Sci.* 19(8): 2194–2207. doi: 10.3390/ijms19082194.
Helander, I. M., E. L. Nurmiaho-Lassila, R. Ahvenainen et al. 2001. Chitosan disrupts the barrier properties of the outer membrane of Gram-negative bacteria. *Int. J. Food Microbiol.* 71(2–3): 235–244. doi: 10.1016/S0168-1605(01)00609-2.
Hewajulige, I. G. N., T. Shiina and N. Nakamura. 2011. Antifungal effects of chitosan with different molecular weights on in vitro growth of anthracnose causing fungi of papaya. In *Proceedings of Horticulture for the Future APHC/AuSHS/NZIAHS Conference*, Lorne, Victoria, Australia, p. 49.
Hewajulige, I. G. N., Y. Sultanbawa and R. W. Wijeratnam. 2009a. Effect of irradiated chitosan treatment on storage life of fruits of two commercially grown papaya (Carica papaya L.) varieties. *J. Nat. Sci. Foundation of Sri Lanka* 37(1): 61–66. doi: 10.4038/jnsfsr.v37i1.458.
Hewajulige, I. G. N., Y. Sultanbawa and R. W. Wijeratnam. 2009b. Mode of action of chitosan coating on anthracnose disease control in papaya. *Phytoparasitica* 37(5): 437–444. doi: 10.1007/s12600-009-0052-5.
Hirano, S. and N. Nagao. 1989. Effect of chitosan, pectic acid, lysozyme and chitinase on the growth of several phytopathogens. *Agric. Biol. Chem.* 53(11): 3065–3066. doi: 10.1080/00021369.1989.10869777.
Hong, S. I., J. H. Han and J. M. Krochta. 2004. Optical and surface properties of whey protein isolate coatings on plastic films as influenced by substrate, protein concentration, and plasticizer type. *J. Appl. Polym. Sci.* 92(1): 335–343. doi: 10.1002/app.20007.
Illum, L. (1998). Chitosan and its use as a pharmaceutical excipient. *Pharma. Res.* 15(9): 1326–1331. doi: 10.1023/a:1011929016601.
Imeri, A. G. and D. Knorr. 1988. Effects of chitosan on yield and compositional data of carrot and apple juice. *J. Food Sci.* 53(6): 1707–1709. doi: 10.1111/j.1365-2621.1988.tb07821.x.
Iriti, M. and F. Faoro. 2009. Chitosan as a MAMP, searching for a PRR. *Plant Signal. Behav.* 4(1): 66–68. doi: 10.1016/j.plaphy.2008.08.002.66.
Jeuniaux, C. and M. F. Voss-Foucart. 1991. Chitin biomass and production in the marine environment. *Biochem. Syst. Ecol.* 19(5): 347–356. doi: 10.1016/0305-1978(91)90051-Z.
Jinasena, D., P. Pathirathna, S. Wickramarachchi et al. 2011. Use of chitosan to control anthracnose on "Embul" banana. In *International Conference on Asia Agriculture and Animal*, Vol. 13, Hong Kong, pp. 56–60.
Kaczmarek, B. and A. Sionkowska. 2018. Chitosan/collagen blends with inorganic and organic additive—A review. *Adv. Polym. Technol.* 37(6): 2367–2376. doi: 10.1002/adv.21912.
Kaya, M., L. Česoniene, R. Daubaras et al. 2016. Chitosan coating of red kiwifruit (*Actinidia melanandra*) for extending of the shelf life. *Int. J. Biol. Macromol.* 85: 355–360. doi: 10.1016/j.ijbiomac.2016.01.012.
Khiareddine, H. J., R. S. R. El-Mohamedy, F. Abdel-Kareem et al. 2016. Variation in chitosan and salicylic acid efficacy towards soil-borne and air-borne fungi and their suppressive effect of tomato wilt severity. *J. Plant Pathol. Micro.* 6(11): 1–10. doi: 10.4172/2157-7471.1000325.
Kittur, F. S., K. R. Kumar and R. N. Tharanathan. 1998. Functional packaging properties of chitosan films. *Eur. Food Res. Technol.* 206(1): 44–47. doi: 10.1007/s002170050211.
Kong, M., X. G. Chen, K. Xing et al. 2010. Antimicrobial properties of chitosan and mode of action: A state of the art review. *Int. J. Food Microbiol.* 144(1): 51–63. doi: 10.1016/j.ijfoodmicro.2010.09.012.
Kulikov, S. N., S. N. Chirkov, A. V. Il'ina et al. 2006. Effect of the molecular weight of chitosan on its antiviral activity in plants. *Appl. Biochem. Microbiol.* 42(2): 200–203. doi: 10.1134/s0003683806020165.
Kumar, M. N. V. R. 2000. A review of chitin and chitosan applications. *React. Funct. Polym.* 46(1): 1–27. doi: 10.1016/S1381-5148(00)00038-9.
Kumar, S., F. Ye and S. Dobretsov. 2017. Plant latex capped colloidal silver nanoparticles: A potent anti-biofilm and fungicidal formulation. *J. Mol. Liq.* 230: 705–713. doi: 10.1016/j.molliq.2017.01.004.
Kumar, S., F. Ye, S. Dobretsov, et al. 2019. Chitosan nanocomposite coatings for food, paints, and water treatment applications. *Appl. Sci.* 9(12): 2409–2436. doi: 10.3390/app9122409.
Leuba, L. and P. Stossel. 1986. Chitosan and other polyamines: Antifungal activity and interaction with biological membranes. In *Chitin in Nature and Technology*. Muzzarelli, R., C. Jeuniaux and G. W. Gooday (eds.). Boston, MA: Springer, pp. 215–222.
Li, X., X. Feng and S. Yang. 2008. Effects of molecular weight and concentration of chitosan on antifungal activity against *Aspergillus niger, Iran. Polym. J.* (English Edition) 17(11): pp. 843–852.
Lin, B., Y., Du, X. Liang et al. 2011. Effect of chitosan coating on respiratory behavior and quality of stored litchi under ambient temperature. *J. Food Eng.* 102(1): 94–99. doi: 10.1016/j.jfoodeng.2010.08.009.
Malerba, M. and R. Cerana. 2015. Reactive oxygen and nitrogen species in defense/stress responses activated by chitosan in sycamore cultured cells. *Int. J. Mol. Sci.* 16(2): 3019–3034. doi: 10.3390/ijms16023019.

Melo, N. F. C. B., B. L. de MendonçaSoares, K. Marques-Diniz et al. 2018. Effects of fungal chitosan nanoparticles as eco-friendly edible coatings on the quality of postharvest table grapes. *Postharvest Biol. Technol.* 139: 56–66. doi: 10.1016/j.postharvbio.2018.01.014.
Meng, X., L. Yang, J. F. Kennedy et al. 2010. Effects of chitosan and oligochitosan on growth of two fungal pathogens and physiological properties in pear fruit. *Carbohydr. Polym.* 81(1): 70–75. doi: 10.1016/j.carbpol.2010.01.057.
Minami, S., M. Oh-Oka and Y. Okamoto. 1996. Chitosan-inducing hemorrhagic pneumonia in dogs. *Carbohydr. Polym.* 29(3): 241–246. doi: 10.1016/0144-8617(95)00157-3.
Moret, A., Z. Muñoz and S. Garcés. 2009. Control of powdery mildew on cucumber cotyledons by chitosan. *J. Plant Pathol.* 91(2): 375–380.
Moussa, S. H., A. A. Tayel, A. S. Alsohim et al. 2013. Botryticidal activity of nanosized silver-chitosan composite and its application for the control of gray mold in strawberry. *J. Food Sci.* 78(10): 1589–1594. doi: 10.1111/1750-3841.12247.
Muzzarelli, R., R. Tarsi, O. Filippini. 1990. Antimicrobial properties of N-carboxybutyl chitosan. *Antimicrob. Agents Chemother.* 34(10): 2019–2023. doi: 10.1128/AAC.34.10.2019.
Nadarajah, K., W. Prinyawiwatkul, K. N. Hong et al. 2006. Sorption behavior of crawfish chitosan films as affected by chitosan extraction processes and solvent types. *J. Food Sci.* 71(2): E33–E39. doi: 10.1111/j.1365-2621.2006.tb08894.x.
Nunes, M. C. D. N. 2008. Impact of environmental conditions on fruit and vegetable quality. *Stewart Postharvest Rev.* 4(4): 1–15. doi: 10.2212/spr.2008.4.4.
Oh, J., S. C. Chun and M. Chandrasekaran. 2019. Preparation and in vitro characterization of chitosan nanoparticles and their broad-spectrum antifungal action compared to antibacterial activities against phytopathogens of tomato. *Agronomy* 9(1): 21. doi: 10.3390/agronomy9010021.
Olawuyi, I. F., J. J. Park, J. J. Lee et al. 2019. Combined effect of chitosan coating and modified atmosphere packaging on fresh- cut cucumber. *Food Sci. Nutr.* 7(1): 1–10. doi: 10.1002/fsn3.937.
Palma-Guerrero, J., I. C. Huang, H. B. Jansson et al. 2009. Chitosan permeabilizes the plasma membrane and kills cells of *Neurospora crassa* in an energy dependent manner. *Fungal Genet. Biol.* 46(8): 585–594. doi: 10.1016/j.fgb.2009.02.010.
Palma-Guerrero, J., J. A. Lopez-Jimenez, A. J. Pérez-Berná et al. 2010. Membrane fluidity determines sensitivity of filamentous fungi to chitosan. *Mol. Microbiol.* 75(4): 1021–1032. doi: 10.1111/j.1365-2958.2009.07039.x.
Parvin, N., M. A. Kader, R. Huque et al. 2018. Extension of shelf-life of tomato using irradiated chitosan and its physical and biochemical characteristics. *Int. Lett. Nat. Sci.* 67: 16–23. doi: 10.18052/www.scipress.com/ilns.67.16.
Petriccione, M., F. Mastrobuoni, M. S. Pasquariello et al. 2015. Effect of chitosan coating on the postharvest quality and antioxidant enzyme system response of strawberry fruit during cold storage. *Foods* 4: 501–523. doi: 10.3390/foods4040501.
Pinotti, A., A. Bevilacqua and N. Zaritzky. 1997. Optimization of the flocculation stage in a model system of a food emulsion waste using chitosan as polyelectrolyte. *J Food Eng.* 32(1): 69–81. doi: 10.1016/S0260-8774(97)00003-4.
Poverenov, E., R. Rutenberg, S. Danino et al. 2014. Gelatin-chitosan composite films and edible coatings to enhance the quality of food products: Layer-by-layer vs. blended formulations. *Food Bioprod. Techol.* 7(11): 3319–3327. doi: 10.1007/s11947-014-1333-7.
Pranoto, Y., V. M. Salokhe and S. K. Rakshit. 2005. Physical and antibacterial properties of alginate-based edible film incorporated with garlic oil. *Food Res. Int.* 38(3): 267–272. doi: 10.1016/j.foodres.2004.04.009.
Qin, C., J. Gao and L. Wang. 2006. Safety evaluation of short-term exposure to chitooligomers from enzymic preparation. *Food Chem. Toxicol.* 44(6): 855–861. doi: 10.1016/j.fct.2005.11.009.
Raafat, D., K. Von Bargen, A. Haas et al. 2008. Insights into the mode of action of chitosan as an antibacterial compound. *Appl. Environ. Microbiol.* 74(12): 3764–3773. doi: 10.1128/AEM.00453-08.
Rabea, E. I., M. E. T. Badawy, C. V. Stevens et al. 2003. Chitosan as antimicrobial agent: Applications and mode of action. *Biomacromolecules* 4(6): 1457–1465. doi: 10.1021/bm034130m.
Reddy, M. V. B., J. Arul, E. A. Barka et al. 1998. Effect of chitosan on growth and toxin production by *Alternaria alternata* f. sp. Lycopersici. *Biocontrol Sci. Technol.* 8(1): 33–43. doi: 10.21273/hortsci.32.3.467f.
Remunan-Lopez, C. and R. Bodmeier. 1997. Mechanical, water uptake and permeability properties of cross-linked chitosan glutamate and alginate films. *J. Controlled Release* 44: 215–225.
Revah Moiseev, S. and P. A. Carroad. 1981. Conversion of the enzymatic hydrolysate of shellfish waste chitin to single cell protein. *Biotechol. Bioeng.* 23(5): 1067–1078. doi: 10.1002/bit.260230514.
Romanazzi, G., E. Feliziani and D. Sivakumar. 2018. Chitosan, a biopolymer with triple action on postharvest decay of fruit and vegetables: Eliciting, antimicrobial and film-forming properties. *Front. Microbiol.* 9(2745): 1–9. doi: 10.3389/fmicb.2018.02745.

Romanazzi, G., F. M. Gabler and D. Margosan. 2009. Effect of chitosan dissolved in different acids on its ability to control postharvest gray mold of table grape Gianfranco. *Phytopathology* 99: 1028–1036. doi: 10.1094/PHYTO-99-9-1028.

Romanazzi, G., F. Nigro and A. Ippolito. 2002. Effects of pre- and postharvest chitosan treatments to control storage grey mold of table grapes. *J. Food Sci.* 67(5): 1862–1867. doi: 10.1111/j.1365-2621.2002.tb08737.x.

Romanazzi, G., F. Nigro and A. Ippolito. 2003. Short hypobaric treatments potentiate the effect of chitosan in reducing storage decay of sweet cherries. *Postharvest. Biol. Techol.* 29: 73–80. doi: 10.1016/S0925-5214(02)00239-9.

Salunkhe, D. K., H. R. Bolin and N. R. Reddy. 1991. *Storage, Processing, and Nutritional Quality of Fruits and Vegetables*, 2nd ed. Boca Raton, FL: CRC Press.

Shahidi, F. and J. Synowiecki. 1991. Isolation and characterization of nutrients and value-added products from snow crab (*Chinoecetes opilio*) and shrimp (*Pandalus borealis*) processing discards. *J. Agric. Food Chem.* 39(8): 1527–1532. doi: 10.1021/jf00008a032.

Shao, X. F., K. Tu and S. Tu. 2012. A combination of heat treatment and chitosan coating delays ripening and reduces decay in "Gala" apple fruit. *J. Food Qual.* 35: 83–92. doi: 10.1111/j.1745-4557.2011.00429.x.

Silva, G. M., W. B. Silva, D. B. Medeiros et al. 2017. The chitosan affects severely the carbon metabolism in mango (*Mangifera indica* L. cv. Palmer) fruit during storage. *Food Chem.* 237: 372–378. doi: 10.1016/j.foodchem.2017.05.123.

Silva, W. B., G. M. Silva, D. B. Santana et al. 2018. Chitosan delays ripening and ROS production in guava (*Psidium guajava* L.) fruit. *Food Chem.* 242: 232–238. doi: 10.1016/j.foodchem.2017.09.052.

Sothornvit, R. and J. M. Krochta. 2014. Plasticizers in edible films and coatings. In *Innovations in Food Packaging*. Han, J. H. (ed.), 2nd ed. Cambridge, MA: Academic Press, Elsevier Ltd., pp. 403–433.

Srinivasa, P. C., R. Baskaran, M. Ramesh et al. 2003. Storage studies of mango packed using biodegradable chitosan film. *Eur. Food Res. Techol.* 215(6): 504–508. doi: 10.1007/s00217-002-0591-1.

Tago, K., Y. Naito, T. Nagata et al. 2007. A ninety-day feeding, subchronic toxicity study of oligo-N-acetylglucosamine in Fischer 344 rats. *Food Chem. Toxicol.* 45(7): 1186–1193. doi: 10.1016/j.fct.2006.12.027.

Tattiyakul, T. and U. Siripatrawan. 2017. Shelf-life extension of banana fruit using a chitosan-containing photocatalyst. *Acta Horticulturae* 1179: 111–118. doi: 10.17660/ActaHortic.2017.1179.17.

Terry, L. A. and D. C. Joyce. 2004. Elicitors of induced disease resistance in postharvest horticultural crops: A brief review. *Posthar. Bio. Technol.* 32(1): 1–13. doi: 10.1016/j.postharvbio.2003.09.016.

Uriarte-Montoya, M. H., J. L. Arias-Moscoso, M. Plascencia-Jatomea et al. 2010. Jumbo squid (*Dosidicus gigas*) mantle collagen: Extraction, characterization, and potential application in the preparation of chitosan-collagen biofilms. *Biores. Technol.* 101(11): 4212–4219. doi: 10.1016/j.biortech.2010.01.008.

Velickova, E., E. Winkelhausen, S. Kuzmanova et al. 2013. Impact of chitosan-beeswax edible coatings on the quality of fresh strawberries (*Fragaria ananassa* cv *Camarosa*) under commercial storage conditions. *LWT - Food Sci. Tech.* 52(2): 80–92. doi: 10.1016/j.lwt.2013.02.004.

Verlee, A., S. Mincke and C. V. Stevens. 2017. Recent developments in antibacterial and antifungal chitosan and its derivatives. *Carbohydr. Polym.* 164: 268–283. doi: 10.1016/j.carbpol.2017.02.001.

Walker-Simmons, M. and C. A. Ryan. 1984. Proteinase inhibitor synthesis in tomato leaves. *Plant. Physiol.* 76(3): 787–790.

Wang, Q., J. H. Zuo, Q. Wang. 2015. Inhibitory effect of chitosan on growth of the fungal phytopathogen, *Sclerotinia sclerotiorum*, and sclerotinia rot of carrot. *Jour. Integr. Agric.* 14(4): 691–697. doi: 10.1016/S2095-3119(14)60800-5.

Wang, X., Y. Du, L. Fan et al. 2005. Chitosan- metal complexes as antimicrobial agent: Synthesis, characterization and Structure-activity study. *Polym. Bull.* 55(1–2): 105–113. doi: 10.1007/s00289-005-0414-1.

Wilson, C. L., A. El-Ghaouth and E. Chalutt. 2010. Potential of induced resistance to control postharvest diseases of fruits and vegetables. *Plant Dis.* 78(9): 837–844. doi: 10.1094/pd-78-0837.

Yin, H., X., Zhao, X. Bai et al. 2009. Molecular cloning and characterization of a *Brassica napus* L. MAP kinase involved in oligochitosan-induced defense signaling. *Plant Mol. Biol. Rep.* 28: 292–301. doi: 10.1007/s11105-009-0152-x.

Young, D. H., H. Köhle and H. Kauss. 1982. Effect of chitosan on membrane permeability of suspension-cultured *Glycine max* and *Phaseolus vulgaris* cells. *Plant Physiol.* 70(5): 1449–1454. doi: 10.1104/pp.70.5.1449.

Zambrano-Zaragoza, M. L., R. González-Reza, N. Mendoza-Muñoz et al. 2018. Nanosystems in edible coatings: A novel strategy for food preservation. *Int. J. Mol. Sci.* 19(3): 705. doi: 10.3390/ijms19030705.

Zhang, D. and P. C. Quantick. 1998. Antifungal effects of chitosan coating on fresh strawberries and raspberries during storage. *J. Hort. Sci. Biotech.* 73(6): 763–767. doi: 10.1080/14620316.1998.11511045.

Zhang, X., G. Xiao, Y. Wang et al. 2017. Preparation of chitosan-TiO_2 composite film with efficient antimicrobial activities under visible light for food packaging applications. *Carbohydr. Polym.* 169: 101–107. doi: 10.1016/j.carbpol.2017.03.073.

Zuppini, A., B. Baldan and R. Millioni. 2003. Chitosan induces Ca 2+ mediated programmed cell death in soybean cells. *New Phytol.* 161: 557–568. doi: 10.1046/j.1469-8137.2003.00969.x.

4 Biobased Coatings
An Alternative Approach in Managing Postharvest Diseases

Smita Rai, Anmol Gupta, Ambreen Bano,
Neelam Pathak and Swati Sharma

CONTENTS

4.1 INTRODUCTION

Fruits and vegetables are an integral part of a healthy diet. They are also highly perishable by nature because they are subjected to decay and injury in postharvest handling as well as during transport. In 2000, postharvest loss of fruits and vegetables was defined as "that weight of wholesome edible product (exclusive of moisture content) that is normally consumed by humans and that has been separated from the medium and sites of its immediate growth and production by deliberate human action with the intention of using it for human feeding but which for any reason fails to be consumed by humans." Postproduction losses in fruits and vegetables can be up to 30% or 40% (Hegazy 2013), especially in developing countries such as India. The spoilage or degradation is caused due to infection by various pathogenic bacteria and fungi. Microbial spoilage affects the cost and acceptability of the fruits, reduces their shelf life, renders them unfit for consumption and can also be a source of risk for human and animal health because of the accumulation of cancerogenic mycotoxins. The high nutrient content in fruits and vegetables, as well as their succulent nature, makes them highly susceptible to infection. The packaging conditions in storage can increase relative humidity and if the temperature is also conducive and in favor of the development of postharvest decay organisms, they can cause extensive damage. Sometimes, the complete package or whole consignment suffers. Some of the common pathogens that cause dry and soft rots in fruits and vegetables are species of *Alternaria*, *Botrytis*, *Diplodia*, *Monilinia*, *Phomopsis*, *Rhizopus*, *Penicillium* and *Fusarium*, and bacteria such as *Erwinia* or *Pseudomonas* (Yahaya and Mardiyya 2019).

Besides low temperature storage, different physical and chemical techniques have been developed such as hot water treatment, fungicide or chemical dipping/spray/coatings, modified atmosphere packaging with inhibitor 1-methylcyclopropene and ethylene absorbents and their combinations (Valdés et al. 2017). The use of chemicals has its own disadvantages, as some fungicides or waxes can even cause cancer. Moreover, there is increasing global concern about environmental pollution as well as pesticide resistance caused due to indiscriminate use of chemicals. Thus, there is a pressing need to find more effective, safe and ecofriendly methods to control postharvest diseases. Physical treatments based on heat and UV irradiation, biological control treatments based on naturally occurring microorganisms and the use of food additives or carbonic acid salts are some promising non-chemical methods for controlling postharvest diseases today (Siddiqui 2018).

Biobased postharvest management systems hold great promise in providing a safer alternative to present chemical-based systems reducing health and environmental hazards. Biobased products are defined by USDA as "commercial or industrial products that are composed in whole, or in significant part, of biological products or renewable domestic agricultural materials or forestry materials" (excluding materials embedded in geological formations and/or fossilized). The use of biobased system involves use of biocoatings or biopackaging. These biocoatings are generally edible in nature and are applied as films, coatings or brushings.

4.2 EDIBLE COATINGS

Edible coatings (EC) have been successful not only in reducing water loss and delaying senescence, but also in decreasing pathogen infection and increasing the shelf life of fruits. Different kinds of edible coatings have been developed (Figure 4.1) from different materials, characterized by moldability (Yousuf et al. 2018). Edible coatings may be formulated as a mixture of biopolymeric base material mixed with crosslinkers, stabilizers, pH regulator, emulsifiers for lipids, antibrowning agents, detergents for ease of spread, Ca or Mg salts for increasing activity and yeast/ bacterial cells and nutrients for their growth. Edible coatings can be engineered to incorporate several active ingredients such as seaweed extract enhancing shelf life and even nutritional and sensory attributes. A new generation of EC are being designed to allow the incorporation of various value-adding additives like antioxidants, vitamins, nutraceuticals and natural antimicrobial agents by nanoencapsulation or layer-by-layer (LbL) assembly. Recently, nanocomposites, nanoemulsions, nanoparticles and nanofibers have been used to form active edible coatings (Ansorena and Ponce 2019). Edible coatings can give the same effect as modified atmosphere packaging (Dhall 2013). Several biopolymeric materials are currently being used, such as polysaccharides (cellulose, chitosan, starch, pectin or their derivatives, seaweed extracts and gums), lipids (fatty acids, waxes, acyl glycerol), hydrocolloids proteins and composites. These materials are the most favored in making edible coatings. These coatings act by various mechanisms like controlling respiratory rate and the exchange of gases/moisture, controlling ethylene production and thus delaying ripening and controlling pathogens and increasing shelf life as well as improving fruit quality (Figure 4.2). These films can be applied by brushing, dipping or spraying on the fruit surface (Cazón et al. 2017).

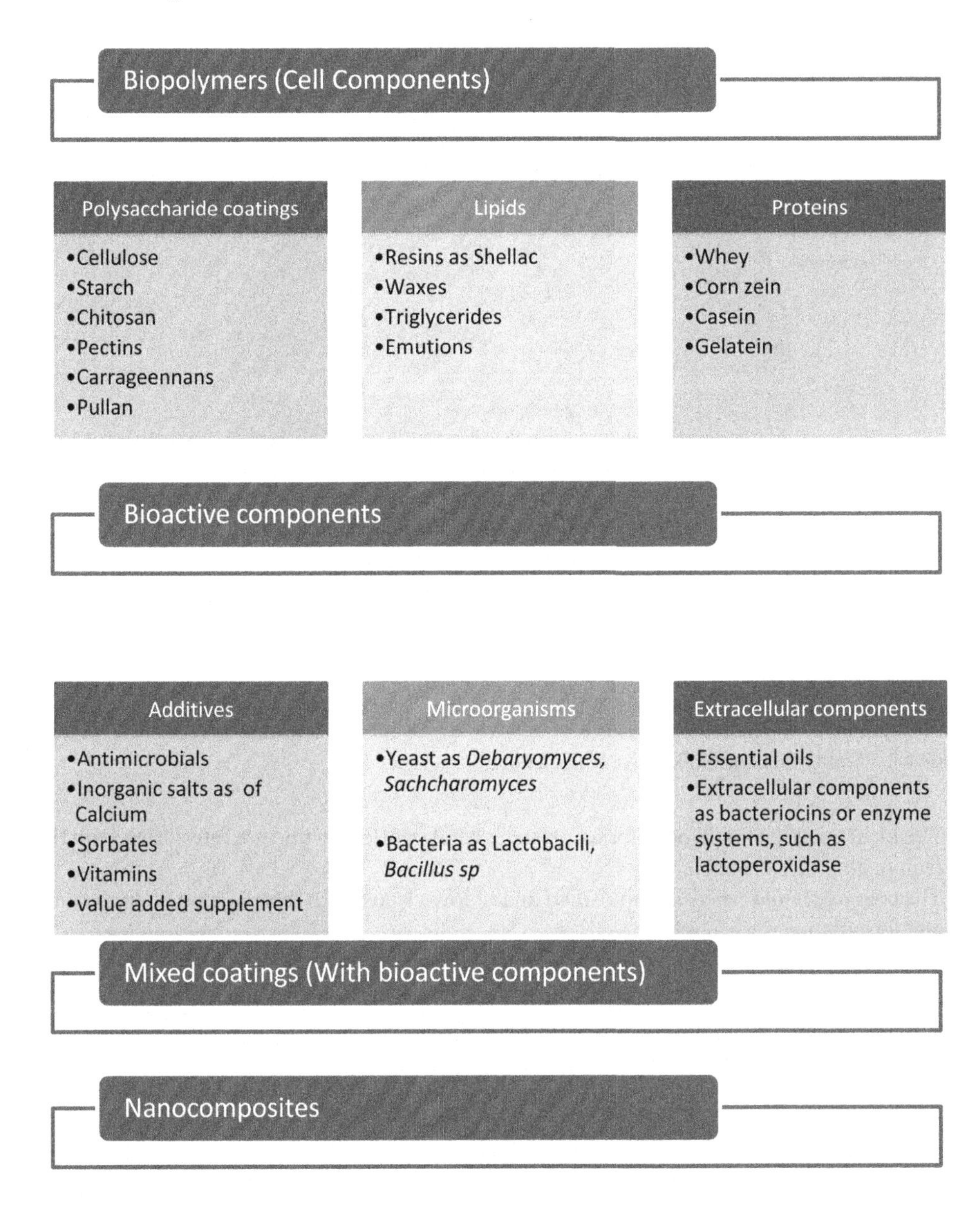

FIGURE 4.1 Types of edible coatings.

4.3 DESIRED PROPERTIES OF EDIBLE COATINGS

The properties of any edible coating rely on the basic molecular structure of the monomers. The properties have been reviewed by Arvanitoyannis and Gorris (1999) in detail. The qualities desirable for edible fruit coatings include:

- The coating should be water-resistant as well as easily spreadable so that it remains intact and covers fruit surface adequately when applied.
- The coating should act as a structural barrier to reduce moisture as well as gaseous permeability.
- The coating should be easily and efficiently dried.

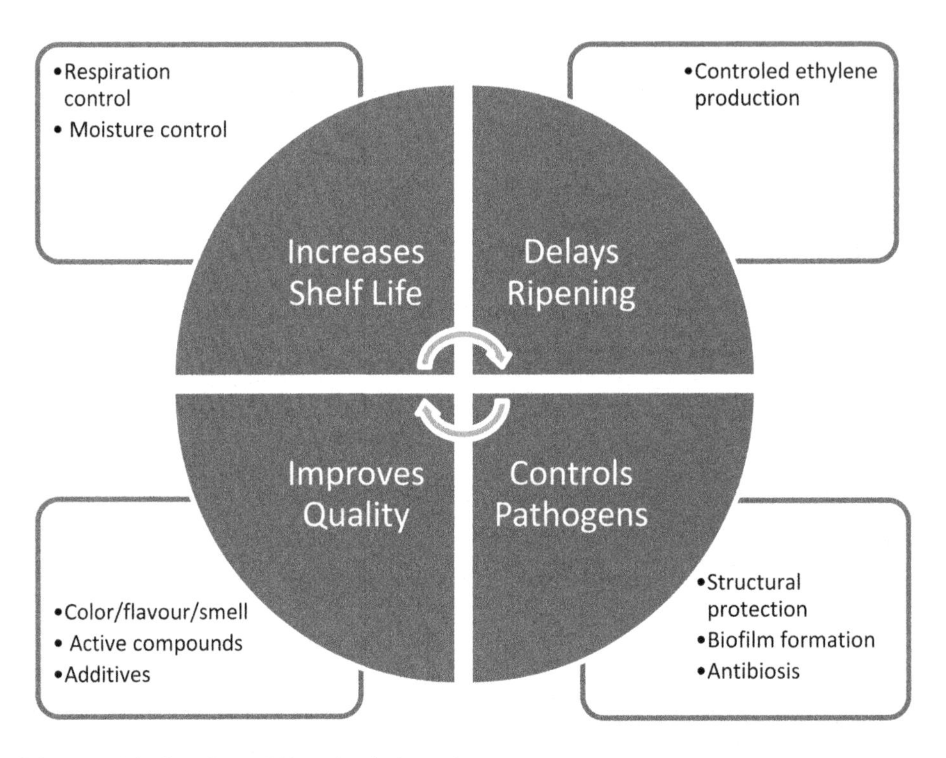

FIGURE 4.2 Main functions of bioactive fruit coatings.

- The ideal coating should be colorless, tasteless and odorless, while not interfering with the fruit quality or its acceptability.
- The coating should be easily emulsified and of low viscosity so there is ease of application and formulation.
- The base material should be easily available and economical.
- The coating should not hinder the incorporation or absorption of additives.
- The coating should be translucent to opaque, but not like glass, and should be capable of tolerating slight pressure.

4.4 BIOPOLYMER-BASED COATINGS

Edible coatings can be made from a variety of polymers, the most commonly available are extracted from marine and agricultural plants and animals.

4.4.1 Polysaccharides

Polysaccharides are the most favored biopolymers for making edible coatings. Fruit coatings have been used to retard moisture loss during storage. Polysaccharides are generally hydrophilic and can act as a moisture barrier to the atmosphere, hence, the moisture content can be maintained (Kester and Fennema 1986). Additionally, polysaccharides form a layer or film on fruit surfaces which, being less permeable to oxygen, acts as a respiratory or gas barrier and helps in prolonging ripening. Polysaccharides like cellulose, starch, gums and chitosan have been used in making edible coatings (Table 4.1). Cellulose, amylose (a component of starch) and chitosan are better at making edible coatings as their linear structure makes them tougher, transparent and flexible as well as resistant to fats and oils. Research on microbial polysaccharides like xanthan, pullan, curdlan or hyaluronic acid, which are more complex in nature, is also gaining interest (Dhall 2013).

TABLE 4.1
Polysaccharide-based Fruit Coatings Used for the Control of Postharvest Pathogens

Study no.	Fruits	Type of coatings	Pathogens on fruit	Controls	References
1	Papaya	1.5% chitosan	*Erwinia mallotivora*	Pathogen control and shelf life increase to 5 weeks	Ali et al. (2012)
2	Mango	1.0% chitosan	–	Pathogen control and shelf life increase to 3 weeks	Eshetu et al. (2019)
3	Guava	0.5, 1% and 2% chitosan	*Botrytis cinerea* *Rhizopus stolonifer*	Pathogen control and shelf life increase to 2 weeks	Hong et al. (2012)
4	Apricot	Acid-soluble chitosan and water-soluble chitosan	*Burkholderia seminalis*	Fungal spoilage	Lou et al. (2011)
5	Cavendish banana	2% chitosan, chitosan of 80% DD*	*Fusarium oxysporum*	Panama disease	Suseno et al. (2014)
6	Longan	Chitosan (0.6–1.5% w/v)	*Lasiodiplodia theobromae*	Inhibited mycelial growth	Apai et al. (2007)
7	Kola nuts	Chitosan	*Aspergillus parasiticus* *Colletotrichum apperessoria* *Cladosporium cladosporioides* *Colletotrichum gloeosporioides*	Gray mold, *Cyanea acuminata* and *Cola nitida*	Idris et al. (2017)
8	Broccoli	Chitosan	*Escherichia coli* O157:H7	Also controls diarrhea, foodborne diseases in humans	Xu et al. (2005) Dutta et al. (2009)
9	Soursop (*Annona muricata*)	1.0% chitosan	*Colletotrichum gloeosporioides*	Anthracnose pathogen control and shelf life increase to 9 days	Ramos-Guerrero et al. (2019)
10	*Citrus sinensis* (L. Osbeck)	1.5% chitosan	*Podosphaera xanthii* *Diplocarpon rosae* Wolf	Black spot, powdery mildew	Etebu and Nwauzoma (2014)
11	Peach	Chitosan (5.0 and 10.0mg ml^{-1})	*Monilinia fructicola*	Brown rot	Li and Yu (2001)
13	Citrus	Chitosan	*Penicillium digitatum* *Penicillium italicum*	Black spot	Chen et al (2019)
17	Strawberries	1% or 1.5% chitosan + 0.5% Calcium gluconate (CaGlu)	Mycelium	Controls mycelium causing severe allergic rhinitis	Hernández-Muñoz et al. (2008)
18	Avocados	Pectin-based edible emulsion coating	*Lasiodiplodia theobromae*	Stem-end rot	Maftoonazad et al. (2007)
19	Apple	Alginate (2% w/v) or gellan-based (0.5% w/v)	*Listeria innocua*	Maintain the overall quality of the minimally processed fruit	Tapia et al. (2008)
20	Pears	N-acetylcysteine and glutathione alginate based (2 %w/w) pectin based (2 %w/w) gellan based (0.5 % w/w)	–	Reduces microbial growth and prevents fresh pears from browning	Oms-Oliu et al. (2008)

*DD = degree of deacetylation

4.4.2 Alginate

Alginates as well as carrageenans can also be used to prepare edible coatings. Alginate (salt of alginic acid) is a linear copolymer of D-mannuronic and L-guluronic acid. Alginate reacts with cations (di-valent and tri-valent) like $Ca^{++}/Fe^{+++}/Mg^{++}$ added as gelling agents to form a coating (Oms-Oliu et al. 2008; Cha and Chinnan 2004).

4.4.3 Cellulose and Derivatives

Cellulose is a highly crystalline structure and is naturally available in abundance. Its water solubility is low but that can be increased by alkali treatment which causes swelling in its crystalline structure. It forms a mechanical or structural barrier around fruits and the property increases with an increase in molecular weight properties (Avena-Bustillos et al. 1993). This can be followed by its conversion to carboxymethyl cellulose (CMC), hydroxypropyl cellulose (HPC), hydroxypropyl methyl cellulose (HPMC) or methyl cellulose (MC) by treatment of chloroacetic acid, propylene oxide or methyl chloride. These modifications can increase its film-forming characteristics. The coatings are generally transparent, tasteless and odorless. They are moderate in strength, flexible and water-soluble. Moreover, they are also resistant to oil and fats as well as moderately permeable to moisture or oxygen, regulating the solute, moisture and oxygen content (Krochta and Mulder-Johnston 1997).

4.4.4 Chitosan

Chitosan, a deacetylated product of chitin, is a generally recognized as safety (GRAS) biopolymer that is currently being used in functional foods. It has excellent nutritional, antioxidant, anti-inflammatory, health-promoting properties and has bioactivities against many bacterial and fungal pathogens (Badawy and Rabea 2011). Chitosan is also used to improve and extend the shelf life of perishables as well as to maintain food quality (Table 4.1). Chitosan has very good emulsifying properties and is used in various oil-in-water emulsions for stabilizing. The properties depend on the molecular structure and degree of polymerization as well as concentration. Due to amino groups, chitosan has a high positive charge when in acidic media. It can be easily used for the encapsulation of additional compounds as nano-/microparticles and can be used for making hydrogels and nanocomposites. Chitosan can be modified structurally for enhancement of its functional properties and activity.

4.4.5 Lipid Coating

Many lipids can be used to form films which act as a barrier to moisture due to their hydrophobic nature. They do not exhibit good mechanical properties but can give gloss to fruit surfaces. Waxes or petroleum-based mineral oils have been in use as fruit glazing, but they are not considered safe for consumption. Lipid-based edible coatings are generally made by using neutral lipids and fatty acids derived from vegetable oils, waxes (beeswax, carnauba wax, candelilla wax, rice bran wax) or shellac/wood resins (Baldwin et al. 1997). Most fatty acids derived from vegetable oils are considered GRAS and have been suggested as substitutes for the petroleum-based mineral oils used in the preparation of edible coatings (Baldwin et al. 1997). Waxes can reduce moisture loss and control gas composition in fruits along with reducing surface abrasion during handling. Hence, waxes have been commercially used as coatings on fresh fruits and vegetables like fruits as apples, citrus and squash, and vegetables such as tomatoes, brinjal, pepper, cucumbers, asparagus, carrots, turnip, radish, okra, potato and sweet potato (Bai et al. 2002).

4.4.6 Wax-based Emulsion Coatings

Wax-based emulsion coatings offer a better alternative than wax coatings as they have excellent moisture/gas barrier properties, but they do not add glaze to the surface. Glycerol and natural fatty acids are used in such wax emulsions as polyglycerols-polystearates (Galus and Kadzińska 2015). They can also be used as a carrier of various essential oils.

4.5 BIOACTIVE COATINGS WITH BIOCONTROL AGENTS

Biological control can be a safer alternative to presently used postharvest fungicides by reducing their harmful side effects on health and the environment as well as the problem of resistance development in the pathogen (Sharma et al. 2012). Many postharvest pathogens have been shown to be controlled in several fruits and vegetables in *in vitro* studies as well as in *in vivo* studies (Sharma et al. 2009). A number of biological control agents have been studied for controlling postharvest diseases of fruits and vegetables (Wilson and Wisniewski 1989; Pang et al. 2002). Many bacteria and filamentous fungi, especially yeasts, have been isolated from plant surfaces. The naturally occurring microflora has been shown to protect fruit against postharvest pathogens (Janisiewiez 1997; Wilson and Chalutz 1989; Roberts 1990).

Some of these living fungicides have been patented and tested on a large scale under commercial conditions (Pusey et al. 1988; Hofstein et al. 1994). The mechanisms of action can involve one or more of the following:

- Forming a biofilm-like structural barrier
- Modification of the microenvironment by altering moisture and gases in fruits
- Competition for nutrients
- Direct or indirect antagonism by
  1. antibiosis
  2. hyperparasitism
  3. use of lytic enzymes as cellulases/chitinase
  4. production of siderophores
- Manipulation of host resistance by induction of phytoalexins

Often, more than one mechanism is involved (Janisiewicz and Korsten 2002; Liu et al. 2013). Investigations on various biocontrol agents concluded that competition by microbial antagonists with pathogens for nutrients and fruit surface is the most important mechanism of biocontrol (Wilson and Wisniewski 1989; Arras et al. 1998; Li et al. 2008). Non-pathogenic microorganisms (especially yeasts) colonize the surface of fruits as well as wounds, they multiply, forming a protective layer, and thereby exhaust the limited nutrients available for pathogens to grow. Yeasts like *Pichia guilliermondii*, *Candida saitoana*, *Rhodotorula glutinis*, *Rhodosporidium paludigenum*, *Kloeckera apiculata*, *Hanseniaspora guilliermondii* and *Metschnikowia andauensis* use competition as the mechanism of action (El-Ippolito et al. 2000; Liu et al. 2010). The role of direct parasitism by microbes is less documented in controlling postharvest pathogens of fruit but some yeasts like *Pichia guilliermondii* have shown potential against *Penicillium italicum* (Arras et al. 1998). Pulcherrimin, a siderophore produced by the yeast *Metschnikowia pulcherrima*, has shown ability to reduce the growth of various postharvest fungal pathogens (Saravanakumar et al. 2008). The activation of reactive oxygen species (ROS) and secretion of lytic enzymes such as chitinases, cellulose or b-1,3 glucanases can also induce host defense responses by increasing ethylene and phenylalanine ammonia lyase or phytoalexins (scoparone and scopoletin).

4.5.1 Yeast Containing Bioactive Coatings

Yeast readily colonize dry surfaces in carpospheres and wounds. They have simple nutritional requirements and hence are easy to grow in bioreactors (Chanchaichaovivat et al. 2007). A list of biocontrol yeast is given in Table 4.2. Along with nutrient and space colonization or biofilm

TABLE 4.2
Yeast/Polymeric Matrix-based Bioactive Fruit Coatings Used for the Control of Postharvest Pathogens

Study no.	Fruits	Microorganism/s used	Polymeric matrix	Pathogens on fruit	References
1	Citrus	*Mucor rouxii*	Cress seeds, olive leaves and senna pods	*Penicillium digitatum* and *Penicillium italicum* (blue mold/green mold)	Tayel et al. (2016)
2	Mandarin orange	*Wickerhamomyces anomalus, Metschnikowia pulcherrima* and *Aureobasidium pullulans*	Locust bean gum	*Penicillium digitatum, Penicillium italicum* (green mold)	Parafati et al. (2016)
3	Grapefruit	*Candida oleophila*	Methylcellulose and hydroxypropyl cellulose	Naturally found	McGuire and Baldwin (1994)
4	Orange	*Candida guilliermondii* and *Debaryomyces* sp.	Methylcellulose, sodium carboxymethylcellulose and hydroxypropyl cellulose	Naturally found	Potjewijd et al. (1995)
5	Strawberry	*Candida laurentii*	Alginate	Naturally found	Fan et al. (2009)
6	Grape	*Candida sake*	Hydroxypropyl methylcellulose, corn starch, sodium caseinate and pea protein	*Botrytis cinerea*	Marín et al. (2016)
7.	Orange	*Wickerhamomyces anomalus*	Sodium alginate and locust bean gum	*Penicillium digitatum*	Aloui et al. (2015)

formation (Parafati et al. 2015), the modes of action of various antagonistic yeasts against different fungi also include production of:

- Antifungal hydrolases (El Ghaouth et al. 2003)
- Defense-related proteins (for induction of metabolism of proteins—certain defense responses cause changes in cell structure, transcription and energy metabolism) (Chan et al. 2007)
- ROS production in host as well as tolerance to ROS
- Pigments (causing iron depletion) (Sipiczki 2006)
- Volatile organic compounds (Parafati et al. 2015)

4.5.2 Bacterial Bioactive Coatings

Bacterial bioactive coatings involve the incorporation of live bacterial cells which can act as biocontrol agents as listed in Table 4.3. Several Bacilli have been studied for their biocontrol effect but as their carposphere competence is less than that of yeast and *Pseudomonas*—they are not preferred as fruit coating. Strains of *B. subtilis* and its antibiotics were found to be effective against *Penicillium* (Leelasuphakul et al. 2008). Siderophores have also been reported to be produced by microbial antagonists. *Pantoea agglomerans* strain CPA-2, a gram-negative bacteria isolated from apple surface, was found to be effective against postharvest pathogens of citrus and pome fruit (Nunes et al. 2002). *Pseudomonas syringae*, another common gram-negative epiphytic bacterium, has been reported to control the growth of postharvest pathogens by production of antibiotics, siderophores (Iacobelliset al. 1992) or biosurfactants (Lindow and Brandl 2003), or inducing resistance in plants (Smith and Metraux 1991).

4.5.2.1 Lactobacilli

Lactic acid bacteria (LAB) are naturally present in plants and also exhibit probiotic properties along with antimicrobial properties (Leroy and De Vuyst 2004; Gerez et al. 2013). Many genera of Lactobacillus, Lactococcus and Lactosphaera have been found to be associated with food products (Vries et al. 2006). As they have GRAS and queries per second (QPS) status, they can be used safely in foods for pathogen control (Martinez et al. 2013). LABs show antagonism by exhibiting competition for nutrients and space resulting in exclusion of pathogens from wounds or entry sites and the alteration of spore membrane permeability (Pawlowska et al. 2012). Their antimicrobial nature also depends on their ability to release antifungal metabolites, which are mainly organic acids (acetic, lactic and propionic acids), proteinaceous compounds (bacteriocins) and certain fatty acids that can alter membrane permeability (Crowley et al. 2013).

4.5.3 Essential Oils

Essential oils like lime, lemongrass, eucalyptus, cinnamon or thyme essential oils, which exhibit antimicrobial properties, have also been incorporated in fruit coatings (Ramos-García et al. 2012; Ding and Lee 2019). Recently, essential oils have been used as combined coatings as shown in a chitosan coating with Mentha oil for mangoes (de Oliveira et al. 2020).

4.6 NANOFORMULATIONS

Nanocomposites, nanofibers and nanoparticles as well as nanoemulsions are currently being experimented with for formulation of fruit coatings, as reviewed by Ansorena and Ponce (2019). Nanoformulations can improve the functionality and performance of the coating. The new

TABLE 4.3
Edible Coatings Incorporating Living Bacteria and Pathogen Inhibition

Study no.	Fruits	Bacteria	Polymeric matrix	Inhibition of pathogens	References
1	Apple and papaya	*BifidobacteriumactisBb-12*	Alginate and gellan	*Escherichia coli* and *Streptococcus pyogenes*	Tapia et al. (2007)
2	Strawberries	*Lactobacillus acidophilus*	Alginate	*Staphylococcus aureus* and *Escherichia coli*	Moayednia et al. (2009)
3	Apple	*Bacillus amyloliquefaciens*	–	*Botryosphaeria dothidea*	Li et al. (2013)

generation of edible coatings not only reduces moisture loss and delays senescence, but also shows increased antimicrobial properties. The coatings are engineered so as to incorporate antioxidants, vitamins or nutraceuticals by nanoencapsulation and layer-by-layer assembly and can also help in their controlled release. Fruits and vegetables are one of the challenges of the postharvest industry (Ansorena and Ponce 2019).

4.7 MIXED COATINGS

Generally, the coatings used are not standalone coatings but they are formulated using a base material or polymeric matrix incorporating different bioactive ingredients. Additives such as MCP or Hexanal can also be added for delaying ripening. Some of the mixed coatings are listed in Table 4.4.

4.8 COMMERCIAL COATINGS

Many edible coatings are currently in the market for use on fresh and fresh-cut fruits and vegetables like citrus, apples, mangoes, pomegranates, papayas, avocados and tomatoes (Olivas et al. 2005; Dhall 2013). Some of them are shown in Table 4.5. They have been reported to decrease weight loss and increase shelf life and fruit quality. Among commercial postharvest coatings in the market, BioSave (*Pseudomonas syringae*) from the United States and "Shemer" (*Metschnikowia fructicola*) from Israel are noteworthy (Droby et al. 2009), whereas in composite coating formulations, Pro-long, TAL-Prolong (Courtaulds Group, London), Natural Shine 9000 (Pace International, Seattle, WA, USA) and Semperfresh (AgriCoat Industries Ltd., Berkshire, UK) are available (Nisperos-Carriedo et al. 1992). Initially, two biobased coatings, Aspire™ (Ecogen, Philadelphia, PA, USA) and YieldPlus (Anchor Yeast, Cape Town, South Africa) were also available, but due to some problems, they are now no longer available (Droby et al. 2009). Aspire contained an active ingredient in form of yeast *Candida oleophila* for the control of postharvest rots of citrus fruit. Aspire, along with thiabendazole (TBZ), was evaluated in commercial cold storage tests and found to reduce decay due to green and blue molds, as well as sour rot due to *Geotrichum candidum* (Droby et al. 1998).

4.9 CONCLUSION

Biobased products hold great promise for providing economical and safer alternatives to present postharvest synthetic fungicides, reducing the health and environmental hazards that they pose. They have shown to increase the shelf life as well as quality of fruits and vegetables. As necrotrophic pathogens invade wounds, their growth is limited due to the structural barriers and their growth is also inhibited by use of various biocontrol antagonists. These coatings can be engineered to encapsulate various value-added components as vitamins, probiotics or nutraceuticals. The coatings can be applied directly to fruit wounds or fruit surfaces can be coated by using dips, drenches or sprayers. Currently, research in the area of functional bioactive coatings is heating up as biobased products are renewable as well as sustainable. Certain products like BioSave are commercially available and a few others are still in line to be released in the market in the future. The efficacy and consistency of the coatings in field conditions as well as on shelf life have to be seriously evaluated before release.

TABLE 4.4
Mixed-based Bioactive Edible Coatings in Fruits and Vegetables Used for the Control of Postharvest Pathogens

Study no.	Fruits	Type of coatings	Pathogens on fruit	References
1	Tomato	Chitosan-based (1%), beeswax (0.1%), oleic acid (1.0%), lime or thyme essential oil (0.1%)	*Rhizopus stolonifer* *Escherichia coli* (black bread mold)	Ramos-García et al. (2012)
2	Mango	Chitosan and beeswax	–	Efendi and Hermawati (2010)
3	Apple	Glucose/chitinase (0.25, 0.5 and 0.75)	–	Nisha et al. (2016)
4	Apple	AgNPs/chitinase (0.25, 0.5 and 0.75)	–	Nisha et al. (2016)
5	Cherry tomato	Chitosan (CHI; 4mg/mL) + *Origanum vulgare* L. essential oil (OVEO;1.25μL/mL)	*Aspergillus niger* *Rhizopus stolonifer* (black bread mold)	Barreto et al. (2016)
6	Strawberries	1% or 1.5% chitosan + 0.5% calcium gluconate (CaGlu)	Mycelium	Muñoz et al. (2008)
7	Pomegranate	Chitosan + locust bean gum (LBG)	*Penicillium digitatum* *Penicillium expansum* *Penicillium herquei* (green rot, green mold or black Spot)	Chen et al. (2019)
8	Cherry tomatoes	Hydroxypropyl methylcellulose (HPMC) + beeswax (BW)	Gray mold *Botrytis cinerea*	Fagundes et al. (2015)
9	Strawberries	Alginate coating +yeast antagonist	Gray mold and blue mold *Cryptococcus laurentii*	Fan et al. (2009)
10	Strawberries	1% or 1.5% chitosan + 0.5% calcium gluconate (CaGlu)	Mycelium	Hernández-Muñoz et al. (2008)

TABLE 4.5
Applications of Edible Coatings in Fresh Fruits and Vegetables

Study no.	Coating material	Composition	Fruit/vegetable	Effect of coating	References
1	Semperfresh™	Sucrose ester-based coating (Sodium salts of CMC, and mixed mono and diglycerides)	Cherries and pears	Reduced color changes, retained acid, increased shelf life and maintained quality	www.paceint.com/product/semperfresh/
			Banana	Decreased ethylene production and delayed chlorophyll loss	
2	Nature-seal™	Cellulose-based edible coating (More than 30 formulations; blends of vitamins and minerals)	Mango and tomato	Delayed ripening	www.natureseal.com
			Pome fruit and carrot	Retard discoloration and carotene loss, and is a barrier for O2 diffusion	Dhall (2013)
3	FreshSeal™	Polyvinyl alcohol, starch and surfactant	Guava	Extended shelf life	Manuel (2005)
4	Prolong™	Sucrose polyester of fatty acid esters	Mango	Retard ripening, reduce weight loss and chlorophyll loss	Dhall. (2013)
		Sodium CMC and mono and diglycerides	Pear	Retention of firmness, green skin color and titratable acidity	
5	Nature-seal™	Cellulose-based edible coating	Tomato	Delayed ripening	Manuel (2005)

ACKNOWLEDGMENT

This chapter bears Integral University Manuscript Number (IU/R&D/2020-MCN000776). The authors are grateful to Integral University, UPCST, UP; DST-FIST and DST Inspire (IF160803), Delhi, India, for economic support.

REFERENCES

Ali, A., M. T. M. Mohamed and Y. Siddiqui. 2012. Control of anthracnose by chitosan through stimulation of defence-related enzymes in Eksotika II papaya (*Carica papaya* L.) fruit. *Journal of Biology and Life Science 3*: 114–126.

Aloui, H., F. Licciardello, F. Khwaldia et al. 2015. Physical properties and antifungal activity of bioactive films containing *Wickerhamomyces anomalus* killer yeast and their application for preservation of oranges and control of postharvest green mold caused by *Penicillium digitatum*. *International Journal of Food Microbiology 200*: 22–30.

Ansorena, M. R. and A. G. Ponce. 2019. Coatings in the Postharvest. In T. J. Gutiérrez (Ed.), *Polymers for Agri-Food Applications*. Cham: Springer, pp. 339–354.

Apai, W., V. Sardsud, P. Boonprasom et al. 2007. Antifungal activity of chitosan coating and its components on Lasiodiplodia theobromae in longan. In *Europe-Asia Symposium on Quality Management in Postharvest Systems-Eurasia 2007 804*, pp. 235–242.

Arras, G., V. D. Cicco, S. Arru et al. 1998. Biocontrol by yeasts of blue mould of citrus fruits and the mode of action of an isolate of *Pichia guilliermondii*. *The Journal of Horticultural Science and Biotechnology 73*(3): 413–418.

Arvanitoyannis, I. and L. G. M. Gorris. 1999. Edible and biodegradable polymeric materials for food packaging or coating. In J. C. Oliveira and F. A. R. Oliveira (Eds), *Processing Foods: Quality Optimization and Process Assessment*. Boca Raton, FL: CRC Press, pp. 357–371.

Avena-Bustillos, R. J., L. A. Cisneros-Zevallos, J. M. Krochta et al. 1993. Optimization of edible coatings on minimally processed carrots using response surface methodology. *Transactions of the ASAE 36*(3): 801–805.

Badawy, M. E. and E. I. Rabea. 2011. A biopolymer chitosan and its derivatives as promising antimicrobial agents against plant pathogens and their applications in crop protection. *International Journal of Carbohydrate Chemistry 2011*: 29.

Bai, J., E. A. Baldwin and R. H. Hagenmaier. 2002. Alternatives to shellac coatings provide comparable gloss, internal gas modification, and quality for 'Delicious' apple fruit. *HortScience 37*(3): 559–563.

Baldwin, E. A., M. O. Nisperos, R. H. Hagenmaier et al. 1997. Use of lipids in edible coatings for food products. *Food Technology 51*(6): 56–62.

Barreto, T. A., S. C. Andrade, J. F. Maciel et al. 2016. A chitosan coating containing essential oil from *Origanum vulgare* L. to control postharvest mold infections and keep the quality of cherry tomato fruit. *Frontiers in Microbiology 7*: 1724.

Cazón, P., G. Velazquez, J. A. Ramírez et al. 2017. Polysaccharide-based films and coatings for food packaging: A review. *Food Hydrocolloids 68*: 136–148.

Cha, D. S. and M. S. Chinnan. 2004. Bipolymer-based antimicrobial packaging: A review. *Critical Reviews in Food Science and Nutrition 44*: 223–237.

Chan, Z., G. Qin, X. Xu et al. 2007. Proteome approach to characterize proteins induced by antagonist yeast and salicylic acid in peach fruit. *Journal of Proteome Research 6*(5): 1677–1688.

Chanchaichaovivat, A., P. Ruenwongsa and B. Panijpan. 2007. Screening and identification of yeast strains from fruits and vegetables: Potential for biological control of postharvest chilli anthracnose (*Colletotrichum capsici*). *Biological Control 42*(3): 326–335.

Chen, C., N. Cai, J. Chen et al. 2019. Clove essential oil as an alternative approach to control postharvest blue mold caused by penicillium italicum in citrus fruit. *Biomolecules 9*(5): 197.

Crowley, S., J. Mahony and D. van Sinderen. 2013. Current perspectives on antifungal lactic acid bacteria as natural bio-preservatives. *Trends in Food Science and Technology 33*(2): 93–109.

De Oliveira, K. A. R., M. C. da Conceição, S. P. A. de Oliveira et al. (2020) Postharvest quality improvements in mango cultivar Tommy Atkins by chitosan coating with *Mentha piperita* L. essential oil. *The Journal of Horticultural Science and Biotechnology 95*(2): 260–272.

De Vries, M. C., E. E. Vaughan, M. Kleerebezem et al. 2006. *Lactobacillus plantarum* survival, functional and potential probiotic properties in the human intestinal tract. *International Dairy Journal 16*(9): 1018–1028.

Dhall, R. K. 2013. Advances in edible coatings for fresh fruits and vegetables: A review. *Critical Reviews in Food Science and Nutrition 53*(5): 435–450.

Ding, P. and Y. L. Lee. 2019. Use of essential oils for prolonging postharvest life of fresh fruits and vegetables. *International Food Research Journal 26*(2): 363–366.

Droby, S., L. Cohen, A. Daus et al. 1998. Commercial testing of Aspire: A yeast preparation for the biological control of postharvest decay of citrus. *Biological Control 12*(2): 97–101.

Droby, S., M. Wisniewski, D. Macarisin et al. 2009. Twenty years of postharvest biocontrol research: Is it time for a new paradigm? *Postharvest Biology and Technology 52*(2): 137–145.

Efendi, D. and H. Hermawati. 2010. The use of bee wax, chitosan and BAP to prolong shelflife of Mangosteen (*Garcinia mangostana* L.) fruit. *Jurnal Hortikultura Indonesia 1*(1): 32–39.

El Ghaouth, A., C. L. Wilson and M. Wisniewski. 2003. Control of postharvest decay of apple fruit with *Candida saitoana* and induction of defense responses. *Phytopathology 93*(3): 344–348.

Eshetu, A., A. M. Ibrahim, S. F. Forsido et al. 2019. Effect of beeswax and chitosan treatments on quality and shelf life of selected mango (*Mangifera indica* L.) cultivars. *Heliyon 5*(1): e01116.

Etebu, E. and A. B. Nwauzoma. 2014. A review on sweet orange (*Citrus sinensis* L Osbeck): Health, diseases and management. *American Journal of Research Communication 2*(2): 33–70.

Fagundes, C., L. Palou, A. R. Monteiro et al. 2015. Hydroxypropyl methylcellulose-beeswax edible coatings formulated with antifungal food additives to reduce Alternaria black spot and maintain postharvest quality of cold-stored cherry tomatoes. *Scientia Horticulturae 193*: 249–257.

Fan, Y., Y. Xu, D. Wang, Zhang et al. 2009. Effect of alginate coating combined with yeast antagonist on strawberry (Fragaria×ananassa) preservation quality. *Postharvest Biology and Technology 53*(1–2): 84–90.

Galus, S. and J. Kadzińska. 2015. Food applications of emulsion-based edible films and coatings. *Trends in Food Science and Technology 45*(2): 273–283.

Gerez, C. L., M. J. Torres, G. F. De Valdez et al. 2013. Control of spoilage fungi by lactic acid bacteria. *Biological Control 64*(3): 231–237.

Hegazy, R. 2013. *Post-harvest Situation and Losses in India*. Accessed online at https://www.researchgate.net/publication/301770292_Postharvest_Situation_and_Losses_in_India.

Hernandez-Munoz, P., E. Almenar, V. Del Valle, D. Velez et al. 2008. Effect of chitosan coating combined with postharvest calcium treatment on strawberry (Fragaria × ananassa) quality during refrigerated storage. *Food Chemistry 110*(2): 428–435.

Hofstein, R., B. Fridlender, E. Chautz et al. 1994. Large-scale production and pilot testing of biological control agents for postharvest diseases. In: Wilson, C. and Wisniewski, M. Eds. *Biological Control of Postharvest Diseases of Fruits and Vegetables*, pp. 89–100. Boca Raton, FL: CRC Press.

Hong, K., J. Xie, L. Zhang et al. 2012. Effects of chitosan coating on postharvest life and quality of guava (*Psidium guajava* L.) fruit during cold storage. *Scientia Horticulturae 144*: 172–178. doi: 10.1016/j.scienta.2012.07.002

Iacobellis, N. S., P. Lavermicocca, I. Grgurina et al. 1992. Phytotoxic properties of *Pseudomonas syringe* epv syringae toxins. *Physiological and Molecular Plant Pathology 40*(2): 107–116.

Idris, M. A., S. B. Sadiq, A. W. Abubakar et al. 2017. Isolation and identification of Phytopathogenic fungi responsible for Kolanuts (*Kola acuminate*) rot in Jimeta modern market, Yola Adamawa state Nigeria. *IJAR 3*(3): 272–274.

Ippolito, A., A. El Ghaouth, C. L. Wilson et al. 2000. Control of postharvest decay of apple fruit by *Aureobasidium pullulans* and induction of defense responses. *Postharvest Biology and Technology 19*(3): 265–272.

Janisiewicz, W. J. and L. Korsten. 2002. Biological control of Post harvest diseases of fruits. *Annual Review of Phytopathology 40*(1): 411–441. doi: 10.1146/annurev.phyto.40.120401.13.

Janisiewiez, W. J. and S. N. Jeffers. 1997. Efficacy of commercial formulation of two biofungicides for control of blue mold and grey mold of apples in cold storage. *Crop Protection 16*: 629–633.

Kester, J. J. and O. R. Fennema. 1986. Edible films and coatings: A review. *Food Technology 40*: 47–59.

Krochta, J. M. and C. D. Mulder-Johnston. 1997. Edible and biodegradable polymer films: Challenges and opportunities. *Food Technology 51*: 61–74.

Leelasuphakul, W., P. Hemmanee and S. Chuenchitt. 2008. Growth inhibitory properties of *Bacillus subtilis* strains and their metabolites against the green mold pathogen (*Penicillium digitatum Sacc.*) of citrus fruit. *Postharvest Biology and Technology 48*(1): 113–121.

Leroy, F. and L. De Vuyst. 2004. Lactic acid bacteria as functional starter cultures for the food fermentation industry. *Trends in Food Science and Technology 15*(2): 67–78.

Li, B. Q., Z. W. Zhou and S. P. Tian. 2008. Combined effects of endo-and exogenous trehalose on stress tolerance and biocontrol efficacy of two antagonistic yeasts. *Biological Control 46*(2): 187–193.

Li, H. and T. Yu. 2001. Effect of chitosan on incidence of brown rot, quality and physiological attributes of postharvest peach fruit. *Journal of the Science of Food and Agriculture 81*(2): 269–274.

Lindow, S. E. and M. T. Brandl. 2003. Microbiology of the phyllosphere. *Appl. Environ. Microbiol. 69*(4): 1875–1883.

Lou, M. M., B. Zhu, I. Muhammad et al. 2011. Antibacterial activity and mechanism of action of chitosan solutions against apricot fruit rot pathogen Burkholderia seminalis. *Carbohydrate Research 346*(11): 1294–1301.

Liu, J., Y. Sui, M. Wisniewski et al. 2013. Utilization of antagonistic yeasts to manage postharvest fungal diseases of fruit. *International journal of Food Microbiology 167*(2): 153–160.

Liu, X., W. Fang, L. Liu et al. 2010. Biological control of postharvest sour rot of citrus by two antagonistic yeasts. *Letters in Applied Microbiology 51*(1): 30–35.

Maftoonazad, N., H. S. Ramaswamy, M. Moalemiyan et al. 2007. Effect of pectin-based edible emulsion coating on changes in quality of avocado exposed to Lasiodiplodiatheobromae infection. *Carbohydrate Polymers 68*(2): 341–349.

Manuel, B., C. Jorge, C. M. Rosalba et al. 2005. Commercial film coatings reduce weight loss and improve appearance of 'Keitt' mango fruits (*Mangifera indica* L.). *HortScience 40*: 994C–994. doi: 10.21273/HORTSCI.40.4.994C

Marín, A., M. Cháfer, L. Atarés et al. 2016. Effect of different coating-forming agents on the efficacy of the biocontrol agent Candida sake CPA-1 for control of *Botrytis cinerea* on grapes. *Biological Control 96*: 108–119.

Martinez, F. A. C., E. M. Balciunas, A. Converti et al. 2013. Bacteriocin production by Bifidobacterium spp. A review. *Biotechnology Advances 31*(4): 482–488.

Moayednia, N., M. R. Ehsani, Z. Emamdjomeh et al. 2009. The effect of sodium alginate concentrations on viability of immobilized *Lactobacillus acidophilus* in fruit alginate coating during refrigerator storage. *Journal Australian Journal of Basic and Applied Sciences 3*(4): 3213–3216.

Nisha, V., C. Monisha, R. Ragunathan et al. 2016. Use of chitosan as edible coating on fruits and in micro biological activity – An ecofriendly approach. *Internatial Journal of Pharmaceutical Science Invention 5*(81): 7–14.

Nisperos-Carriedo, M. O., E. A. Baldwin and P. E. Shaw. 1992. Development of an edible coating for extending postharvest life of selected fruits and vegetables. *Proceedings Annual Meeting Florida State Horticultural Society 104*: 122–125.

Nunes, C., J. Usall, N. Teixido et al. 2002. Control of *Penicillium expansum and* Botrytis cinerea on apples and pears with the combination of Candida sake and *Pantoea agglomerans*. *Journal of Food Protection 65*(1): 178–184.

Olivas, G. I. and G. V. Barbosa-Cánovas. 2005. Edible coatings for fresh-cut fruits. *Critical Reviews in Food Science and Nutrition 45*(7–8): 657–670.

Oms-Oliu, G., R. Soliva-Fortuny and O. Martín-Belloso. 2008. Edible coatings with antibrowning agents to maintain sensory quality and antioxidant properties of fresh-cut pears. *Postharvest Biology and Technology 50*(1): 87–94.

Pang, X, Z. Q. Qunzhang and X. M. Huang. 2002. Biological control of post harvest disease of fruits and vegetables. *Journal of Tropical Subtropical Botany 10*: 186–192.

Parafati, L., A. Vitale, C. Restuccia et al. 2015. Biocontrol ability and action mechanism of food-isolated yeast strains against *Botrytis cinerea* causing post-harvest bunch rot of table grape. *Food Microbiology 47*: 85–92.

Parafati, L., A. Vitale, C. Restuccia et al. 2016. The effect of locust bean gum (LBG)-based edible coatings carrying biocontrol yeasts against *Penicillium digitatum* and *Penicillium italicum* causal agents of post-harvest decay of mandarin fruit. *Food Microbiology 58*: 87–94.

Pawlowska, A. M., E. Zannini, A. Coffey et al. 2012. "Green preservatives": Combating fungi in the food and feed industry by applying antifungal lactic acid bacteria. In J. Henry (Ed.), *Advances in Food and Nutrition Research*. Cambridge, MA: Academic Press, Vol. 66, pp. 217–238.

Potjewijd, R., M. O. Nisperos, J. K. Burns et al. 1995. Cellulose-based coatings as carriers for *Candida guillermondii* and *Debaryomyces* sp. in reducing decay of oranges. *HortScience 30*(7): 1417–1421.

Pusey, P. L., M. W. Hotchkiss, H. T. Dulmage et al. 1988. Pilot tests for commercial production and application of *Bacillus subtilis* (B-3) for postharvest control of peach brown rot. *Plant Disease 72*(7): 622–626.

Ramos-García, M., E. Bosquez-Molina, M. Hernández-Romano et al. 2012. Use of chitosan-based edible coatings in combination with other natural compounds, to control *Rhizopus stolonifer* and Escherichia coli DH5α in fresh tomatoes. *Crop Protection 38*: 1–6.

Ramos-Guerrero, A., R. R. González-Estrada, G. Romanazzi et al. 2019. Effects of chitosan in the control of postharvest anthracnose of soursop (*Annonam uricata*) FRUIT. *Revista Mexicana de IngenieríaQuímica 19*(1): 99–108.

Roberts, R. G. 1990. Postharvest biological control of gray mold of apple by *Cryptococcus laurentii. Phytopathology 80*(6): 526–530.

Sharma, N., S. Sharma and B. Prabha. 2012. Postharvest biocontrol–new concepts and application. In B. Venkateswarlu, A. K. Shanker, C. Shanker, and M. Maheswari (Eds), *Crop Stress and its Management: Perspectives and Strategies*. Dordrecht: Springer, pp. 497–515.

Sharma, R. R., D. Singh and R. Singh. 2009. Biological control of postharvest diseases of fruits and vegetables by microbial antagonists: A review. *Biological control 50*(3), 205–221.

Siddiqui, M. W. (Ed.). 2018. *Postharvest Disinfection of Fruits and Vegetables*. Cambridge, MA: Academic Press.

Sipiczki, M. 2006. *Metschnikowia* strains isolated from botrytized grapes antagonize fungal and bacterial growth by iron depletion. *Applied and Environmental Microbiology 72*(10): 6716–6724.

Smith, J. A. and J. P. Métraux. 1991. *Pseudomonas syringe* epv. *syringae* induces systemic resistance to *Pyricularia oryzae* in rice. *Physiological and Molecular Plant Pathology 39*(6): 451–461.

Suseno, N., E. Savitri, L. Sapei et al. 2014. Improving shelf-life of cavendish banana using chitosan edible coating. *Procedia Chemistry 9*: 113–120.

Tapia, M. S., M. A. Rojas-Graü, A. Carmona et al. 2008. Use of alginate-and gellan-based coatings for improving barrier, texture and nutritional properties of fresh-cut papaya. *Food Hydrocolloids 22*(8): 1493–1503.

Tayel, A. A., S. H. Moussa, M. F. Salem et al. 2016. Control of citrus molds using bioactive coatings incorporated with fungal chitosan/plant extracts composite. *Journal of the Science of Food and Agriculture 96*(4): 1306–1312.

Valdés, A., M. Ramos, A. Beltrán et al. 2017. State of the art of antimicrobial edible coatings for food packaging applications. *Coatings* 7(4): 56. doi: 10.3390/coatings7040056

Wilson, C. L. and E. Chalutz. 1989. Postharvest biological control of Penicillium rots of citrus with antagonistic yeasts and bacteria. *Scientia Horticulturae 40*(2): 105–112.

Wilson, C. L. and M. E. Wisniewski. 1989. Biological control of postharvest diseases of fruits and vegetables: An emerging technology. *Annual Review of Phytopathology 27*(1): 425–441.

Yahaya, S. M. and A. Y. Mardiyya. 2019. Review of post-harvest losses of fruits and vegetables. *Biomedical Journal of Scientific and Technical Research* 13(4): 10192–10200. doi: 10.26717/BJSTR.2019.13.002448

Yousuf, B., O. S. Qadri and A. K. Srivastava. 2018. Recent developments in shelf-life extension of fresh-cut fruits and vegetables by application of different edible coatings: A review. *LWT 89*: 198–209.

5 Non-chemical Management of Postharvest Diseases of Mango

P. K. Shukla, Tahseen Fatima, Gundappa and Nidhi Kumari

CONTENTS

5.1 INTRODUCTION

Mango (*Mangifera indica* L.) is one of the most popular fruits in the world. It is an evergreen tree, grown throughout subtropical and tropical regions of the world (Alemu et al., 2014b) and is indigenous to north-east India and north Burma and is considered to have originated in the Indo-Burma region (De Candolle, 1904). It belongs to the Anacardiaceae family which has 75 genera and over 700 species (Lizada, 1993). Mango's delicious taste and high caloric value make it one of the most liked fruits on the international market (Diedhiou et al., 2007). It is the world's fifth most important fruit crop, which is cultivated in 56.81 million hectares with global fruit production of about 50.64 million tonnes (FAO, 2017). India is the world's largest mango producer with around 56 per cent share in global production. The other major mango producing countries are China, Mexico, Thailand, Indonesia, Pakistan, Nigeria, Philippines, Brazil, Egypt and Haiti (Swart, 2010). It is grown in almost all parts of India over an area of 2.26 million hectares with 21.82 million tonnes of production (NHB Database, 2017–2018). The tree grows luxuriantly in sandy, loamy, clayey, well-drained and aerated soil with a pH range of 5.5–7.5 (Young et al., 1965; Kadman et al., 1976). Mango is drought resistant but can withstand occasional flooding as well (Singh, 1960). Rainfall and temperature are two important environmental factors crucial for optimum flowering and fruit set in mango, however, their influence varies with cultivars (Schaffer et al., 1994).

Dry weather before blossoming and a temperature range of 24–30 °C is required for profuse flowering; however, during fruit development, if sufficient water is provided, the tree can withstand up to 48 °C. Low temperature hampers plant growth and results in flower deformation as well as loss of pollen viability and may cause death of young trees (Popenoe, 1957; Issarakraisila et al., 1992), however, it has been reported that older trees can endure up to −4 °C for a few hours with limited damage (Crane and Campbell, 1994). In India, Dashehari, Chaunsa, Langra, Kesar, Alphonso and Totapuri are the most appreciated mango cultivars due to their wide range of adaptability, high nutritive value, delicious taste and excellent flavour. The fruit is used in both raw and ripe forms. Raw mango fruits of local cultivars are used for making various traditional products like pickle, chutney, amchur, murabba, aamras and so on, while cultivars like Totapuri and Alphonso are used in various types of drinks. Along with fruits, other parts of the mango tree like bark and leaves have been utilized in various types of treatments or preventatives for a long time because of its medicinal properties like anti-parasitic, anti-diabetic, anti-inflammatory, anti-oxidant, laxative and stomachic (Kalita, 2014). The ripe fruit possesses fattening, diuretic and laxative properties and also aids in human digestion.

The mango crop has been reported to suffer from a large number of diseases at various growth, development and production stages. The diseases are responsible for loss in plant growth, yield and fruit quality (Prakash et al., 1996, Prakash, 2004; Diedhiou et al., 2007; Shukla et al., 2018a,b). Various preharvest and postharvest diseases caused by fungi and bacteria are responsible for the immense loss to mango production. Among those, important preharvest diseases are wilt, powdery mildew, anthracnose, blossom blight, dieback/twig blight, malformation, sooty mould and shoulder browning; postharvest diseases include anthracnose, stem end rot, Aspergillus rot and Alternaria rot. Mango fruits are highly susceptible to infection during the period between harvest and consumption because of high moisture content and the nutrient reserve in fruits. Mango fruits, being highly perishable, are marketed immediately after harvest. Postharvest losses in mango have been reported up to 17–36 per cent (Haggag, 2010). The mango crop is the host of many pathogens, among which fungi are the major agents throughout the world (Diedhiou et al., 2007).

Postharvest management means handling of fruits after harvest to prolong disease-free storage life, maintain freshness and an attractive appearance. To deliver quality fruits to market and ultimately to the consumer, proper postharvest management is necessary. Care must be taken at every stage like harvesting, de-sapping, grading, packaging, storage and transport. Postharvest management should be considered as a second production operation to protect fruits from diseases and add value to the fruits as a basic means for effective marketing.

Prevention is a better and more effective tool of postharvest disease management. Chemicals have mainly been used to prevent and manage postharvest diseases; however, the residue results in environmental contamination, and therefore poses a potential health hazard (Alemu et al., 2014a). It is necessary to adopt non-chemical methods instead of chemical ones, particularly those applied one month before harvest for postharvest disease management. Therefore, management of postharvest diseases below the economic threshold using economically viable, ecologically safe and easily operational non-chemical procedures is obligatory. This can be achieved by adopting integrated disease management (IDM) strategies (Parakash, 412004; Ploetz, 2009). However, chemicals are an important component of disease management in IDM, when integrated at the right time for protection of the crop at critical stages. The important postharvest diseases are discussed in detail in the following section.

5.2 ANTHRACNOSE

The disease is caused by *Colletotrichum gloeosporioides* (Penz.) Penz and Sacc, which produces symptoms on every tender part of the tree including fruitlets and mature fruits. It is the most serious postharvest disease, reported to be a key constraint in almost all mango growing regions of the world where rain and high humidity prevails during fruit development and maturity (Akem, 2006; Haggag, 2010; Shukla et al., 2016, Shukla, 2017), therefore, it is a major limitation in export trade expansion (Sangeetha and Rawal, 2008). Fruits at all growth stages are prone to the disease and bear typical disease symptoms (Yenjit et al., 2004). Sunken necrotic lesions on the fruit surface and mummified fruits are typical symptoms (Rivera-Vargas et al., 2006). The postharvest fruit rot is the most damaging and economically important phase of the disease. It directly reduces the number of marketable fruits resulting in huge losses. The postharvest rot is directly associated with the preharvest phase where infection remains symptomless and quiescent before harvest during fruit development but causes decay and rotting upon fruit ripening after harvest (Haggag, 2010).

5.2.1 Symptoms

Anthracnose disease symptoms include blossom blight, typical shot hole spots on leaves and spots or lesions of different sizes on fruits. Infection on the panicles starts as small black or dark-brown spots which later on may enlarge, coalesce and kill the flowers, severely reducing fruit set. On leaves, anthracnose infections start as small, round or angular, brown to black spots. Infected young fruits develop black spots and later on shrivel and finally drop off. Depending on the prevailing weather conditions, the disease may vary in severity from slight to heavy infection. Postharvest anthracnose appear as round brown to black lesions with an indefinite border on the fruit surface then followed by the development of circular lesions on the ripening fruit which increase in size rapidly and may cover the entire fruit surface in severe cases. Sometimes, lesions of different sizes can coalesce and cover extensive areas of the fruit, typically in a tearstain pattern, developing from the basal towards the distal end of the fruit (Arauz, 2000; Giblin et al., 2010; Shukla et al., 2016). Initially, lesions are restricted to the peel, but in later stages, the fungus penetrates into the pulp. In advanced stages of infection, the fungus produces acervuli and abundant orange to salmon pink masses of conidia on the lesions (Akem, 2006).

5.2.2 Epidemiology

Environmental parameters *viz*., temperature, humidity, rain, misty conditions or heavy dews especially at the time of blossoming, initiation of new flushes and fruit maturity greatly influence the extent of infection and damage. The infection is mostly favoured by temperature ranging from 20 to 30 °C, with the optimum being 25 °C (Arauz, 2000). Continuous wet weather during flowering and fruit maturity causes severe infection. Availability of free water on the fruit surface as well as relative humidity above 95 per cent for 12 hours is essential for conidial germination, which are primary and secondary inoculum sources of infection, but during ripening, if storage temperature reaches up to 40 °C, it favours severe rotting (Prakash et al., 1996).

5.2.3 Management

Postharvest anthracnose can be successfully managed by integrating pre- and postharvest treatments and their application at critical stages (Ploetz, 2009). The first critical stage is after fruit set during their early development phase. At this stage, fruits are highly susceptible to infection under high relative humidity. Use of fungicides at this stage ensures crop protection and the elimination of residue in mature fruits, although at maturity, fungicides cannot be recommended. An ideal anthracnose management strategy includes both pre- and postharvest management practices.

5.2.3.1 Preharvest Measures

5.2.3.1.1 Cultural Practices

Preharvest cultural practices like the application of fertilizers, manures, growth regulators and canopy management influence fruit physiology, chemical composition and the extent of infection in fruits (Prakash, 2004). Well managed orchards with over 30 per cent light penetration and free wind blow have minimum inoculum build-up. Since the development of the disease depends on wetness over the fruit surface, new orchards should be established in areas where the dry season coincides with the fruit development phase (Arauz, 2000).

5.2.3.1.2 Resistant Varieties

Use of resistant cultivars is the most economical and environmentally friendly approach of plant disease management. Although resistance of commercial cultivars has not been consistent against anthracnose at different locations. All commercial cultivars are susceptible to disease, however, susceptibility varies among them (Akem, 2006).

5.2.3.1.3 Chemical Control

Since *C. gloeosporioides* remains active on the tree and on fallen fruits or leaves throughout the year except during winter, inoculum is available to cause fresh infections on tender tissue under favourable conditions. Excellent disease control has been achieved using fungicides dithiocarbamate, mancozeb, febran in the field (Akem, 2006). Benzimidazoles, tinidazole and prochloraz are the commonly used fungicides for mango after infection activity. Besides benomyl applied as an eradicant spray following infection periods (Arauz, 2000), it is also sprayed along with protectants to impede the development of the resistant pathogen population. Carbendazim, thiophanate methyl, mancozeb and difenoconazole have also been found effective, however, none of these fungicides should be sprayed during the last 30 days before harvesting to avoid residual toxicity.

5.2.3.2 Postharvest Measures

Postharvest anthracnose management measures are applied to reduce the level of infection on the fruits. Keeping in view the set standards regarding fruit quality and fungicide residues in fruits, no chemical method can be applied at this stage.

5.2.3.2.1 Biological Management

Relatively little research has been conducted on the biological control of anthracnose. *Bacillus licheniformis*, a gram-positive bacterium, has been reported to reduce anthracnose incidence and was successfully evaluated as a biocontrol agent in South Africa (Govender et al., 2005).

Senghor et al. (2007) also reported that *Bacillus subtilis* significantly reduced anthracnose incidence in storage. Gram-negative bacteria and some other amendments have also been employed by some workers for biological management of mango anthracnose (Vivekananthana et al., 2004).

Similarly, Kefialew and Ayalew (2008) reported that different isolates of bacteria and yeast antagonists significantly reduced anthracnose severity on fruits artificially inoculated with *C. gloeosporioides*. Those investigations indicated that *in vivo* reduction in disease severity was obtained when fruits were inoculated with the bacterium 24 hours prior to inoculation with the fungus, but not when fruits were inoculated with the pathogen first. Since, at the time of harvest, fruits mostly have symptomless infection, field application of bacterial bioagents could not be popularized.

5.2.3.2.2 Physical Method

Hot water treatment (HWT), which includes dipping of fruits in hot water with or without fungicides at 50–55 °C for up to 5 minutes (Alemu, 2014b) just after harvesting, is an age-old and successful practice that efficiently reduces anthracnose development on fruits on its own (Uddin et al., 2018). HWT is an effective, easy to apply and simple practice still used in many mango producing countries. However, the effectiveness of HWT in reducing anthracnose development is greatly influenced by the initial infection level as well as storage conditions. In cases of high disease pressure, the potency of HWT can be greatly enhanced by combining it with fungicides (Kefialew and Ayalew, 2008). HWT should not be delayed beyond 2 days after harvesting and must be transferred in ambient cool water after treatment followed by air drying before packing.

5.2.3.2.3 Use of Plant Products

Since increased awareness regarding the harmful effects on environmental and health hazards posed by fungicides, the use of plant extracts and essential oils has gained interest in crop management schedules due to their apparently safe nature and antimicrobial activities. These are extracts are composed of secondary metabolites mixtures (Gottlieb et al., 2002). Few people have studied the potential of different plant extracts and essential oil *viz*., castor oil, eucalyptus oil, lemongrass oil, garlic bulb extract, *Azadirachta indica*, *Zingiber officinale*, *Lantana* and *Curcuma longa* leaf extract in suppressing anthracnose development and postharvest rots of mango fruits (Hasabnis and D'Souza, 1987; Chauhan and Joshi, 1990; Singh and Thakur, 2003; Duamkhanmanee, 2008; Alemu et al., 2014a). Besides anthracnose control, these extracts also positively influence mango's storage quality by better retaining total soluble solids and sugar content.

5.3 STEM END ROT

Stem end rot is another important postharvest disease responsible for severe postharvest losses in mango production worldwide (Ni et al., 2012). The disease incidence is more severe in orchards with older trees and if fruits are subjected to prolonged storage (Cooke et al., 2009). Several fungi, *viz. Lasiodiplodia theobromae*, *Aspergillus niger*, *Phomopsis mangiferae* or *Dothiorella dominicana* and *Colletotrichum gloeosporioides* have been reported to be associated with this disease (Cooke et al., 2009). The major causal organism of stem end rot, *Botryodiplodia theobromae* is a wound/secondary pathogen and a saprophyte, capable of living on different hosts or substrates at relatively high temperatures. It has been found to be associated with postharvest rots of a wide range of vegetables.

5.3.1 Disease Symptoms

The pathogen remains dormant on the unripe fruit, upon ripening, the fruit becomes brown, soft and starts rotting at the pedicel and peduncle tissues, which eventually can cover the entire fruit surface with steel-grey coloured mycelium in extreme cases (Govender, 2005; Prakash, 2004). Injuries to fruits may lead to lesion formation even away from the stem end. Adjacent healthy fruits also get infected after coming in physical contact with these decaying fruits. The symptoms may vary according to the causal agent. *P. mangiferae* results in formation of firmer lesions with defined margin, which spread comparatively slower than those caused by other stem end rot fungi. The fruiting bodies formed by *P. mangiferae* are dark, pinhead-sized, whereas *C. gloeosporioides* produces pink spore masses (Prakash, 2004; Cooke et al., 2009).

5.3.2 Epidemiology and Disease Cycle

The fungi remain within the branches and twigs, causing latent infection. The endophytic hyphae colonize the inflorescence and reach the fruit's stem end several weeks after flowering but do not extend into the fruit until harvesting (Govender and Korsten 2006; Cooke et al., 2009). The stem end rot causing fungi may also harbour in soil and tree litter from where the fruit also gets infected after harvest (Cooke et al., 2009). The pathogen is suspected to invade mango trees either through natural openings or wounds and further causing latent infection in the fruit by entering through stem ends. The conidia are dispersed and released with free water under high humidity. Conidia present in rainwater and the pathogen's fruiting structure on the mango litter are the sources of inoculum. Black pycnidia and perithecia in the stomata on cankered limbs of trees bearing dried up fruits are the overwintering structures of the pathogen.

5.3.3 Management

5.3.3.1 Preharvest

Disease incidence can be managed by reducing inocula load and infection in twigs by adopting proper orchard management year-round. Latent infection in branches and twigs can be reduced by pruning, orchard sanitation and fungicide spraying. Practices *viz.*, resistant or tolerant cultivar planting, avoiding injuries which are the entry points of the pathogen, and fungicide sprays can contribute to reducing the deposition of pre-harvest inocula. Application of *Bacillus licheniformis* and covering fruit with polyethylene or paper bags have also been reported to reduce fruit rot incidence (CABI, 2005).

5.3.3.2 Postharvest

Irradiation, HWT and controlled atmosphere with regulated O_2 and CO_2 concentrations coupled with low temperature during storage contribute greatly in minimizing postharvest losses (Govender, 2005; Cooke et al., 2009). A warm water dip *B. licheniformis* amended with fungicide prochloraz in reduced concentration effectively controls fruit rots (CABI, 2005; Cooke et al., 2009).

5.4 BACTERIAL BLACK SPOT (CANKER)

In general, it is not considered as an important postharvest disease of mango fruits, however, the fruits severely infected with *Xanthomonas campestris* pv. *mangiferaeindicae* may suffer rotting when subjected to high humidity during storage. The tissue injured due to bacterial infection also allows other saprophytic fungi to grow. Its incidence in new orchards can be avoided by planting bacteria-free grafts. Spraying of copper-based bactericides has also been recommended during the rainy season to avoid preharvest infection in fruits.

5.5 OTHER FRUIT ROTS

After harvest, rotting of mango fruits is common, this may be due to anthracnose and stem end rot or due to infection caused by other pathogenic or infestation by other saprophytic fungi over the fruit's injured portion (Prakash et al., 2011). Such rots can be avoided by proper sorting and removal of injured fruits from the batch. HWT, mentioned for anthracnose, may also help in managing such rots (Mansour et al., 2006). Javadpour et al. (2018) demonstrated a significant decrease in mango fruit rot after treatment with essential oils of *Thymus vulgaris* (1000 µl/l), *Artemisia persica* (1000 µl/l) and *Rosmarinus officinalis* (500 µl/l). The description of various types of rots is briefly mentioned below.

5.5.1 Black Rot

Caused by *Aspergillus niger* V. Tiegh, *A. variecolor* (Berk and Br.) Thorn and Raper, *A. nidulans* (Eidorn) Wint, *A. fumigatus* Fres., *A. flavus* Link and *A. chevalieri* (Mang.) Thorn. and Church. After harvest, fungus invades the fruits through wounds or cut ends and infection starts in the form of greyish to pale brown soft, sunken spots which later on coalesce into dark brown to black lesions and spread rapidly. Later on, profuse fungal growth develops over these lesions. The disease development intensifies at 30–36 °C. Careful handling at all stages to avoid mechanical injury and sap burn damage to the fruits can significantly reduce black rot incidence.

5.5.2 Botryosphaeria Rot

Caused by *Botryosphaeria ribis* Gross. and Dugg. Initially, the fruit tissue around the spot softens followed by brown discoloration of affected skin and darkening of the flesh beneath. Dark, sunken, oval to round or sometimes elliptical lesions with minute black bodies and white to grey mouldy growth in advance stages accompanied with shallow decay of the flesh beneath are formed. Variable symptoms are produced by *Botryosphaeria ribis* and therefore sometimes confused with previously described stem end rot, hence the association of *B. ribis* must be confirmed with culturing and microscopic examination. Incidence of *Botryosphaeria* rot can be reduced by harvesting fruits with stalks, maintaining strict orchard hygiene and good trees.

5.5.3 Brown Rot

Caused by *Lasiodiplodia theobromae* (Pat.) Griffon and Moubl. Presence of dark brown to black necrotic patches without any defined margin near the stalk or anywhere on a wounded fruit is a characteristic symptom. A heavily infected fruit turns dark, brittle, light weighted with disintegrated and off flavoured internal pulp. Incidence of brown rot can be reduced by careful handling of fruit to avoid bruising/wounding.

5.5.4 Stem End Soft Rot

The disease is caused by *Ceratocystis paradoxa* (Dade) Moreau and results in heavy postharvest losses. Pale, soft and watery, irregular lesions develop near the stem end which later darken due to spore formation and subsequently cover the entire fruit. Rotting is accompanied by a rancid odour. By proper handling to prevent fruit injuries, avoiding packing of wet fruits and following proper hygienic practices, incidence of stem end soft rot can be reduced.

5.5.5 Pestalotiopsis Rot

Caused by *Pestalotiopsis versicolor* (Speg.) Stey and *P. mangiferae* (P. Henn.) Stey. The symptoms are characterized with the appearance of small spots which slowly increase in size; the brown spots later turn into dark brown spots with a greyish white centre and black dots of acervuli over them. As the disease advances to the stalk end, the fruit drop off the trees.

5.5.6 Charcoal Rot

The causal agent of charcoal rot is *Macrophomina phaseolina* (Maubl.) Ashby which usually infects the mango fruits through cut ends. Black irregular necrotic leathery patches usually originating from the distal end of the fruits are characteristic symptoms. Rotting and brown discoloration of tissue generally takes place from the fruit's lower side. As the infection advances under humid conditions, the entire fruit surface as well as the internal pulp turns dark brown and completely rots.

Incidence of charcoal rot can be reduced by maintaining field hygiene and removing diseased fruits as well as careful handling of harvested fruits to avoid soil contamination and bruising.

5.5.7 Alternaria Rot

The fruit rot is caused by *Alternaria tenuissima* (Fr.) Wiltshire and *A. alternata* (Fr: Fr.) Keissl. Initially, there are water soaked circular and subcircular spots which later become black with a sunken centre that enlarges in size. Dark brown spores may grow on lesions under humid conditions. Reddish patches on the flesh below the lesions are clearly visible when epicarp is removed.

5.5.8 Phoma Rot

Phoma mangiferae P. Henn. produces black to brown spots which may appear on the entire fruit surface. The outer skin remains hard but the internal portion rots quickly. In advance stages, dark pycnidia are formed over such rotten tissue.

5.5.9 Macrophoma Rot

Water soaked, brownish, circular lesions which later on become irregular in shape, cover the entire fruit surface and are covered by deep brown mould are produced by *Macrophoma mangiferae* Hingorani and Sharma on mango. Minute black bodies are formed in the fruit skin of infected fruits.

5.5.10 Rhizopus Rot

Rhizopus arrhizus Fisch and *R. oryzae* Went and Gerlings commonly cause soft decay which emits a very peculiar and unpleasant foul odour. Profuse development of coarse white spore heads mould strands giving rise to globular white spore heads (sporangia), which later on turn black and are easily visible to the naked eye takes place.

5.5.11 Cladosporium Rot

Caused by *Cladosporium herbarum* (Pers.) Link and *C. cladosporioides* (Fres) de Vries. Lesions covered with a white mould are associated with injury on fruits which later gives rise to a velvety mat of dark green spores. Unlike other rots, the lesions remain limited in area at the surface, but decay penetrates deeply towards the stone. The rotted area is dark brown to black and remains shallow.

5.5.12 Fusarium Rot

The losses due to fusarium rot are more common in the rainy season. The causal agent is *Fusarium oxysporum* Schld. Like most postharvest rots, infection in this case also takes place through the stem end or other injuries. Appearance of large, dark brown, irregular water-soaked areas on the fruits. Soon after, soft rot is accompanied by a putrescent odour and profuse growth of dull pink fungal colonies can be seen.

5.5.13 Rhizoctonia Rot

Rhizoctonia rot is caused by *Rhizoctonia solani* Kühn, a fungus which thrives well under warm-wet conditions and is characterized by the development of brown to black, irregular lesions on the fruits. The lesion development is followed by hard brown rotting on the upper portion with a clear line of distinction between healthy and affected tissue rotting with abundant sclerotial bodies of the fungus. In advanced stages, the fungus also penetrates inside fruits, which are close to the ground.

5.5.14 Mucor Rot

Caused by *Mucor subtilissimus* Oud, a wound parasite. White mould development may take place at the stem end or at any place on the injured fruit surface. Yellowish mould bearing black spore heads (sporangia) covers the water-soaked lesions. The decay is accompanied by a yeasty odour. Infection is escalated with fruit fly infestation.

5.5.15 Grey Mould Decay

Grey mould decay is caused by the fungus *Botrytis cinerea* Pers. A light brown firm decay, which is later on covered by grey or brown fungal growth that contains a mass of spores. The pathogen spreads readily by contact with adjacent fruits, giving rise to larger nests of diseased fruit in packed containers.

5.5.16 Blue Mould

Caused by *Penicillium expansum* link ex Gray. A soft brown decay with white powdery spore masses, which becomes blue when mature. The decay on the skin and flesh develops at the stem end or at the wound which later on may spread to the entire fruit surface.

5.5.17 Hendersonia Rot

Pale brown circular lesions with cracked centre caused by *Hendersonia creberrima* Sydow and Butler appear on the fruit. Decayed tissue is soft and moist.

5.5.18 Watery soft rot

This is a common postharvest disease during the rainy season under subtropical conditions. It is caused by *Sclerotinia sclerotiorum* Lib. de Bary. Like other rots, the tissue softens and fluid is released. Under humid conditions, there is copious development of a bright white cottony mould spreading onto neighbouring fruits, which later gives rise to numerous firm black bodies (sclerotia).

5.5.19 Sclerotium Fruit Rot

This rot is caused by *Sclerotium rolfsii* Sacc. and is more severe in warm moist conditions. It is common in mango fruits harvested at relatively late stages and subsequently heaped on the soil. The fungus persists in the soil in the form of sclerotia and hyphal strands. Therefore, infection usually occurs near the soil surface. Rotted tissue is light brown to pinkish, soft but not watery. In early stages, the lesions have sharply defined margins. Later, rotting becomes intrusive, with a substantial amount of internal tissue becoming water-soaked yet firm and hard. In humid conditions, there is profuse white mould development, which eventually gives rise to small (1–2 mm) spherical sclerotia, white at first and later turning to brown.

5.5.20 Yeasty Rot

Caused by *Saccharomyces* spp. Such rots are associated with ripe fruits. Yeasty fungi are able to invade via wounds or injuries and cause fermentation in the flesh, which turns light yellow. Pockets of gas may be formed and juice bubbles out, eventually leaving the fruit interior in a spongy and fibrous form. It is characterized by exudation of the frothy juice having an odour of fermentation. To reduce further infection, fruits should promptly be discarded and destroyed. Cold storage is effective in arresting development of rot. Strict field hygiene should be maintained with crop rotation if possible. Harvested fruits should be kept in clean places to avoid contamination.

5.6 FACTORS AFFECTING DISEASE DEVELOPMENT

5.6.1 Preharvest Conditions and Postharvest Disease Incidence

Improper orchard management often leads to low yield and poor fruit quality. Some factors like microclimate, agrochemicals, nutrition, management systems and maturity of the fruits are responsible for fruit quality parameters and their susceptibility to diseases (Prakash, 2004; Poll et al., 1996; Mattheis and Fellman, 1999; Léchaudel and Joas, 2006; Lalel et al., 2003; Huong, 2008). Light is an essential factor for proper growth and development of trees and fruits. The green leaves absorb the sunlight to induce carbohydrate and sugar synthesis, which is translocated to the buds, flowers and fruits. Because of its crucial role, canopy management should aim for maximal utilization of light by regulating growth to give optimal productivity with quality fruit production. Pruning enhances the quantity of sunlight intercepted by trees because tree shape regulates the exposure of the leaf area to incoming radiation. Consequently, better light penetration into the tree canopy improves tree growth, productivity, yield and fruit quality (Ahmad et al., 2006). Therefore, it also helps to avoid microclimate conditions, which are beneficial for disease and pest infestation. Economical aspects of obtaining the required canopy architecture and convenience in carrying out cultural practices should also be considered to prevent postharvest losses (Srivastava, 2007).

5.6.2 Temperature and Fruit Ripening and Postharvest Diseases

Low temperature is an effective means to prolong the storage and shelf life of fruits, which reduces both respiration and ethylene synthesis (Lee et al., 1996; Crane et al., 2009). Lower temperature helps in the reduction of all metabolic activities and biochemical reactions. Mangoes are quite sensitive to cold temperatures, which can cause chilling injuries at temperatures below 13 °C for mature green mangoes and below 10 °C for partially ripe mangoes (Kader and Mitcham, 2008; Johnson and Hofman, 2009).

5.7 CONCLUSION

Postharvest diseases of mango cause severe losses worldwide. The mature green fruits are mostly symptom free, but during ripening and storage, symptoms appear that can result in considerable losses. To manage such diseases, knowledge and understanding of the pathogen's biology, the conditions aggravating disease development and the economics, efficacy and market acceptability of management practices are necessary. Integrated disease management, including both chemical-based and non-chemical-based approaches, must be preferred to avoid residual toxicity and maintain quality standards of fruits.

REFERENCES

Ahmad, S., Z. A. Chatha, M. A. Nasir et al. 2006. Effect of pruning on the yield and quality of Kinnow fruit. *Journal of Agriculture and Social Sciences* 2: 51–53.

Akem, C. N. 2006. Mango anthracnose disease: Present status and future research priorities. *Journal of Plant Pathology* 5(3): 266–273.

Alemu, K., A. Ayalew and K. Woldetsadic. 2014a. Effect of aqueous extracts of some medicinal plants in controlling anthracnose disease and improving postharvest quality of mango fruit. *Persian Gulf Crop Protection* 3(3): 84–92.

Alemu, K., A. Ayalew and K. Woldetsadic. 2014b. Evaluation of antifungal activity of botanicals for post-harvest management of mango anthracnose (*Colletotrichum gloeosporioides*). *International Journal of Life Science* 8(1): 1–6.
Arauz, L. F. 2000. Mango anthracnose: Economic impact and current options for integrated management. *Plant Disease* 84(6): 600–611.
CABI. 2005. *Crop Protection Compendium* (2005 ed.). Wallingford, UK: CAB International.
Chauhan, H. I. and H. N. Joshi. 1990. Evaluation of phyto extracts for control of mango anthracnose. In *Proceedings of Symposium of Botanical Pesticides in Integrated Pest Management: Proceedings of National Symposium Held on January 21–22, 1990*. Rajahmundry: Central Tobacco Research Institute, pp. 455–459.
Cooke, T., D. Persley and S. House. 2009. *Diseases of Fruit Crops in Australia*. Clayton, Australia: CSIRO Publishing, pp. 157–173.
Crane, J. H. and C. W. Campbell. 1994. *The Mango. Fruit Crops Fact Sheet FC-2*. Gainesville, FL: Florida Cooperative Extension Service.
Crane, J. H., S. Salazar-Garcia, T. S. Lin et al. 2009. Crop production: Management. In *The Mango: Botany, Production and Uses* (2nd ed.), R. E. Litz (ed.). Wallingford, UK: CAB International, pp. 432–483.
De Candolle, A. 1904. *Origin of Cultivated Plants*, 2nd ed., K. Paul, Trench, Trubner, p. 468. London: International Scientific Series.
Diedhiou, P. M., N. Mbaye, A. Dramé et al. 2007. Alteration of postharvest diseases of mango *Mangifera indica* through production practices and climatic factors. *African Journal of Biotechnology* 6(9): 1087–1094.
Duamkhanmanee, R. 2008. Natural Eos from lemon grass (*Cymbopogon citrates*) to control postharvest anthracnose of mango fruit. *International Journal of Biotechnology* 10(1): 104–108.
FAO. 2017. FAOSTAT online database at http://www.fao.org/ default.htm.
Giblin F. R., L. M. Coates and J. A. G. Irwin. 2010. Pathogenic diversity of avocado and mango isolates of *Colletotrichum gloeosporioides* causing anthracnose and pepper spot in Australia. *Australasian Plant Pathology* 39: 50–62.
Gottlieb, O. R., M. R. Borin and N. R. Brito. 2002. Integration of ethno-botany and photochemistry: Dream or reality? *Photochemistry* 60: 145–152.
Govender, V. 2005. *Evaluation of Biological Control Systems for the Control of Mango Postharvest Diseases* (M.Sc. thesis). Pretoria, South Africa: University of Pretoria, p. 126.
Govender, V. and L. Korsten. 2006. Evaluations of different formulations of *Bacillus licheniformis* in mango pack house trials. *Biological Control* 37: 237–242.
Govender, V., L. Korsten and D. Sivakumar. 2005. Semi-commercial evaluation of *Bacillus licheniformis* to control mango postharvest diseases in South Africa. *Postharvest Biology and Technology* 38: 57–65.
Haggag, W. M. 2010. Mango diseases in Egypt. *Agriculture and Biology Journal of North America* 1(3): 285–289.
Hasabnis S. N. and T. F. D'Souza. 1987. Use of natural products in the control of the storage rot in Alphanso mango fruits. *Journal of Maharashtra Agricultural Universities* 12: 105–6.
Huong P. T. 2008. Some initial results of improvement of neglected mango orchards in Coc Lac hamlet, Yen Chau district, Son La Province. *Journal of Science and Development* 2: 105–109.
Issarakraisila, M., J. A. Considine and D. W. Turner. 1992. Seasonal effects on floral biology and fruit set of mangoes in warm temperature region of Western Australia. *Acta Horticulturae* 321: 626.
Javadpour, S., A. Golestani, S. Rastegar et al. 2018. Postharvest control of *Aspergillus niger* in mangos by means of essential oil. *Advances in Horticultural Science* 32(3): 389–398.
Johnson G. I. and P. J. Hofman. 2009. Postharvest technology and quality treatments. In *The mango: Botany, Production and Uses* (2nd ed.), R. E. Litz (ed.). Wallingford, UK: CAB International, pp. 529–605.
Kader, A. and B. Mitcham. 2008. Optimum procedures for ripening mangoes. In *Fruit Ripening and Ethylene Management*. Davis, CA: University of California Postharvest Technology Research Center, pp. 47–48.
Kadman, A., S. Gazit and G. Ziv. 1976. Selection of mango rootstocks for adverse water and soil conditions in arid areas. *Acta Horticulturae* 57: 81–88.
Kalita, P. 2014. An overview on *Mangifera indica*: Importance and its various pharmacological action. *PharmaTutor* 2(12): 72–76.
Kefialew, Y. and A. Ayalew. 2008. Postharvest biological control of anthracnose (*Colletotrichum gloeosporioides*) on mango (*Mangifera indica*). *Postharvest Biology and Technology* 50: 8–11.
Lalel, H. J. D., Z. Singh and S. C. Tan. 2003. The role of ethylene in mango fruit aroma volatiles biosynthesis. *Journal of Horticultural Science and Biotechnology* 78: 485–494.
Léchaudel, M. and J. Joas. 2006. Quality and maturation of mango fruits of cv. 'Cogshall' in relation to harvest date and carbon supply. *Australian Journal of Agricultural Research* 57: 419–26.

Lee, L., J. Arult, R. Lenckit et al. 1996. A review on modified atmosphere packaging and preservation of fresh fruits and vegetables: Physiological basis and practical aspects–Part I. *Packaging Technology and Science* 9: 1–17.

Lizada, C. 1993. Mango. In *Biochemistry of Fruit Ripening*. G. B. Seymour, Taylor J. E. and Tucker, G. A. (eds.). London, UK: Chapman and Hall, pp. 255–271.

Mansour, F. S., S. A. Abd-El-Aziz and G. A. Helal. 2006. Effect of fruit heat treatment in three mango varieties on incidence of postharvest fungal disease. *Journal of Plant Pathology* 88(2): 141–148.

Mattheis, J. P. and J. K. Fellman. 1999. Preharvest factors influencing flavor of fresh fruit and vegetables. *Postharvest Biology and Technology* 15: 227–232.

NHB Database. 2017. *National Horticulture Board, Department of Horticulture and Cooperation, Government of India*. New Delhi: Aristo Printing Press.

Ni, H., H. Yang, R. Chen et al. 2012. New Botryosphaeriaceae fruit rot of mango in Taiwan: Identification and pathogenicity. *Botanical Studies* 53: 467–478.

Ploetz, R. C. 2009. *Management of the Most Important Pre- and Postharvest Disease*. Homestead, FL: University of Florida, TREC, Homestead, Department of Plant Pathology.

Poll, L., A. Rindom, T. B. Toldam-Anderson et al. 1996. Availability of assimilates and formation of aroma compounds in apples as affected by the fruit: Leaf ratio. *Physiologia Plantarum* 97: 223–227.

Popenoe, J. and W. G. Long. 1957. Evaluation of starch contents and specific gravity as measure of maturity of Florida mangoes. *Proceedings of the Florida State Horticultural Society* 70: 272–274.

Prakash, O. 2004. Diseases and disorders of mango and their management. In *Diseases of Fruits and Vegetables Volume I: Diagnosis and Management*. S. A. M. H. Naqvi (ed.), pp. 511–619. Dordrecht, Netherlands: Springer.

Prakash, O., A. K. Misra and R. Kishun. 1996. Some threatening diseases of mango and their management. In *Management of Threatening Plant Diseases of National Importance*. V. P. Agnihotri, A. K. Sarbhoy, D. V. Singh (eds.). New Delhi: Malhotra Publishing House, pp. 179–205.

Prakash, O., A. K. Misra and P. K. Shukla. 2011. Post harvest diseases of mango and their management. In *Proceeding Global Conference Augmenting Production and Utilization of Mango: Biotic and Abiotic Stresses Organized by CISH at Lucknow during 21–24 June* 2011, pp. 137–144. Lucknow: SDHS, CISH Productions.

Rivera-Vargas, L. I., Y. Lugo-Noel, R. J. McGovern et al. 2006. Occurrence and distribution of *Colletotrichum* spp. on mango (*Mangifera indica* L.) in Puerto Rico and Florida, USA. *Journal of Plant Pathology* 5: 191–198.

Sangeetha, C. G. and R. D. Rawal. 2008. Nutritional Studies of *Colletotrichum gloeosporioides* (Penz.) Penz. and Sacc. the incitant of mango anthracnose. *World Journal of Agricultural Sciences* 4(6): 717–720.

Schaffer, B., A. W. Whiley and J. H. Crane. 1994. Mango. In *Handbook of Environmental Physiology of Fruit Crops, Vol. II: Subtropical and Tropical Crops*. B. Schaffer and P. C. Andersen (eds.). Boca Raton, FL: CRC Press, pp. 165–197.

Senghor, A. L., W. J. Liang and W. C. Ho. 2007. Integrated control of *Colletotrichum gloeosporioides* on mango fruit in Taiwan by the combination of *Bacillus subtilis* and fruit bagging. *Biocontrol Science and Technology* 17(8): 865–870.

Shukla, P. K., A. K. Bhattacharjee and A. Dixit. 2018a. Efficacy of difenoconazole against shoulder browning disease of mango (*Mangifera indica* L.) and its residue analysis for safety evaluation in mango fruit. *Indian Phytopathology* 71(1): 147–151.

Shukla, P. K., T. Adak and Gundappa. 2017. Anthracnose disease dynamics of mango orchards in relation to humid thermal index under subtropical climatic condition. *Journal of Agrometeorology* 19(1): 56–61.

Shukla, P. K., T. Adak, A. K. Misra et al. 2016. Appraisal of shoulder browning disease of mango (*Mangifera indica* L.) in subtropical region of India. *Journal of Mycology and Plant Pathology* 46(1): 38–46.

Shukla, P. K., S. Varma, T. Fatima et al. 2018b. First report on wilt disease of mango caused by *Ceratocystis fimbriata* in Uttar Pradesh. *Indian Phytopathology* 71(1): 135–142.

Singh, D. and A. K. Thakur. 2003. Effect of pre harvest sprays of fungicides and calcium nitrate on postharvest rot of kinnow in low temperature storage. *Plant Disease Research* 18: 9–11.

Singh, L. B. 1960. *The Mango*. London, UK: Leonard Hill.

Srivastava, K. K. 2007. *Canopy Management of Fruit Crops*. Lucknow: International Book Distributing Co., pp. 1–25.

Swart, G. 2010. *Epidemiology and Control of Important Postharvest Diseases in Mangoes in South Africa*. Midrand, South Africa: Southern African Society for Plant Pathology.

Uddin, M. N., S. H. T. Shefat, M. Afroz et al. 2018. Management of anthracnose disease of mango caused by *Colletotrichum gloeosporioides*: A review. *Acta Scientific Agriculture* 2(10): 169–177.

Vivekananthana, R., M. Ravia, D. Saravanakumara et al. 2004. Microbially induced defence related proteins against postharvest anthracnose infection in mango. *Crop Protection* 23: 1061–1067.
Yenjit, P., W. Intanoo, C. Chamswarng et al. 2004. Use of promising bacterial strains for controlling anthracnose on leaf and fruit of mango caused by *Colletotrichum gloeosporioides. Walailak Journal of Science and Technology*4 (2): 56–69.
Young, T. W., R. C. Koo and J. T. Miner. 1965. Fertilizer trials with Kent mangoes. *Proceedings of the Florida State Horticultural Society* 78: 369–375.

6 Ecofriendly Combating Natural Essential Oils against Postharvest Phytopathogenic Fungi

Holistic Overview

Abhishek Tripathi and Afroz Alam

CONTENTS

6.1 INTRODUCTION

For a long time, marketable antimicrobial operators have been connected as an approach to oversee sustenance decay or pollution. These days, user worries toward engineered additives have brought about an expanding consideration on different common antimicrobials, for example, essential oils. Sweet-smelling and remedial plant-based essential oils and their derivatives show antimicrobial potential against an extensive variety of microbial contaminants (Basim et al. 2000; Iacobellis et al. 2004; Tripathi and Kumar 2007; Sharma and Tripathi 2008a; Pandey et al. 2014; Sonker et al. 2014; Gormez et al. 2016; Alam et al. 2017). Usually, essential oils are hydrophobic aromatic fluids with a wobbly and oily nature. These are present in various plant parts like flowers, leaves, shoots, bark, seeds and roots. These essential oils are very helpful as a flavoring agent, aroma enhancers, beauty care products, sustenance increasing substances, cleansers, plastics gums and scents. Furthermore, inquisitiveness about their application regarding their characteristic utility against microbial agents is mounting owing to the extensive range of activities, innate origins and the largely accepted "generally recognized as safe" (GRAS) grade of these essential oils. Nowadays, essential oils are repeatedly explored and utilized for their antimicrobial potential (Burt 2004; Nedorostova et al. 2009; Tripathi et al. 2013c), antioxidative character (Mimica-Dukic et al. 2003), antifungal agency (Sharma and Tripathi 2008a), antiulcer compounds (Dordevic et al. 2007), anti-inflammatory nature, as repellent, insecticidal and antifeedant (Pandey et al. 2014a), cytotoxic agency (Sylvestre et al. 2007), anthelmintic nature (Inouye et al. 2001), antiviral potential (Maurya et al. 2005), immunomodulatory, antinociceptive, larvicidal, ovicidal, anesthetic and molluscicidal properties, as well as for their use as food preservatives (Mediratta et al. 2002; Jantan et al. 2003; Abdollahi et al. 2010; Ghelardini et al. 2001; Fico et al. 2004; Pandey et al. 2013; Ukeh and Mordue 2009;

Pandey et al. 2017). Essential oils obtained from plants having aromatic and medicinal characteristics are considered efficient in opposition to various agents distressing stored food products, for instance, human pathogenic fungi, insects and bacteria. Essential oils derived from *Clausena pentaphylla*, *Chenopodium ambrosioides*, *Ocimum sanctum*, *Mentha arvensis* and *Tarchonanthus* are sensitive to contact and act as fumigant compounds (Pandey et al. 2017; Matasyoh et al. 2007).

6.2 ESSENTIAL OILS AND FUNCTIONS OF THEIR ACTIVE CONSTITUENTS

The preponderance of aromatic and medicinal plants holds a volatile and unstable odoriferous assortment of compounds that can be extracted as essential oils (Figure 6.1 and 6.2). Usually, these plants generate an extensive array of characteristic secondary metabolites such as alcoholic compounds (menthol, geraniol, linalool), terpenoids, aldehydes (citral, benzaldehyde, cinnamaldehyde, carvonecamphor), acidic compounds (myristic, benzoic, cinnamic acids), ketonic bodies (eugenol, thymol) and phenols (anethole, ascaridole). Among these compounds, terpenoids (e.g., oxygen-containing hydrocarbons), terpenes (limonene, pinene, myrcene, *p*-cymene, terpinene) and aromatic

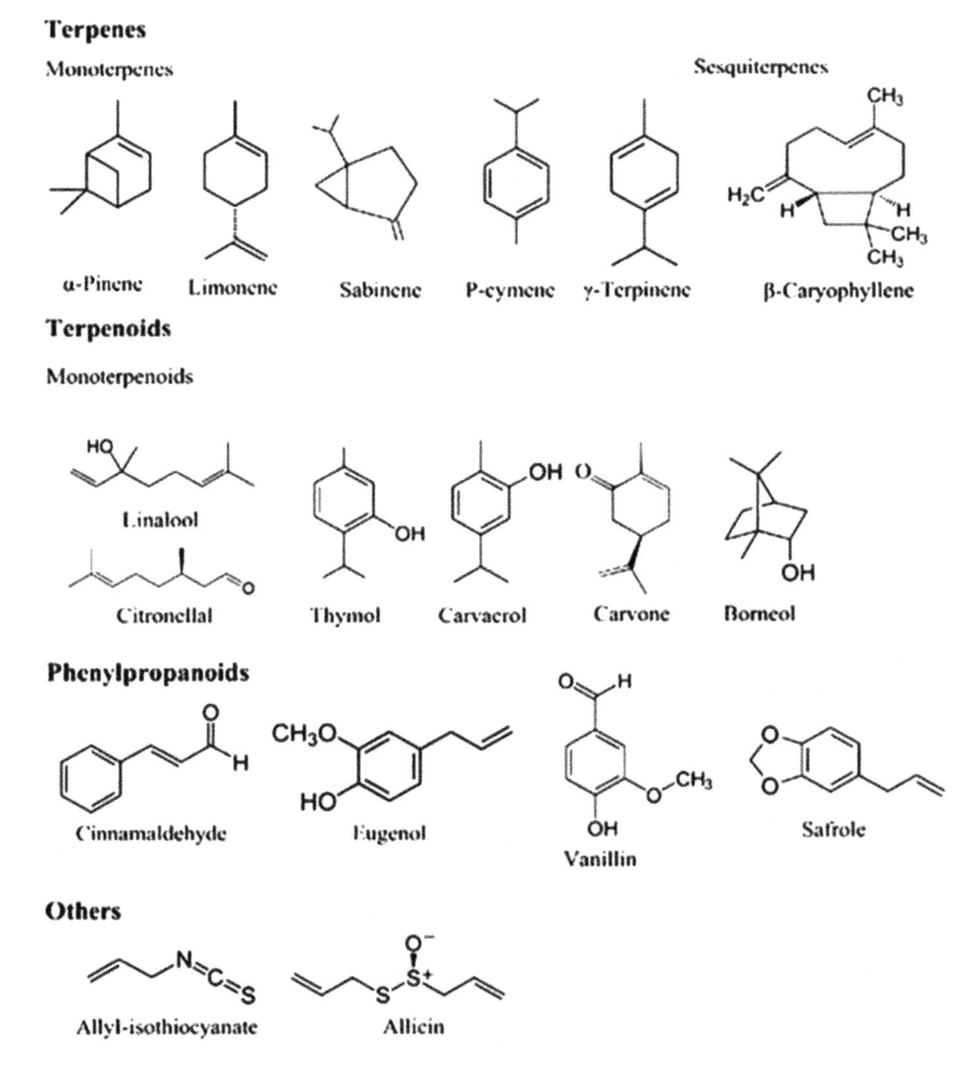

FIGURE 6.1 Chemical structures of essential oil constituents.

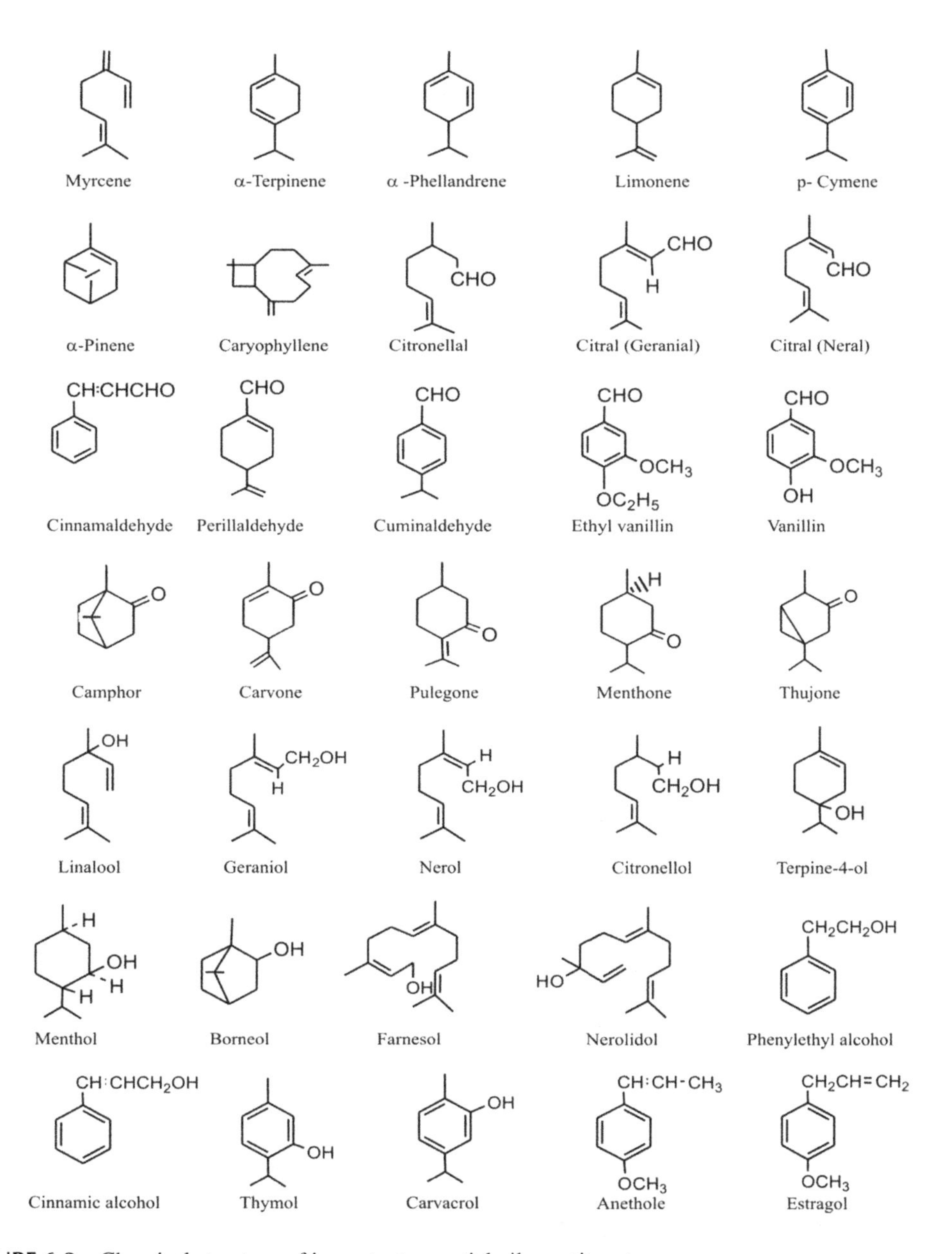

FIGURE 6.2 Chemical structure of important essential oil constituents.

phenols (e.g., carvacrol, thymol, safrole, eugenol) are known to have foremost roles in the composition of a range of essential oils (Koul et al. 2008; Figure 6.3). Offshoots of aromatic polyterpenoids and terpenoids are synthesized by shikimic acid and mevalonic acid pathways, respectively (Bedi et al. 2008). Terpenoids are among a massive collection of secondary metabolites produced by plants with medicinal and aromatic properties and encompass an imperative task in conferring opposition to contaminants. Monoterpenoids have an antimicrobial character which results in an unsettling increase and spreading out of microorganisms as well as an obstruction in physiological and biochemical procedures of bacteria and fungi (Marei et al. 2012; Burt 2004). A number of botanical ingredients such as carvone, azadirachtin, ascaridole, methyl eugenol, menthol, limonene, volkensin and toosendanin have been reported to operate in opposition to numerous microbial pathogens and pests (Isman 2006; Sharma and Tripathi 2008a; Pandey et al. 2017). Furthermore,

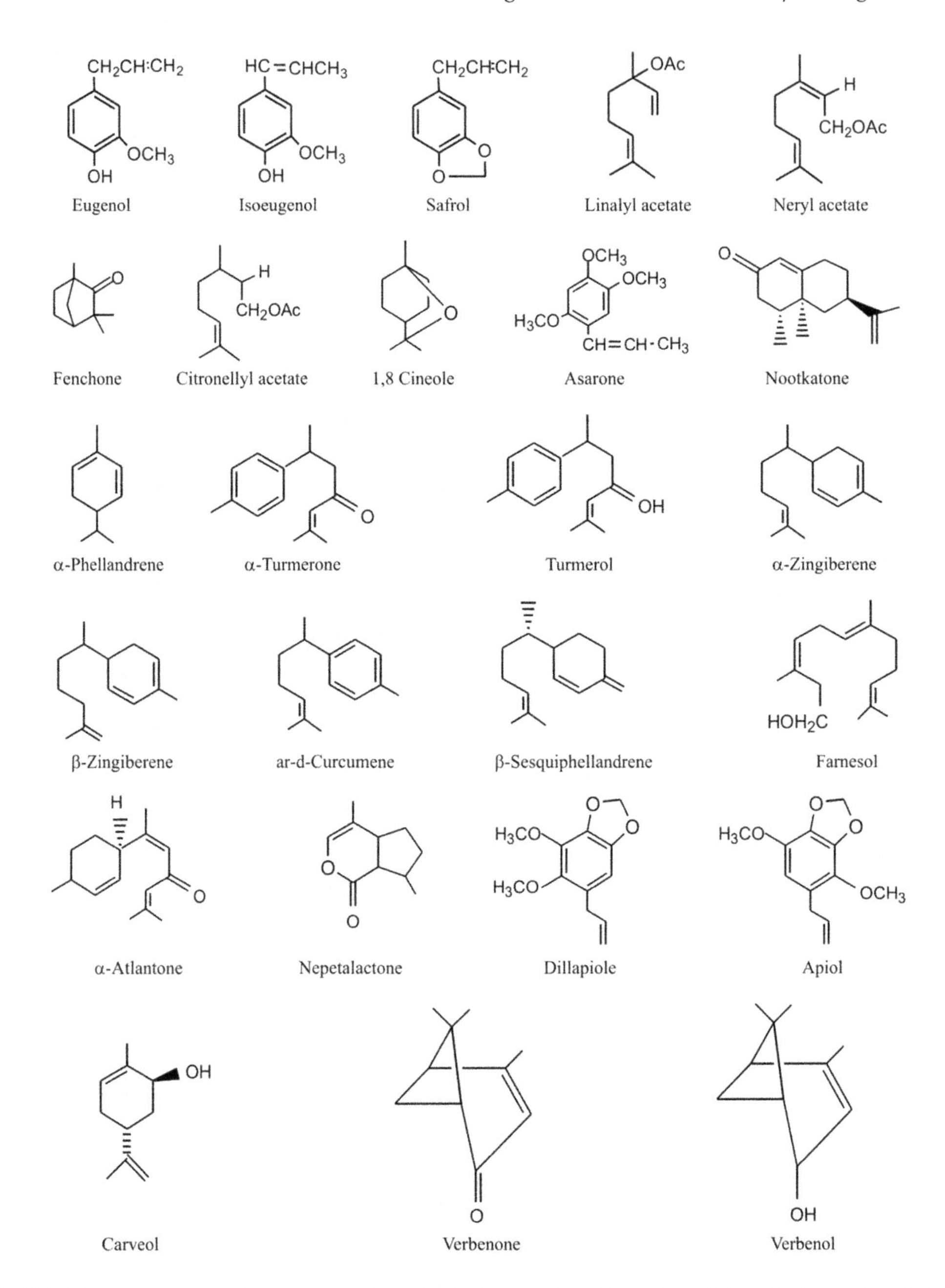

FIGURE 6.3 Active secondary compounds that exist in essential oils. (Adapted from Hyldgaard et al. 2012.)

many of them possess an antimicrobial and insecticidal nature and can be accountable for superior taste or noxious characteristics. Fungi, namely *Neurospora sitophila*, *Aspergillus flavus* and *Penicillium digitatum* have been reported to be completely restricted by essential oil obtained from *Cymbopogon citratus* (Shukla 2009; Sonker et al. 2015). Essential oils obtained from *Cymbopogon citratus*, *Nigella sativa* and *Pulicaria undulata* restrain the expansion of *Bacillus subtilis*, *Pseudomonas aeruginosa*, *Staphylococcus aureus* and *Escherichia coli*. Essential oil from *Citrus sinensis* showed vigorous activity against *Aspergillus niger* (Sharma and Tripathi 2008a). Likewise, essential oils obtained from *Artemisia*, *Acorus*, *Chenopodium*, *Curcuma*, *Clausena*, *Cymbopogon*, *Foeniculum*, *Eupatorium*, *Lippia*, *Hyptis*, *Syzygium*, *Ocimum*, *Putranjiva*, *Cinnamon* and *Vitex* spp. are identified for their well-defined antimicrobial characteristics (Pandey et al. 2017). The antibacterial nature of essential oils and their derivatives in opposition to foodborne microbes and their safety toward food (Burt 2004) could prove to be a substitute to the usual fungicides and bactericides (Perricone et al. 2015).

This chapter will deal with diverse aspects concerning plant-derived essential oils and their utility in fungal pathogenicity in controlling postharvest phytopathogenic fungi, especially during the postharvest storage of fruits.

Many studies have been completed regarding essential oil mitigation abilities against postharvest phytopathogens (Sharma and Tripathi 2006; Alam et al. 2017). Consequently, many essential oils have been validated as inhibitory agents to restrain the growth of postharvest fungal strains in the *in vitro* state (Sharma and Tripathi 2006).

The flora of this planet represents a massive pool of unique fungitoxic compounds that could serve as valuable ecofriendly alternatives to conventional and hazardous synthetic fungicides. These plant-derived essential oils have shown remarkable action against an extensive range of various phytopathogenic fungi (Sharma and Tripathi 2006). The added benefit of essential oils is the bioactive nature of their volatile phase, an attribute that accords them the potential as fumigants for postharvest protection against various microbial contaminants. Additionally, numerous unique constituents of essential oils display a sturdy antifungal nature (Sharma and Tripathi 2006; Alam et al. 2017). In plant systems, these compounds possibly play a defensive function against potent phytopathogenic microorganisms. Currently, several diverse constituents of essential oils are utilized as flavoring agents, food preservatives and in perfumery industries (Alam et al. 2017). Owing to their amazing effectiveness as antimicrobial agents, these plant-derived compounds are progressively employed at a large scale beyond their innate source in controlling various fungal diseases (Sharma and Tripathi 2006).

Studies in the past have found that essential oils from steam distillation of many diverse tropical and temperate plants possess high potential as fungicides even at lower concentrations that can control, combat or kill fungi at the same effectiveness as synthetic antimicrobial compounds (Sharma and Tripathi 2006); moreover, they are also effective as surface disinfectants (Tripathi et al. 2013c).

6.3 POTENCY OF ESSENTIAL OILS AGAINST STORAGE FUNGI

Fungal pathogens are considered as the most lethal destroyers of various food supplies, *viz*., vegetables, pulses, cereals and fruits, through characteristic release of mycotoxins, which makes food products unfit for human consumption and detrimental to human health since these strains harm the dietetic value of foodstuff (Paranagama et al. 2003; Pandey et al. 2017). Spoilage of stored food commodities is unceasing trouble in various climatic conditions during the storage process. The Food and Agriculture Organization has already stated that foodborne pathogenic fungal pathogens and their noxious metabolites have the potential to create a quantitative and qualitative deterioration of up to 35% of overall agricultural and horticultural food commodities throughout the globe (Alam et al. 2017). An alarming decline of food quality, color and consistency, as well as a reduction in the nutrient content along with worsening in the physiological and biochemical characteristics of various food products, is caused due to contamination of fungal pathogens in food products (Dhingra et al. 2001). Fungi release mycotoxins when they enter the host and start their lifecycle, which is the cause of devastating famines, especially in developing nations (Alam et al. 2017). In context to molds, food contagion by *Aspergillus, Alternaria, Rhizopus, Penicillium* and *Fusarium* spp. is of immense implication owing to the linked health vulnerabilities and food-driven infections (Pandey et al. 2014). For this reason, during storage and shipment, application of essential oils could be a lucrative move to avoid fungal growth in order to fight, combat and kill fungal pathogens and check the huge food losses during storage.

In the recent past, throughout the world, the antifungal properties of essential oils have been accepted as an important feature (Lopez-Reyes et al. 2013). The antifungal effect of these essential oils leads to the breakdown of fungal hyphae due to the presence of mono- and sesquiterpene compounds in these essential oils. Furthermore, compounds of essential oils can liquefy in cell membranes and cause membrane puffiness, thereby changing the function of the membrane (Dorman and Deans 2000).

In addition, the lipophilic assets of essential oils are accountable for their antifungal action, as this provides them the capacity to pierce the cell walls and affect the functioning of enzymes concerned with cell wall synthesis, therefore, changing the morphological characteristics of the attacking fungi (Cox et al. 2000). Antifungal activity on growth and morphogenesis of *Aspergillus niger* by an application of essential oil from *Citrus sinensis* led to a loss of cytoplasm in fungal hyphae, the hyphal wall and its diameter becoming markedly thinner and distorted, cell wall disruption and ultimately death of the pathogen (Sharma and Tripathi 2008a).

The tabulated compilation reviewing the investigations into essential oils tested for antifungal action as a probable additive to food storage procedure is shown in Table 6.1.

6.4 EFFECTIVENESS OF ESSENTIAL OILS IN FOOD SAFEGUARDING

Research into the effectiveness of essential oils in the maintenance of food supplies in order to augment shelf life has been lucratively carried out in the recent past. Several investigators have used essential oils, either in untainted or formulation form, to improve the shelf life of food supplies in diverse storage containers such as those prepared with tin, cardboard, glass, natural fabrics or polyethylene, and shown experiential noteworthy enrichment of shelf life (Tripathi and Kumar 2007; Pandey et al. 2014). Previous studies have shown that some essential oil constituents such as citronella, citral, citronellol, farnesol, eugenol and nerol could guard chili fruits and seeds from fungal contamination for more than 6 months (Sharma and Tripathi 2008a). Essential oil derived from *Ageratum conyzoides* effectively controlled the decaying of mandarins by blue mold and increased the shelf life until the 30th day. Later on, Anthony et al. (2003) also examined essential oils of *Cymbopogon flexuosus, C. nardus* and *Ocimum basilicum* and observed that these could appreciably control anthracnose disease in bananas with an extended shelf life up to the 21st day. Essential oil obtained from *Cymbopogon flexuosus* (20 µL/mL) is capable of shielding fruit rotting in *Malus pumila* for up to 3 weeks (Shahi et al. 2003). Essential oils of *Putranjiva roxburghii* as a fumigant were found to be effective against *A. niger* and *A. flavus* infecting groundnuts during storage and improved the shelf life of groundnut by up to 6 months by protecting them against fungal infections (Tripathi and Kumar 2007). The application of *Cymbopogon pendulus* essential oil as a fumigant has increased groundnut shelf life by 6–12 months (Pandey et al. 2017), thus being more effectual than *Pinus roxburghii* essential oil. These disparities in effectiveness of essential oils may be connected to the extraction of oils from different species of plants, as well as to their phytochemical composition, dose level and type of storage container. *Thymus capitata* (0.1%) and *Citrus aurantifolia* (0.5%) oils were reported to reduce disease occurrence in papaya fruit while oil of cinnamon (0.3%) extended the shelf life of banana by up to 28 days by reducing fungal disease incidence (Maqbool et al. 2010). Enhancement of the shelf life of *Buchanania* is based on the seed dressing and fumigation with *Ocimum canum* oil (1 µL/mL) (Singh et al. 2011; Tripathi et al. 2013b). Likewise, *Chenopodium ambrosioides* and *Clausena pentaphylla* oils, when used as fumigants for natural fabric bags and glass containers, were effective in guarding pigeon pea seeds from infestations of *Aspergillus flavus, A. niger, A. ochraceus*, and *A. terreus* infection for more than 6 months. Powdery formulations of *Chenopodium pentaphylla* and *C. ambrosioides* oils were also found effective to protect harvested seeds of pigeon pea for up to 6 months (Pandey et al. 2017). Oil obtained from *Artemisia nilagirica* when used as a fumigant in cardboard was found effective in improving the shelf life of table grapes by up to 9 days (Sonker et al. 2015). Likewise, *Lippia alba* (bushy matgrass) oil, when used as an air dosage (spray) treatment in glass containers, inhibited fungal propagation and production of aflatoxin in *Vigna radiata* and improved its shelf life by more than 6 months (Pandey et al. 2016).

TABLE 6.1
Findings on Fungitoxicity of Essential Oils of Higher Plants against Phytopathogens

Study no.	Investigators	Plants studied	Test fungi	Observations
1.	Behura et al. (2000)	*Curcuma longa* (rhizome and leaves)	Five rice pathogens *viz., Rhizoctonia solani, Trichoconiella padwickii, Helminthosporium oryzae, Fusarium moniliforme* and *Curvularia lunata*	The oil showed antifungal activity
2.	Shukla et al. (2000)	*Citrus sinensis*	*Aspergillus flavus, Penicillium italicum* and *Alternaria alternata*	Oil fungitoxic at 0.5% conc.
3.	Pandey et al. (2000)	*Eucalyptus* species	*Fusarium solani, Rhizopus arrhizus* and *Alternaria alternata*	Maximum inhibitory for *F. solani*
4.	Deena and Thoppil (2000)	*Lantana camara*	Eight fungal pathogens	Possessed good antifungal activity
5.	Babu et al. (2000)	*Cymbopogon martinii*	*Alternaria solani*	Inhibition at 0.3% conc.
6.	Mishra et al. (2000)	*Eucalyptus rostrata*	*Fusarium moniliforme*	Effective at dose of 3000 ppm
7.	Mishra et al. (2000)	*Cinnamomum zeylanicum* (bark)	*Stachybotrys chartarum*	Oil showed fungitoxicity at 400 ppm.
8.	Daferera et al. (2000)	Some Greek aromatic plants	*Penicillium digitatum*	Good antifungal activity was found
9.	Mishra and Tripathi (2001)	*Caesulia axillaris* (leaves)	*Aspergillus flavus* and *A. niger*	The MIC* of the oil against *A. flavus* and *A. niger* was found to be 1000 ppm.
10.	Liu et al. (2001)	*Artemisia princeps* (leaves) and *Cinnamomum camphora* (seeds)	*Gaeumannomyces graminis* var. *tritici*	Mycelial growth of *Gaeumannomyces graminis* var. *tritici* was inhibited 100% at 50 gml^{-1}
11.	Singh and Pant (2001)	*Pogostemon plectranthoides*	Seven phytopathogens	Showed activity
12.	Srivastava and Singh (2001)	*Murraya koenigii*	*Rhizoctonia bataticola, Helminthosporium oryzae* and *Rhizoctonia solani*	Very good activity against *R. bataticola* and *H. oryzae*
13.	Singatwadia and Katewa (2001)	*Cymbopogon martini* and *Cymbopogon citratus*	Seven fungal pathogens	Showed good activity
14.	Kumar and Tripathi (2002)	*Cuminum cyminum* (seeds)	*Aspergillus flavus* and *Aspergillus niger*	The MIC of the oil was 100 ppm and a exhibited broad range of activity
15.	de Billerbeck et al. (2001)	*Cymbopogon nardus*	*Aspergillus niger*	Mycelium growth completely inhibited at 800 mg/l. Ultrastructural modifications were also recorded

(Continued)

TABLE 6.1 (CONTINUED)
Findings on Fungitoxicity of Essential Oils of Higher Plants against Phytopathogens

Study no.	Investigators	Plants studied	Test fungi	Observations
16.	Sagar et al. (2002)	*Tagetes patula*	*Aspergillus niger, Penicillium funiculosum, Fusarium solani, Rhizomucor* sp. and *Trichoderma viridae*	Oil conc. of 1600 and 3200 ppm fully inhibited the growth of fungi
17.	Dubey et al. (2002)	*Ocimum gratissimum, Zingiber cassumunar* and *Cymbopogon citratus*	*Aspergillus flavus*	Oil showed antifungal activity and M1C ranging from 500 to 1300 ppm
18.	Baranowska et al. (2002)	*Pinus* species	*Fusarium culmorum, Fusarium poae* and *Fusarium solani*	*Pinus ponderosa* showed 100% inhibition against *F. culmorum* and *F. solani* at 2% conc.
19.	El-Shazly et al. (2002)	*Senecio aegyptius* var. *discoideus*	*Candida albicans* and *Aspergillus flavus*	Showed significant level of antifungal activity
20.	Ramezani et al. (2002)	*Eucalyptus citriodora*	*Rhizoctonia solani* and *Helminthosporium oryzae.*	Complete inhibition of *R. solani* and *H. oryzae* was observed at 10 and 20 ppm
21.	Beg and Ahmad (2002)	*Syzygium aromaticum*	*Alternaria alternata, Fusarium chlamydosporum, Helminthosporium oryzae* and *Rhizoctonia solani*	Highly sensitive at conc. 100 μl/well
22.	Siddiqui and Tripathi (2002)	*Seseli indicum*	*Penicillium italicum* and 14 other fungi	The MIC of the oil was 1000 ppm
23.	Garg and Jain (2003)	*Curcuma longa* (rhizome)	*Aspergillus niger, Aspergillus fumigatus, Curvularia lunata, Fusarium psidi* and *Rhizopus oryzae*	The oil exhibited good to moderate activity against *C. lunata, A. niger* and *A. fumigatus* at a dilution of 1:1000
24.	Mishra et al. (2003)	*Acorus calamus* and *Hedychium spicatum* (rhizome and fruits)	*Helminthosporium oryzae* and *Fusarium moniliforme*	The oil completely inhibited the growth of both test fungi at lowest doses 0.5 × 103 ml/l and 1.0 × zl/l
25.	Parcha et al. (2003)	*Dysoxylum malabaricum* (leaves)	*Aspergillus flavus, Aspergillus niger, Penicillium* sp. and *Candida albicans*	The oil showed MIC at a dilution of 1/600, 1/700, 1/200 and l/1000, respectively
26.	Bouchra et al. (2003)	Seven Moroccan Labiatae plants	*Botrytis cinerea*	*Origanum compactum* and *Thymus glandulosus* greatly inhibited the growth
27.	Duru et al. (2003)	Three *Pistacia* species	*Pythium ultimum, Rhizoctonia solani* and *Fusarium sambucinum*	Significant inhibition was found against *Rhizoctonia solani* and *Fusarium sambucinum.*
28.	Paranagama et al. (2003)	*Cymbopogon citratus*	*Aspergillus flavus*	Oil was fungicidal at 1.0 mg/ml concn.
29.	Singh et al. (2003)	*Curcuma longa* (rhizome)	Ten storage fungi	The oil showed 100% toxicity at 3000 ppm.

(Continued)

TABLE 6.1 (CONTINUED)
Findings on Fungitoxicity of Essential Oils of Higher Plants against Phytopathogens

Study no.	Investigators	Plants studied	Test fungi	Observations
30.	Daferera et al. (2003)	Oregano, thyme, *Dictamnus*, lavender, rosemary, sage and pennyroyal oil	*Botrytis cinerea, Fusarium* sp. and *Clavibacter michiganensis* subsp. *michiganensis*	All were inhibited with concentration range of 85–300 µg/ml
31.	Benkeblia (2004)	*Allium cepa* and *Allium sativum*	*Aspergillus niger, Penicillium cyclopium* and *Fusarium oxysporum*	*F. oxysporum* showed the lowest sensitivity toward EO** extracts, whereas *A. niger* and *P. cyclopium* were significantly inhibited, particularly at low concentrations (50 and 100 ml/l)
32.	Nguefack et al. (2004)	*Cymbopogon citratus, Monodora myristica, Ocimum gratissimum, Thymus vulgaris* and *Zingiber officinalis*	*Fusarium moniliforme, Aspergillus flavus* and *Aspergillus fumigatus*	Showed activity between 800 ppm and 2500 ppm
33.	Sokmen et al. (2004)	*Thymus spathulifolius*	19 test fungi	Strongly inhibited the growth except four fungal species
34.	Palhano et al. (2004)	Lemongrass oil	*Colletotrichum gloeosporioides*	Inhibited growth with 0.75 mg/ml conc.
35.	Tripathi et al. (2004)	*Mentha arvensis, Ocimum canum* and *Zingiber officinale*	*Penicillium italicum*	Absolute fungitoxicity against test fungi
36.	Rasooli et al. (2005)	*Thymus eriocalyx* and *Thymus x-porlock*	*Aspergillus niger*	MFC were 250 and 500 ppm, respectively of *Thymus*. Oil also showed irreversible damage to cell wall, cell membrane and cellular organelles
37.	Dob et al. (2006)	*Thymus algeriensis*	*Fusarium oxysporum* f.sp. *albedinis* and *Mucor ramaniamus*	MIC was 1.0 µl/ml
38.	Sharma and Tripathi (2006)	*Citrus sinensis*	10 postharvest pathogens	Fungicidal in the 700 to 1000 ppm range. In the case of *Aspergillus niger*, treatment of oil leads to the distortion and thinning of hyphal wall and the reduction in hyphal diameter and absence of conidiophores
39.	Ozcan et al. (2006)	*Foeniculum vulgare* ssp. *piperitum*	*Alternaria alternata, Fusarium oxysporum*, and *Rhizoctonia solani*	The oils exerted varying levels of antifungal effects on *Alternaria alternata, Fusarium oxysporum* and *Rhizoctonia solani*

(*Continued*)

TABLE 6.1 (CONTINUED)
Findings on Fungitoxicity of Essential Oils of Higher Plants against Phytopathogens

Study no.	Investigators	Plants studied	Test fungi	Observations
40.	Dubey et al. (2007)	*Eupatorium cannabinum*	*Botryodiploidia theobromae* and *Colletotrichum gloeosporioides*	MIC was 1000 μg/ml against both pathogens
41.	Babu et al. (2007)	*Curcuma longa*	*Fusarium oxysporum f. sp. dianthi*, *Alternaria dianthi*, *Fusarium oxysporum* f. sp. *gladioli* and *Curvularia trifolii* f. sp. *gladioli*	Oil showed better antifungal activity over other oil samples at each of the tested concentrations
42.	Kumar et al. (2007)	*Chenopodium ambrosioides*	*Aspergillus flavus*, *Aspergillus niger*, *Aspergillus fumigatus*, *Botryodiplodia theobromae*, *Fusarium oxysporum*, *Sclerotium rolfsii*, *Macrophomina phaseolina*, *Cladosporium cladosporioides*, *Helminthosporium oryzae* and *Pythium debaryanum*	Oil exhibited broad fungitoxic spectrum at 100 μg/ml
43.	Kishore et al. (2007)	Clove oil, cinnamon oil and five essential oil components (citral, eugenol, geraniol, limonene and linalool)	14 phytopathogenic fungi	Citral completely inhibited the growth of *Alternaria alternata*, *Aspergillus flavus*, *Curvularia lunata*, *Fusarium moniliforme*, *Fusarium pallidoroseum* and *Phoma sorghina* in paper disc agar diffusion assays
44.	Szczerbanik et al. (2007)	Spearmint, tea tree, pine and cinnamon oils	Postharvest pathogens	Spearmint and tea tree oils controlled the growth of *Botrytis cinerea*, *Fusarium solani*, *Colletotrichum* sp., *Geotrichum candidum*, *Rhizopus oryzae*, *Aspergillus niger* and *Cladosporium cladosporioides* more effectively than pine or cinnamon oil
45.	Hadian et al. (2007)	*Artemisia khorasanica*	*Tiarosporella phaseolina*, *Fusarium moniliforme*, *Fusarium solani* and *Rhizoctonia solani*	The oil was effective and showed fungistatic activity against *Tiarosporella phaseolina*, *Fusarium moniliforme* and *Fusarium solani*, whereas against *Rhizoctonia solani*, it exhibited high fungicidal activity
46.	Tripathi et al. (2008)	*Hyptis suaveolens*	Wide range of postharvest fungal strains	Essential oil exhibited remarkable antifungal activity (*in vivo* and *in vitro*)
47.	Sharma and Tripathi (2008a)	*Citrus sinensis*	*Aspergillus niger*	Essential oil exhibited death and decay of fungal hyphae
48.	Sharma and Tripathi (2008b)	*Hyptis suaveolens*	*Fusarium oxysporum* f.sp. *gladioli*	Essential oil, UV-C and hot water treatment showed protection of corm rot of gladiolus
49.	Tripathi et al. (2009)	*Hyptis suaveolens*	*Fusarium oxysporum* f.sp. *gladioli*	Essential oil exhibited remarkable antifungal activity

(Continued)

TABLE 6.1 (CONTINUED)
Findings on Fungitoxicity of Essential Oils of Higher Plants against Phytopathogens

Study no.	Investigators	Plants studied	Test fungi	Observations
50.	Sharif et al. (2010)	*Cestrum nocturnum* L.	Wide range of fungal strains	Essential oil exhibited remarkable antifungal activity (*in vivo* and *in vitro*)
51.	Tian et al. (2011)	*Cicuta virosa L. var. latisecta Celak*	Wide range of fungal strains	Essential oil exhibited remarkable antifungal activity (*in vivo* and *in vitro*)
52.	Mohammadi Aminifard (2012)	*Prunus persica*	*Botrytis cinerea*	Essential oil of this plant found to be very effective in inhibiting the expansion of fungus
53.	Lopez-Reyes et al. (2013)	*Ocimum basilicum* and *Rosmarinus officinalis*	*Monilinia laxa* and *Botrytis cinerea*	The obtained oil was found bioactive as antimicrobial agent *in vivo*
54.	Tripathi et al. (2013a)	*Hyptis suaveolens*	Broad range of fungal strains	The obtained oil was found bioactive as antimicrobial agent *in vivo*
55.	Mancini et al. (2014)	*Origanum vulgare ssp. hirtum*	*Fusarium solani*, *Botrytis cinerea* and *Rhizoctonia solani*	Essential oil of this plant found to be very effective in inhibiting the expansion of fungal strains
56.	Elshafie et al. (2015a, b)	*Origanum vulgare L.*	Peach specific fungal strains like *Botrytis cinerea*	Essential oil of this plant found to be very effective in inhibiting the expansion of fungal strains
57.	Samah et al. (2016)	*Artemisia herba-alba*	Broad spectrum	The obtained oil was found bioactive as an antimicrobial agent
58.	Gakuubi et al. (2016)	*Tagetes minuta* L.	*Botrytis cinerea*	The obtained oil was found effective against fungal spread during storage
59.	Nazzaro et al. (2017)	*Commiphora myrrha* and *Hedychium spicatum*	Wide range of postharvest fungal strains like *Fusarium solani*, *Botrytis cinerea* and *Rhizoctonia solani*	The obtained essential oils have shown significant antifungal activity
60.	Jagana et al. (2018)	*Syzygium aromaticum* (L.) Merr. et Perry	*Colletotrichum musae*	Complete inhibition of fungal growth was reported with clove oil at all the concentrations tested (0.5, 1.0 and 2.0%)
61.	yournews.com/2019	Peppermint (*Mentha × piperita*)	*Fusarium sporotrichioides*	Provide protection against *Fusarium* infection in important cereal crops, including barley, maize, oats, rice and wheat

*MIC =
**EO = essential oil

6.5 CONCLUDING REMARKS

Synthetic fungicides are known as hazardous to the environment. Therefore, in the near future, these chemicals must be strictly regulated by governments. This may lead to a growing demand for plant origin products as biological plant protection materials.

Approved data and scientific publications in the past have revealed that plant essential oils/extracts are biodegradable. These natural products do not cause any environmental risks like that of synthetic chemical fungicides and pesticides.

Overall examinations related to various essential oils have motivated analysts to center their enthusiasm toward the investigation of plant antimicrobials. It is evident that the utilization of fundamental oils and their subsidiaries has been broadly portrayed and essential oils have been utilized against a wide range of pathogens.

The present review gives a concise outline of various essential oils, their dynamic constituents and their potential against various storage fungi and other microorganisms. This indicates that essential oils explore various possibilities of antimicrobial properties and shows their normal maintainability when utilized as potential biocontrol operators against fungal pathogens. Hence, it is evident that these natural essential oils possess the potential of biocontrol activity that ought to be investigated further due to their capability to secure sustenance. Likewise, due to the fumigating nature of essential oils, its antimicrobial action ought to have a promising GRAS status in mammalian frameworks.

The United States Food and Drug Administration (USDA) declared a few essential oils and their constituents (e.g., carvone, carvacrol, cinnamaldehyde, thymol, linalool, citral, limonene, eugenol, limonene and menthol) to be GRAS and endorsed them as flavor or nourishment additive substances. Basic oil applications are advancing as a method for coordinating pathogens into sustenance compartments; for instance, fumigants that can be valuable in common texture and cardboard holders and even compartments made of wooden sheets. A few essential oils can be utilized as light showers and used as a fumigant itself. Various essential oils and their constituents are dynamic against microorganisms and restrict their development on the host, and they can be conveyed from normally accessible crude materials; by and large, they are fairly minimal effort medications.

In this review, it is understood that it is conceivable to create systems for sustenance product insurance without the utilization, or with decreased use, of commercial synthetic bactericides and fungicides. In spite of the fact that accessible data demonstrates that essential oils possess biodegradable properties and have a constrained impact on non-target living beings with low dimensions of mammalian harmfulness, they still need to be brought to the mainstream. There, support and business have some downsides, for instance, cost viability. Regardless, there are multitudinous potential employments of fundamental oils and more research is expected to address the issues of the food industry moving toward the utilization of green innovation.

Regarding explicit imperatives, the viability of these materials misses the mark when contrasted with manufactured pesticides, in spite of the fact that there are explicit pathogen settings where control proportional to that with regular items has been observed. Essential oils additionally require, to some degree, more prominent application rates (as high as 1% dynamic fixing) and may require reapplications when utilized out-of-entryways.

There are extra difficulties in the commercial utilization of plant essential oil-based natural pesticides such as the accessibility of adequate amounts of plant material, institutionalization and refinement of pesticide items, security of innovation (licenses) and administrative endorsement (Isman 2006). Numerous essential oils might be rich and accessible all year because of their utilization in fragrance, food and refreshment enterprises, substantial scale commercial use of essential oil-based pesticides could require more prominent generation of specific oils. Also, as the concoction profile of plant species can normally shift by relying upon geographical, hereditary, climatic, yearly or occasional elements, pesticide makers must find a way to guarantee that their items will perform reliably. The majority of this requires significant expense and smaller organizations may not contribute the required finances except if there is a high likelihood of recouping their expenses through some type of market selectiveness. Finally, when these issues are tended to, administrative endorsement is required. Appropriately, administrative endorsement keeps on being an obstruction to commercialization and will probably continue being a hindrance until administrative frameworks are changed in accordance with ecofriendly approaches (Isman and Machial 2006).

Indeed, pesticides and fungicides derived from the essential oils of plants have a few vital advantages but, owing to their unpredictable character, there is also a faint chance of causing harm to the ambient environment. However, this will be always lower than synthetic pesticides. Populations of predators and naturally occurring parasitoids and pollinators will be less affected, in light of the insignificant remaining action, of the use of essential oil-based insecticides harmonized with the environment. It is additionally clear that obstruction will grow steadily to essential oil-derived pesticides after assumptions from the puzzling mix of ingredients in a lot of these oils.

Eventually, in developing nations, the source plants of significant essential oils are sometimes found to be endemic and the attained pesticides, due to their effects, have to be included in postharvest management. Normally, these pesticides will locate their most noteworthy commercial application in urban pest control, general wellbeing, veterinary wellbeing, vector control vis-à-vis human health and packing products. In agribusiness, these pesticides will be most helpful for insured crops, high-esteem push crops and inside natural nourishment generation frameworks where elective pesticides are accessible. Consequently, there is an open door for changing buyer inclination toward the following: utilization of "common" over engineered items; the presence of and development in specialty markets, where quality could easily compare with value; the solid development sought after for essential oils and plant separates; the potential to expand the scope of accessible items including new item advancement through biotechnology; and the creation of fundamental oils and other useful plant extracts from minimal effort creating nations.

REFERENCES

Abdollahi, A., A. Hassani, Y. Ghosta et al. 2010. Study on the potential use of essential oils for decay control and quality preservation of table grapes. *Journal of Plant Protection Research* 50:45–52.

Alam, A., A. Tripathi, V. Sharma et al. 2017. Essential oils: A novel consumer and eco-friendly approach to combat postharvest phytopathogens. *Journal of Advances in Biology and Biotechnology* 11(1):1–16.

Anthony, S., K. Abeywickrama and S. W. Wijeratnam. 2003. The effect of spraying essential oils *Cymbopog onnardus*, *C. flexuosus* and *Ocimum basilicum* on post-harvest diseases and storage life of Embul banana. *Journal of Horticultural Science and Biotechnology* 78:780–85.

Babu, G. D. K., V. Shanmugam, S. D. Ravindranath et al. 2007. Comparison of chemical composition and antifungal activity of *Curcuma longa L.* leaf oils produced by different water distillation techniques. *Flavour and Fragrance Journal* 22(3):191–96.

Babu, S., K. Seetharaman, R. Nanda Kumar et al. 2000. In vitro antifungal activity of *Palmarosa* (*Cymbopogon martini* (Roxb.) *Watson*) oil against *Alternaria solani. Indian Perfumer* 44(4):275–77.

Baranowska, M. K., M. Mardarowicz, L. P. Wiwart et al. 2002. Antifungal activity of the essential oil from some species of genus Pinus. *Zeitschrift für Naturforschung* 57c:478–82.

Basim, H., O. Yegen and W. Zeller. 2000. Antibacterial effect of essential oil of *Thymbra spicata L. var. spicata* on some plant pathogenic bacteria. *Zeitschrift furPflanzenkr Pflanzenschutz* 107:279–84.

Bedi, S. Tanuja, and S. P. Vyas. 2008. *A Hand Book of Aromatic and Essential Oil Plants Cultivation, Chemistry, Processing and Uses*. Jodhpur: AGROBIOS Publishers.

Beg, A. Z. and I. Ahmad. 2002. In vitro fungitoxicity of the essential oil of *Syzygium aromaticum. World Journal of Microbiology and Biotechnology* 18:313–15.

Behura, C., P. Ray, C. C. Rath et al. 2000. *Journal of Essential Oil Bearing Plants* 3(2):79–84.

Benkeblia, N. 2004. Antimicrobial activity of essential oil extracts of various onions (*Allium cepa*) and garlic (*Allium sativum*). *Lebensmittel-Wissenschaft und-Technologie* 37:263–68.

Bouchra, C., M. Achouri, L. M. I. Hassani et al. 2003. Chemical composition and antifungal activity of essential oils of seven Moroccan Labiatae against *Botrytis cinerea* Pers: Fr. *Journal of Ethnophatmacology* 89:165–69.

Burt, S. 2004. Essential oils: their antibacterial properties and potential applications in food—A review. *International Journal of Food Microbiology* 94:223–53.

Cox, S., C. Mann, J. Markham, et al. 2000. The mode of antimicrobial action of the essential oil of *Melaleuca alternifolia* (tea tree oil). *Journal of Applied Microbiology* 88:170–75.

Daferera, D. J., B. N. Ziogas and M. G. Polissiou. 2000. GC-MS analysis of essential oils from some Greek aromatic plants and their fungitoxicity on *Penicillium digitatum*. *Journal of Agricultural and Food Chemistry* 48(6):2576–81.

Daferera, D. J., B. N. Ziogas and M. G. Polissiou. 2003. The effectiveness of plant essential oils on the growth of *Botrytis cinerea, Fusarium* sp. and *Clavibacter michiganensis* sub sp. michiganensis. *Crop Protection* 22:39–44.

de Billerbeck, V. G., C. G. Roques, J. M. Bessiere, et al. 2001. Effects of *Cymbopogon nardus* (L.) W. Watson essential oil on the growth and morphogenesis of *Aspergillus niger*. *Canadian Journal of Microbiology* 47:9–17.

Deena, M. J. and J.E. Thoppil. 2000. Antimicrobial activity of essential oil of *Lantana camara*. *Fitoterapia* 71:453–55.

Dhingra, O. D. and R. A. Netto Coelho. 2001. Reservoir and non reservoir hosts of bean wilt pathogen, *Fusarium oxysporum* f. sp. phaseoli. *Journal of Phytopathology* 149:463–67.

Divya Jagana, D., Y. R. Hegde and R. Lella. 2018. Bioefficacy of essential oils and plant oils for the management of Banana anthracnose—A major post-harvest disease. *International Journal of Current Microbiology and Applied Sciences* 7(4):2359–65.

Dob, T., D. Dahmane, T. Benabdelkader et al. 2006. Studies on the essential oil composition and antimicrobial activity of *Thymus algeriensis* Boiss. Et Reut. *The International Journal of Aromatherapy* 16:95–100.

Dordevic, S., S. Petrovic, S. Dobric et al. 2007. Antimicrobial, anti-inflammatory, anti-ulcer and antioxidant activities of *Carlina acanthifolia* root essential oil. *Journal of Ethnopharmacology* 109:458–63.

Dorman, H. and S. Deans. 2000. Antimicrobial agents from plants: Antibacterial activity of plant volatile oils. *Journal of Applied Microbiology* 88:308–16.

Dubey, A. K., N. Kumar and N. N. Tripathi. 2002. Caesulia axillaris—Volatile herbal fungitoxicants. *Journal of Basic and Applied Mycology* 1:68–73.

Dubey, R. K., R. Kumar, N. K. Dubey. 2007. Evaluation of *Eupatorium cannabinum* Linn. oil in enhancement of shelf life of mango fruits from fungal rotting. *World Journal of Microbiology and Biotechnology* 23:467–73.

Duru, M. E., A. Cakir, S. Kordali, et al. 2003. Chemical composition and antifungal properties of essential oils of three Pistacia species. *Fitoterapia* 74:170–76.

Elshafie, H. S., E. Mancini, I. Camele et al. 2015a. In vivo antifungal activity of two essential oils from Mediterranean plants against post-harvest brown rot disease of peach fruit. *Industrial Crop Production* 66:11–15.

Elshafie, H. S., E. Mancini, L. De Martino et al. 2015b. Antifungal activity of some constituents of *Origanum vulgare L.* essential oil against post-harvest disease of peach fruit. *Journal of Medicinal Food* 18(8):929–34.

El-Shazly, A., G. Doral and M. Wink. 2002. Chemical composition and biological activity of the essential oil of *Senecio aegyptius var. discoideus Boiss. Zeitschrift für Naturforschung* 57c:434–39.

Fico, G., L. Panizzi, G. Flamini et al. 2004. Biological screening of *Nigella elamascena* for antimicrobial and molluscicidal activities. *Phytotherapy Research* 18:468–70.

Gakuubi, M. M., W. Wanzala, J. M. Wagacha et al. 2016. Bioactive properties of *Tagetes minuta L.* (Asteraceae) essential oils: A review. *American Journal of Essential Oils and Natural Products* 4(2):27–36.

Garg, S. C. and R. K. Jain. 2003. Anethole, a potential insecticide from *Illicium verum* Hook. F., against two stored product insects. *Indian Perfumer* 47(2):199–202.

Ghelardini, C., N. Galeotti and G. Mazzanti. 2001. Local anesthetic activity of monoterpenes and phenylpropanes of essential oils. *Planta Medica* 67:564–66.

Gormez, A., S. Bozari, D. Yanmis et al. 2016. The use of essential oils of *Origanum rotundifolium* as antimicrobial agent against plant pathogenic bacteria. *Journal of Essential Oil-Bearing Plants* 19:656–63.

Hadian, J., T. Ramak-Masoumi, M. Farzaneh et al. 2007. Chemical compositions of essential oil of *Artemisia khorasanica* Podl. and its antifungal activity on soil-born phytopathogens. *Journal of Essential Oil-Bearing Plants* 10(1):53–9.

Hyldgaard, M., T. Mygind and R. L. Meyer. 2012. Essentials oils in food preservation: Mode of action, synergies and interactions with food matrix components. *Frontiers in Microbiology* 3:12.

Iacobellis, N. S., P. L. Cantore, A. D. Marco et al. 2004. Antibacterial activity of some essential oils. Management of plant diseases and arthropod pests by BCAsIOBC/wprs. *Bulletin* 27:223–26.

Inouye, S., T. Takizawa and H. Yamaguchi. 2001. Antibacterial activity of essential oils and their major constituents against respiratory tract pathogens by gaseous contact. *Journal of Antimicrobial Chemotherapy* 47:565–73.

Isman, M. B. 2006. Botanical insecticides, deterrents and repellents in modern agriculture and an increasingly regulated world. *Annual Review of Entomology* 51:45–66.
Isman, M. B. and C. M. Machial. 2006. Pesticides based on plant essential oils: from traditional practice to commercialization. In: M. Rai and M. C. Carpinella (eds.), *Naturally Occurring Bioactive Compounds*. Amsterdam, Netherlands: Elsevier B.V.
Jantan, I., W. O. Ping, S. D. Visuvalingam et al. 2003. Larvicidal activity of the essential oils and methanolic extracts of Malaysian plants on *Aedes aegypti*. *Pharmaceutical Biology* 41:234–36.
Kishore, G. K., S. Pande and S. Harish. 2007. Evaluation of essential oils and their components for broad-spectrum antifungal activity and control of late leaf spot and crown rot diseases in peanut. *Plant Disease* 91(4):375–79.
Koul, O., S. Walia and G. S. Dhaliwal. 2008. Essential oils as green pesticides: Potential and constraints. *Biopesticides International* 4:63–84.
Kumar, R., A. K. Mishra, N. K. Dubey et al. 2007. Evaluation of *Chenopodium ambrosioides* oil as a potential source of antifungal, antiaflatoxigenic and antioxidant activity. *International Journal of Food Microbiology* 115(2):159–64.
Kumar, N. and N. N. Tripathi. 2002. Fungal infestation in groundnut seeds during storage and their control by essential oil of *Putranjiva roxburghii* wall. *Journal of Indian Botanical Society* 81:127–32.
Liu, C. H., A. K. Mishra, B. He et al. 2001. Composition and antifungal activity of essential oils from *Artemisia princeps* and *Cinnamomum camphora. International Pest Control* 43(2):72–74.
Lopez-Reyes, J. G., D. Spadaro, A. Prelle et al. 2013. Efficacy of plant essential oils on postharvest control of rots caused by fungi on different stone fruits in vivo. *Journal of Food Protection* 76(4):631–39.
Mancini, E., I. Camele, H. S. Elshafie, et al. 2014. Chemical composition and biological activity of the essential oil of *Origanum vulgare* ssp. hirtum from different areas in the southern Apennines (Italy). *Chemistry and Biodiversity* 11(4):639–51.
Maqbool, M., A. Ali and P. G. Alderson. 2010. Effect of cinnamon oil on incidence of anthracnose disease and post-harvest quality of bananas during storage. *International Journal of Agriculture and Biology* 12:516–20.
Marei, G. K., M. A. Rasoul and S. A. Abdelgaleil. 2012. Comparative antifungal activities and biochemical effects of monoterpenes on plant pathogenic fungi. *Pesticide Biochemistry and Physiology* 103(1):56–61.
Matasyoh, J. C., J. J. Kiplimo, N. M. Karubiu et al. 2007. Chemical composition and antimicrobial activity of essential oil of *Tarchonanthus camphorates*. *Food Chemistry* 101:1183–87.
Maurya, S., P. Marimuthu, A. Singh et al. 2005. Antiviral activity of essential oils and acetone extracts of medicinal plants against papaya ring spot virus. *Journal of Essential Oil-Bearing Plants* 8:233–38.
Mediratta, P. K., K. K. Sharma and S. Singh. 2002. Evaluation of immunomodulatory potential of *Ocimum sanctum* seeds oil and its possible mechanism of action. *Journal of Ethnopharmacology* 80:15–20.
Mimica-Dukic, N., B. Bozin, M. Sokovic et al. 2003. Antimicrobial and antioxidant activities of three Mentha species essential oils. *Planta Medica* 69:413–19.
Mishra, D., P. K. Jaiswal, D. Mishra et al. 2000. Control of Fusarium rot of tomato by essential oils of *Eucalyptus rostrata* Schlecht. *Indian Perfumer* 44(2):79–82.
Mishra, D., C. O. Samuel and S. C. Tripathi. 2003. Evaluation of some essential oils against seed-borne pathogen of rice. *Indian Phytopathology* 56(2):212–13.
Mishra, D. and S. C. Tripathi. 2001. International Symposium on Frontiers of Fungal Diversity and Diseases in South East. Asia, Dept. of Botany, D.D.U. Gorakhpur University, Gorakhpur, Feb. 9–11, p. 96.
Mohammadi, M. and M. H. Aminifard. 2012. Effect of essential oils on postharvest decay and some quality factors of peach *(Prunus persica* var. Redhaven). *Journal of Biological and Environmental Science* 6(17):147–53.
Nazzaro, F., F. Fratianni, R. Coppola et al. 2017. Essential oils and antifungal activity. *Pharmaceuticals (Basel)* 10(4):86.
Nedorostova, L., P. Kloucek, L. Kokoska et al. 2009. Antimicrobial properties of selected essential oils in vapour phase against food borne bacteria. *Food Control* 20:157–60.
Nguefack, J., V. Leth, P. H. Amvam Zollo et al. 2004. Evaluation of five essential oils from aromatic plants of Cameroon for controlling food spoilage and mycotoxin producing fungi. *International Journal of Food Microbiology* 94:329–34.
Ozcan, M. M., J. C. Chalchat, D. Arslan et al. 2006. Comparative essential oil composition and antifungal effect of bitter fennel (*Foeniculum vulgare* ssp. piperitum) fruit oils obtained during different vegetation. *Journal of Medicinal Food* 9(4): 552–61.
Palhano, F. L., T. T. B. Vilches, R. B. Santos et al. 2004. Inactivation of *Colletotrichum gloeosporioides* spores by high hydrostatic pressure combined with citral or lemongrass essential oil. *International Journal of Food Microbiology* 95:61–6.

Pandey, A. K., P. Kumar, P. Singh et al. 2017. Essential oils: Sources of antimicrobials and food preservatives. *Frontiers in Microbiology* 7:1–14.
Pandey, A. K., U. T. Palni and N. N. Tripathi. 2014a. Repellent activity of some essential oils against two stored product beetles *Callosobruchus chinensis L.* and *C. maculates F.* (*Coleoptera:Bruchidae*) with reference to *Chenopodium ambrosioides L.* for the safety of pigeonpea seeds). *Journal of Food Science and Technology* 51:4066–71.
Pandey, A. K., M. K. Rai and S. Qureshi. 2000. In vitro antimycotic effect of volatile oils of Eucalyptus species. *Indian Perfumer* 44(2):75–7.
Pandey, A. K., P. Singh, U. T. Palni et al. 2013. Application of *Chenopodium ambrosioides* Linn. essential oil as botanical fungicide for the management of fungal deterioration in pulses. *Biological Agriculture and Horticulture* 29:197–208.
Pandey, A. K., P. Singh, N. N. Tripathi. 2014. Chemistry and bioactivities of essential oils of some Ocimum species: An overview. *Asian Pacific Journal of Tropical Biomedicine* 4(9):682–94.
Pandey, A. K., N. Sonker and P. Singh. 2016. Efficacy of some essential oils against *Aspergillus flavus* with special reference to Lippia alba oil an inhibitor of fungal proliferation and aflatoxin b1 production in green gram seeds during storage. *Journal of Food Science* 81: 928–34.
Paranagama, P. A., K. H. T. Abeysekera, K. Abeywickrama et al. 2003. Fungicidal and anti-aflatoxigenic effects of the essential oil of Cymbopogon citrates (DC.) Stapf. (lemongrass) against *Aspergillus flavus* Link. Isolated from stored rice. *Letters in Applied Microbiology* 36:1–5.
Parcha, V., M. Gahlot and B. K. Bajaj. 2003. Antimicrobial activity of essential oil of *Dysoxylum malabaricum. Indian Perfumer* 47(2):183–87.
Perricone, M., E. Arace, M. R. Corbo et al. 2015. Bioactivity of essential oils: Are view on their interaction with food components. *Frontiers in Microbiology* 6:76.
Ramezani, H., H. P. Singh, D. R. Batish et al. 2002. Fungicidal effect of volatile oils from *Eucalyptus citriodora* and its major constituent citronellal. *New Zealand Plant Protection (Arable Entomology and Pathology)* 55:327–30.
Rasooli, I., M. B. Rezaei and A. Allameh. 2005. Growth inhibition and morphological alterations of *Aspergillus niger* by essential oil from *Thymus eriocalyx* and *T. x-porlock. Food Control* 17:359–64.
Sagar, D. V., S. N. Naik and P. Vasudevan. 2002. Antifungal activity of *Tagetes patula* essential oil. *Indian Perfumer* 46(3):269–72.
Samah, L., A. Meliani, S. Benmimoune, et al. 2016. Essential oil composition and antimicrobial activity of *Artemisia herba—Alba Asso* grown in Algeria. *Medicinal Chemistry* 6:435–39.
Shahi, S. K., M. Patra, A. C. Shukla et al. 2003. Use of essential oil as botanical-pesticide against post-harvest spoilage in Malus pumilo fruits. *BioControl* 48:223–32.
Sharif, M., R. Atiqur, A. Yunus et al. 2010. Inhibition of plant pathogens in vitro and in vivo with essential oil and organic extracts of *Cestrum nocturnum L. Pesticide Biochemistry and Physiology* 96:86–92.
Sharma, N. and A. Tripathi. 2008a. Effects of *Citrus sinensis* (*L.*) Osbeck epicarp essential oil on growth and morphogenesis of *Aspergillus niger* (*L.*) Van Tieghem. *Microbiological Research* 163:337–44.
Sharma, N. and A. Tripathi. 2008b. Integrated management of postharvest Fusarium rot of gladiolus corms using hot water, UV-C and *Hyptis suaveolens* (L.) Poit. essential oil. *Postharvest Biology and Technology* 47:246–54.
Sharma, N. and A. Tripathi. 2006. Fungitoxicity of the essential oil of *Citrus sinensis* on postharvest pathogens. *World Journal of Microbiology and Biotechnology* 22:587–93.
Shukla., A. C. 2009. Volatile oil of *Cymbopogon pendulus* as an effective fumigant pesticide for the management of storage-pests of food commodities. *National Academy Science Letters* 32:51–59.
Shukla, A. C., S. K. Shahi and A. Dixit. 2000. Epicarp of *Citrus sinensis*: A potential source of natural pesticides. *Indian Phytopathology* 53(4):468–71.
Siddiqui, N. and S. C. Tripathi. 2002. Fungitoxicity study of essential oil of *Seseli indicum. Indian Perfumer* 46(3):205–10.
Singatwadia, A. and S. S. Katewa. 2001. In vitro studies on antifungal activity of essential oil of *Cymbopogon martini* and *Cymbopogon citratus. Indian Perfumer* 45(1):53–5.
Singh, D. P. and A. K. Pant. 2001. Chemical composition and biological activity of essential oil of *Pogostemon plectranthoides* Desf. *Indian Perfumer* 45(1):35–8.
Singh, G., I. P. S. Kapoor, S. K. Pandey et al. 2003. Curcuma longa-chemical, antifungal and antibacterial investigation of rhizome oil. *Indian Perfumer* 47(2):173–78.
Singh, P., A. K. Pandey, N. Sonker et al. 2011. Preservation of *Buchnania lanzan* Spreng. seeds by *Ocimum canum* Sims. essential oil. *Annals of Plant Protection Sciences* 19:407–10.

Sokmen, A., M. Gulluce, A. H. Akpulat et al. 2004. The in vitro antimicrobial and antioxidant activities of the essential oil and methanolic extracts of endemic *Thymus spathulifolius*. *Food Control* 15:627–34.
Sonker, N., A. K. Pandey, P. Singh et al. 2014. Assessment of *Cymbopogon citratus* (DC.) Stapf essential oil as herbal preservatives based on antifungal, antiaflatoxin and antiochratoxin activities and in vivo efficacy during storage. *Journal of Food Science* 79: 628–34.
Sonker, N., A. K. Pandey and P. Singh. 2015. Efficiency of *Artemisia nilagirica* (Clarke) Pamp essential oil as a mycotoxicant against postharvest mycobiota of table grapes. *Journal of the Science of Food and Agriculture* 95:1932–39.
Srivastava, S. and R. P. Singh. 2001. Antifungal activity of the essential oil of *Murraya koenigii* (*L.*) Spreng. *Indian Perfumer* 45(1): 49–51.
Sylvestre, M., A. Pichette, S. Lavoie et al. 2007. Composition and cytotoxic activity of the leaf essential oil of *Comptonia peregrine* (*L*). Coulter. *Phytotherapy Research* 21:536–40.
Szczerbanik, M., J. Jobling, S. Morris et al. 2007. Essential oil vapours control some common postharvest fungal pathogens. *Australian Journal of Experimental Agriculture* 47(1):103–9.
Tian, J., X. Q. Ban, H. Zeng et al. 2011. Chemical composition and antifungal activity of essential oil from *Cicuta virosa L.* var. latisecta Celak. *International Journal of Food Microbiology* 145:464–70.
Tripathi, A., N. Sharma and V. Sharma. 2008. *Hyptis suaveolens* (*L.*) Poit: A source of potential fungitoxic essential oil against post-harvest pathogens. *Biochemical and Cellular Archives* 8(1):1–6.
Tripathi, A., N. Sharma and V. Sharma. 2009. In vitro efficacy of *Hyptis suveolens L.* (Poit) essential oil on growth and morphogenesis of *Fusarium oxysporum* f.sp. gladioli (Massey) Snyder & Hansen. *World Journal of Microbiology and Biotechnology* 25:503–12.
Tripathi, A., N. Sharma, V. Sharma et al. 2013a. Bioactivity of essential oil of *Hyptis suaveolens* (*L.*) Poit. against pathogenic microflora. *Global Journal of Applied Agricultural Research* 3(1):45–48.
Tripathi, A., N. Sharma, V. Sharma et al. 2013b. A review on conventional and non-conventional methods to manage post-harvest diseases of perishables. *Researcher* 5(6):6–19.
Tripathi, A., N. Sharma, V. Sharma et al. 2013c. Integrated eco-friendly management of Fusarium corm rot and yellows by sowing hot water, UV-C and/or essential oil treated gladiolus corms in soil solarized and/or essential oil fumigated experimental fields. *International Journal of Horticultural and Crop Science Research* 3(1):51–63.
Tripathi, N. N. and N. Kumar. 2007. Putranjiva roxburghii oil—A potential herbal preservative for peanuts during storage. *Journal of Stored Products Research* 43:435–42.
Tripathi, P., N. K. Dubey, R. Banerji et al. 2004. Evaluation of some essential oils as botanical fungitoxicants in management of post-harvest rotting of fruits. *World Journal of Microbiology and Biotechnology* 20:317–21.
Ukeh, D. A. and A. J. Mordue. 2009. Plant based repellents for the control of stored product insect pests. *Biopesticides International* 5:1–23.

7 Entomopathogenic Biopesticides
Opportunities and Challenges

N. Sharma, A.S. Bhandari and P.K. Shukla

CONTENTS

7.1 INTRODUCTION

Agriculture and forests are important resources in sustaining global economic, environmental and social systems. For this reason, a global challenge is to secure high and quality yields and to make agricultural produce environmentally compatible. The latest report launched by the UN Food and Agriculture Organization (FAO, 2019) estimates that around 14% of the world's food is lost after harvesting and before reaching the retail level through on-farm activities, storage and transportation. The damage and destruction inflicted on crops by pathogens and pests has had a serious impact on farming and agricultural practices for a long time. These biotic factors include animals, bacteria, birds, fungi, insects, nematodes, viruses and weeds. Roughly, it has been calculated that nearly 50,000 species of fungi, 10,000 species of insects, 15,000 species of nematodes and 1800 species of weeds destroy food and fiber crops worldwide.

Devi (2015) has quoted the amount of global crop losses due to pests between one-third and one-half of the attainable crop production with crop losses in developing countries at the higher end. In developing countries, pests, weeds and diseases destroy about 40% of crops while they are still in the field and 6–7% after harvest. In Africa and Asia, preharvest losses are estimated at 50%. Insects cause the highest crop damage followed by plant pathogens and weeds. In India alone, 40% of the crop yield is lost, equivalent to 30 million tons of food grains. In recent years, the qualitative and quantitative deterioration and damage to harvested products amounts to more than 10–90% due to diseases caused by pests and pathogens with an average of 35–40% for all potential food and fiber crops (Abang et al., 2014).

Until the 1950s, insect pests were not of major concern in crop production. There were some instances of pest outbreaks, particularly grasshoppers and locusts, pink bollworm, cotton boil weevil, leaf defoliators and stem borers in standing crops. Stored grains were usually infested by storage pests such as *Callosobruchus chinensis* (pulse beetle), *Cryptolestes ferrugineus* (rusty grain beetle), *Oryzaephilus surinamensis* (saw-toothed grain beetle), *Rhyzopertha dominica* (lesser grain borer), *Sitophilus granarius* (granary weevil), *Sitophilus oryzae* (rice weevil), *Tribolium castaneum* (red flour beetle), *Trogoderma granarium* (Kharpa beetle) and *Liposcelis corrodens* (mites) belonging to the order Coleoptera.

The strategies employed for managing these losses due to pathogens and pests mainly relied on various physical, cultural and chemical treatments. Although cultural methods were found to be effective, they are handicapped by their low and slow efficacy. The primary strategy employed to eliminate insect pests in harvested and stored products was by using chemical pesticides. Consequently, the use of chemical pesticides in agriculture has been an integral part of crop production in many regions, often at very high levels and in an unscientific pattern of application (Devi, 2010).

7.2 PESTICIDES IN USE

Use of pesticides is an agricultural management technique that enables producers to check further development of pests and pathogens (Mengistie et al., 2017). Their role in augmenting agricultural output has been well perceived and they have been considered as an essential input in agricultural production (Yadav and Dutta, 2019). Among the pest control chemicals, insecticides dominate the industry with 65% utilization, followed by herbicides (16%), fungicides (15%) and others (4%). This pattern is different from the global pattern where herbicides form the major share (44%), followed by fungicides (27%), insecticides (22%) and others (7%). The herbicide sector in India is the one that has shown the fastest growth, mainly due to rising farm wages, thus making manual weed control costly (Devi, 2011).

A recent survey has shown that Europe was the largest consumer of diverse pesticides valued at $12,850 million in 2008; approximately 31.7% of the world's total. The six largest national agrochemical markets are in France, Italy, Spain, the UK, Germany and Turkey. France alone accounts for 26.2% of agrochemical sales; Italy comes second at 19.8%. Among others, Surinam, Malta, Columbia, Palestine, Japan, Korea, Chile and China are the countries where more than 10 kg $ha^{-1}yr^{-1}$ pesticides are used.

The Nordic countries, on the other hand, consume comparatively fewer pesticides, which include insecticides, herbicides and fungicides that are employed in modern agriculture to control pests and increase crop yield. In both developed and developing countries, the use of chemical pesticides has increased dramatically during the last few decades.

According to a survey conducted by IDAO (Interational Data Analysis Organization) of Ziway and Meki districts, in Ethiopia, about 530,441 kg of insecticide and 50,957 kg of fungicide were applied by 13,889 smallholder vegetable farmers (Mengistie et al., 2017).

As per the Standing Committee on Chemicals and Fertilizers of India, pesticide production in India during 2011–2012 stood at 68,490 tons. About 60 companies produce technical pesticides in the country and about 500 units produce their formulations. The total value of annual production of pesticides in the country is about 8000 crores, out of which pesticides worth 6000 crores are consumed in the country and the rest are exported. There are 256 registered pesticide products in India.

According to a study by The Research and Markets, the Indian pesticide market was worth INR 197 billion in 2018 and is further projected to reach a value of INR 316 billion by 2024, growing at a compound annual growth rate (CAGR) of 8.1% during 2019–2024. Though crop losses due to pest attack in India are reported to be very high, the intensity of pesticide consumption in the country is one of the lowest in the world (291.2 grams/ha) compared with the US (4.5kg/ha), Japan (11kg/ha), China (14kg/ha) and the world average of 3 kg per ha. The worldwide consumption of pesticides is about two million tons per year, out of which India's share is only 3.75%. The area under pesticide use in India is only 9% of the total cultivated land (16.7 million hectares).

There is overwhelming evidence that some of these chemicals do pose potential risks to the ecosystem in general and human beings in particular (Devi, 2010). Chemical pesticides tend to persist in soil and surface water bodies as well as groundwater for long periods, thus imposing serious health threats for humans and animals. It is estimated that around 800,000 people in developing countries have died due to pesticides since the onset of the green revolution. Nearly 20,000 people in developing countries die each year because of pesticide consumption through their food (Bhardwaj and Sharma, 2013). An analysis by PAN India revealed that more than 115 pesticides out of the 275 are highly hazardous (Yadav and Dutta, 2019).

7.3 RESISTANCE TO CHEMICALS

Over the last five decades, chemical insecticides have been the backbone of insect control; producers and cultivators had to rely heavily on conventional groups of insecticides, such as DDT, BHC (organochlorines), aldrin, dieldrin, endosulfan (cyclodienes), monocrotophos, quinalphos, chlorpyriphos, profenophos, dimethoate, phosalone, metasystox, acephate, phorate, methyl parathion (organophosphates), carbosulfan, carbaryl, thiodicarb, methomyl (carbamates), cypermethrin, deltamethrin, fenvalerate, λ-cyhalothrin (pyrethroids), chlordimeform and amitraz (formamidines) (Khan et al., 2015).

Fumigants such as cynogen, ethyl formate, methyl bromide, phosphines and sulfuryl fluorides were quick enough to attack and kill all life stages of insects in a community or in storage structures. Pesticide application has often been adopted as a risk avoidance strategy, where the chances of pest incidence/critical pest population are often wrongly perceived to be on the high side. Later, the indiscriminate use of these chemicals in a highly unscientific manner, starting from the choice of chemical to application practices, timing and even necessity led to a negative impact (Devi, 2009). The private operators in developing countries, who handle retail sales, generally do not have any formal training or information on these aspects, inadvertently leading to a rise in incidence of pesticide resistance in insect species and weeds, thus threatening our ability to harness these pests. More than 500–550 arthropods have evolved resistance to at least one pesticide in use. The concerns on the negative externalities of chemical pesticides across the world have resulted in increasing awareness on pesticide use, especially in socio-economically advanced societies.

Fumigation is one of the most effective methods for the prevention of losses of stored products due to insect pest infestation but studies emphasizing an upsurge in fumigation resistance in stored products have been reported. Insect resistance to phosphine was reported to be so high in Australia (Nayak et al., 2013) and India (Kaur et al., 2015) that it caused control failures of the pests. The reason behind this resistance is said to be the common practice of fumigation in unsealed silos, farm storages and bulk handling storage resulting in frequent exposure of insect populations to sublethal dosages. Repeated exposure to such sublethal dosages may allow the rare individuals with a new resistance gene to survive treatment and they may continue breeding, passing on their acquired characteristics, while on the other hand, normal and susceptible insects get killed.

Methyl bromide, another efficient fumigant, has been held as a major contributor to ozone depletion and had been banned worldwide. By 2014, populations of at least 590 insect species were found to demonstrate resistance to nearly 300 insecticidal compounds (Belinato and Martins, 2016).

Shankarganesh et al. (2012) observed resistance against conventional as well as new chemical insecticides for quinalphos, monocrotophos, lindane, endosulfan, benzene hexachloride, avermectins, spinosad, fipronil, indoxacarb and chitin synthesis inhibitors.

7.4 BIOPESTICIDES: A POSSIBLE ALTERNATIVE

With the success achieved so far, the last century has been a witness to an increase as well as a backlash against chemical pesticides. Despite many years of effective control by agrochemical insecticides, numerous factors are threatening the effectiveness and continued use of these agents. These include the development of insecticide resistance use-cancellation and/or de-registration of some insecticides due to human health complications. Serious effects on biodiversity, non-target organisms and the food chain have emerged; compounded further by indiscriminate and excessive use of products that have had a substantial impact on the environment. Consequently, beneficial species have been lost and residual problems have increased, with a subsequent impact on groundwater contamination and resistance in pests. The high level uses of these "poisons by design" are prevalent among farmers with low levels of education. In rural areas of developing countries, nearly three million farmers suffer from mild poisoning resulting in approximately 180,000 fatalities annually because of incorrect perception, lack of knowledge and regulation among farmers (Ozkara et al., 2016).

Therefore, an eco-friendly alternative is needed. Improvement in pest control strategies represents one method to generate higher quality and greater quantity of agricultural products. Therefore, there is a need to develop biopesticides which are effective, biodegradable and do not leave any harmful effects on the environment (Usta, 2013). The use of biopesticides based on microorganisms and their products has proven to be highly effective, species specific and eco-friendly in nature, leading to their adoption in pest management strategies all around the world.

The current growth of the chemical pesticide market is about 1–2% per year, while growth in microbial pest control is about 10–20% per year (Harris, 2009).

The microbial biopesticide market constitutes about 90% of total biopesticides and there is ample scope for further development in agriculture and public health. North America uses the largest percentage of the biopesticide market share at 44%, followed by the EU and Oceania with 20% each, South and Latin American countries with 10% and about 6% in India and other Asian countries. In terms of sales of biopesticides, the global market in 2007 was US$672 million and projected at US$1000 million for 2010. As of 2007/2008, estimates of microbial biopesticide sales were US$396 million at the end-user level although these estimates are likely to be just a fraction of the total usage of such products, owing to the lack of information available on the non-commercial use of such products in some regions.

The recent report by Microbial Pesticides Market (2019) showed that the global market for these pesticides was valued at US$1944.3 million in the year 2018 and is forecasted to reach a value of US$4753.1 million in the year 2024 with a CAGR of 16.3% for the forecast period of 2019–2024 (www.mordorintelligence.com).

Although biopesticides represent only 3% of the global crop protection business, its growth is at the rate of 10% per year. Of the total global biopesticide market, mycoinsecticide is second (27%) to *Bacillus thuringiensis* products (Rai et al., 2014). This growth is an indication that bio-insecticides will soon play an important role in insect pest management. The biopesticide sector is experiencing a significant growth and many discoveries are being developed into new biological pesticide products that are fueling a growing global market (Ruiu, 2018).

Biological pesticides, or biopesticides, represent a range of bio-based substances, which act against invertebrate pests with different mechanisms of action. According to the technical definition provided by the United States Environmental Protection Agency, they can be categorized as follows: (i) microbial entomopathogens; (ii) naturally occurring biochemicals that act through nontoxic mechanisms; and (iii) PIP/plant-incorporated protectants deriving from genetically engineered plants (Biopesticides, 2018).

In agricultural ecosystems, employing living organisms and natural products enhances the properties of natural ecosystem components to counteract the biotic and reproductive potential of pests as the growth of harmful insect and other invertebrate populations is favored by an oversimplification of living communities. Biological control methods based on the use of predators and parasitoids, natural enemies (Kenis et al., 2017) and pest disease agents (Kaya and Vega, 2012) may restore a lost ecological balance. Particularly bacteria, fungi, baculoviruses and nematodes represent the basic concept of invertebrate pathogens.

In the last few decades, numerous experimentations and investigations conducted in the academic and industrial context have led to the discovery, development and market launch of several microbial biopesticides (Marrone, 2014). Although mycoinsecticides have their limitations, which led to the withdrawal of products such as Boverin (former USSR), Mycar, Taerain and CornGaurd (USA) and Bio 1020 (Germany) from the market, their potential as biocontrol agents remains promising. The interest in this specific field of study is internationally fostered by recently revised legislative frameworks, like the European Pesticide Regulation (EC) No. 1107/2009, encouraging the use of safer pesticides with less environmental impact (Villaverde et al., 2014).

7.5 ENTOMOPATHOGENIC BIOPESTICIDES

Due to the magnitude of these damages and various other reasons, indiscriminate use of chemicals has seen both qualitative and quantitative changes over the past five decades in insect pest problems and their management. Insects, like other organisms, are susceptible to a variety of diseases caused by different groups of microorganisms (Kachhawa, 2017). Microbial pathogens consist of disease-causing organisms, which are disseminated in the pest population in large quantities in a manner like the application of chemical pesticides. These pathogens are exploited for biological control of insect pests through introductory or inundated applications (Pathak et al., 2017). Microbial pathogens of insects are intensively investigated to develop eco-friendly pest management strategies in agriculture.

Several microorganisms like bacteria, baculoviruses, fungi, protozoa and nematode-associated bacteria have been found to possess antagonistic properties to be utilized against various pests (Dara, 2017). Following a few decades of successful use of the entomopathogenic bacterium *Bacillus thuringiensis* and a few other microbial species, recent academic research and industrial efforts have led to the discovery of new microbial species and strains and their specific toxins and virulence factors; many of these have, therefore, been developed into commercial products.

Bacterial entomopathogens include species of *Bacillus, Serratia, Pseudomonas, Yersinia, Burkholderia, Chromobacterium, Streptomyces* and *Saccharopolyspora*, while fungal genera comprise different strains *of Beauveria bassiana, B. brongniarti, Metarhizium anisopliae, Verticillium, Lecanicillium, Hirsutella, Paecilomyces* and *Isaria* species. Baculoviruses are species specific and refer to niche products active against chewing insects, especially Lepidopteran caterpillars. Entomopathogenic nematodes (EPNs) mainly include species in the genera *Heterorhabditis* and *Steinernema* while *Nosema* and *Vairimorpha* are the commonly used protozoa.

7.5.1 Bacteria

Different bacterial species have, for a long time, been the object of studies investigating their pathogenic relationship with invertebrates, especially insects (Mnif and Ghribi, 2015). The bacteria that are used as biopesticides can be divided into four categories: crystalliferous spore formers (*Bacillus thuringiensis*), obligate pathogens (*Bacillus popilliae*), potential pathogens (*Serratia marcescens*) and facultative pathogens (*Pseudomonas aeruginosa*). Out of these, the spore forming ones has been most widely adopted for commercial use because of their safety and effectiveness.

Most of the insect pathogenic bacteria occur in the families Bacillaceae, Enterobacteriaceae, Micrococcaceae, Pseudomonadaceae and Streptococcaceae, with maximum attention to *Bacillus* spp., as microbial control agents. This group of entomopathogens is well represented by *Bacillus thuringiensis* (Bt), the most studied and commercially used bacterial species (Table 7.1).

Subspecies of this bacterial species in use as biopesticides include *B. thuringiensis tenebrionis* (targeting Colorado potato beetle and elm leaf beetle larvae), *B. thuringiensis kurstaki* (targeting a variety of caterpillars), *B. thuringiensis israelensis* (targeting mosquito, black fly and fungus gnat larvae) and *B. thuringiensis aizawai* (targeting wax moth larvae and various caterpillars, especially the diamondback moth caterpillar).

A similar insecticidal behavior is associated with the *Lysinibacillus sphaericus* species group that acts against mosquitoes and black flies. Insecticidal toxins have been found in entomopathogenic species belonging to the *Paenibacillus* genus. Another bacterium in the same bacterial family but showing a wider spectrum of pesticidal activity is *Brevibacillus laterosporus*, a species characterized by a swollen sporangium containing a spore with a canoe-shaped parasporal body attached to one side (Ruiu, 2013; Marche et al., 2017). This bacterium holds several virulence factors (Marche et al., 2018) and its use in integrated management of different pests has been proposed (Ruiu et al., 2011, 2014). Many details of this process are still not understood, and the action seems to be more complex as indicated by the existence of novel receptors and signal transduction pathways induced within the host following intoxication (Ffrench-Constant et al., 2006). This could lead to altering the activation of midgut proteases, resulting in differences in the toxin structure that could affect binding to the peritrophic membrane, thereby accounting for host specificity (Hurst et al., 2000).

Several bacterial products based on different species and strains are in use worldwide (Table 7.2). Among these, Gammaproteobacteria represents a heterogeneous group of species including several entomopathogens like the endosymbionts of insecticidal nematodes *Photorhabdus, Xenorhabdus* and *Serratia* species, whose insecticidal action is a toxin mediated process (Ffrench-Constant et al., 2006; Hurst et al., 2000).

The same group includes the non-spore forming species *Yersinia entomophaga* producing the toxin complex Yen-Tc, containing toxins and chitinases (Landsberg et al., 2011), and *Pseudomonas entomophila* which has a toxin secretion system (Vodovar et al., 2006), both acting by ingestion.

Beta proteobacteria represent another class including species with significant potential as biocontrol agents. An insecticidal strain of *Burkholderia rinojensis* has recently been discovered and developed into a product acting by ingestion and contact against diverse chewing and sucking insects and mites (Cordova-Kreylos et al., 2013). The insecticidal action relies on different metabolites

TABLE 7.1
Commercially Used Bacterial Species

Bt variety	Target pest
Bacillus moritai	Diptera
B. popilliae	Japanese beetle grubs
B. sphaericus	Mosquito larvae
B. thuringiensis subsp. aizawai	Moth larvae
B. thuringiensis subsp. israelensis	Mosquito and blackflies
B. thuringiensis subsp. kurstaki	Lepidopteran larvae
B. thuringiensis subsp. tenebrionis	Colorado potato beetle
B. thuringiensis subsp. galleriae	Lepidopteran larvae

Source: Kunimi (2007) and Kabaluk et al. (2010).

TABLE 7.2
Commercial Products Based on Entomopathogenic Bacteria

Active substances	Commercial names*	Main targets
Bacillus thuringiensis aizawai	Able-WG, Agree-WP, Florbac, XenTari	Armyworms, diamondback moth
Bacillus thuringiensis kurstaki	Biobit, Cordalene, Costar-WG, Crymax-WDG, Deliver, Dipel, Foray, Javelin-WG, Lepinox Plus, Lipel, Rapax	Lepidoptera
Bacillus thuringiensis israelensis	Teknar, VectoBac, Vectobar	Mosquitoes and blackflies
Bacillus thuringiensis tenebrionis	Novodor, Trident	Colorado potato beetle
Bacillus thuringiensis sphaericus	VectoLex, VectoMax	Mosquitoes
Burkholderia spp.	Majestene, Venerate	Chewing and sucking insects and mites; nematodes
Saccharopolyspora spinosa	Tracer™ 120	Conserve insects
Chromobacterium subtsugae	Grandevo	Chewing and sucking insects and mites
Bacillus firmus	BioNemagon	Nematodes

Source: Ruiu, 2018.
*Different products may refer to different microbial strains. The representative trade names are those shown on the relevant company websites to which reference should be made for details.

and the commercial product is based on heat-killed cells and spent fermentation media. Another commercially successful beta propteobacterium is a strain of *Chromobacterium subtsugae* whose metabolites show a broad-spectrum insecticidal activity against different species of Lepidoptera, Hemiptera, Coleoptera and Diptera (Martin et al., 2007).

Actinobacteria include different *Streptomyces* species producing a variety of insecticidal toxins, such as the macrocyclic lactone derivatives acting on the insect peripheral nervous system (Copping and Menn, 2000). Within the same phylum, *Saccharopolyspora spinosa* produces potent and broad-spectrum insecticidal toxins known as spinosyns, whose natural and semi-synthetic derivatives have had good commercial success (Kirst, 2010).

7.5.2 Fungi

Recently, there has been a renewed interest in fungal pathogens of insects because of their potentials as biocontrol agents. Entomopathogenic fungi (EPF) are naturally occurring microorganisms that play an important role in insect population dynamics in a natural ecosystem. Their activities play a significant role in the stability of insect population dynamics in natural ecosystems. Consequently, more than 750 species of fungi, mostly from hyphomycetes and entomophthorales were found to possess pathogenic action against insects, many of them offering great potential for pest management (Ruiu, 2018). Globally, more than 171 products have been developed from these fungi on at least 12 species.

The groundbreaking field trials with EPF started with a Russian microbiologist, Elie Metchnikoff in 1888, who first described *Entomophthora anisopliae* near Odessa in Ukraine from infected larvae of the wheat cockchafer *Anisoplia austriaca*, and later on, from *Cleonus punctiventris*; renamed as *Metarhizium anisopliae* by Sorokin and more recently named *as M. brunneum* (Vega et al., 2009).

Metarhizium causes a disease known as "green muscardine" in insect hosts because of the green color of its conidial cells. Since then, researchers around the world have identified more than 750 fungi species that are pathogenic to insects of different orders. From these fungi, more than 171 mycoinsecticides have been produced using propagules from at least 12 fungal species.

Entomopathogenic fungi include a variety of genera and species acting against diverse targets showing varying degrees of specificity. These are commensals, obligate or facultative or symbionts of insects. Common fungal entomopathogens include species belonging to Chytridiomycota, Zygomycota, Oomycota, Ascomycota and Deuteromycota. Some of the most widely used species include *Beauveria bassiana, Fusarium pallidoroseum, Metarhizium anisopliae, Nomuraea rileyi, Paecilomyces farinosus* and *Verticillium lecanii.* Other fungal species, whose action is associated with the production of insecticidal metabolites and commercially exploited worldwide for pest management include *Hirsutella* spp. and *Isaria* spp. (Ruiu, 2018). Many of them, which have been commercialized globally, are presented in Table 7.3.

The genus *Beauveria* contains at least 49 species out of which approximately 22 are considered pathogenic. *Beauveria bassiana*, representing one of the most used fungal bio-insecticides, was the first example of microbial control of insects at the end of the 19th century. *Beauveria bassiana*, a white muscardine fungus, is the most historically important of the commonly used fungi in this genus. Originally known as *Tritirachium shiotae*, this fungus was renamed after the Italian lawyer and scientist Agostino Bassi, who first implicated it as the causative agent of a white (later yellowish or occasionally reddish) muscardine disease in domestic silkworms (Furlong and Pell, 2005). Within the same genus, *B. bassiana* and *B. brongniartii* strains showing varying levels of virulence against diverse targets are now used as active substances in diverse formulations (Zimmermann, 2007a). Recent studies have highlighted the potential of some *B. bassiana* and *B. brongniartii* strains as endophytes in biological control applications (Vidal and Jaber, 2015; McKinnon et al., 2017). The commercial mycoinsecticide "Boverin" based on *B. bassiana* with reduced doses of trichlorophon has been used to suppress the second-generation outbreaks of *Cydia pomonella* L. Higher insect mortality was detected when *B. bassiana* along with sublethal concentrations of insecticides were applied to control Colorado potato beetle (*Leptinotarsa decemlineata*), attributing higher rates of synergism between two agents (Usta, 2013).

Metarhizium anisopliae represents another well-exploited fungal species with diverse strains offering a variety of toxins and virulence factors produced by this fungus (Schrank and Vainstein, 2010), demonstrating insecticidal properties against a wide range of targets (Zimmermann, 2007b).

TABLE 7.3
Fungi in Use as Biopesticides

Fungi	Target
Alternaria cassiae	Sickle pod (weed)
Aschersonia aleyrodes	Whitefly
Beauveria bassiana (muscardine fungus).	Colorado potato beetle, corn root worm, citrus root weevil, cotton bollworms, coffee berry borer, codling moth, Japanese beetle, pod borer, mango mealy bug, boll weevil, cotton leaf hopper, Chinch bug, yellow stem borer, rice leaf folder, brown plant hopper, etc.
Colletotrichum gloeosporioides	Northern joint vetch (weed)
Fusarium lateritium	Velvet leaf (weed)
Hirsutella thompsonii	Phytophagous mites
Metarhizium anisopliae	Spittlebug, sugarcane hopper, rhinoceros beetle, termite, locust, grasshoppers, etc.
Nomuraea rileyi	*Helicoverpa armigera, Spodoptera litura, Trichoplusiani ni* and *Achaea janata*
Pandora delphacis	Brown plant hopper and green leaf hopper of rice
Phytophthora palmivora	Milk weed vine (weed)
Verticillium lecanii	Aphid, whiteflies and scales

Source: Pawar and Singh 1993 and Zimmermann, 1993.

Lecanicillium (Verticillium lecanii) is a widely distributed fungus, which can cause large epizootics in tropical and subtropical regions, as well as in warm and humid environments (Nunez et al., 2008). In the 1970s, *V. lecanii* was developed to control whitefly and several aphid species, including green peach aphids (*Myzus persicae*) for use in greenhouse chrysanthemums. Kim et al. (2008) reported that *V. lecanii* was an effective biological control agent against *Trialeurodes vaporariorum* in South Korean greenhouses. This fungus attacks nymphs and adults and sticks to the leaf underside by means of a filamentous mycelium (Nunez et al., 2008). This entomopathogen was found to be effective against cotton aphids and powdery mildew (Kim et al., 2008).

Nomuraea rileyi, another potential entomopathogen, is a dimorphic hyphomycete that can cause epizootic death in various insects. It has been shown that many insect species belonging to Lepidoptera including *Spodoptera litura* and some belonging to *Coleoptera* are susceptible to *N. rileyi*. The host specificity of *N. rileyi* and its eco-friendly nature encourage its use in pest management against several insect hosts such as *Anticarsia gemmatalis, Bombyx mori, Heliothis zea, Plathypena scabra, Pseudoplusia* and *Trichoplusia ni*.

Isaria fumosorosea (Paecilomyces fumosoroseus) is one of the most important natural enemies of whiteflies worldwide, and causes the sickness called "Yellow Muscardine" (Nunez et al., 2008). Strong epizootic potential of this fungus against *Bemisia* and *Trialeurodes* spp. in both greenhouse and open field environments has been demonstrated. The ability of this fungus to grow extensively over the leaf surface under humid conditions is a characteristic that certainly enhances its ability to spread rapidly through whitefly populations (Wraight et al., 2000). Nunez et al. (2008) reported that *P. fumosoroseus* is best for controlling the nymphs of whitefly. These fungi cover the body of whitefly with mycelial threads and stick them to the underside of the leaves. The nymphs show a "feathery" appearance and are surrounded by mycelia and conidia (Nunez et al., 2008).

Among the species already use in formulated mycoinsecticides are *Beauveria bassiana, B. brongniartii, Cladosporium oxysporum, Hirsutella thompsonii, Lecanicillium* spp. (previously *Verticillium lecanii*) and *Metarhizium anisopliae*.

Several mycotoxins like Beauvericin, Beauverolides, Bassianolide (by *B. bassiana, V. lecanii, Paecilomyces* spp.) and Destruxins A, B, C, D, E, F (by *M. anisopliae*) are produced during pathogenesis and these act like poison to the insects. After the death of the insects, the fungus breaks open the integument and forms aerial mycelia and sporulation on the cadavers.

Streptomyces are reported to produce toxins that act against insects (Kekuda et al., 2010). About 50 such compounds have been reported as active against various insect species belonging to Lepidoptera, Homoptera, Coleoptera, Orthoptera and mites (Kaur et al., 2014). The most active toxins are actinomycin A, cycloheximide and novobiocin. Spinosyns are commercially available biopesticidal compounds that were originally isolated from the actinomycetes *Saccharopolyspora spinosa* and are found to be active against dipterans, hymenopterans, siphonaterans and thysanopterans, but are less active against coleopterans, aphids and nematodes (Bacci et al., 2016).

Currently, the largest single microbial control program using fungi involves the use of *M. anisopliae* for control of spittlebugs (Cercopidae) in South American sugarcane and pastures (Iwanicki et al., 2019). The application of *B. bassiana* for the control of pine moth *Dendrolimus* spp. in China probably represents one of the largest uses of a biocontrol agent, with over one million hectares of pine forest (Maina et al., 2018). *B. bassiana* strain Bb-147 (Abrol, 2017) is registered on maize in Europe for the control of the European corn borer and the Asiatic corn borer which normally requires specific environmental conditions (i.e., temperature, relative humidity).

A limitation of fungal-based biopesticides is their action by contact and a strict range of conditions for conidia and spore germination (Ahmad et al., 2019). Improved products are to be developed employing endophytic strains targeting insects after their penetration inside the plant. Another aspect to be considered before applying fungal entomopathogens on the crop is the lack of fungicide residues used against phytopathogenic fungi (Koul, 2011). Most commercial products are based on suspensions of conidia (Table 7.4).

TABLE 7.4
Commercial Products Based on Entomopathogenic Fungi

Active substances	Commercial names*	Main targets
Beauveria bassiana	Bio-Power, Biorin/Kargar, Botanigard, Betal, Daman, Naturalis, Nagestra, Beauvitech-WP, Bb-Protec, Racer, Mycotrol, Ostrinol	Corn borer, insects and mites
Beauveria brongniartii	Bas-Eco, Engerlingsplz	Scrab beetle larva, *Helicoverpa armigera*, berry borer, root grubs
	No-Mite, Mycohit	Spider mites
Isaria fumosorosea	No-Fly WP, PFR-97	Whitefly
Metarhizium anisopliae	Biomet/Ankush, Bio-Magic, Devastra, Green Muscle, Kalichakra, Novacrid, Bioblast, Met52/ BIO1020 granular	Pacer beetles, caterpillar pests, grasshoppers, termites
Metarhizium brunneum	Attracap	*Agriotes* spp.
Metarhizium flavoviride	BioGreen	Red-headed cockchafer
Paecilomyces lilacinus	Bio-Nematon, MeloCon Mytech-WP, Paecilo	Plant pathogenic nematodes
Paecilomyces fumosoroseus	Bioact WG, No-Fly WP, Paecilomite	Insects, mites, nematodes, thrips
Verticillium lecanii	Bio-Catch, Mealikil, Bioline/Verti-Star	Mealy bugs, sucking insects
Lecanicillium muscarium	Mycotal	White flies
Lecanicillium lecanii	Lecatech-WP, Varunastra	Aphids, leafminers, mealybugs, scale insects, thrips, whiteflies
Lecanicillium longisporum	Vertalec	Aphids
Myrothecium verrucaria	DiTera	Nematodes

*Different products may refer to different microbial strains. The representative trade names are those shown on the relevant company websites to which reference should be made. *Source*: Ruiu, 2018.

7.5.3 Viral Biopesticides

There are more than 1600 different viruses, which infect 1100 species of insects and mites. Over 700 insect-infecting viruses have been isolated, mostly from Lepidoptera (560) followed by Hymenoptera (100), Coleoptera, Diptera and Orthoptera. A special group of viruses, known as baculovirus, in the family Baculoviridae, represents DNA viruses establishing pathogenic relationships with invertebrates (Clem and Passarelli, 2013; Haase et al., 2015), and accounts for more than 10% of all insect pathogenic viruses inhabiting about 100 insect species. Globally, these viruses are widely used for the control of vegetable and field crop pests and are effective against plant-chewing insects.

The virus infectivity is associated with the production of crystalline occlusion bodies, containing infectious particles, within the host cell.

Based on the morphology of these occlusion bodies, Baculoviruses (Rohrmann, 2011) are divided into two main groups: nucleopolyhedroviruses (NPVs), in which these bodies are polyhedron-shaped and develop in cell nuclei, and granuloviruses (GVs), in which these bodies are granular-shaped. A different, double-stranded RNA virus family (Reoviridae), presents polyhedron-shaped occlusion bodies in the cell cytoplasm (CPVs).

Viruses used for insect control are the DNA-containing baculoviruses, NPVs, GVs, iridoviruses, parvoviruses, polydnaviruses and poxviruses, and the RNA-containing reoviruses, cytoplasmic polyhedrosis viruses, nodaviruses, picrona-like viruses and tetraviruses. However, the main categories used in pest management have been NPVs and GVs.

The first well-documented introduction of baculovirus into the environment, that resulted in effective suppression of a pest, occurred accidentally before World War II. Along with a parasitoid imported to Canada to suppress spruce sawfly *Diprion hercyniae*, an NPV specific for spruce sawfly was introduced and since then no control measures have been required against this hymenopteran species. Given the close relationship and specificity with the host, the name of the entomopathogenic virus includes the initial of the host name. For instance, LdMNPV refers to *Lymantria dispar multicapsid nucleopolyhedrovirus*.

Sandoz Inc., in 1975, introduced the first relatively broad range baculovirus insecticide Elcar™; a preparation of *Heliothis zea* NPV (HzSNPV) that can infect many species belonging to genera *Helicoverpa* and *Heliothis*. HzSNPV provided control of not only cotton bollworm, but also of pests belonging to these genera that attack soybean, sorghum, maize, tomato and beans; however, in 1982, the company decided to discontinue the production. The resistance to many chemical insecticides, including pyrethroids, revived the interest in HzSNPV and the same virus was registered under the name GemStar™, a product of choice for biocontrol of *Helicoverpa armigera*. Countries with large areas of such crops like cotton, pigeon pea, tomato, pepper and maize, for example, India and China, introduced special programs for the reduction of this pest by biological means.

In the early 1980s in Brazil, employing baculovirus *Anticarsia gemmatalis* nucleopolyhedrovirus (AgMNPV) as a biopesticide to control the velvetbean caterpillar in soybean turned out to be extremely successful. Since then, over two million ha of soybean has been treated with this virus annually.

Their use has had a substantial impact against gypsy moths, pine sawflies, Douglas fir tussock moths and pine caterpillars in forest habitats. Codling moth is controlled by *Cydia pomonella* GVs on fruit trees (Lacey and Unruh, 2005) and potato tuber worm by *Phthorimaea operculella* GVs in stored tubers (Rondon and Gao, 2018). Virus-based products are also available for cabbage moths, corn earworms, cotton leaf worms and bollworms, beet armyworms, celery loopers and tobacco budworms (Table 7.5).

7.5.4 Nematodes

Entomopathogenic nematodes used for insect control infect only insects or related arthropods such as weevils, gnats, white grubs and various species of the Sesiidae family (Lewis and Clarke, 2012). These fascinating organisms suppress insects in cryptic habitats such as soil-borne pests and stem borers (Table 7.6).

Entomopathogenic nematode species in the genera *Heterorhabditis* and *Steinernema* act as obligate parasites and because of their mutualistic symbiosis with insect pathogenic bacteria in the genera *Photorhabdus* and *Xenorhabdus,* respectively, possess significant insecticidal potential (Gozel and Gozel, 2016). Various species include *Steinernema carpocapsae, S. riobrave, Steinernema glaseri, Steinernema scapterisci, Heterorhabditis bacteriophora* and *Heterorhabditis megidis*.

7.5.5 Protozoans

Protozoa are taxonomically subdivided into several phyla, some of which contain entomopathogenic species. Microsporidia are ubiquitous, obligatory intracellular parasites that are disease agents for several insect species. Two genera, *Nosema* and *Vairimorpha*, have some potential as they attack lepidopteran and orthopteran insects and seem to kill hoppers more than any other insect (Kachhawa, 2017).

7.6 MODE OF INFECTION

Studies have demonstrated that entomopathogens produce insecticidal toxins, which are an important tool in pathogenesis. A microbial toxin or a biological toxin material is derived from a microorganism and their pathogenic effect on target pests is species specific. Generally, the toxins produced by microbial pathogens, which have been identified, are peptides, but they vary greatly in terms of

TABLE 7.5
Commercial Products Based on Entomopathogenic Viruses

Active substances	Commercial names*	Main targets
Helicoverpa zea nucleopolyhedrovirus	Heligen	*Helicoverpa* spp. and *Heliothis virescens*
Spodoptera litura nucleopolyhedrovirus	Biovirus–S, Somstar-SL	*Spodoptera litura*
Adoxophyes orana granulovirus (AoGV)	Capex	Summer fruit tortrix moth
Cryptophlebia leucotreta granulovirus	Cryptex	False codling moth (*Thaumatotibia leucotreta*)
Helicoverpa armigera nucleopolyhedrovirus (HearNPV)	Biovirus–H, Helicovex, Helitec, Somstar-Ha	African cotton bollworm (*Helicoverpa armigera*), corn earworm (*H. zea*), *H. virescens*, *H. punctigera*
Helicoverpa zea nuclearpolyhedrosis virus	Gemstar	*Heliothis* and *Helicoverpa* sp.
Plutella xylostella (GV)	Plutellavex	*Plutella xylostella*
Spodoptera littoralis nucleopolyhedrovirus (SpliNPV)	Littovir	African cotton leaf worm
Lymantria dispar multiple nucleopolyhedrovirus (LdMNPV)	Gypchek	*Lymantria dispar*
Cydia pomonella granulovirus (CpGV)	CYD-X, Madex, Carpovirusine	*Cydia pomonella*
Neodiprion abietis nucleopolyhedrovirus (NeabNPV)	*Neodiprion abietis* NPV	*Neodiprion abietis*
Spodoptera exigua nucleopolyhedrovirus (SeNPV)	Spexit, Spod-X	*Spodoptera exigua*

*Different products may refer to different microbial strains. The representative trade names are those shown on the relevant company websites to which reference should be made for details.
Source: Ruiu, 2018.

TABLE 7.6
Entomopathogenic Nematodes as Biopesticides

Name of nematode	Host
Steinernema glaseri	White grubs (scarabs, especially Japanese beetle, *Popillia* sp.), banana root borer
S. kraussei	Black vine weevil, *Otiorhynchus sulcatus*
S. carpocap	Turf grass pests—billbugs, cutworms, armyworms, sod webworms, chinch bugs, crane flies. Orchard pests, ornamental and vegetable pests—codling moth, banana moth, cranberry girdler, dogwood borer and other clearwing borer species, black vine weevil, peach tree borer, shore flies
S. feltiae	Fungus gnats (*Bradysia* spp.), shore flies, western flower thrips
S. scapterisci	Mole crickets (*Scapteriscus* spp.)
S. riobrave	Citrus root weevils (*Diaprepes* spp.) mole crickets
Heterorhabditis bacteriophora	White grubs (scarabs), cutworms, black vine weevil, flea beetles, corn root worm, citrus root weevils
H. megidis	Weevils
H. indica	Fungus gnats, root mealybug, grubs
H. marelatus	White grubs (scarabs), cutworms, black vine weevil
H. zealandica	Scarab grubs

Source: Tofangsazi et al 2015.

structure, toxicity and specificity (Rai et al., 2014). The effect by microbial entomopathogens occurs by invasion through the integument or gut of the insect, followed by multiplication of the pathogen resulting in the death of the host, for example, insects.

7.6.1 Bacteria

Bacterial pesticides act by a variety of mechanisms; they may act by inhibiting the development, feeding, growth or reproduction of a pathogen/pest and as competitors or inducers of host resistance in plants. The insecticidal activity of bacterium relies on toxins and virulence factors, some of which are produced and released by the cell during the vegetative phase of growth (Jurat-Fuentes et al., 2010). The biosynthesis of crystal toxins (Cry and Cyt) is associated with parasporal bodies produced during the sporulation phase. Different toxin gene sequences result in different affinity with insect midgut receptors, so that diverse insecticidal protein toxins characterize different strains and strain-specific insecticidal properties (Pigott and Ellar, 2007). After being ingested, these toxins specifically bind to insect midgut receptors, thus triggering a pore-forming process that alters the epithelial membrane permeability that determines consequent disruption of the intestinal barrier functions and eventual bacterial septicemia leading to insect death (Bravo et al., 2007). Cry protoxins are ingested (Ruiu, 2013; Ruiu et al., 2011) and then solubilized, releasing a protease resistant biologically active endotoxin, before being digested by protease of the gut to remove amino acids from its C– and N–terminal ends. The C-terminal domain of the active toxin binds to specific receptors on the brush border membranes of the midgut followed by insertion of the hydrophobic region of the toxin into the cell membrane (Ruiu et al., 2014). This creates a disruption in the osmotic balance because of the formation of transmembrane pores and ultimately cell lysis occurs in the gut wall, leading to leakage of the gut contents that ultimately results in the senescence of the target pest. Some types of toxins need more alkaline pH while others respond more to neutral pH. A similar mechanism is associated with the *Lysinibacillus sphaericus* species group that acts against mosquitoes and black flies through the production of complementary crystal proteins BinA and BinB, and mosquitocidal toxins Mtx (Charles et al., 2000). Insecticidal toxins showing high homology with Bt Cry toxins have been found in entomopathogenic species belonging to the *Paenibacillus* genus.

7.6.2 Fungi

These entomopathogenic fungi act by a multiplicity of mechanisms for pathogenesis. Their pathogenic action depends on contact causing infection and killing the sucking insect pests such as aphids, thrips, mealy bugs, whiteflies, scale insects, mosquitoes and all types of mites (Maina et al., 2018). During vegetative growth, the fungus may produce and release a variety of metabolites, favoring its growth or acting as virulence factors or toxins.

Mostly, entomopathogenic fungi (EPFs) are opportunistic, soil-borne fungal species from hyphomycetes group and usually cause insect mortality by nutritional deficiency, destruction of tissues and by the release of toxins. The entry of entomopathogenic fungi through the insect cuticle occurs by a synergistic combination of mechanical pressure and enzymatic degradation. Cuticle degrading enzymes such as chitinase, lipase and protease play an important role in the pathogenicity of these organisms on insects and facilitate the breakdown of insect cuticle for penetration of the fungal germ tube into the insect body.

Due to a combined enzymatic and mechanical action, the fungus penetrates the host body and the mycelium develops internally, often producing different types of conidia or spores colonizing the host. The infection process normally starts with the germination of conidia or spores that have encountered the host cuticle. These fungi attack the host via the integument or gut epithelium and establish their conidia in the joints and integument. When the fungal conidia encounter the host, they attach themselves to the cuticle through hydrophobic mechanisms and germinate to form germ tubes in favorable conditions (Inglis et al., 2001).

During this process, the fungus produces specialized infection structures that include penetration pegs and/or appressoria, which enable the growing hyphae to penetrate the host integument (Ortiz-Urquiza and Kehyani, 2013).

The germ tube penetrates the cuticle aided by the action of other enzymes such as metalloid proteases and amino peptidases (Bidochka and Small, 2005). Once inside the insect, the fungi develop as hyphal bodies that disseminate through the haemocoel and invade diverse fatty bodies, malpighian tubules, mitochondria, muscle tissues and hemocytes, leading to insect mortality within 3–14 days after infection (Kachhawa, 2017).

The infection is triggered by the germination of a conidium or spore, requiring specific environmental conditions (i.e., temperature and relative humidity). To ensure spread in the environment, new conidia or spores will be produced outside the infected host. Before this stage, the host affected by both the biochemical and mechanical action of the fungus normally dies. Some species, such as *B. bassiana* and *M. anisopliae*, cause muscardine disease after the cadaver or the dead body is mummified and covered by mycelial growth (Patrick and Kaskey, 2012).

7.6.3 Viruses

The mechanism of viral pathogenesis is through replication of the virus in the nucleus or in the cytoplasm of the target cell. Baculoviruses act orally against insects, and the first infection normally takes place after ingestion of contaminated food. Ingested occlusion bodies within the midgut environment release specific types of virions, called occlusion-derived viruses (Slack and Basil, 2007).

Numerous virions of NPVs are occluded within each occlusion body to develop polyhedra. However, the GV virion is occluded in a single small occlusion body to generate granules. Infected nuclei can produce hundreds of polyhedra and thousands of granules per cell. The expression of viral proteins occurs in three phases, namely: (1) early phase, i.e., 0–6 h post-infection; (2) late phase, i.e., 6–24 h post-infection; and 3) very late phase, i.e., up to 72 h post-infection. It is at the late phase that virions assemble as the 29 kDa-occlusion body protein is synthesized.

These occlusion derived viruses (ODVs) interact directly with the membrane of microvillar epithelial cells through the action of their envelope proteins (i.e., protein infectivity factors or PIFs). Within the nucleus of infected midgut cells, a second type of virion, called budded viruses (BVs), are produced, ensuring the successive spread of the virus throughout the host. As the infection spreads, the dead insect body progressively liquefies, favoring the dispersal of virus particles in the environment (Williams et al., 2017). Viral infections are also able to induce behavioral changes in the hosts, affecting their gene expression mechanisms (Katsuma et al., 2012). These can create enzootics, deplete the pest populations and ultimately create a significant impact on the economic threshold of the pest (Harrison and Hoover, 2012).

Viral biopesticides are usually only active against a narrow host spectrum and after their application on plant surfaces; baculovirus occlusion bodies (OBs) are rapidly inactivated by solar ultraviolet (UV) radiation, particularly in the UV-B range of 280–320 nm (Akhanaev et al., 2017). UV inactivation can be controlled by using plastic greenhouse structures that reduce the intensity of UV-B (280–315 nm) readings by >90% compared with external readings leading to an increase in the prevalence of infection in larvae. However, their efficacy can further be improved by the use of formulations that include stilbene-derived optical brighteners, which enhance the susceptibility to NPV infection by disrupting the peritrophic membrane (Shapiro and Argauer, 2001), inhibiting sloughing (Williams et al., 2017) or virus-induced apoptosis of insect midgut cells (Harrison and Hoover, 2012). Due to their mode of action, commercially available baculovirus-based products are only active against chewing insects, especially Lepidopteran caterpillars. Because of the low stability of baculovirus formulations in the environment and their high production costs related to the need to reproduce them within their host, their use in biological pest management is limited to specific niche market segments (Sun, 2015).

7.6.4 Nematodes

Normally, insect pathogenic nematodes enter the host through natural openings like the oral cavity, anus or spiracles and release their symbiotic bacteria in the hemocoel, resulting in bacterial proliferation which is accompanied by the release of toxins and virulence factors that weaken the host and the production of metabolites that favor the creation of a suitable environment for nematode reproduction (Dar, 2019). *Heterorhabditis* spp. harbor *Photorhabdus* and *Steinernema* spp. carry *Xenorhabdus* sp., producing an insecticidal toxin complex (Tc), including different subunits that show toxicity against insects by ingestion (Ffrench-Constant et al., 2007). The only insect-parasitic nematode possessing an optimal balance of biological control attributes are insecticidal nematodes in the genera *Steinernema* and *Heterorhabditis*, which attack the hosts as infective juveniles (IJs) (Ruiu, 2018). IJs are free-living organisms, which enter the hosts through the mouth, anus, spiracles or cuticle and can release their bacterial symbionts within 24–48 h (Ffrench-Constant et al., 2007). Nematodes can complete up to three generations within the host, after which, the IJs leave the cadaver to find new hosts (Shapiro-Ilan et al., 2012). A variety of improved *in vivo* and *in vitro* methods for nematode production at small and large-scale have been developed. The quality of the final formulation plays a major role in the efficacy of nematode-based biological control applications against pests (Shapiro-Ilan et al., 2012).

A variety of products based on different nematode species are commercialized worldwide, which target specific pest species and market segments (Table 7.7).

7.6.5 Protozoan

Protozoans infect a wide range of insect hosts in natural environments. Although these pathogens can kill their insect hosts, many are more important for their chronic, debilitating effects. One important and common consequence of protozoan infection is a reduction in the number of offspring produced by infected insects. Although protozoan pathogens play a significant role in the natural limitation of insect populations, few appear to be suited for development as insecticides (Henry, 2003).

Two types of transmission were reported during the study of *Nosema pyrausta*, a microsporidium infecting the European corn borer, *Ostrinia nubilalis.* In horizontal transmission, it was found that a spore eaten by the insect larva germinates in the midgut, extrudes a polar filament and injects sporoplasm into a midgut cell. The sporoplasm reproduces and then forms more spores, which can infect other tissues. Spores in infected midgut cells are sloughed into the gut lumen and are eliminated along with feces to the maize plant. These spores remain viable and are consumed during larval feeding so that the infection cycle is repeated in midgut cells of the new host (Koul, 2011). If a female larva is infected, *Nosema* is passed to the filial generation by vertical transmission. As the infected larva develops through an adult, the ovarian tissue and developing oocytes become infected with *N. pyrausta.* The embryo is infected within the yolk and when larvae hatch, they are infected with *N. pyrausta.*

Microsporidian infections in insects are thought to be common and responsible for naturally occurring, low to moderate insect mortality, taking days or weeks to harm their hosts (Kermani et al., 2013). Frequently, they reduce host reproduction or feeding rather than killing the pest outright. Some microsporidia are being investigated as microbial insecticides, and at least one is available commercially, but the technology is new, and work is needed to perfect the use of these organisms (Usta, 2013). Although they infect a wide range of pests naturally and induce chronic and debilitating effects that reduce the target pest populations, the use of protozoan pathogens as biopesticide agents has not been very successful.

7.7 RESISTANCE TO MICROBIALS

Among the various groups of microbial pathogens, development of resistance has been most frequently reported in the case of *B. thuringiensis* (Downes et al., 2010). Within the last few years, at

TABLE 7.7
Selection of Commercial Products Based on Entomopathogenic Nematodes

Active substances	Commercial names*	Main targets
Steinernema carpocapsae	Capsanem, Carpocapsae-System, Exhibitline SC, Optinem-C, NemaGard, Nemastar, NemaTrident-T, NemaRed, Nemasys C, Palma-Life	Borer beetles, caterpillars, crane fly, moth larvae, *Rhynchophorus ferrugineus*, Tipulidae
S. feltiae	Entonem, NemaShield, NemaTrident-F, Nemapom, Nemaplus, Nemaflor, NemaFly, Nemafrut, Nemasys F, Nematrip, Nematech-SSP, NemaTrident-S, Nemax-F, Nemycel, Steinernema-System, Optinem-F	*Bradysia* spp., *Chromatomyia syngenesiae*, *Phytomyza vitalbae*, soil-dwelling pests, codling moth larvae, sciarids, thrips
S. kraussei	Kraussei-System	Vine weevil larvae
Heterorhabditis bacteriophora	Larvanem, Nemaplant, NemaShield-HB, Nematop, Nematech-H NemaTrident-H, NemaTrident-C, Nema-green, Optinem-H	*Otiorhynchus* spp., chestnut moths, black vine weevil, soil-dwelling beetle larvae, *Melolontha melolontha*, caterpillars, cutworms, leafminers
H. downesi	NemaTrident-CT	Black vine weevil
Otiorhynchus sulcatus, *Phasmarhabditis hermaphrodita*	Slugtech-SP	Molluscs

Source: Ruiu, 2018.
*Different products may refer to different microbial strains. The representative trade names are those shown on the relevant company websites to which reference should be made for details.

least 16 insect species have been identified that exhibit resistance to *B. thuringiensis* endotoxins. Reports of development of resistance in field populations of *Plutella xylostella* are from the countries where *Bacillus thuringiensis* is extensively used, i.e., China, Japan, Philippines, Malaysia, India and North America. The field-evolved insect resistance to *B. thuringiensis* crops and various aspects related to resistance are more prominent in lepidopterans (Huang et al., 2011). Field-evolved resistance has been documented in noctuids such as *Spodoptera frugiperda* and *H. zea* (Tabashnik et al., 2009).

Factors associated with field resistance are the failure to use high dose *B. thuringiensis* cultivars and a lack of a sufficient refuge as demonstrated by implementation of the high dose/refuge insect resistance management strategy that has been successful in delaying field resistance to Bt crops (Huang et al., 2011).

Genetic engineering is an important and useful tool to resolve and avoid this problem of resistance, where microbial genes from *B. thuringiensis* were incorporated and transferred to plants resulting in transgenics. To date, we have *B. thuringiensis* cotton and *B. thuringiensis* maize available in 13 and nine countries, respectively, grown on 42.1 million ha of land (Shelton et al., 2008). The development of such transgenics was seen as a panacea in terms of microbial control of pests; however, field resistance in *H. zea* as a result of an increase in the frequency of resistance alleles is alarming (Tabashnik et al., 2008).

Gene pyramiding is another approach used to try and address the emerging resistance problem (Sushmita et al., 2016). Pyramiding (Manyangariwa et al., 2006) means the stacking of multiple genes so that more than one toxin is expressed in the transgenic plant. However, gene pyramiding needs to be sustainable and no cross or multiple resistances should occur. The problem of developing multiple resistances cannot be completely ignored as in the end they would render such strategies ineffective. Asymmetrical cross-resistance between *B. thuringiensis* toxins Cry1Ac and Cry2Ab in pink boll worm (Downes et al., 2010; Tabashnik et al., 2009) suggests that it is important to incorporate the potential effects of such cross-resistance in resistance management plans to help sustain the efficacy of pyramided *B. thuringiensis* crops. Current evidence suggests that gene pyramiding may not be a sustainable strategy *per se*; therefore, other management strategies such as crop rotation, use of predators and parasitoids and refugia might be considered and needs to be incorporated in the management plans (Zhao et al., 2003).

Transgenic plants that control insects via RNA interference (RNAi) are going to be a reality soon (Baum et al., 2007; Vogel et al., 2019), which can help in minimizing the drawbacks of resistance and will further broaden the scope of transgenics. Some recent studies (Samuel et al., 2019) have demonstrated that toxin-binding proteins such as cadherin facilitate toxin oligomerization and thus modify and promote the toxicity of *B. thuringiensis* toxin, which in turn can prevent the resistance in comparison with the standard *B. thuringiensis* toxins. The studies showed that cadherin gene silencing with RNAi in *Manduca sexta* reduces the toxicity of *B. thuringiensis* toxin Cry1Ab. The toxins that had cadherin deletion mutations killed cadherin-silenced *M. sexta* and *B. thuringiensis* resistant *Pectinophora gossypiella* (Soberon et al., 2007).

The first documented instance of field resistance to a commercially applied baculovirus (Eberle and Jehle, 2006) in the field was found in Europe where *Cydia pomonella* GV is one of the main components of codling moth control. *C. pomonella* GV in apple orchards has led to a high degree of resistance in some populations (Frisch et al., 2007).

Apparently, this is either the result of the overuse of the product or the predominant control strategy applied.

However, there does not seem to be any reported examples of field development of resistance to entomopathogenic fungi or nematodes (Shelton et al., 2007). However, there is evidence to demonstrate the existence of natural resistance mechanisms in insects against fungi (Wilson et al., 2001) and nematodes (Kunkel et al., 2004), suggesting that resistance to these pathogens cannot be completely ignored.

7.8 BENEFITS OF MICROBIAL BIOPESTICIDES

Invertebrate pathogenic microorganisms employed as active substances in pest management are recognized as generally safe for the environment and non-target species in comparison with synthetic chemicals. This is in relation to the specificity of their mode of action, limiting their efficacy against one or a narrow range of pest species (Kaya and Vega, 2012). Microbial pesticides are non-toxic and non-pathogenic to non-target organisms and the safety offered is their greatest strength. Action of microbials is specific to a single group or species of pests, therefore, they do not directly affect beneficial animals such as predators and parasitoids.

Biological pesticides exhibit a multisite action, which hinders the development of resistant pests, fostering their use in resistance management programs. Accordingly, the employment of microbials in combination or in rotation with conventional pesticides are encouraged (Musser et al., 2006). A good efficacy against their targets can be achieved by employing these biopesticides as standalone products in organic farming. Reduced preharvest interval and the lack of significant residues on crops are the additional advantages of biopesticides. Microbial pesticides can be used in many habitats where chemical pesticides have been prohibited. Such habitats include recreational and urban areas, lakes and stream borders of watersheds and near homes and schools in agricultural settings.

Residues of microbial pesticides have the potential to control vectors, are non-hazardous and are safe all the time, even close to the harvesting period of the crop. Some pathogenic microbes can establish in a pest population or its habitat and provide control during subsequent seasons or pest generations.

The method of application in the field and the formulation features play a key role in the performance of any biopesticide. With the aim of enhancing efficacy, proprietary technologies have been developed by industry, maximizing the effects on the target and improving product application features (Satinder et al., 2006).

7.9 DISADVANTAGES OF MICROBIAL PESTICIDES

Owing to the specificity of the action, microbes may control only a portion of the pests present in a field and may not control other types of pests present in treated areas, which can cause continuous damage. As heat, UV light and desiccation reduces the efficacy of microbial pesticides, the delivery systems become an important factor. Special storage and formulation procedures are necessary. Shelf life is a major constraint, and given this short shelf life and because of their pest specificity, markets are limited. The development, registration and production costs cannot be spread over a wide range of pest control sales; for example, insect viruses are not widely available. Some insects develop resistance to several insect pathogens. Resistance management will have to be practiced, as it is with chemical pesticides. This trend is in line with the implementation of legislative frameworks fostering the registration and use of environmentally friendly products in different world regions. In this context, premarket authorization remains an important factor that slows down this innovation process, even if it is a necessary and indispensable tool to guarantee health safety.

7.10 FUTURE PERSPECTIVES

Over the last five decades, agriculturists/producers/cultivators have had to rely on conventional synthetic chemical insecticides for insect pest control. The availability of many broad-spectrum chemical pesticides is declining as a result of the evolution of resistance and legislation. Resistance has been reported for all insecticide classes either for one or more key pest species. The increasingly serious problems associated with broad-spectrum synthetic pesticides have led to the need for effective biodegradable pesticides with greater selectivity. This includes the development of novel solutions against new targets or the introduction of new technologies that enhance the efficacy of already available active substances. Since the establishment of the fact that microbes pathogenic to insects can be a key component in the fight against insect pests, academic and industrial investment in the biopesticide sector is experiencing a significant growth and many discoveries are being developed into new biopesticidal products that enlarge the global market. The availability of biopesticides acting against diverse crop pests is essential to ensure the management of agro-ecosystems respecting the environment and human health. The growing demand from farmers is accompanied by an increasing market offer of newly introduced and improved products that can be used alone, in rotation or in combination with conventional chemicals.

To date, several microbial insecticides have been developed and are being used against many insect pests of economic importance in several countries (Bailey et al., 2010). Nonetheless, new strains of microbes, which are pathogenic to insects are being discovered, a situation, which presents a brighter future for the use of entomopathogens in insect pest management. Nevertheless, it is still far behind synthetic chemicals in efficacy and popularity. While acknowledging limitations, one can still argue that the use of microbial pesticide is likely to rise if research is focused on it, improving its performance under challenging environmental conditions, with formulations that will increase persistence, shelf life, ease of application, pathogen virulence and a wider spectrum of action.

Advanced molecular studies on insect-microbial community diversity are also opening new frontiers for the development of innovative pest management strategies (Abdelfattah et al., 2018; Malacrinò et al., 2018). On the other hand, recent findings are contributing to foster a deeper understanding of the insect-microbial interactions within the plant ecosystem (Bennett et al., 2018).

Modern legislative frameworks require following criteria and principles of integrated pest management (IPM) in agro-ecosystems and are further fueling a significantly expanding market. Added to this are the efforts made by scientists working in the field of invertebrate pathology, whose studies aim to give light to new and increasingly effective microbial derived active substances.

Owing to some of the early successes and the continuing growth of the biopesticide market, expectations for the performance of microbial biopesticides have been high, however, there are many challenges that will need to be overcome. There is an immediate requirement to investigate the ecological relevance vis-à-vis the use of microbial biopesticides. Some recent studies revealed that the pattern and impact of these toxins varies from species to species, depending on the ecosystem, the route of exposure and the non-Bt control against which effects are quantified. As such, the effect of microbial biopesticides on microbial communities must be carefully monitored. In fact, there is a need for well-defined selection criteria and a complete process description for the development of a microbial pest control product. For a commercial microbial product, three specific criteria for selection are required: toxicity, production efficiency and safety of the product. That means while screening, process toxicity of the product will be relative to dose rate, mode of action, speed of kill, host range, sensitivity to abiotic factors and persistence. Second, mass production will be a critical criterion and should be a high-yield-oriented process. Third, safety of the product will be essential in relation to registration requirements and costs involved.

The current regulatory guidelines are inadequate, while information on the uptake of microbial control strategies must be collated and shared with the rest of the world.

One way to streamline and speed up product registration processes is through international harmonization of regulatory framework, for example, data requirements, fees, timelines, criteria for approval and risk assessments. Indeed, major steps have been taken to increase both the harmonization and transparency of data requirements and the procedures for risk assessment at OECD, North American and European Union levels (Kabaluk et al., 2010). Although, while harmonization is desirable, it should be kept in mind that microbial agents have a wide range of mechanisms of action, and because their properties are generally poorly understood relative to chemical pesticides, regulatory assessment frameworks must retain a degree of flexibility and reliance on expert opinion in order to comply with the "intra and interspecific variation of microorganisms and their constituents" (Mensink and Scheepmaker, 2007). Furthermore, to implement local production schemes in developing countries, intervention at the national and international level will be important.

An important question, however, remains that "when and where" the use of microbials as biopesticides is appropriate and required. As soon as any new idea floats in the market, researchers, scientists and biocontrol companies start developing a product without any specific well-developed plan, whereas the approach should be to develop a product to solve a problem or to grasp an opportunity. Thus, it is essential to make a detailed plan such as which pest to target, specific crop, region, the time at which the pest or the problem occurred, what are the available solutions, acceptable costs and market potential (Usta, 2013). If these aspects are considered in detail, a potential microbial product could be obtained. Therefore, certain steps should be taken into consideration in order to obtain a good microbial pest control product. These would be: (i) the collection of isolates and identification of the perfect isolate; (ii) laboratory screening for efficacy; (iii) assessment of production efficiency; (iv) mode-of-action and toxicological properties; (v) glass house trials; and (vi) evaluation of efficacy under commercial conditions. If all these factors are considered, success is inevitable; perseverance to develop such products will be rendered less risky.

In order to increase the utility of microbial pathogens in the Ecologically Based Invasive Plant Management (EBIPM) program, systematic surveys are required in different agroecological zones to identify naturally occurring pathogens. Detailed studies are necessary on the properties, mode of action and pathogenicity of such organisms. Ecological studies on the dynamics of diseases in insect populations are necessary because environmental factors play a significant role in disease outbreaks and ultimate control of the pests. It is expected that with the recent advancements in microbial research coupled with dedicated efforts from extension specialists, farmers, pest management regulators and the general public, microbial biopesticides could play a prominent role in future EBIPM and Area Wide Pesst Management (AWPM) programs (Koul, 2011). As mentioned above, structured project plans are required to achieve the goal. The roadmap to successful development and commercialization of a microbial pest control product is amply illustrated in new flow diagrams which provide details of various phases involved and output information leading to consecutive steps for decision making, and ultimately, market potential (Ravensberg, 2011).

REFERENCES

Abdelfattah, A., A. Malacrinò, M. Wisniewski et al. 2018. Metabarcoding: A powerful tool to investigate microbial communities and shape future plant protection strategies. *Biological Control* 120: 1–10.

Abang, A. F., C. M. Kouame, M. Abang et al. 2014. Assessing vegetable farmer knowledge of diseases and insect pests of vegetable and management practices under tropical conditions. *International Journal of Vegetable Science* 20(3): 240–253. doi: 10.1080/19315260.2013s.800625

Abrol, D. P. 2017. *Technological Innovations in Integrated Pest Management: Biorational and Ecological Perspective. (e-Book).* Scientific Publishers, p. 454.

Akhanaev, Y. B., I. A. Belousova, N. I. Ershov et al. 2017.Comparison of tolerance to sunlight between spatially distant and genetically different strains of *Lymantria dispar* nucleopolyhedrovirus. *PLoS ONE* 12(12): e0189992. doi: 10.1371/journal.pone.0189992.

Ahmad, S., Z. Khan, A. Khan et al. 2019. Biopesticides – Its prospects and limitation: An overview. In V. K. Gupta, A. K. Verna, and G. D. Singh (eds.). *Perspective in Animal Ecology and Reproduction.* New Delhi, India: Astral International (P) Ltd., pp. 296–314.

Bacci, L., D. Lupi, S. Savoldelli et al. 2016. A review of spinosyns, a derivative of biological acting substances as a class of insecticide with a broad range of action against many insect pests. *Journal of Entomological and Acarological Research* 48(1): 40–52. doi: 10.4081/jear2016.5653

Bailey, K. L., S. M. Boyetchko and T. Langle. 2010. Social and economic drivers shaping the future of biological control: A Canadian perspective on the factors affecting the development and use of microbial biopesticides. *Biological Control* 52: 221–229.

Baum, J. A., T. Bogaert, W. Clinton et al. 2007. Control of coleopteran insect pests through RNA interference. *Nature Biotechnology* 25: 1322–1326.

Belinato, T. A. and A. J. Martins. 2016. Insecticide resistance and fitness cost. In *Insecticides Resistance.* S. Trdan (ed.). Rijeka, Croatia: IntechOpen.

Bennett, A. E., P. Orrell, A. Malacrinò et al. 2018. Fungal-mediated above–belowground interactions: The community approach, stability, evolution, mechanisms, and applications. In *Aboveground–Belowground Community Ecology. Ecological Studies (Analysis and Synthesis).* T. Ohgushi et al. (eds.). Cham, Switzerland: Springer, Vol. 234, pp. 85–116.

Bhardwaj, T. and J. P. Sharma. 2013. Impact of pesticides application in agricultural industry: An Indian scenario. *International Journal of Agricultural Science and Food Technology* 14(8): 817–822.

Bidochka, M. J. and C. Small. 2005. Phylogeography of *Metarhizium*, an insect pathogenic fungus. In *Insect-Fungal Associations.* F. E. Vega and M. Blackwell (eds.). New York: Oxford University Press Inc., pp. 28–49.

Bravo, A., S. S. Gill and M. Soberon. 2007. Mode of action of *Bacillus thuringiensis* Cry and Cyt toxins and their potential for insect control. *Toxicon* 49: 423–435.

Charles, J. F., M. H. Silva-Filha and C. Nielsen-LeRoux. 2000. Mode of action of *Bacillus sphaericus* on mosquito larvae: Incidence on resistance. In *Entomopathogenic Bacteria: From Laboratory to Furrow Application.* J. F. Charles et al. (eds.). London, UK: Kluwer Academic Publishers, pp. 237–252.

Clem, R. J. and A. L. Passarelli. 2013. Baculoviruses: Sophisticated pathogens of insects. *PLoS Pathogens* 9: e1003729.

Copping, G. L. and J. J. Menn. 2000. Biopesticides: A review of their action, applications and efficacy. *Pest Management Science* 56: 651–676.

Cordova-Kreylos, A. L., L. E. Fernandez, M. Koivunen et al. 2013. Isolation and characterization of *Burkholderia rinojensis* sp. nov. a non-*Burkholderia cepacia* complex soil bacterium with insecticidal and miticidal activities. *Applied and Environmental Microbiology* 79: 7669–7678.

Dar, A. S., Z. H. Khan, A. A. Khan et al. 2019. Biopesticides- its prospects and limitations: An overview. In V. K. Gupta, A. K. Verma, and G. D. Singh (ed.). *Perspective in Animal Ecology and Reproduction.* New Delhi, India: Astral International (P) Ltd., pp. 296–314.

Dara, S. K. 2017. Entomopathogenic microorganisms: Modes of action and role in IPM. *UCANR e-Journal of Entomology and Biologicals.* UC ANR Blogs, May 20, 2017.

Devi, P. I. 2009. Health risk perceptions, awareness and handling behavior of pesticides by farm workers. *Agricultural Economics Research Journal* 22(9): 263–268.

Devi, P. I. 2010. Pesticides in agriculture—A boon or a curse? A case study of Kerala. *Economic and Political Weekly* 45: 26–27.

Devi, P. I. 2011. Is *Farm Labour Compensated for Occupational Risk? An Attempt Employing Hedonic Wage Model. Project Report.* Thrissur, Kerala: Kerala Agricultural University, p. 42.

Devi, P. I. 2012. Dynamics of farm labour use—An empirical analysis. *Agricultural Economics Research Review* 25(2): 317.

Devi, P. I. 2015. *Supply Side Analysis of Pesticide Markets in Kerala: Evidences from Retail Traders. A Research Report*. Thrissur, Kerala: Kerala Agricultural University, p. 56

Downes, S., R. J. Mahon, L. Rossiter et al. 2010. Adaptive management of pest resistance by *Helicoverpa* species (Noctuidae) in Australia to the Cry2Ab Bt toxin in Bollgard II cotton. *Evolution Applications* 3: 574–584.

Eberle, K. E. and J. A. Jehle. 2006. Field resistance of codling moth against *Cydia pomonella* granulovirus (CpGV) is autosomal and incompletely dominant inherited. *Journal of Invertebrate Pathology* 93: 201–206.

FAO. 2019. *The State of Food and Agriculture. Moving Forward on Food Loss and Waste Reduction*. Rome License: CC BY-NC-SA 3.0 IGO. Rome, Italy: FAO.

French-Constant, R. H., A. Dowling and N. R. Waterfield. 2007. Insecticidal toxins from *Photorhabdus* bacteria and their potential use in agriculture. *Toxicon* 49: 436–451.

Ffrench-Constant, R. and N. Waterfield. 2006. An ABC guide to the bacterial toxin complexes. *Advances in Applied Microbiology* 58: 169–183.

Frisch, E., K. Undorf-Spahn and J. Kienzle. 2007. Codling moth granulosivirus: First indication of variations in the susceptibility of local codling moth populations. *IOBC/ WPRS Bulletin* 30: 181–186.

Furlong, M. J. and K. J. Pell. 2005. Interactions between entomopathogenic fungi and arthropod natural enemies. In *Insect-Fungal Associations: Ecology and Evolution*. F. E. Vega and M. Blackwell (eds.). Oxford, UK: Oxford University Press, pp. 51–73.

Gozel, U. and C. Gozel 2016. Entomopatogenic mematodes in pest management. In *Integrated Pest Management(IPM): Environmentally Sound Pest Management*. H. K. Gill and G. Goyal (eds.). Rijeka, Croatia: IntechOpen. Available from: https://www.intechopen.com/books/integrated-pest-management-ipm-environmentally-sound-pest-management/entomopathogenic-nematodes-in-pest-management

Harris, A. K. 2009. Available from: http://www.farmchemicalsinternational.com/magazine/2000

Harrison, R. and K. Hoover. 2012. Baculoviruses and other occluded insect viruses. In F. E. Vega and H. K. Kaya (eds.), *Insect Pathology* (2nd ed.). London, UK: Academic Press, pp. 73–131.

Haase, S., A. Sciocco-Cap and V. Romanowski. 2015. Baculovirus insecticides in Latin America: Historical overview, current status and future perspectives. *Viruses* 7: 2230–2267.

Henry, J. E. 2003. Natural and applied control of Insects by Protozoa. *Annual Review of Entomology* 26(1): 49–73.

Huang, F., D. A. Andow and L. L. Buschman. 2011.Success of the high dose /refuge resistance management strategy after 15 years of Bt crop use in North America. *Entomologia Experimentalis et Applicata* 140: 1–16.

Hurst, M. R., T. R. Glare, T. A. Jackson et al. 2000. Plasmid-located pathogenicity determinants of *Serratia entomophila*, the causal agent of amber disease of grass grub, show similarity to the insecticidal toxins of *Photorhabdus luminescens*. *Journal of Bacteriology* 182: 5127–5138.

Iwanicki, N. S., A. A. Pereira, A. B. R. Z. Botelho et al. 2019. Monitoring of field application of *Metarhizium anisopliae* in Brazil revealed high molecular diversity of *Metarhizium* spp. in insects, soil and sugarcane roots. *Scientific Reports* 9: 4443. doi: 10.1038/s41598-019-38594-8

Inglis, G. D., M. Goettel, T. Butt et al. 2001. Use of hyphomycetous fungi for managing insect pests. In *Fungi as Biocontrol Agents: Progress, Problems and Potentials*. T. M. Butt et al. (eds.). Wallingford, UK: CABI Publishing, pp. 23–70.

Jurat-Fuentes, J. L. and T. A. Jackson. 2012. Bacterial Entomopathogens. In *Insect Pathology* (2nd ed.). F. E. Vega and H. K. Kaya (eds.). London, UK: Academic Press, pp. 265–349.

Kabaluk, J., A. M. Svircev, M. S. Goette et al. 2010. *The Use and Regulation of Microbial Pesticides in Representative Jurisdictions Worldwide*. IOBC Global, p. 99. www.IOBC-global.org

Kachhawa, D. 2017. Microorganisms as a biopesticides. *Jour. of Entomology Zoology Studies* 5(3): 468–473.

Katsuma, S., Y. Koyano, W. Kang et al. 2012. The baculovirus uses a captured host phosphatase to induce enhanced locomotory activity in host caterpillars. *PLoS Pathogens* 8: e1002644. doi: 10.1371/journal.ppat.1002644

Kaur, R., M. Subbarayalu, R. Jagadeesan et al. 2015. Phosphine resistance in India is characterised by a dihydrolipoamide dehydrogenase variant that is otherwise unobserved in eukaryotes. *Heredity* 115(3): 188–194

Kaur, T., A. Vasudev, S. K. Sohal et al. 2014. Insecticidal and growth inhibitory potential of *Streptomyces hydrogenans* DH16 on major pest of India, *Spodoptera litura*(Fab.) (Lepidoptera: Noctuidae). *Microbiology* 14: 227. doi: 10.1186/s12866-014-0227-1

Kaya, H. K. and F. E. Vega. 2012. Scope and basic principles of insect pathology. In *Insect Pathology* (2nd ed.). F. E. Vega and H. K. Kaya (eds.). London, UK: Academic Press, pp. 1–12.

Kekuda, T. R., K. S. Shobha and R. Onkarappa. 2010. Potent insecticidal activity of two Streptomyces species isolated from the soils of Western ghats of Agumbe, Karnataka. *Journal of Natural Pharmaceuticals* 1(1): 29–32.

Kermani, N., Z.-A. Abu-hassan, H. Dieng et al. 2013. Pathogenecity of *Nosema* sp. (Microsporidia) in the Diamondback Moth, *Plutella xylostella* (Lepidoptera: Plutellidae). *PloS ONE* 8(5): e62884. doi: 10.1371/journal.pone.0062884

Kenis, M., B. P. Hurley, A. E. Hajek et al. 2017. Classical biological control of insect pests of trees: Facts and figures. *Biological Invasions* 19: 3401–3417.

Khan, M. A., Z. Khan, A. Wasim et al. 2015. Insect pest resistance: An alternative approach for crop protection. In *Genetic Modification of Crop Plants: Issues and Challenges*. K. R. Hakeem (ed.). Basel, Switzerland: Springer International Publishing, p. 257.

Kim, J. J., M. S. Goettel and D. R. Gillespie. 2008. Evaluation of *Lecanicillium longisporum*, Vertalec® for simultaneous suppression of cotton aphid, *Aphis gossypii*, and cucumber powdery mildew, *Sphaerotheca fuliginea*, on potted cucumbers. *Biological Control* 45: 404–409.

Kirst, H. A. 2010. The spinosyn family of insecticides: Realizing the potential of natural products research. *The Journal of Antibiotics* 63: 101–111.

Koul, O. 2011. Microbial biopesticides: Opportunities and challenges. CAB Reviews: *Perspectives in Agriculture, Veterinary Science, Nutrition and Natural Resources* 6(056): 1–26.

Kunimi, Y. 2007. Current status and prospects on microbial control in Japan. *Journal of Invertebrate Pathology* 95: 181–186.

Kunkel, B. A., P. S. Grewal and M. F. Quigley. 2004. A mechanism of acquired resistance against an entomopathogenic nematode by *Agrotis ipsilon* feeding on perennial ryegrass harboring a fungal endophyte. *Biological Control* 29: 100–108.

Landsberg, M. J., S. A. Jones, R. Rothnagel et al. 2011. 3D structure of the *Yersinia entomophaga* toxin complex and implications for insecticidal activity. *Proceedings of the National Academy of Sciences of the United States of America* 108: 20544–20549.

Lewis, E. E. and D. J. Clarke 2012. Nematode parasites and entomopathogens. In *Insect Pathology* (2nd ed.). F. E. Vega and H. K. Kaya (eds.). London, UK: Academic Press, pp. 395–424.

Lacey, L. and T. Unruh. 2005. Biological control of codling moth (*Cydia pomonella*, Lepidoptera: Tortricidae) and its role in integrated pest management, with emphasis on entomopathogens. *Vedalia* 12: 33–60.

Maina, U. M., I. B. Galadima, F. M. Gambo et al. 2018. A review on the use of entomopathogenic fungi in the management of insect pests of field crops. *Journal of Entomology and Zoology Studies* 6(1): 27–32.

Malacrinò, A., O. Campolo, R. F. Medina. 2018. Instar- and host-associated differentiation of bacterial communities in the Mediterranean fruit fly *Ceratitis capitata*. *PLoS ONE* 13: e0194131.

Marche, M. G., M. E. Mura, G. Falchi et al. 2017. Spore surface proteins of *Brevibacillus laterosporus* are involved in insect pathogenesis. *Scientific Reports* 7: 43805.

Marche, M. G., S. Camiolo and A. Porceddu. 2018. Survey of *Brevibacillus laterosporus* insecticidal protein genes and virulence factors. *Journal of Invertebrate Pathology* 155: 38–43.

Manyangariwa, W., M. Turnbill, G. S. McCutcheon et al. 2006. Gene pyramiding as a Bt resistance management strategy: How sustainable is this strategy. *African Journal of Biotechnology* 5: 781–785.

Marrone, P. G. 2014. The market and potential for biopesticides. In *Biopesticides: State of the Art and Future Opportunities*. A. D. Gross et al. (eds.). Washington, DC: American Chemical Society, pp. 245–258.

Martin, P. A. W., D. Gundersen-Rindal, M. Blackburn et al. 2007. *Chromobacterium subtsugae sp. nov.*, a betaproteobacterium toxic to Colorado potato beetle and other insect pests. *The International Journal of Systematic and Evolutionary Microbiology* 57: 993–999.

McKinnon, A. C., S. Saari, M. E. Moran-Diez et al. 2017. *Beauveria bassiana* as an endophyte: A critical review on associated methodology and biocontrol potential. *BioControl* 62: 1–17.

Mengistie, B. T., A. P. J. Mol and P. Oosterveer. 2017. Pesticide use practices among smallholder vegetable farmers in Ethiopian Central Rift Valley. *Environment, Development and Sustainability* 19: 301–324. doi: 10.1007/s10668-015-9728-9

Mensink, B. J. W. G. and J. W. A. Scheepmaker. 2007. How to evaluate the environmental safety of microbial plant protection products: A proposal. *Biocontrol Science and Technology* 17: 3–20.

Mnif, I. and D. Ghribi. 2015. Potential of bacterial derived biopesticides in pest management. *Crop Protection* 77: 52–64. doi: 10.1016/j.cropro.2015.07.017

Musser, F. R., J. P. Nyrop and A. M. Shelton. 2006. Integrating biological and chemical controls in decision-making: European corn borer (Lepidoptera: Crambidae) control in sweet corn as an example. *Journal of Economic Entomology* 99: 1538–1549.

Nayak, M. K., J. C. Holloway, R. N. Emery et al. 2013. Strong resistance to phosphine in the rusty grain beetle, *Cryptolestes ferrugineus* (Stephens) (Coleoptera: Laemophloeidae): Its characterization, a rapid assay for diagnosis and its distribution in Australia. *Pest Management Science* 69: 48–53.

Nunez, E. J. Iannacone, and H. G. Omez. 2008. Effect of two entomopathogenic fungi in controlling *Aleurodicus cocois* (Curtis, 1846) (Hemiptera: Aleyrodidae). *Chilean Journal of Agricultural Research* 68: 21–30.

Ortiz-Urquiza, A. and N. O. Keyhani. 2013. Action on the surface: Entomopathogenic fungi versus the insect cuticle. *Insects* 4: 357–374.

Ozkara, A., D. Akyil and M. Konuk. 2016. Pesticides, environmental pollution, and health. In *Environmental Health Risk-Hazardous Factors to Living Species*. M. L. Larramendy and S. Soloneski (eds.). Rijeka, Croatia: IntechOpen.

Pathak, D. V., R. Yadav and M. Kumar. 2017. Microbial pesticides: Development, prospects and popularization in India. In *Plant-Microbe Interaction in Agro-Ecological Perspectives*. H. B. Singh et al. (eds.). Singapore: Springer Nature Singapore Pvt. Ltd, pp. 455–471.

Patrick, W. and J. Kaskey. 2012. Biopesticide: Killer bugs for hire. *Bloomberg Business Week*. https://www.bloomberg.com/news/articles/2012-07

Pawar, A. D. and B. Singh. 1993. Prospects of botanicals and biopesticides. In *Botanical and Biopesticides*. New Delhi: Westville Publishing House.

Pigott, C. R. and D. J. Ellar. 2007. Role of receptors in *Bacillus thuringiensis* crystal toxin activity. *Microbiology and Molecular Biology Reviews* 71: 255–281.

Rai, D., V. Updhyay, P. Mehra et al. 2014. Potential of entomophogenic fungus as biopesticides. *Indian Journal of Science and Technology* 2(5): 7–13.

Ravensberg, W. J. 2011. *Roadmap to the Successful Development and Commercialization of Microbial Pest Products for Control of Artropods*. Dordrecht, Netherlands: Springer.

Rohrmann, G. F. 2011. *Baculovirus Molecular Biology* (2nd ed.). Bethesda, MD: National Library of Medicine (US), National Center for Biotechnology Information. Available online: http://www.ncbi.nlm.nih.gov/books/NBK49500/ (accessed on 29 August 2019).

Rondon, S. I. and Y. Gao. 2018. *The Journey of the Potato Tuberworm Around the World*. Rijeka, Croatia: IntechOpen.

Ruiu, L., A. Satta and I. Floris. 2011. Comparative applications of azadirachtin- and *Brevibacillus laterosporus*-based formulations for house fly management experiments in dairy farms. *Journal of Medical Entomology* 48: 345–350.

Ruiu, L. 2013. *Brevibacillus laterosporus*, a pathogen of invertebrates and a broad-spectrum antimicrobial species. *Insects* 4: 476–492.

Ruiu, L., A. Satta and I. Floris. 2014. Administration of *Brevibacillus laterosporus* spores as poultry feed additive to inhibit house fly development in feces: A new eco-sustainable concept. *Poultry Science* 93: 519–526.

Ruiu, L. 2018. Microbial biopesticides in agroecosystems. *Agronomy* 8: 235. doi: 10.3390/agronomy8110235

Samuel, C. C., M. Y. Rafii, S. I. Ramlee et al. 2019. Marker-assisted selection and gene pyramiding for resistance to bacterial leaf blight disease of rice (*Oryzae sativa* L.). *Biotechnology and Biotechnological Equipment* 33(1): 440–445. doi: 10.1080/13102818.2019.1584054

Satinder, K. B., M. Verma, R. D. Tyagi et al. 2006. Recent advances in downstream processing and formulations of *Bacillus thuringiensis* based biopesticides. *Process Biochemistry* 41: 323–342.

Shankarganesh, K., S. Walia, S. Dhingra et al. 2012. Effect of dihydrodillapiole on pyrethroid resistance associated esterase inhibition in an Indian population of *Spodoptera litura* (Fabricius). *Pesticide Biochemistry and Physiology* 102: 86–90.

Schrank, A. and M. H. Vainstein. 2010. *Metarhizium anisopliae* enzymes and toxins. *Toxicon* 56: 1267–1274.

Shelton, A. M., J. Romeis and G. G. Kennedy. 2008. IPM and GM, insect-protected plants: Thoughts for the future. In *Genetically Modified Crops within IPM Programs*. A. M. Shelton and G. G. Kennedy (eds.). New York: Springer, pp. 419–429.

Shelton, A. M., P. Wang, J.-Z. Zhao et al. 2007. Resistance to insect pathogens and strategies to manage resistance: An update. In *Field Manual of Techniques in Invertebrate Pathology*. L. Lacey and H. K. Kaya (eds.). Dordrecht, Netherlands: Springer, pp. 793–811.

Shapiro, M. and R. Argauer. 2001. Relative effectiveness of selected stilbene optical enhancers of the beet armyworm (Lepidoptera: Noctuidae) nuclear polyhedrosis virus. *Journal of Economic Entomology* 94(2): 339–343. doi: 10.1603/0022-0493-94.2.339

Shapiro-Ilan, D. I., R. Han and C. Dolinksi. 2012. Entomopathogenic nematode production and application technology. *Journal of Nematology* 44: 206–217.

Slack, J. and A. Basil. 2007. The baculoviruses occlusion-derived virus: Virion structure and function. *Advances in Virus Research* 69: 99–165. doi: 10.1016/S0065-3527(06)69003-9

Soberon, M., L. Pardo-Lopez, I. Lopez et al. 2007. Engineering modified Bt toxins to counter insect resistance. *Science* 318: 1640–1642.

Sun, X. 2015. History and current status of development and use of viral insecticides in China. *Viruses* 7: 306–319.

Sushmita, K., B. Ramesh, D. Pattanayak et al. 2016. Gene pyramiding: A strategy for insect resistance management in Bt transgenic crops. *IJBT* 15(3): 283–291.

Tabashnik, B. E., A. J. Gassmann, D. W. Crowder et al. 2008. Insect resistance to Bt crops: Evidence versus theory. *Nature Biotechnology* 26: 199–202.

Tabashnik, B. E., G. C. Unnithan, L. Masson et al. 2009.Asymmetrical cross-resistance between *Bacillus thuringiensis* toxins Cry1Ac and Cry2Ab in pink bollworm. *Proceedings of National Academy of Sciences United State of America* 106: 11889–11894.

Tabashnik, B. E., J. B. J. Van Rensburg and Y. Carriere. 2009. Field evolved insect resistance to Bt crops: Definition, theory, and data. *Journal of Economic Entomology* 102: 2011–2025.

Tofangsazi, N., S. P. Arthurs and R. M. G. Davis. 2015. *Entomopathogenic Nematodes (Nematoda: Rhabditida: Families Steinernematidae and Heterorhabditidae)*. One of a series of the Entomology and Nematology Department. Gainesville, FL: UF/IFAS Extension, pp. 1–5.

United States Environmental Protection Agency. 2017. Biopesticides. Available online: www.epa.gov/pesticides/biopesticides.

Usta, C. (2013). Microorganisms in biological pest control — A review (Bacterial toxin application and effect of environmental factors). In *Current Progress in Biological Research*. M. Silva-Opps (ed.). Rijeka, Croatia: IntechOpen.

Vega, F. E., M. S. Goettel, M. Blackwell et al. 2009. Fungal entomopathogens: New insights on their ecology. *Fungal Ecology* 2: 149–159.

Vidal, S. and L. R. Jaber. 2015. Entomopathogenic fungi as endophytes: Plant-endophyte-herbivore interactions and prospects for use in biological control. *Current Science* 109: 46–54.

Villaverde, J. J., B. Sevilla-Morán, P. Sandín-España et al. 2014. Biopesticides in the framework of the European Pesticide Regulation (EC) No. 1107/2009. *Pest Management Science* 70: 2–5.

Vodovar, N., D. Vallenet, S. Cruveiller et al. 2006. Complete genome sequence of the entomopathogenic and metabolically versatile soil bacterium *Pseudomonas entomophila. Nature Biotechnology* 24: 673–679.

Vogel, E., D. Santos, L. Mingels et al. 2019. RNA interference in insects: Protecting beneficials and controlling pests. *Frontiers in Physiology* 9: 1912. doi: 10.3389/fphys.2018.01912

Williams, T., C. Virto, R. Murillo et al. 2017. Covert infection of insects by baculoviruses. *Frontiers in Microbiology* 8: 1337.

Wilson, K., S. C. Cotter, A. F. Reeson et al. 2001. Melanism and disease resistance in insects. *Ecology Letters* 4: 637–649.

Wraight, S. P., R. I. Carruthers, S. T. Jaronski et al. 2000. Evaluation of the entomopathogenic fungi *Beauveria bassiana* and *Paecilomyces fumosoroseus* for microbial control of the silver leaf whitefly, *Bemisia argentifolii. Biological Control* 17: 203–217.

Yadav, S. and S. Dutta. 2019. A study of pesticide consumption pattern and farmer's perceptions towards pesticides: A case of Tijara Tehsil, Alwar (Rajasthan). *International Journal of Current Microbiology and Applied Sciences* 8(4): 96–104. doi: 10.20546/ijcmas.2019.804.012

Zhao, J.-Z., J. Cao, Y. Li et al. 2003. Transgenic plants expressing two *Bacillus thuringiensis* toxins delay insect resistance evolution. *Nature Biotechnology* 21: 1493–1497.

Zimmermann, G. 1993.The entomopathogenic fungus *Metarhizium anisopliae* and its potential as a biocontrol agent. *Pesticide Science* 37: 375–379.

Zimmermann, G. 2007a. Review on safety of the entomopathogenic fungi *Beauveria bassiana* and *Beauveria brongniartii. Biocontrol Science and Technology* 17: 553–596.

Zimmermann, G. 2007b. Review on safety of the entomopathogenic fungus *Metarhizium anisopliae. Biocontrol Science and Technology* 17: 879–920.

8 Bioactive Plant Metabolites as Stored Product Protectants

Amritesh C. Shukla

CONTENTS

8.1 INTRODUCTION

Over the past few decades, there has been much work on the extraction, isolation and identification of an extensive range of plant-based bioactive natural products that in some way affect the behavior, development and/or reproduction of agricultural pests/insects. Secoy and Smith (1983) recorded 677 different species of plants in 131 families suitable for use in pest control. Grainge et al. (1986) produced a suitable list of 1,600 plant species and Ahmed et al. (1984) reported about 2,000 species. Yang and Tang (1988) reported that in China, different parts or extracts of 276 plant species are used as pesticides. Jacobson (1990) in a survey reported that almost 1,500 plant species from 175 plant families act as insect feeding deterrents. In controlling stored product insects, Talukder (1995) listed 43 plant species as insect repellents, 21 plants as insect feeding deterrents, 47 plants as insect toxicants, 37 plants as grain protectants, 27 plants as insect reproduction inhibitors and seven plants as insect growth and development inhibitors. Machial et al. (2010) reported 17 plant metabolites as insecticidal against *Choristoneura rosaceana* Harris, and *Trichoplusia ni* Hubner. Akhtar et al. (2015) reported the antifeedant activity of the essential oil of *Azadirachta indica*, *Melia azedarach*, *Colocynthis citrullus*, *Nicotiana tabacum* and *Eucalyptus camaldulensis* against *Tribolium castaneum*, *Rhyzopertha dominica* and *Trogoderma granarium*. Saeidi and Pezhman (2018) investigated insecticidal properties of essential oils of *Mentha piperita*, *Mentha pulegium*, *Zataria multiflora* and *Thymus daenensis* against *Bruchus lentis* and *Callosobruchus maculatus*. Further, Brari and Kumar (2019) reported the plant metabolites of *Rabdosia rugosa* (Wall. ex Benth), *Zanthoxylum armatum* (DC.), *Artemisia maritima* (L.) and *Colebrookea oppositifolia* (Sm.) as antifeedant against four stored product insect pests viz., *Tribolium castaneum* (Herbst.), *Sitophilus oryzae* (L.),

Stegobium paniceum (L.) and *Plodia interpunctella* (Hubner). These findings indicate the increasing attempts to replace synthetic insecticides with low cost, locally available, ecofriendly pest control methods have been undertaken, at the global level.

Pesticidal plants are utilized in two main ways:

1. The first approach is the isolation, identification and chemical synthesis of active compounds. If feasible, these compounds or their active derivatives are synthesized and marketed by the chemical industry.
2. The second approach is suitable for farmers in developing countries and organic farming. Plant tissues or crude products of the plant, such as aqueous or organic solvent extracts, are used directly. These practices are labor intensive, but are often economically and ecologically sound, and do not require sophisticated technology.

8.2 CLASSIFICATION OF BIOACTIVE PLANT METABOLITES

Depending upon the physiological activities of the insects, plant products and their components can be classified into six different categories viz., attractants, repellents, feeding deterrents/antifeedants, toxicants, reproduction inhibitors and growth retardants (Jacobson 1982; Rajashekar et al., 2012).

8.2.1 ATTRACTANTS

Plant products and their components that cause insects to make oriented movements toward their source are known as insect attractants. They influence both gustatory (taste) and olfactory (smell) receptors or sensilla. Isothiocyantes from seeds of *Crucifera* and sugar and molasses and terpenes from bark with pheromones are natural attractants for various insects of Cruciferae and bark beetles. Onion propyl mercapton from Umbelliferae and phenylacetaldehyde from flowers of *Araujia sericifera* are attracted carrot fly (*Psila rosae*) and Lepidoptera, respectively. Insect attractants can be used in three ways for the control of insects. In sampling or monitoring insect populations to assess the extent of infestation, the measure of control to be adapted is decided by using, for example, insecticide-coated traps, poison baits or distracting insects from normal mating, aggregation feeding or oviposition. They do not kill the insects, therefore, they do not disturb ecosystems. They can be used to misguide the insects to the wrong oviposition sites whereby their number will go down by starvation or by producing unfertilized eggs. They cannot be relied on as a sole control measure used only in an integrated control program (Arora et al., 2012).

8.2.2 REPELLENT

Repellents are desirable chemicals offering protection with minimal impact on the ecosystem by driving away insect pests from treated materials by stimulating olfactory or other insect receptors. According to Dethier et al. (1960), an insect repellent is a chemical stimulus, which causes the insect to make oriented movements away from the source of stimulus. Further, repellents from plant origin are considered safe in pest control operations as they minimize pesticide residues and ensure the safety of people, food, the environment and wildlife (Talukder et al., 2004). The plant extracts, powders and essential oils from different bioactive plants were reported as repellents against different economically important stored product insects (Tripathi et al., 2000; Owusu, 2001; Khan and Gumbs, 2003; Boeke et al., 2004; Talukder et al., 2004; Garcia et al., 2007; Rajendran and Sriranjini, 2008; Tripathi and Upadhya, 2009; Ko et al., 2009; Shaaya and Kostjukovsky, 2009). The essential oil of *Artemisia annua* was found as a repellent against *Tribolium castaneum* and *Calloso-bruchus maculatus* (Tripathi et al., 2000). However, the essential oil of *Melia azedarach* (Bakain), *Azadirachta indica* (Neem) and *Datura stramonium* (Datura) were recorded for repellent activity against *Tribolium castaneum*, *Rhyzopertha dominica* and *Trogoderma granarium*

(Huang et al., 2000; Rajendran and Sriranjini, 2008; Batish et al., 2008; Sahaf and Moharramipour, 2008; Cosimi et al., 2009; Islam et al., 2009; and Hanif et al., 2016).

8.2.3 Feeding Deterrents/Antifeedants

Plant products and their pesticides inhibit feeding or disrupt insect feeding by rendering the treated materials unattractive or unpalatable (Talukder, 2006; Rajashekar et al., 2012). Antifeedants are sometimes referred to as "feeding deterrents" and are of great value in protecting stored commodities from insects. Insects remain on treated food indefinitely and they eventually starve to death without eating (Talukder and Howse, 2000; Talukder et al., 2004; Talukder 2005). Some naturally occurring antifeedants, which have been characterized, include glycosides of steroidal alkaloids, aromatic steroids, hydroxylated steroid meliantriol, triterpene hemiacetal etc. (Talukder and Howse, 2000; Lee, et al., 2004; Talukder et al., 2004; Ko et al., 2009). The screening of several medicinal herbs showed that the root bark of *Dictamnus dasycarpus* possessed significant feeding deterrence against two stored product insects (Liu et al., 2002). Chaudhary et al. (2017) and Ghoneim and Hamadah (2017) pointed out that azadirachtin, which is s prominent constituent of neem, was established as a pivotal insecticidal ingredient. Abdullah et al. (2017) reported that 1,8-cineole found in Galangal essential oil exhibited antifeedant activity, repellent activity and toxicity effect toward termites. Jose and Sujatha (2017) revealed that terpenoids, coumarin and phenols, present in the methanol extracts of *Gliricidia sepium* exhibited significant antifeedant activity. Similarly, Brari and Kumar (2019) reported the essential oils of *Rabdosia rugosa*, *Zanthoxylum armatum*, *Artemisia maritima* and *Colebrookea oppositifolia* had effects against four stored product insect pests viz., *Tribolium castaneum*, *Sitophilus oryzae*, *Stegobium paniceum* and *Plodia interpunctella*.

8.2.4 Toxicants

Some botanical pesticides are toxic and cause death to stored product insects (Padin et al., 2013). Toxicants are specific types of chemicals that directly kill insects and are also referred to as insecticides. As an insecticide, it is a stomach poison because it must be ingested to be effective (Isman, 2006). Worldwide reports on the toxicity of different plant derivates show that many plant products are toxic to stored product insects (Tripathi et al., 2000; Shukla et al., 2001; Channoo et al., 2002; Park et al., 2003; Boeke et al., 2004; Talukder et al., 2004; Islam and Talukder, 2005; Talukder, 2006; Obeng-Ofori, 2007; Shaaya and Kotjukovsky, 2009; Obeng-Ofori, 2010; Shukla, 2009; Shukla, 2012; Bouguerra et al., 2017; Germinara et al., 2017; Lucia et al., 2017; Papanastasiou et al., 2017; Qari et al., 2017; Trivedi et al., 2017; Wu et al., 2017; Zhao et al., 2017).

Essential oil of *Lavandula angustifolia* exhibited good fumigant and contact toxicity against granary weevil adults. Two major constituents of the essential oil of garlic, *Allium sativum*, methyl allyl disulfide and diallyl trisulfide were found as potent contact toxicant, fumigant and feeding deterrent against *Sitophilus zeamais* and *Tribolium castaneum* (Huang et al., 2000). The essential oil vapors distilled from anise, cumin, eucalyptus, oregano and rosemary were also reported as a fumigant and caused 100% mortality of the eggs of *Tribolium confusum* and *Ephestia kuehniella* (Tunc et al., 2000). In addition, a strong repellent activity is able to disrupt granary weevil orientation to an attractive host substrate (Germinara et al., 2017). Trivedi et al. (2017) demonstrated fumigant toxicity against the stored grain pest *Callosobruchus chinensis*. The essential oils of cinnamon, clove, rosemary, bergamot and Japanese mint showed potential to be developed as possible natural fumigants or repellents for control of the pulse beetle. Lucia et al. (2017) found that the mortality of adults and eggs for head lice associated with the use of (geraniol, citronellol, 1,8-cineole, linalool, α-terpineol, nonyl alcohol, thymol, menthol, carvacrol and eugenol) essential oils. Bouguerra et al. (2017) showed that *Thymus vulgaris* essential oil exhibited significant activity and could be considered as a potent natural larvicidal agent against *Culex pipiens*. Zhao et al. (2017) indicated that the essential oil of *Echinops grijsii* roots and the isolated thiophenes have an excellent potential for use in the control of *Aedes albopictus*, *Anopheles sinensis*, and *C. pipiens* pallens larvae and could be used in the search for new, safer and more effective natural compounds as larvicides. Wu et al. (2017) observed the toxicity and repellent activities of the rhizomes of *Zingiber zerumbet* (L.) Smith (Zingiberaceae) essential oil, which contains the component α-caryophyllene against cigarette beetles (*Lasioderma serricorne*). Papanastasiou et al. (2017) showed the toxicity of limonene, linalool and α-pinene on adult Mediterranean fruit flies. Qari et al. (2017) showed DNA damage due to alterations in the enzymatic system (acetylcholinesterase, acid phosphatase, alkaline phosphatase, lactate dehydrogenase and phenol oxidase), total protein and DNA concentration after treatment with essential oils of *Citrus aurantium*, *Eruca sativa*, *Zingiber officinale* and *Origanum majorana* against *Rhyzopertha dominica*.

8.2.5 Reproduction Inhibitors/Chemosterilants

Ground plant parts, extracts, oils and vapors also suppressed the fecundity and fertility of many insects (Singh et al., 2016). Some of the potential plant metabolites/products commonly used as reproduction inhibitors of insect pests are Azadirachtin (stored grain pests, aphids, caterpillars and mealybugs); Asarone (*Sitophilus oryzae*, *Lasioderma serricorne* and *Callosobruchus chinensis*); Nicotine (caterpillars, aphids and thrips); Pyrethrum (beetles, caterpillars, aphids, leafhoppers, spider mites, bugs and cabbage worms); Rotenone (beetles, bugs, aphids, spider mites and carpenter ants); Ryania (beetle, lace bugs, aphids and squash bug); Sabadilla (caterpillars, leaf hoppers, thrips, stink and squash bugs); Spinosads (caterpillars, leaf miners and foliage-feeding beetles). Further, many researchers reported that plant parts, oils or extracts mixed with grains reduced insect oviposition, egg hatchability, post-embryonic development and progeny production (Isman et al., 2008; Shaaya and Kotjukovsky, 2009; Ahtar et al., 2010). Reports also indicated that plant derivatives including essential oils caused mortality of insect eggs (Tunc et al., 2000; Lee et al., 2004; Ogendo et al., 2008; Rajendran and Sriranjini, 2008; Isman et al., 2008; Shaaya and Kotjukovsky, 2009; Ahtar et al., 2010; Isman, 2015; Tak and Isman, 2015; Stevenson et al., 2017).

8.2.6 Insect Growth and Development Inhibitors/Growth Retardants

Plant products and pesticides that show deleterious effects on the growth and development of insects, reducing the weight of larva, pupa and adult stages and lengthening the development stages are called growth retardants (Talukder, 2006). Plant derivatives also reduced the survival rates of larvae, pupae and adult emergence as well as inhibiting the development of eggs and immature stages inside grain kernels (Isman, 2000; Tripathi et al., 2000; Koul et al., 2008; Lee et al., 2004; Rajendran and Sriranjini, 2008; Ahtar et al., 2010; Isman et al., 2008; Isman et al., 2010; Isman, 2015; Tak and Isman 2015; Stevenson et al., 2017).

8.3 MANAGEMENT OF STORAGE INSECT PESTS

8.3.1 Synthetics Used for Pest Management

During the past few decades, the application of synthetic pesticides to control agricultural pests has been a standard practice. However, with growing evidence regarding detrimental effects of many of the conventional pesticides on health and environment, a requirement for safer means of pest management has become crucial (Isman and Akhtar, 2007; Shukla 2012; Stevenson et al., 2017). Despite numerous and ongoing research being conducted with new grain protectants, both synthetic and natural ones, only a few have been adopted to be used as grain protectants. Daglish (2006) discussed the barriers under biological, technical, legal and commercial categories as to why the adoption of new grain protectants is not widespread.

At the beginning of the new millennium, only two fumigants were in wide use in the world: phosphine and methyl bromide. Methyl bromide was already phased out, although critical uses were still allowed for some consumption awaiting alternatives, exceptions being for quarantine and pre-shipment treatment. There are various reasons for the disappearance of dozens of fumigants. First, there were health reasons (suspected or alleged carcinogens), flammability, no food registration, lack of interest, strict limitations on fumigant re-registration and so on. The restrictions on the use of fumigants have posed new global challenges to the food and chemical industry and have resulted in energized efforts to develop and register new fumigants as an alternative, primarily to methyl bromide (Navarro, 2006; Ducom, 2006). There are several newly developed fumigants or new fumigant formulations such as sulfuryl fluoride (Bell, 2005; Tsai et al., 2006; Chayapraser et al., 2006), carbonyl sulfide (Navarro, 2006; Ducom, 2006), propylene oxide (Navarro et al., 2004; Isikber et al., 2006), methyl iodide (Greech et al., 1996), ozone (Mason et al., 1999), ethyl formate (Annis et al., 2000), cyanogens (Yong and Trang, 2003) and ethanedinitrile (Navarro, 2006; Ducom, 2006). Some of these fumigants suffer from limitations and may only be used for treatment of a particular type of commodity or for application in a specific situation. Ethyl formate can be used as a promising candidate for the fumigation of stored food commodities especially dried fruits; carbon disulfide (an old fumigant still in use) for protection of seed materials; and carbonyl sulfide for grain fumigation (Navarro, 2006; Ducom, 2006). Global challenges in the research and development of new fumigants and technology of fumigation are in developing fumigants that will successfully replace highly effective and cheap phosphine and methyl bromide (Navarro, 2006; Ducom, 2006; Isman, 2006; Shaaya and Kostjukovsky, 2009; Isman, 2015).

8.3.2 Botanicals Used for Pest Management

Since time immemorial, plant materials have been used as a kind of natural protectant to protect stored grains. Neem plant parts, i.e., leaves, crushed seeds, powdered fruits and oil are the most traditional examples in this regard (Talukder et al., 2004; Islam and Talukder, 2005; Isman et al., 2008; Akhtar et al., 2008; Miresmailli and Isman, 2014; Tak and Isman, 2015; Wanna and Ngoen, 2019). Yadav and Bhatnagar (1987) reported that dried leaves of *Azadirachta indica* have been mixed with stored grains for protection against insects for a long time. In parts of Eastern Africa, leaves of some plants have traditionally been mixed as grain protectants (Isman 2006; Miresmailli and Isman, 2014; Stevenson et al., 2017). The grain protectant potential of different plant derivatives, including plant oils against major stored product pests, were also found to be very promising (Shukla et al., 2001; Lee et al., 2004; Tripathi and Upadhya, 2009). Worldwide reports have shown that leaf, bark, seed powder or oil extracts of plants mixed with stored grains reduced oviposition rate and suppressed adult emergence of stored product insects and reduced seed damage rates (Keita et al., 2001; Tapondjou et al., 2002; Talukder et al., 2004; Isman, 2006; Shaaya and Kotjukovsky, 2009; Ko et al., 2009; Tripathi and Upadhya, 2009; Hanif et al., 2016).

Further, the use of botanical pesticides has been emerging as one of the primary ways to protect crops and their products as well as the environment from pesticide pollution, which is a global problem (Isman, 2000, 2006, 2008, 2015). When extracted from plants, these chemicals are referred to collectively as "botanicals." Botanical insecticides possess a spectrum of properties including insecticidal activity, repellence to pests, antifeedancy, insect growth regulation and toxicity to nematodes, mites, snail, slugs and other pests of agricultural importance. Also, they possess antifungal, antiviral and antibacterial properties against pathogens. Generally, botanicals degrade more rapidly than most conventional (synthetic) pesticides, and so are considered relatively environmentally benign and less likely to kill beneficial insects and mites than insecticides with longer residual activity. Since most of them generally degrade within a few days, and sometimes within a few hours, these insecticides must be applied more often. More frequent application, plus higher costs of production usually make botanicals more expensive to use than synthetic insecticides (Shukla 2012; Isman, 2015). Among botanicals, plant volatile essential oils (EOs) are the most frequently studied as pesticides for pest and disease management (Isman, 2000; Shukla et al., 2001; Pascual, 2003; Lee et al., 2003, 2004; Rozman et al., 2007; Tripathi and Upadhya, 2009; Ko et al., 2009; Isman 2015).

However, essential oils, besides a large-scale demonstration of their efficacy and penetration, need a lot of research in order to determine their toxicological and safety data prior to registration (Daglish, 2006). Also, as with other groups of insecticides, the potential use of natural EOs in stored grain insect pest management depends on many factors. Isman (2006, 2008, 2015) tried to outline the challenges to and opportunities for the development and commercialization of new botanical insecticides and other natural insecticides. He believed, in spite of mostly favorable toxicology and minimal environmental impact and efficacy, botanicals and other natural insecticides need to fulfill many other considerations for successful commercialization and use. However, he believes that this group of insecticides may find a place in applications where there is a greater tolerance for the presence of insects and a focus is placed on environmental safety.

According to Rajendran and Sriranjini (2008), although in laboratory tests with adult insects some of the plant extracts have shown significant insect toxicity, their physical properties such as high boiling point, high molecular weight and very low vapor pressure are barriers for application in large-scale fumigations. The authors believe that plant products have the potential for small-scale treatments and space fumigations. Still, there is a lack of data for single or multiple components of essential oils on sorption, tainting and residues in food commodities. Also, the requirements for the registration of plant products may be another barrier (Rajendran and Sriranjini, 2008).

8.4 ESSENTIAL OILS FOR THE MANAGEMENT OF STORED GRAIN INSECT PESTS

The concentrations of natural EOs and their active components needed for effective fumigation have been studied by many researchers. In order to enable the comparison of toxicity data, we analyzed the reports that presented the doses of EOs in volumes, mostly in μg L^{-1} or μl L^{-1}, published during the last few years.

Tunc et al. (2000) tested the ovicidal activity of essential oil vapors distilled from anise *Pimpinella anisum* (L.), cumin *Cuminum cyminum* (L.), eucalyptus *Eucalyptus camaldulensis* (Dehnh.), oregano *Origanum syriacum* (L.) var. *bevanii* and rosemary *Rosmarinus officinalis* (L.) against the confused flour beetle, *Tribolium confusum* (du Val.) and the Mediterranean flour moth, *Ephestia kuehniella* (Zeller). The exposure to vapors of essential oils from anise and cumin resulted in 100% mortality of the eggs. At a concentration of 98.5 μl L^{-1} of anise essential oil, the LT_{99} values were 60.9 and 253.0 h for *E. kuehniella* and *T. confusum*, respectively. For the same concentration of the essential oil of cumin, the LT_{99} value for *E. kuehniella* was 127.0 h.

Sánchez-Ramos and Castanera (2000) found that the vapor of natural monoterpenes pulegone, eucalyptol, linalool, fenchone, menthone, α-terpinene and γ-terpinene at the concentration of 14 μl L^{-1} or below generated 90% mortality of the mobile stages of *Tyrophagus putrescentiae* (Schrank).

Lee et al. (2001) examined the fumigant toxicity of different essential oils toward rice weevil, *S. oryzae*. The essential oil from eucalyptus contained 1,8-cineole (81.1%), limonene (7.6%) and α-pinene (4.0%). The oil generated LD_{50} = 28.9 μl L^{-1} air; 1,8-cineole was more active (LD_{50} = 23.5 μl L^{-1} air) than limonene and α-pinene. Benzaldehyde (LD_{50} = 8.65 μl L^{-1} air) occurring in peach and almond kernels also had a potent fumigant toxicity toward rice weevils.

Papachristos and Stamopoulos (2002) assessed the toxicity of vapors of essential oils from *Lavandula hybrida* (Reverch.), *R. officinalis* and *Eucalyptus globulus* (Lab.) against the larvae and pupae of *Acanthoscelides obtectus* (Say.). The essential oil vapors were toxic to all immature stages tested with LC_{50} values ranging between 0.6 and 76 μl L^{-1} air, depending on oil and development stages.

Lee et al. (2003) evaluated the fumigant toxicity of 20 naturally occurring monoterpenoids against *S. oryzae*, *T. castaneum*, *O. surinamensis*, the house fly, *Musca domestica* L. and the German cockroach, *Blattella germanica* L. Cineole, *l*-fenchone and pulegone at 50 μg ml^{-1} air caused 100% mortality in all five species tested.

Lee et al. (2004) studied the potent fumigant toxicity of 42 essential oils and found out that six of them extracted from *Eucalyptus nicholii* (Maiden and Blakely), *Eucalyptus codonocarpa* (Blakely and McKie), *Eucalyptus blakely* (Maiden), *Callistemon sieberi* (F.Muell.), *Melaleuca fulgens* (R.Br.) and *Melaleuca armillary* (R.Br.) were toxic to *S. oryzae*, *R. dominica* and *T. castaneum*. The LD_{50} and LD_{95} against the adults of *S. oryzae* were between 19.0 to 30.6 and 43.6 to 56.0 μg ml^{-1} air, respectively. The LD_{95} of 1,8-cineole for *S. oryzae* was 47.9, for *R. dominica* 30.4 and for *T. castaneum* 21.0 μg ml^{-1} air.

Prajapati et al. (2005) evaluated the insecticidal, repellent and oviposition-deterrent activity of essential oils extracted from ten medicinal plants against *Anopheles stephensi* (Liston), *Aedes aegypti* (L.) and *Culex quinquefasciatus* (Say.). The essential oil of *Pimpinella anisum* (L.) showed toxicity against the fourth instar larvae of *A. stephensi* and *A. aegypti* with equivalent LD_{95} values of 115.7 μg ml^{-1}, whereas it was 149.7 μg ml^{-1} against *C. quinquefasciatus* larvae. Essential oils of *Zingiber officinale* and *Rosmarinus officinalis* were found to be ovicidal and repellent, respectively, toward the three mosquito species.

Ketoh et al. (2005) studied the effectiveness of the essential oil extracted from *Cymbopogon schoenanthus* (L.) against all development studies of *Callosobruchus maculatus* (Fab.). At the highest concentration tested (33.3 μl L^{-1}), all adults of *C. maculatus* were killed within 24 h of exposure to the oil and the development of newly laid eggs and neonate larvae was also inhibited.

Tapondjou et al. (2005) investigated the toxicity of cymol and essential oils of *Cupressus sempervirens* (L.) and *Eucalyptus saligna* (Sm.) against *S. zeamais* and *T. confusum*. *Eucalyptus* oil was more toxic than *Cupressus* oil to both insect species (LD_{50} = 0.36 μl cm^{-2} for *S. zeamais* and 0.48 μl cm^{-2} for *T. confusum)* on filter paper discs and was more toxic to *S. zeamais* on maize (LD_{50} = 38.05 μl per 40 g grain).

Ketoh et al. (2006) assessed the insecticidal activity of crude essential oils extracted from *Cymbopogon schoenanthus* (L.) and of its main constituent, piperitone, on different developmental stages of *C. maculatus*. Piperitone was more toxic to adults with a LC_{50} value of 1.6 μl L^{-1} vs. 2.7 μl L^{-1} obtained with the crude extract.

Wang et al. (2006) investigated repellent and fumigant activity of essential oil from mugwort *Artemisia vulgaris* (L.) to *T. castaneum*. At 8.0 μl mL^{-1}, mortality of adults reached 100%, but with 12-, 14- and 16-day larvae, mortalities were 49%, 53% and 52%, respectively. At dosages of 10, 15 and 20 μl L^{-1} air and a 96-h exposure period, mortality of eggs reached 100%. No larvae, pupae and adults were observed following a 60 μl L^{-1} dosage.

Choi Won-Sik et al. (2006) determined the toxicity of volatile components of thyme, sage, eucalyptus and clove bud against mushroom sciarid, *Lycoriella mali* (Fitch); *α*-pinene was the most toxic fumigant compound found in thyme essential oil (LD_{50} = 9.85 μl L^{-1} air) followed by *β*-pinene (LD_{50} = 11.85 μl L^{-1} air) and linalool (LD_{50} = 21.15 μl L^{-1} air). The mixture of *α*- and *β*-pinene exhibited stronger fumigant toxicity than *α*- or *β*-pinene itself against mushroom fly adults.

Negahban et al. (2007) determined the content of essential oil extracted from *Artemisia sieberi* (Besser). The oil contained camphor (54.7%), camphene (11.7%), 1,8-cineol (9.9%), *β*-thujone (5.6%) and *α*-pinene (2.5%).The mortality of seven-day-old adults of *C. maculatus*, *S. oryzae*, and *T. castaneum* increased with concentration from 37 to 926 μl L^{-1} and with exposure time from 3–24 h. A concentration of 37 μl L^{-1} and an exposure time of 24 h were sufficient to obtain 100% kill of the insects. *C. maculatus* was significantly more susceptible than *S. oryzae* and *T. castaneum*.

Rozman et al. (2007) investigated the toxicity of 1,8-cineole, camphor, eugenol, linalool, carvacrol, thymol, borneol, bornyl acetate and linalyl acetate against adults of *S. oryzae*, *R. dominica and T. castaneum*. The most sensitive species was *S. oryzae*, followed by *R. dominica*. *T. castaneum* was highly tolerant of the tested compounds; 1,8-cineole, borneol and thymol were highly effective against *S. oryzae* when applied for 24 h at the lowest dose (0.14 μl L^{-1}). For *R. dominica*, camphor and linalool were highly effective and produced 100% mortality in the same conditions. Against *T. castaneum*, no oil compounds achieved more than 20% mortality after exposure for 24 h, even with the highest dose (139 μl L^{-1}). However, after 7 days exposure, 1,8-cineole produced 92.5% mortality, followed by camphor (77.5%) and linalool (70.0%).

Stamopoulos et al. (2007) tested the vapor form of monoterpenoids terpinen-4-ol, 1,8-cineole, linalool, *R*-(+)-limonene and geraniol against different stages of *T. confusum*. The LC_{50} values ranged between 1.1 and 109.4 μl L^{-1} air for terpinen-4-01, 4 and 278 μl L^{-1} air for (*R*)-(+)-limonene and 1,8-cineole 3.5 and 466 μl L^{-1} air were the most toxic to all stages tested, followed by linalool with LC_{50} values ranging between 8.6 and 183.5 μl L^{-1} air, while the least toxic monoterpenoid tested was geraniol with LC_{50} values ranging between 607 and 1627 μl L^{-1} air.

Korunic and Rozman (2008) carried out three different experiments with 1,8-cineole. The authors conducted experiments in order to determine the efficacy of 509 g m^{-3} of cineole against different developmental stages of *S. oryzae*, *R. dominica* and *Cryptolestes ferrugineus* (Steph.) in wheat grain in the space 50% filled up with grain. Apparently, an applied dose of 50 g m^{-3} was not sufficient for effective control of younger developmental stages of *S. oryzae*, *R. dominica* and even *C. ferrugineus*, the most sensitive species tested. In the second experiment, the authors tested the effective concentration of cineole against adults of *S. oryzae*, *R. dominica*, *T. castaneum* and *C. ferrugineus* in the space 50% filled up with wheat applying cineole in the concentration range of 50, 100, 150, 200 and 250 g m^{-3}. The 100% mortality of *C. ferrugineus* was obtained with 50 g m^{-3} (lowest applied concentration). However, 100% mortality of *R. dominica* was obtained with a concentration of 150 g m^{-3} and 100% mortality of *S. oryzae* and *T. castaneum* at a concentration of 250 g m^{-3}. In the third experiment, the concentration of 50 g m^{-3} cineole in spaces differently filled up with wheat (empty space, 50% and 95% filled up) was assessed against the same four species. This concentration in empty space induced nearly 100% mortality in all four tested insect species. However, fumigation in a space 50% filled up with wheat, cineole was absolutely effective against *C. ferrugineus* only, with 50% to 60% efficacy against rice weevil and lesser grain borer, and only 11% against red flour beetle. In space 95% filled up with wheat, mortality of rusty grain beetle was 88%, rice weevil 34%, lesser grain borer 64% and red flour beetle only 4.5%.

Ebadollahi and Mahboubi (2011) reported the strong insecticidal toxicity of the essential oil of *Azilia eryngioides* against *Sitophilus granarius* and *Tribolium castaneum* (Herbst). Fumigation bioassays revealed that *A. eryngioides* oil had a strong insecticidal activity on adult test insects that were exposed to 37.03, 74.07, 111.11 and 148.14 μL L^{-1} to estimate mean lethal time (LT_{50}) values. Mortality increased as concentration and exposure time increased and reached 100% at the 39-h exposure time and concentrations higher than 111.11 μL L^{-1}. Another experiment was designed to determine the mean lethal concentration at the 24-h exposure time (LC_{50}), and these values indicated that *S. granarius* was more susceptible than *T. castaneum*.

Bossou et al. (2015) explored the fumigant toxicity of essential oils of *Cymbopogon citratus*, *Cymbopogon giganteus*, *C. schoenanthus* and *Eucalyptus citriodora*, against *Tribolium castaneum*. The findings recorded the LC_{50} values as 4.2 mL L^{-1} air, 2.3 mL L^{-1} air, 2.1 mL L^{-1} air and 2.0 mL L^{-1} air, as well as mortalities of 100%, 82%, 75% and 72%, respectively.

Aref et al. (2016) reported the insecticidal efficacy of the essential oil of *Eucalyptus floribundi* on the adult of *Rhyzopertha dominica* and *Oryzaephilus surinamensis*. The findings show that LC_{50} of *E. floribundi* essential oil was obtained for *R. dominica* and *O. surinamensis* as 34.39 and 43.54 μl l^{-1} air, respectively. LT_{50} of *R. dominica* at 100, 200 and 500 μl l^{-1} air concentrations were 4.28, 3.92 and 3.13. However, LT_{50} for *O. surinamensis* at the same concentrations was calculated as 4.90, 4.44 and 2.92. The experiment results showed that increasing the concentration of essential oil and the exposure time led to increases in mortality (Rajashekar et al., 2010).

Bett et al. (2016) reported the fumigant toxicity of *Cupressus lusitanica* and *Eucalyptus saligna* leaf essential oils against *T. castaneum*, *A. obtectus*, *Sitotroga cerealella* and *S. zeamais*. The findings show that the essential oil of *C. lusitanica* caused 90.6 and 100% mortality of adult *S. cerealella* and *A. obtectus*, however, the essential oil of *E. saligna*, at 15 μl L^{-1} air, caused 94.7 and 100% mortality for *A. obtectus* and *S. cerealella*, respectively, 24 h post-fumigation. Further, it was recorded that *C. lusitanica* oil was highly toxic with LC_{50} values of 4.08 and 4.71 μl L^{-1} air against

A. obtectus and *S. cerealella*, respectively, 24 h post-fumigation. The *E. saligna* leaf essential oil was moderately toxic with LC_{50} values of 6.71 and 7.02 and μl L^{-1} air for *S. cerealella* and *A. obtectus*, respectively, 24 h post-fumigation. However, *C. lusitanica* at a concentration of 20 μl L^{-1} air was more toxic to *S. zeamais and T. castaneum* with LC_{50} values of 13.54 and 15.28 μl L^{-1} air, respectively, 168 h post-fumigation.

De Souza et al. (2016) reported the fumigant activity of essential oils of *Ocimum basilicum*, *Citrus aurantium*, *Mentha spicata* and *Croton pulegiodorus* against *R. dominica*. The findings show that out of all the four essential oils, *O. basilicum*, exhibited strong fumigant toxicity against *R. dominica* adults, with a LC_{50} value of 17.67 μl L^{-1} air and LC_{100} value of 27.15 μl L^{-1} air.

Saeidi and Pezhman (2018) recorded the insecticidal activity of the essential oil of *Mentha piperita*, *Mentha pulegium*, *Zataria multiflora* and *Thymus daenensis* against two stored product beetles: *Bruchus lentis and Callosobruchus maculatus*. The LC_{50} value of *M. piperita* after 24 hours for *B. lentis* and *C. maculatus* was 14.62 and 13.70 μl L^{-1} air, respectively, while the values of LC_{50} were 92.32 and 95.80 for *M. pulegium*, 58.43 and 99.94 for *Z. multiflora* and 63.97 and 65.55 μl L^{-1} air for *T. daenensis*, respectively.

Idouaarame et al. (2018) reported the insecticidal activity of *Artemisia arborescens*, *Artemisia herba alba*, *Cupressus sempervirens*, *Eucalyptus camaldulensis*, *Tanacetum annuum* against *Rhyzopertha dominica*, *Sitophilus oryzae* and *Tribolium castaneum*. The observations show that the essential oil of *E. camaldulensis* proved to be the most toxic oil against *R. dominica* with 100% mortality after the first day of exposure at concentrations 0.21 and 0.11 μl/cm^3. Further, the essential oils of *A. arborescencs* and *A. herba alba* showed mortality of 100% at a concentration of 0.21 μl/cm^3. Other essential oils showed relatively low insecticidal activity (20% of mortality) at 0.053 μl/cm^3. However, in the case of *S. oryzae*, the essential oil of *E. camaldulensis* has a very significant insecticidal activity with a mortality of 100% at a concentration of 0.21 μl cm^{-3} after 1 day of exposure.

Similarly, in another study, Abd El-Salam et al. (2019) investigated the fumigant and toxic activity of some aromatic oils against *Oryzaephilus surinamensis*, and recorded that *O. basilicum* oil at the concentration of 80.0 ml/625 cm achieved good results for stored products.

The results clearly indicated that the use of plant metabolites/botanical products achieved good and high mortality percentages against various stored insect pests.

8.5 CHALLENGES TO THE UTILIZATION OF PLANT-BASED PESTICIDES

The successful utilization of plant products/botanicals can be a potential source for the management of the world's destructive pests and diseases, as well as reducing erosion, deforestation, desertification and perhaps even reducing the human population by acting as a spermicide (Obeng-Ofori, 2010). Besides this, plant metabolites/products can also increase the income of rural farmers and promote safety and quality of food and life in general (Obeng-Ofori, 2007; Rajashekar and Shivanandappa, 2010). But there are several challenges and limitations that must be overcome before their use as potential pesticides. The major constraints include:

1. Lack of experience and appreciation of the efficacy of botanicals for pest control. There are still doubts as to the effectiveness of plant-derived products (both "home-made" and commercial products) due to their slow action and lack of rapid knock-down effect.
2. Economic uncertainties occasioned by the seasonal supply of seeds, the perennial nature of most botanical trees and changes in potency with location and time with respect to geographical limitations.
3. Handling difficulties as there is no method for mechanizing the process of collecting, storing or handling the seeds or leaves or flowers from some of the perennial trees.
4. Instability of the active ingredients when exposed to direct sunlight.

5. Rapid degradation, although desirable in some respects, creates the need for more precise timing or more frequent applications.
6. Data on the effectiveness and long-term (chronic) mammalian toxicity are unavailable for some botanicals and tolerances for some have not been established.
7. Unavailability of data related to plant-based products/botanical pesticides and large-scale trials.
8. Difficulty of registration and patenting of natural products and a lack of standardization of botanical pesticide products.
9. Competition with synthetic pesticides through aggressive advertising by commercial pesticide dealers.
10. Commercial-formulated botanicals are more expensive than synthetic insecticides and are not as widely available.

8.6 CONCLUSION

Losses in stored products and agricultural commodities occur due to inappropriate harvesting, transportation, storage and distribution. Losses also occur during marketing because the shelf life of these products is governed by water content, respiratory rate and exogenous factors such as microbial growth, temperature, relative humidity and atmospheric compositions. Further, stored product losses of agricultural produce can be minimized and their storage life can be greatly increased by careful implementation of green technology (Maia and Moore, 2011). However, systematic analysis of each phase of commodity production and postharvest processing are important steps in identifying an appropriate strategy for reducing stored product losses. Future research must be focused on conducting large-scale trials to prove the feasibility of combination treatments of EOs and synergists. Cost benefit analysis also needs to be conducted in order to implement EO application as a green pesticide.

REFERENCES

Abd El-Salam AME, SAE Salem and RS Abdel-Rahman. (2019). Fumigant and toxic activity of some aromatic oils for protecting dry dates from *Oryzaephilus surinamensis* (L.) (Coleoptera: Silvanidae) in stores. *Bulletin of the National Research Centre* 43: 63–68.

Abdullah F, P Subramanian, H Ibrahim et al. (2017). Chemical composition, antifeedant, repellent, and toxicity activities of the rhizomes of galangal, *Alpinia galanga* against Asian subterranean termites, *Coptotermes gestroi* and *Coptotermes curvignathus* (Isoptera: Rhinotermitidae). *Journal of Insect Science* 15(7): 2015. doi: 10.1093/jisesa/ieu175

Ahmed S, C Mitchell and YR Saxena. (1984). Renewable resource utilization for agriculture and rural development and environmental protection: Use of indigenous plant material for pest control by limited resource farmers. *Planning Workshop, Botanical Pest Control Protect*, International Rice Research Institute, Los Banos, Philippines, pp. 1–29.

Akhtar S, M Hasan, M Sagheer et al. (2015). Antifeedant effect of essential oils of five indigenous medicinal plants against stored grain insect pests. *Pakistan Journal of Zoology* 47(4): 1045–1050.

Akhtar Y, YR Yeoung and MB Isman. (2008). Comparative bioactivity of selected extracts from Meliaceae and some commercial botanical insecticides against two noctuid caterpillars *Trichoplusia ni* and *Pseudaletia unipuncta*. *Phytochemistry Reviews* 7: 77–88.

Akhtar Y, Y Yu, MB Isman et al. (2010). Dialkoxybenzene and dialkoxyallylbenzene feeding and oviposition deterrents against the cabbage looper, Trichoplusia ni: Potential insect behavior control agents. *Journal of Agricultural and Food Chemistry* 58: 4983–4991.

Annis PC, JE Graver and S Van. (2000). Ethyl formate – a fumigant with potential for rapid action. In *2000 Annual International Research Conference on Methyl Bromide Alternatives and Emissions Reduction*. Orlando, FL, pp. 70-1–70-3.

Aref SP, O Valizadegan and ME Farashiani. (2016). The insecticidal effect of essential oil of *Eucalyptus floribundi* against two major stored product insect pests; *Rhyzopertha dominica* (F.) and *Oryzaephilus surinamensis* (L.). *TEOP* 19(4): 820–831.

Arora R, B Singh and AK Dhawan. (2012). *Theory and Practice of Integrated Pest Management*. Jodhpur: Scientific Publishers.

Batish DR, HP Singh, RK Kohli et al. (2008). Eucalyptus essential oil as natural pesticide. *Forest Ecology and Management* 256(12): 2166–2174.

Bell CH. (2005). Factors affecting the efficacy of sulfuryl fluoride as a fumigant. In *Proceedings of the 9th International Working Conference on Stored Product Protection*. Campinas, Sao Paulo, Brazil, pp. 519–526.

Bett PK, AL Deng, JO Ogendob et al. (2016). Chemical composition of *Cupressus lusitanica* and *Eucalyptus saligna* leaf essential oils and bioactivity against major insect pests of stored food grains. *Industrial Crops and Products* 82: 51–62.

Boeke SJ, IR Baumgart, AV Huis et al. (2004). Toxicity and repellence of African plants traditionally used for the protection of stored cowpea against *Callosobruchus maculatus*. *Journal of Stored Products Research* 40(4): 423–438.

Bossou AD, E Ahoussi, E Ruysbergh et al. (2015). Characterization of volatile compounds from three *Cymbopogon* species and *Eucalyptus citriodora* from Benin and their insecticidal activities against *Tribolium castaneum*. *Industrial Crops and Products* 76: 306–317.

Bouguerra N, FT Djebbar and N Soltani. (2017). Algerian *Thymus vulgaris* essential oil: Chemical composition and larvicidal activity against the mosquito *Culex pipiens*. *International Journal of Mosquito Research* 4(1): 37–42.

Brari J and V Kumar. (2019). Antifeedant activity of four plant essential oils against major stored product insect pests. *International Journal of Pure and Applied Zoology* 7(3): 41–45.

Channoo C, S Tantakom, S Jiwajinda et al. (2002). Fumigation toxicity of eucalyptus oil against three stored-product beetles. *Thai Journal of Agricultural Science* 35(3): 265–272.

Chaudhary S, RK Kanwar, A Sehgal et al. (2017). Progress on *Azadirachta indica* based biopesticides in replacing synthetic toxic pesticides. *Frontiers in Plant Science* 8: 610.

Chayapraser W, DE Maier, KE Ileleyi et al. (2006). Real-time monitoring of a flour mill fumigation with sulfuryl fluoride. In *Proceedings of the 9th International Working Conference on Stored Product Protection*. Campinas, Sao Paulo, Brazil, pp. 541–550.

Choi W-S, B-S Park, Y-H Lee et al. (2006). Fumigant toxicities of essential oils and monoterpenes against *Lycoriella mali* adults. *Crop Protection* 25(4): 398–401.

Cosimi S, E Rossi, PL Cioni et al. (2009). Bioactivity and qualitative analysis of some essential oils from Mediterranean plants against stored-product pests: Evaluation of repellency against *Sitophilus zeamais* Motschulsky, *Cryptolestes ferrugineus* (Stephens) and *Tenebrio molitor* (L.). *Journal of Stored Products Research* 45: 125–132.

Daglish GJ. (2006). Opportunities and barriers to the adoption of new grain protectants and fumigants. In *Proceedings 9th International Working Conference on Stored Product Protection*, Sao Paulo, Brazil, pp. 209–216.

De Souza VN, CRF De Oliveira, CHC Matos et al. (2016). Fumigation toxicity of essential oils against *Rhyzopertha dominica* (F.) in stored maize grain. *Revista Caatinga, Mossoró* 29(2): 435–440.

Dethier VG, LB Browne and CN Smith. (1960). The designation of chemicals in terms of the responses they elicit from insects. *Journal of Economic Entomology* 35(1): 134–136.

Ducom PJF. (2006). The return of the fumigants. In *Proceedings of the 9th International Working Conference on Stored Product Protection*. Campinas, Sao Paulo, Brazil. pp. 510–516.

Ebadollahi A and M Mahboubi. (2011). Insecticidal activity of the essential oil isolated from *Azilia eryngioides* (pau) Hedge Et Lamond against two beetle pests. *Chilean Journal of Agricultural Research* 71(3): 406–411.

Garcıa M, A Gonzalez-Coloma, OJ Donadel et al. (2007). Insecticidal effects of *Flourensia oolepis* Blake (Asteraceae) essential oil. *Biochemical Systematics and Ecology* 35: 181–187.

Germinara GS, MG Distefano, LD Acutis et al. (2017). Bioactivities of *Lavandula angustifolia* essential oil against the stored grain pest *Sitophilus Ghoneim* and *Hamadah granaries*. *Bulletin of Insectology* 70(1): 129–138.

Ghoneim K and K Hamadah. 2017. Antifeedent activity and detrimental effect of Nimecidine (0.03% Azadirachtin) on the nutitional perforormance of Egyptian cotton leafworm *Spodoptera littoralis* Boisd. (Noctuidae: Lepidoptera). *Bio. Bulletin* 3: 39–55.

Grainge M, S Ahmed, WC Mitchell et al. (1986). *EWC/UH Database*. East-West Center, Honolulu: Resource System Institute, p. 249.

Greech NM, HD Ohr and JJ Sims. (1996). Methyl iodide as a soil fumigante. U.S Patent 5,518,692, U.S. Patent and Trade-mark Office.

Hanif CMS, MU Hasan, M Sagheer et al. (2016). Insecticidal and repellent activities of essential oils of three medicinal plants towards insect pests of stored wheat. *Bulgarian Journal of Agricultural Science* 22(3): 470–476.

Huang Y, SX Chen and SH Ho. (2000). Bioactivities of methyl allyl disulfide and diallyl trisulfide from essential oil of garlic to two species of stored-product pests, *Sitophilus zeamais* (Coleoptera: Curculionidae) and *Tribolium castaneum* (Coleoptera: Tenebrionidae). *Journal of Economic Entomology* 93(2): 537–543.

Idouaarame S, AA Abdel-hamid, M Elfarnini et al. (2018). Insecticidal activity of essential oils from five Moroccan plants on three insect pests of stored cereals. *GSC Biological and Pharmaceutical Sciences* 4(2): 52–57.

Isikber AA, MH Alma, M Kanat et al. (2006). Fumigant toxicity of essential oils from *Laurus nobilis* and *Rosmarinus officinalis* against all life stages of *Tribolium confusum*. *Phytoparasitica* 34(2): 167–177.

Islam MS and FA Talukder. (2005). Toxic and residual effects of *Azadirachta indica*, *Tagetes erecta* and *Cynodon dactylon* extracts against *Tribolium castaneum*. *Journal of Plant Diseases and Protection* 112(6): 594–601.

Islam MS, MM Hasan, W Xiong et al. (2009). Fumigant and repellent activities of essential oil from Coriandrum sativum (L.) (Apiaceae) against red flour beetle *Tribolium castaneum* (Herbst) (Coleoptera: Tenebrionidae). *Journal of Pest Science* 82: 171–177.

Isman MB. (2000). Plant essential oils for pest and diseases management. *Crop Protection* 19: 603–608.

Isman MB. (2006). Botanical insecticides, deterrents, and repellents in modern agriculture and an increasingly regulated world. *Annual Review of Entomology* 51: 45–66.

Isman MB. (2008). Botanical insecticides: For richer, for poorer. *Pest Management Science: Formerly Pesticide Science* 64(1): 8–11.

Isman MB. (2015). A renaissance for botanical insecticides? *Pest Management Science* 71(12): 1587–1590.

Isman MB and Y Akhtar. (2007). Plant natural products as a source for developing environmentally acceptable insecticides. In I Ishaaya, R Nauen and AR Horowitz (eds.), *Insecticides Design Using Advanced Technologies*, pp. 235–248. Switzerland: Springer.

Isman MB, S Miresmailli and C Machial. (2010). Commercial opportunities for pesticides based on plant essential oils in agriculture, industry and consumer products. *Phytochemistry Reviews* 10: 197–204. doi: 10.1007/s11101-010-9170-4.

Isman MB, JA Wilson and R Bradbury. (2008). Insecticidal activities of commercial Rosemary Oils (*Rosmarinus officinalis*) against larvae of *Pseudaletia unipuncta* and *Trichoplusia ni* in relation to their chemical compositions. *Pharmaceutical Biology* 46(1–2): 82–87.

Jacobson M. (1982). Plants, insects, and man their interrelationship. *Economic Botany* 36(3): 346–354.

Jacobson M. (1990). *Glossary of Plant-Derived Insect Deterrents*. Boca Raton, FL: CRC Press, Inc., p. 213.

Jose S and K Sujatha. (2017). Antifeedant activity of different solvent extracts of *Gliricidia sepium* against third in star larvae of *Helicoverpa armigera* (Hubner) (Lepidoptera: Noctuidae). *International Journal of Advanced Research in Biological Sciences* 4(4): 201–204.

Keita SM, C Vincent, JP Schmit et al. (2001). Efficacy of essential oil of *Ocimum basilicum* L. and *O. gratissimum* L. applied as an insecticidal fumigant and powder to control *Callosobruchus maculatus* (Fab.) (Coleoptera: Bruchidae). *Journal of Stored Products Research* 37: 339–349.

Ketoh GK, K Honore, IA Glitho et al. (2006). Comparative effects of *Cymbopogon schoenanthus* essential oil and piperitone on *Callosobruchus maculatus* development. *Fitoterapia* 77(7–8): 506–510.

Ketoh GK, K Honore, IA Koumaglo et al. (2005). Inhibition of *Callosobruchus maculatus* (F.) (Coleoptera: Bruchidae) development with essential oil extracted from *Cymbopogon schoenanthus* L. Spreng. (Poaceae), and the wasp *Dinarmus basalis* (Rondani) (Hymenoptera: Pteromalidae). *Journal of Stored Products Research* 41(4): 363–371.

Khan A and FA Gumbs. (2003). Repellent effect of ackee (*Blighia sapida* Koenig) component fruit parts against stored-product insect pests. *Tropical Agricultural* 80(1): 19–27.

Kheradmand K, S Beynaghi, S Asgari et al. (2015). Toxicity and repellency effects of three plant essential oils against twi-spotted spider mite, *Tetranychus urticae* (Acari: Tetranychidae). *Journal of Agricultural Sciences and Technology* 17: 1223–1232.

Ko K, W Juntarajumnong and A Chandrapatya. (2009). Repellency, fumigant and contact toxicities of *Litsea cubeba* (Lour.) Persoon against *Sitophilus zeamais* Motschulsky and *Tribolium castaneum* (Herbst). *Kasetsart Journal (Natural Sciences)* 43: 56–63.

Korunic Z and V Rozman. (2008). Fumigacija cineolom *in vitro* (Fumigation with cineole essential oil *in vitro)*. In *Proc. of Croatian Sem DDD and ZUPP*, Sibenik, Croatia, April 2–4.

Koul O, S Walia and GS Dhaliwal. (2008). Essential oils as green pesticides: Potential and constraints. *Biopesticides International* 4: 63–88.

Lee BH, PC Annis, F Tumaaliia et al. (2004). Fumigant toxicity of essential oils from the Myrtaceae family and 1, 8-cineole against 3 major stored-grain insects. *Journal of Stored Products Research* 40: 553–564.

Lee BH, WS Choi, SE Lee et al. (2001). Fumigant toxicity of essential oils and their constituent compounds towards the rice weevil, *Sitophilus oryzae* (L.). *Crop Protection* 20: 317–320.

Lee S, CJ Peterson and JR Coats. (2003). Fumigation toxicity of monoterpenoids to several stored product insects. *Journal of Stored Products Research* 39(1): 77–85.

Liu ZL, YJ Xu, J Wu et al. (2002). Feeding deterrents from *Dictamnus dasyca- rpus* Turcz against two stored-product insects. *Journal of Agricultural and Food Chemistry* 50(6): 1447–1450.

Lucia A, AC Toloza, E Guzmán et al. (2017). Novel polymeric micelles for insect pest control: Encapsulation of essential oil monoterpenes inside a triblock copolymer shell for head lice control. *Peer-Reviewed Journal* 5: e3171. doi: 10.7717/peerj.3171.

Machial CM, I Shikano, M Smirle et al. (2010). Evaluation of the toxicity of 17 essential oils against *Choristoneura rosaceana* (Lepidoptera: Tortricidae) and *Trichoplusia ni* (Lepidoptera: Noctuidae). *Pest Management Science* 66(10): 1116–1121.

Maia MF and SJ Moore. (2011). Plant-based insect repellents: A review of their efficacy, development and testing. *Malaria Journal* 10: 1–15.

Mason LJ, CA Strait, CP Woloshuk et al. (1999). Controlling stored grain insects with ozone fumigation. In *Proceedings of the 7th Int Working Conference on Stored Product Protection Beijing*, China, pp. 536–547.

Miresmailli S and MB Isman. (2014). Botanical insecticides inspired by plant–herbivore chemical interactions. *Trends in Plant Science* 19(1): 29–35.

Navarro S. (2006). New global challenges to the use of gaseous treatments in stored products. In *Proceedings of the 9th International Working Conf on Stored Product Protection*. Campinas, Sao Paolo, Brazil, pp. 495–516.

Navarro S, AA Iskber, S Finkelman et al. (2004). Effectiveness of short exposures of propylene oxide alone and in combination with low pressure or carbon dioxide against *Tribolium castaneum* (Herbst) (Coleoptera: Tenebrionidae). *Journal of Stored Products Research* 40: 197–205.

Negahban M, S Moharramipour and F Sefidkon. (2007). Fumigant toxicity of essential oils and their constituent compounds towards the rice weevil, *Sitophilus oryzae* (L.). *Journal of Stored Products Research* 43(2): 123–128.

Obeng-Ofori D. (2007). The use of botanicals by resource poor farmers in Africa and Asia for the protection of stored agricultural products. *Stewart Postharvest Review* 3(6): 1–8.

Obeng-Ofori D. (2010). Residual insecticides, inert dusts and botanicals for the protection of durable stored products against pest infestation in developing countries. In *Stored Products Protection, Proceedings of the 10th International Working Conference on Stored Product Protection* MO Carvalho, PG Fields, CS Adler et al. (eds.), pp. 774–788, Julius-K¨uhn- Archiv 425.

Ogendo JO, M Kostjukovsky, U Ravid et al. (2008). Bioactivity of *Ocimum gratissimum* L. oil and two of its constituents against five insect pests attacking stored food products. *Journal of Stored Products Research* 44: 328–334.

Owusu EO. (2001). Effect of some Ghanaian plant components on control of two stored-product insect pests of cereals. *Journal of Stored Products Research* 37(1): 85–91.

Padin SB, C Fuse, MI Urrutia et al. (2013). Toxicity and repellency of nine medicinal plants against Tribolium castaneum in stored wheat. *Bulletin of Insectology* 66(1): 45–49.

Papachristos DP and DC Stamopoulos. (2002). Toxicity of vapors of three essential oils to the immature stages of *Acanthoscelides obtectus* (Say) (Coleoptera: Bruchidae). *Journal of Stored Products Research* 38(4): 365–373.

Papanastasiou SA, EMD Bali, CS Ioannou et al. (2017). Toxic and hormetic-like effects of three components of citrus essential oils on adult Mediterranean fruit flies (*Ceratitis capitata*). *PLoS One* 12(5): e0177837. doi: 10.1371/journal.pone.0177837

Park IK, SG Lee, DH Choi et al. (2003). Insecticidal activities of constituents identified in the essential oil from leaves of *Chamaecyparis obtusa* against *Callosobruchus chinensis* (L) and *Sitophilus oryzae* (L). *Journal of Stored Products Research* 39: 375–384.

Pascual-Villalobos MJ. (2003). Volatile activity of plant essential oils against stored product beetle pests. In *Proceedings of the 8th International Working Conference on Stored Product Protection, York*. Oxon, UK: CAB International, pp. 648–650.

Prajapati V, AK Tripathi, KK Aggarwal et al. (2005). Insecticidal, repellent and oviposition-deterrent activity of selected essential oils against *Anopheles stephensi*, *Aedes aegypti* and *Culex quinquefasciatus*. *Bioresource Technology* 96(16): 1749–1757.

Qari SH, AH Nilly, AH Abdel-Fattah et al. (2017). Assessment of DNA damage and biochemical responses in *Rhyzopertha dominica* exposed to some plant volatile oils. *Journal of Pharmacology and Toxicology* 12: 87–96. doi: 10.3923/jpt.2017.87.96.

Rajashekar Y, N Bakthavatsalam and T Shivanandappa. (2012). Botanicals as grain protectants. *Psyche*: 1–13. Doi: 10.1155/2012/646740.

Rajashekar Y, N Gunasekaran and T Shivanandappa. (2010). Insecticidal activity of the root extract of *Decalepis hamiltonii* against stored-product insect pests and its application in grain protection. *Journal of Food Science and Technology* 47: 310–314.

Rajashekar Y and T Shivanandappa. (2010). A novel natural insecticide molecule for grain protection. In MO Carvalho, PG Fields, CS Adler et al. (eds.),*Stored Products Protection, Proceedings of the 10th International Working Conference on Stored Product Protection* pp. 913–917, Julius-K¨uhn- Archiv 425.

Rajendran R and V Sriranjini. (2008). Plant products as fumigants for stored-product insect control. *Journal of Stored Product Research* 44: 126–135.

Rozman V, I Kalinovic and Z Korunic. (2007). Toxicity of naturally occurring compounds of Lamiaceae and Lauraceae to three stored-product insects. *Journal of Stored Products Research* 43(4): 349–355.

Saeidi K and H Pezhman. (2018). Insecticidal activity of four plant essential oils against two stored product beetles. *Entomology, Ornithology and Herpetology* 7: 213. doi: 10.4172/2161-0983.1000213

Sahaf BZ and S Moharramipour. (2008). Fumigant toxicity of *Carum copticum* and Vitex pseudo-negundo essential oils against eggs larvae and adults of *Callosobruchus maculatus*. *Journal of Pest Science* 81: 213–220.

Sánchez-Ramos I and P Castañera. (2000). Acaricidal activity of natural monoterpenes on *Tyrophagus putrescentiae* (Schrank), a mite of stored food. *Journal of Stored Products Research* 37(1): 93–101.

Secoy DM and AE Smith. (1983). Use of plants in control of agricultural and domestic pests. *Economic Botany* 37(l): 28–57.

Shaaya E and M Kostjukovsky. (2009). The potential of biofumigants as alternatives to methyl bromide for the control of pest infestation in grain and dry food products. In A Kirakosyan and PB Kaufman (eds.), *Recent Advances in Plant Biotechnology* (Part 4), Switzerland: Springer. pp. 389–403.

Shukla AC. (2009). Volatile essential oil of *Cymbopogon pendulus* as an effective fumigant pesticide for the management of storage-pests of food commodities. *National Academy of Science Letters* 32(1&2): 51–59.

Shukla AC. (ed.) (2012). *Plant Constituents and their Mechanism of Action as Pesticide*, Germany: Lambert Academic Publishers. p. 303.

Shukla AC, SK Shahi, A Dikshit et al. (2001). Plant product as a fumigant for the management of stored product pests. In *Proc. CAF-2000, USA* Donahaye, EJ, Navarro, S and Leesch, JG (eds.), pp. 125–132.

Singh S, DK Sharma, RS Gill. (2016). Evaluation of three plant oils for the control of lesser grain borer, *Rhyzopertha dominica* (Fabricius) in stored wheat. *Journal of Insect Science* 29: 162–169.

Stamopoulos DC, P Damos and G Karagianidou. (2007). Bioactivity of five monoterpenoid vapors to *Tribolium confusum* (du Val) (Coleoptera: Tenebrionidae). *Journal of Stored Products Research* 43(4): 571–577.

Stevenson PC, MB Isman and SR Belmain. (2017). Pesticidal plants in Africa: A global vision of new biological control products from local uses. *Industrial Crops and Products* 110: 2–9.

Tak JH and MB Isman. (2015). Enhanced cuticular penetration as the mechanism for synergy of insecticidal constituents of rosemary essential oil in *Trichoplusia ni*. *Scientific Reports* 5: 12690.

Talukder FA. (2005). Insects and insecticide resistance problems in post-harvest agriculture. In *Proc. Intern. Conf. Postharvest Technology and Quality Management in Arid Tropics*. Sultan Qaboos University, pp. 207–211.

Talukder FA. (2006). Plant products as potential stored product insect management agents-A mini review. *Emirates Journal of Food and Agriculture* 18(1): 17–32. doi: 10.9755/ejfa

Talukder FA and PE Howse. (1995). Evaluation of *Aphanamixis polystachya* as repellents, antifeedants, toxicants and protectants in storage against *Tribolium castaneum* (Herbst). *Journal of Stored Products Research* 31: 55–61.

Talukder FA and PE Howse. (2000). Isolation of secondary plant compounds from *Aphanamixis polystachya* as feeding deterrents against adults *Tribolium castaneum* (Coleoptera: Tenebrionidae). *Journal of Plant Diseases and Protection* 107: 498–504.

Talukder FA, MS Islam, MS Hossain et al. (2004). Toxicity Effects of botanicals and synthetic insecticides on *Tribolium castaneum* (Herbst) and *Rhyzopertha dominica* (F.). *Bangladesh Journal of Environmental Sciences* 10(2): 365–371.

Tapondjou LA, C Adler, H Bouda et al. (2002). Efficacy of powder and essential oil from *Chenopodium ambrosioides* leaves as post-harvest grain protectants against six-stored product beetles. *Journal of Stored Products Research* 38(4): 395–402.

Tapondjou AL, C Adler, DA Fontem et al. (2005). Bioactivities of cymol and essential oils of *Cupressus sempervirens* and *Eucalyptus saligna* against *Sitophilus zeamais* Motschulsky and *Tribolium confusum* du Val. *Journal of Stored Products Research* 41(1): 91–102.

Tripathi AK and S Upadhyay. (2009). Repellent and insecticidal activities of *Hyptis suaveolens* (Lamiaceae) leaf essential oil against four stored-grain coleopteran pests. *International Journal of Tropical Insect Science* 29(4): 219–228.

Tripathi AK, V Prajapati, KK Aggarwal et al. (2000). Repellency and toxicity of oil from *Artemisia annua* to certain stoaed-product beetles. *Journal of Economic Entomology* 93(l): 43–47.

Trivedi A, N Nayak and J Kumar. (2017). Fumigant toxicity study of different essential oils against stored grain pest *Callosobruchus chinensis*. *Journal of Pharmacognosy and Phytochemistry* 6(4): 1708–1711.

Tsai WT, L Mason and KE Ileleyi. (2006). A Preliminary Report of Sulfuryl Fluoride and Methyl Bromide Fumigation of Flour Mills. In *Proceedings of the 9th International Working Conference on Stored Product Protection*. Campinas, Sao Paolo, Brazil, pp. 595–599.

Tunç I, BM Berger, F Erler et al. (2000). Ovicidal activity of essential oils from five plants against two stored-product insects. *Journal of Stored Products Research* 36(2): 161–168.

Wang J, F Zhu, XM Zhou et al. (2006). Repellent and fumigant activity of essential oil from *Artemisia vulgaris* to *Tribolium castaneum* (Herbst) (Coleoptera: Tenebrionidae). *Journal of Stored Products Research* 42(3): 339–347.

Wanna R and P Ngoen. (2019). Efficiency of Indian borage essential oil against cowpea bruchids. *International Journal of Geomate* 16(56): 129–134.

Wu Y, S Guo, D Huang et al. (2017). Contact and repellant activities of zerumbone and its analogues from the essential oil of *Zingiber zerumbet* (L.) Smith against Lasioderma serricorne. *Journal of Oleo Science* 66(4): 399–405. doi: 10.5650/jos.ess16166.

Yadava SRS and KN Bhatnagar. (1987). A preliminary study on the protection of stored cowpea grains against pulse beetle, by indigenous plant products. *Pesticides* (Bombay) 21(8): 25–29.

Yang RZ and CS Tang. (1988). Plants used for pest control in China: A literature review. *Economic Botany* 42(3): 376–406.

Yong LR and LV Trang. (2003). Cyanogen: A possible fumigant for flour/rice mills and space fumigation. In *Proceedings of the 8th International Working Conference on Stored Product Protection, York.*, Oxon, UK: CAB International, pp. 651–653.

Zhao MP, QZ Liu, Q Liu. (2017). Identification of larvicidal constituents of the essential oil of *Echinops grijsii* roots against the three species of mosquitoes. *Molecules* 22: 205. doi: 10.3390/molecules22020205

9 Mycotoxins, Mycotoxicosis and Managing Mycotoxin Contamination

A Review

Abhishek Tripathi and Afroz Alam

CONTENTS

9.1 INTRODUCTION

One of the critical issues facing humanity today is ensuring the availability of safe food and feed worldwide. By using several new technologies, the yield of agricultural and horticultural products has increased many times, but the net yield at the same level is hard to attain due to various natural contaminants. One such group of natural contaminants, mycotoxins, poses a challenge because they exist in an extensive range of food products and are radically different in their chemical structures and the symptoms caused in humans and animals. It has been found that numerous agricultural products, particularly carbohydrate rich food products, are more susceptible to fungal attack and subsequent growth (Kendra and Dyer 2007).

Apart from food products, in developing countries, spoilage and contamination of drugs by fungi is another major problem, which has garnered the attention of scientists due to the amplification of mycotoxins leading to serious quality deterioration (Guan et al. 2008).

9.2 MYCOTOXINS

The growth of fungi on various postharvest products and crude drugs not only deteriorates their quality but also adds toxins. The fungi that are known to produce toxic chemicals are found in almost all major taxonomic groups. Toxic chemicals produced by fungi are collectively referred to as mycotoxins and the diseases associated with the ingestion of mycotoxin contaminated food/feed stuffs are referred to as mycotoxicosis (Bennett and Klich 2003). These are characterized as toxic metabolites produced by a varied assemblage of fungal strains that taint agricultural crops before harvest or during postharvest storage. The range of these harmful substances is reasonably extensive (Zychowski et al. 2013). Mycotoxins are not only detrimental to human and animal health; they also account for huge monetary losses. These losses hurt all the participants *viz.*, production animals/animal producers, grain handlers and distributors, processors of crops and consumers (Rodrigues et al. 2011). At present, about 400 mycotoxins have been recognized and characterized, however, the major attention is with those mycotoxins that are toxic and carcinogenic in nature (Iqbal et al. 2013). The exposure of humans to mycotoxins may be as a result of eating either plant-based foodstuffs that are contaminated with toxins or possess potent mycotoxins and their metabolites (CAST 2003) or direct exposure to air and dust containing mycotoxins (Jarvis 2002).

Mycotoxins represent a group of unrelated chemical compounds produced as secondary metabolites by certain toxigenic isolates of fungi (Ismaiel and Papenbrock 2015). Mold produces several hundred toxic metabolites. In recent years, much emphasis has been given on mycotoxins in human food and animal feed. Feeding on contaminated food can be harmful and lethal to living organisms, causing abnormalities at the cellular and physiological level (Fliege and Metzler 1999).

Consuming a mycotoxin-contaminated diet may provoke acute and long-term chronic effects in humans and animals, with teratogenic, carcinogenic and estrogenic or immune-suppressive effects (Berek et al. 2001). A report was compiled regarding the reduction in the performance of farm animals (ruminants, pigs and poultry) after consuming mycotoxin contaminated feed (AFSSA 2009). Some consequences of eating mycotoxin-contaminated animal feed include reduced feed intake, poor feed conversion, diminished body weight gain, increased incidence of disease (due to immune-suppression) and reduced reproductive capacity (Morgavi and Riley 2007; Pestka 2007; Voss et al. 2008).

The multiplicity of mycotoxin structures also induces miscellaneous toxic effects. For instance, the aflatoxin structure permits the formation of DNA adducts with guanine, inducing cancerous cell formation (Bren et al. 2007). The lactone ring of aflatoxins is accountable for its toxicity. Fumonisins inhibit ceramide synthase (Soriano et al. 2005), inducing adverse effects on the sphinganin/sphingosin ratio. The deamination of fumonisin B_1 induces a loss of toxicity, indicating that amines play a role in fumonisin toxicity (Pagliuca et al. 2005). Deoxynivalenol (DON) and T-2 toxin induce apoptosis in hemopoietic progenitor cells and immune cells (Parent-Massin 2004). They also inhibit protein, Ácido Desoxirribonucleico (ADN) and Ácido Ribonucleico (ARN) synthesis (Richard 2007). The epoxy structure of trichothecenes induces their toxicity (Sundstøl Eriksen et al. 2004). Zearalenone (ZEA), thanks to its conformation, is able to mimic 17-estradiol and bind to estrogen receptors, disrupting fertility and reproduction ability (Gaumy et al. 2001; Iqbal et al. 2014b).

Ergotism is one of the oldest and well-known mycotoxicosis. Aflatoxin research had a dramatic beginning. Before 1960, few mycologists of the USSR and Japan had studied mycotoxins and mycotoxicosis but in 1960, the abrupt death of about 100,000 young turkeys in various poultry farms in England changed the situation. The disease was of unknown etiology, so it was named as "Turkey-X disease" (Blount 1961). Investigations revealed that the common ingredient of feed in various farms was the incorporation of imported Brazilian peanut meal, which was heavily contaminated with light green mold, later this was identified as *Aspergillus flavus*. Bueno et al. (2007) studied the isolated aflatoxin B_1, B_2, G_1 and G_2 from Brazilian peanut meal. In 1969, an outbreak of aflatoxicosis in Karnataka (India) was reported, which caused the death of more than 2200 chicks (Gopal et al. 1969). Reddy and Raghvender (2007) reported aflatoxin production related to host-pathogen interaction. Three important outbreaks due to aflatoxic hepatitis and enteroergotism have been reported in India (Kumar et al. 2016). Bhatt and Krishnamachari (1978) have also reported an outbreak of mycotoxicosis due to toxic substances of *Fusarium* species occurring in Jammu and Kashmir (India).

Among all the mycotoxins, so far identified, aflatoxins have received maximum attention because of their major toxicological characteristics. The physico-chemical character of major aflatoxins has been extensively studied (Akande et al. 2006; Astoreca et al. 2007a, 2007b; Abbas et al. 2010; Agriopoulou et al. 2016). The toxicity of major aflatoxin types in decreasing order is $B_1 > B_2 > G_1 > G_2 > M_1 > M_2$ (Wu et al. 2009).

Citrinin is also an important mycotoxin produced by *Penicillium citrinum. Aspergillus flavus* and *Aspergillus parasiticus* are important toxigenic molds producing aflatoxin. They are a part of the storage flora on commodities stored under inadequate conditions. Ciegler et al. (1972) have reported that pH, moisture and temperature are three important factors that govern the production of mycotoxins. Roy and Chourasia (1990) worked on aflatoxin production on *Piper longum* fruit at different temperatures.

Every kind of food and feed commodity is contaminated with different levels of mycotoxins (EC 2006). In parts of India, 100% of maize samples (Ciegler et al. 1977); cattle feed and poultry feeds (Iqbal et al. 2014a); spices (Bullerman et al. 2002); fruit and dry fruits (Reddy et al. 2010); sunflower seeds (Bilgrami 1984); tomato (Vashisth et al. 2013); mustard seeds (Baig and Fatima 2017) cotton sesame and castor (Aït Mimoune et al. 2016) were found to be contaminated with mycotoxins. However, the literature related to mycotoxin contamination in crude herbals or plant parts is still fragmentary (Bullerman et al. 2002).

Drug samples were highly contaminated because of prolonged storage. Roy and Chourasia (1990) reported mycotoxin (aflatoxin B_1) in 44 samples of five root drugs *viz., Achyranthes aspera* (0.75 μg/g), *Acorus calamus* (1.10 μg/g), *Adhatoda vasica* (0.67 μg/g), *Elerodendrum serratum* (0.51 μg/g) and *Picrorrhiza kurroa* (0.11 μg/g). Researchers worked on aflatoxin producing abilities of *A. flavus* which were isolated from eight drug samples viz., *Rauwolfia serpentina, Emblica officinalis, Azadirachta indica, Datura metel, Abrus precatorius, Holarrhena antidysenterica* and *Strychnos nuxvomica*. It was also found that not all isolates of *A. flavus* are toxigenic (Aït Mimoune et al. 2016). Roy et al. (1988) isolated 158 isolates of *A. flavus* from numerous drug samples, out of which only 49 were found to be toxigenic. Roy and Kumari (2008) reported maximum natural incidence of aflatoxin B_1 in *Acacia nilotica* followed by *Caesalpinia crista* and *Nelumbo nucifera*. The incidence of mycotoxins was also recorded on drug samples used to cure kidney and skin disorders (Pinotti et al. 2016).

Reif and Metzger (1995) examined aflatoxin B_1, B_2, G_1 and G_2 in naturally contaminated common medicinal herbs and plant extracts; 55 samples of 11 drugs manufactured by different registered pharmaceutical industries were found contaminated with either one or more mycotoxins, indicating that contamination present in crude samples would easily be transmitted to finished herbal drugs. The level of mycotoxin contamination was significantly higher than the human tolerance level (Roy et al. 1988). Mycologists analyzed crude samples (15 seeds, six fruits, two barks and ten roots) for aflatoxin, ochratoxin, citrinin and zearalenone contamination. Out of these, only aflatoxin was quantified (Reif and Metzger 1995). It ranged from 0.11–1.25, 0.18–0.71, 0.14–1.51 and 0.11–1.13 μg/g in seed, fruit, bark and root drug samples, respectively. Khan and Singh (2000) reported mycotoxins in medicinal seeds of forest origin. Roy et al. (1988) studied the natural occurrence of aflatoxin in some drug plant parts and reported 0.18 and 0.62 μg/g concentrations of aflatoxin B_1 from *Terminalia chebula* and *Terminalia bellerica*, respectively. Roy and Kumari (2008) reported 1.51, 1.19 and 1.21 μg/g concentration of aflatoxin B_1 from *Emblica officinalis, Terminalia chebula* and *Terminalia bellerica*, respectively, whereas, in 2003, their concentration were 1.02–1.87, 0.64–1.46 and 1.37–2.34 µg/g, respectively. Singh (2003) analyzed mycotoxins in Triphala and its constituents. The range of aflatoxin in naturally infested *Emblica officinalis* was 0.13–0.75 µg/g, 0.08–0.14 µg/g in *Terminalia bellerica* and 0.13–0.75 µg/g in *Terminalia chebula* and it was 0.98-1.83 µg/g in Triphala, under natural conditions.

The threat of mycotoxin contamination of herbal drugs due to molds first emerged in 1976 as a result of a collaborative study conducted by Udagava in Japan. The first report on the natural occurrence of mycotoxin in crude herbal drugs was published from India in 1988 (Roy et al. 1988). Biodeterioration and mycotoxin contamination have drawn worldwide attention to control quality and maintain therapeutic potentials of plant drugs. The mycotoxin present in crude samples can be easily transmitted to powder/finished herbal drugs (Singh et al. 2001).

One of the most dangerous effects, during the postharvest stages of perishable vegetables and fruits and particularly of seeds worsening by fungal strains, is the stimulation of mycotoxicosis, i.e., diseases of animals and humans caused by ingesting fungal-infected fodder and foods that turn out to be toxic substances called mycotoxins (Kralj and Prosen 2009).

As stated earlier, the first contagion of mycotoxicosis was reported as ergotism before 1700 which was associated with the utilization of ergot. Later, the second such report was related to the occurrence of stachybotryotoxicosis in horses and of alimentary toxic aleukia in humans during the 1930s. The extent of mycotoxin was recognized during World War II, when consumption of mold infected grains caused skin necrosis, hemorrhage, kidney and liver failure and the subsequent death of several humans and animals in Russia and adjacent areas. Analogous symptoms were also observed in horses that were fed on moldy hay.

The third incident was from England, where the eruption of aflatoxicosis in 1960 was caused due to feeding aflatoxin tainted peanut meal (Halver 1965). After this alarming incidence, intensive research on mycotoxins was undertaken globally. Consequently, at present many research groups are involved in extensive research related to aflatoxin (aflatoxicosis) and mycotoxins (mycotoxicosis).

Mycotoxins present a ubiquitous menace to the health of humankind and animals, remarkably, not only do these mycotoxins have the ability to cause diseases at higher concentrations, but their low concentrations are more dangerous to the health of humans and animals if they are continuously ingested through food and fodder. In developing countries, the latter situation is more prevalent.

Mycotoxins are produced as metabolic products by fungi which are capable of producing severe or persistent toxicity (*viz*., mutagenic, carcinogenic and teratogenic) on animals and probably in humans depending upon their dosages and exposure levels. The symptoms appearing as a result of these toxic compounds are considered as a syndrome called mycotoxicosis.

At present, more than 100 mycotoxins are identified which originate mainly from the fungal genera: *Aspergillus*, *Penicillium* and *Fusarium*.

The chemical structures of some significant mycotoxins are revealed in Figure 9.1.

Aspergillus, *Penicillium*, *Fusarium* and *Stachybotrys* spp. are the most frequently occurring fungi that are known to cause mycotoxicosis, and a few of them may result in severe illness and even death. *Aspergillus* and *Penicillium* spp. usually generate their toxins in stored seeds, commercially processed foods and feeds or hay, even though infection in seeds usually occurs in the field. *Fusarium* spp. mainly infect bulb/corm and produce toxins in grains; the infection usually occurs in the field or just after storage. *Stachybotrys* spp. colonizes on straw, hay or other cellulose rich products and generates toxins in fodder.

The mycotoxins produced by different fungi usually differ in their chemical structures, the chemical composition which depends upon the substratum on which they are generated and the ambient environment under which they are grown. They are also variable in their effects on different animals and human beings, and in the extent to cause toxicity (Firsvad and Thrane 1996).

Though various fungi are known to produce toxins that are closely related to mycotoxins, however, the major mycotoxins that are produced by the major fungi and their characteristic properties are listed in the following sections.

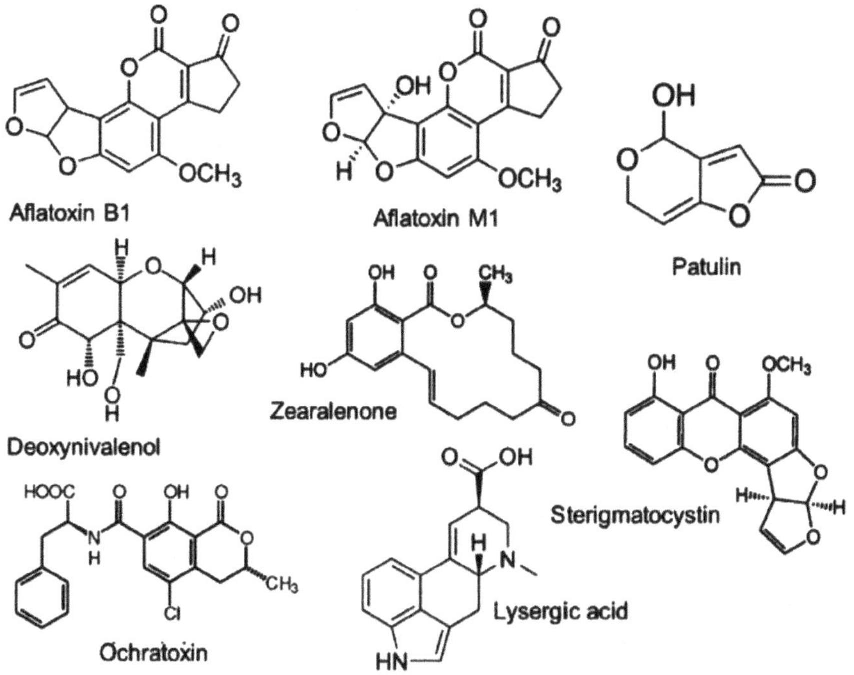

FIGURE 9.1 Chemical structures of selected mycotoxins.

9.2.1 *Aspergillus* Toxins: Aflatoxins

Aflatoxin is among the most common and widely known mycotoxins. Actually, the name "aflatoxin" is an amalgamated word derived from "*A. flavus* toxin." The discovery of aflatoxin has revolutionized research on molds and mold metabolites because of its potency as a toxin and carcinogen. Aflatoxin production has been reported in a large number of fungi, but for many years since the discovery of aflatoxins, *Aspergillus flavus* and *A. parasiticus* were the only two species reliably reported to accumulate aflatoxins. Besides these two species, later *A. nomius*, *A. toxicarius*, *A. pseudotamarii*, *A. flavus* var. *columnaris*, *A. flavus* var. *parvisclerotigenus*, *A. zhaoqingensis* and *A. bombycis* from section *Flavi* were also reported as aflatoxin producing species. Other species in *Aspergillus* and one of its teleomorphs *Emericella*, but also species in *Monocillium*, *Chaetomium*, *Bipolaris* and *Humicola* are able to produce sterigmatocystin, a precursor of the aflatoxins.

However, aflatoxin was discovered in the section *Ochraceorosei* in *Aspergillus* (*A. ochraceoroseus* and *A. rambellii*) and in three species of *Emericella*: *E. astellata*, *E. venezuelensis* and *E. olivicola*. These two *Aspergillus* species of section *Ochraceorosei* and three species of *Emericella* accumulate both aflatoxin B_1 and sterigmatocystin. Whereas the species in section *Flavi* only accumulate aflatoxins and are particularly efficient producers of 3-O-methylsterigmatocystin. These five species, with a possible exception of *E. olivicola*, are of significance for food safety (Ismaiel and Tharwat 2014).

The members of *Aspergillus* section *Flavi* are the important aflatoxin producers in foods and feed stuffs. Of these, *A. flavus* and *A. parasiticus* are by far the most important, with *A. parvisclerotigenus*, *A. nomius*, *A. toxicarius*, *A. pseudotamarii* and *A. bombycis* being important aflatoxin producers in special situations.

Aflatoxins are found primarily in all nuts, especially groundnuts, pistachio and Brazil nuts, copra, cottonseed, rice, wheat, maize, sorghum, pulses, figs and oilseed cakes because the fungus *Aspergillus flavus* has an association with these crops, i.e., the fungus grows in the plants, as a commensal, and has a head start when invading seeds and producing toxins.

Unprocessed vegetable oils made from these fungal affected nuts or seeds more often contain the mycotoxin, aflatoxin. However, during the refining process, aflatoxin is destroyed, therefore refined oils are safe.

Within a range, i.e., a non-toxic concentration (<50 ppb), these aflatoxins are not harmful, but at a high level (approx. 500 ppb), they are dangerous. In recent years, a fairly high percentage (more than 30%) of the corn harvest over huge areas have more than 100 ppb of aflatoxin, i.e., five-fold greater than the permitted limit for humans and in feed for animals. In peanuts, cotton seed, fishmeal, Brazil nuts and so on, or nuts grown in warm and humid regions, aflatoxin is produced at alarming high concentrations (>1000 ppb) and causes typically chronic and sporadically acute mycotoxicosis.

Aflatoxins are bis-furano-isocoumarin derivatives produced in a polyketide pathway by many strains of *Aspergillus flavus* and *A. parasiticus*. Various types of aflatoxins are of analytical interest (Figure 9.2). Among these, four aflatoxins are found in food and the remaining two as metabolites in animal milk. Among the different types of aflatoxins produced by *A. flavus* strains, aflatoxin B1 is the most toxic, mutagenic and carcinogenic type (Ismaiel and Tharwat 2014; Iqbal et al. 2014a).

Aflatoxins include a variety of derivatives, the most important being aflatoxin B_1 and B_2 generated by *A. flavus*, and occasionally isolated from maize, while aflatoxin G_1 and G_2 have been found in *A. parasiticus* regularly isolated from groundnut. Among them, aflatoxins B_1 and G_1 are most often aggregated in toxic concentrations (Hendrickse 1997). From the perspective of community

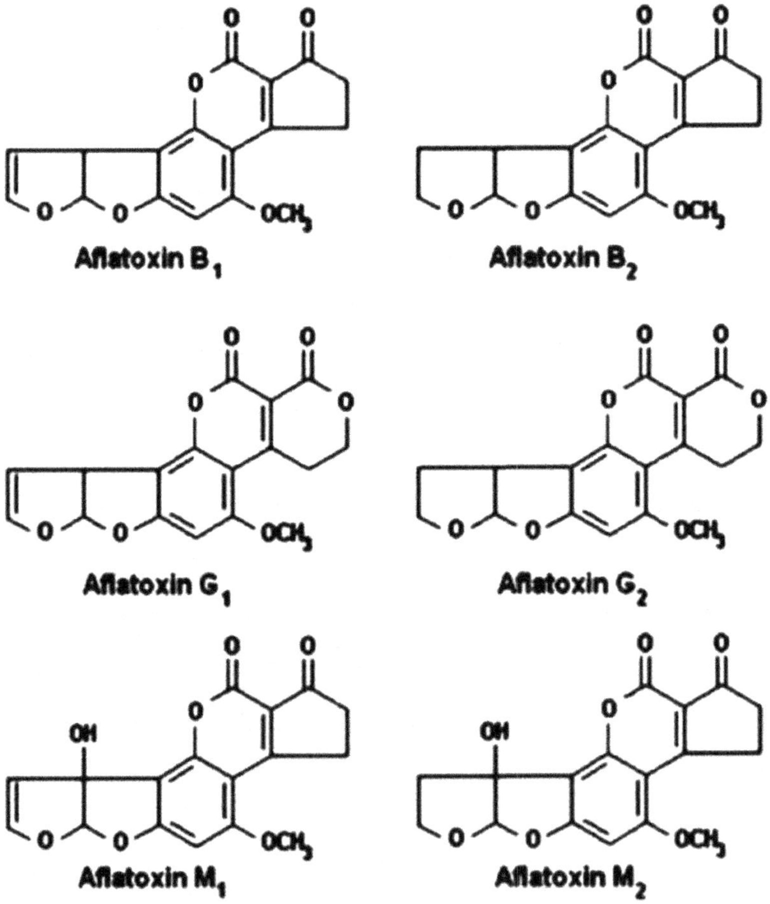

FIGURE 9.2 Chemical structure of the six aflatoxins.

health, the derivative M was reported in the milk of those cows that were fed on contaminated feed containing aflatoxin B_2.

Aflatoxin B_1, B_2, G_1 and G_2 are those toxins which fluoresce as blue (B) or green (G) color under ultraviolet light and are distinguishable by thin layer chromatography (TLC). The lone structural difference between B and G is the addition of oxygen in the cyclopentanone ring.

Aflatoxin M_1 and M_2 correspond to the toxin B_1 and B_2, which have been metabolized in lactating creatures. Their discovery in milk led to the term "M." The apparent structural dissimilarity between B and M is the addition of the hydroxyl group.

The analytical methods for aflatoxin comprise TLC, high performance liquid chromatography (HPLC) and enzyme-linked immunosorbent assay (ELISA). TLC is the most widely used method.

Aflatoxins are carcinogenic in numerous animals and aflatoxins are subject to light deprivation. Consequently, all analytical materials must be sufficiently kept away from light and standard aflatoxin solutions are stocked up using amber colored vials or aluminum foil.

The symptoms of mycotoxicosis caused by aflatoxins in mammals, and apparently humans, vary widely with the exacting toxin, animal species, dosage, age of animal and so on. For instance, young ducklings and turkeys fed on high dosages of aflatoxin became brutally ill and ultimately died. Expecting cows, calves, fattening pigs, mature cattle and sheep fed with low dosages of aflatoxins for long periods were found to exhibit weakening, debilitation, intestinal bleeding, nausea, reduced growth, denial of feed, tendency to other infectious diseases and may potentially abort. Furthermore, most ingested aflatoxins are taken up by the liver and in some experiments it was found that even at low dosages (20 ppb), the animals developed liver cancer (Gupta et al. 2018).

9.2.1.1 Other *Aspergillus* Toxins: Ochratoxin

In addition to aflatoxins, species of *Aspergillus* also produce other toxins in infected grains. The same or similar toxins are also produced in grains infected by strains of *Penicillium*. The most important such toxins are ochratoxins. Ochratoxin was first isolated and characterized during a routine laboratory screening test designed to detect toxic fungal products in foodstuffs in South Africa (Creppy 1999). Ochratoxin causes degeneration and necrosis of the liver and kidney, along with several other symptoms, in domestic animals. Some ochratoxins can persist in the fleshy tissue of animals and can be transmitted to humans through the food chain (Xiao et al. 1996).

Ochratoxins affect protein synthesis and inhibit ATP production. The toxicity of ochratoxins is associated with their isocoumarin moiety (EFSA 2006). It was shown that ochratoxin A, B and C were found to be extremely toxic to ducklings. This toxin is predominantly produced by *Aspergillus ochraceus*, *Penicillium viridicatum* and other *Penicillium* spp. (Van Der Merwe et al. 1965; JECFA 2001; Pfohl-Leszkowicz and Manderville 2007).

Ochratoxin A is a mycotoxin that was discovered to be a metabolite of *Aspergillus ochraceus* in 1965 during a large screen of fungal metabolites designed specifically to identify new mycotoxins (Horie 1995). Subsequent studies revealed that a variety of mold fungus species, including *Aspergillus carbonarius*, *A. niger* and *Penicillium verrucosum* were able to produce ochratoxins (Gil-Serna et al. 2009). Recently, *A. westerdijkiae* and *A. steynii*, two new species from the *Aspergillus* section *Circumdati* have been split from *A. ochraceus* and are reported to be stronger ochratoxin A producers than *A. ochraceus* (Lund and Frisvad 2003). *Penicillium nordicum* and *P. verrucosum* are known to produce ochratoxin A, and have been frequently isolated from cereal crops, meat products and cheese varieties (Beardall and Miller 1994; Bogs et al. 2006). Ochratoxin A has been classified as a possible human carcinogen (group 2B) because of its widespread occurrence in a large variety of agricultural commodities and the potential health risks, mainly toward humans, by the International Agency for Research on Cancer (Alsberg and Black 1913). Given the known human exposure and the abundance of toxicological data from animal studies, the European Union Scientific Committee has recommended ochratoxin A levels below 5 μgkg^{-1} of body weight per day (Bui-Klimke and Wu 2015). In the European Union, some regulatory limits have already been introduced for levels of ochratoxin A in food products such as raw cereal grains (5 μgkg^{-1}), products derived from cereals (3 $\mu g kg^{-1}$), dried fruits (10 $\mu g kg^{-1}$), roasted coffee and coffee products (5 $\mu g kg^{-1}$), grape juice (2 $\mu g kg^{-1}$) (EC No 123/2005) and for all types of wine (2 $\mu g kg^{-1}$) (amended Regulation EC No. 466/2001).

The most studied mycotoxin producing plant pathogenic genus is *Aspergillus* spp. This genus contaminates a wide range of produce including nuts, cereals, sugarcane, beans and sugar beet during storage (e.g., *Aspergillus* spp.). A species, namely *Aspergillus flavus*, is known to infect groundnuts and maize and researches are currently working on it, while other species of *Aspergillus* (e.g., *A. niger*, *A. carbonarius* and *A. parasiticus*) have received less attention in research related to their control on commodities (Abdallah et al. 2018).

9.2.2 *Penicillium* Toxins

Yellowed-rice toxins, primarily citreoviridin, citrinin and luteoskyrin, are all produced by species of *Penicillium* growing in stored rice, barley, corn and dried fish. They cause toxicosis associated with various diseases, nervous and circulatory disorders and degeneration of the kidneys and liver.

9.2.2.1 Penicillic Acid

The mycotoxin penicillic acid (3-methoxy-5-methyl-4-oxo-2,5-hexadienoic acid), a secondary metabolite of *Penicillium puberulum*, was first isolated and named during the investigation of the possible connection between the incidence of pellagra and mold deterioration of maize (Munkvold and Desjardins 1997). These investigators showed that when *P. puberulum* was grown on corn-meal mush or on Raulin's medium, appreciable quantities of penicillic acid were formed (Oxford et al., 1942). They described the general properties of penicillic acid but did not determine its molecular constitution. Oxford and Raistrick (1935) found that cultures of *Penicillium cyclopium* produced relatively large amounts of penicillic acid and this finding enabled Birkinshaw et al. (1936) to prepare sufficient penicillic acid for a successful determination of its molecular constitution by analytical methods. In order to avoid confusion, it should be remembered that so far as is known at present, there is no relationship between penicillic acid and Fleming's penicillin, except the purely fortuitous choice of names for these two substances by their respective discoverers (Macri et al. 2002). Penicillic acid's hazardous effects, carcinogenic nature and antibiotic activity make it a cause of concern (Vella et al. 1995; He et al. 2004; Ezzat et al. 2007; Nonaka et al. 2015).

9.2.2.2 Citrinin

Citrinin was first isolated by Hetherington and Raistrick from a culture of *Penicillium citrinum* and is a benzopyran compound (Bragulat et al. 2008). Several other fungal species within the three genera, *Penicillium* (*P. expansum*, *P. verrucosum*), *Aspergillus* (*A. terreus*) and *Monascus* (*M. ruber*) were also found to produce citrinin. Citrinin contaminates maize, wheat, rye, barley, oats and rice. Citrinin has antibiotic properties against gram-positive bacteria, but it has never been used as a drug due to its high nephrotoxicity. The major target organ of citrinin toxicity are the kidneys, but liver and bone marrow have also been reported as other target organs (Moake et al. 2005; Tanaka et al. 2007).

9.2.2.3 Patulin

Patulin (4-hydroxy-4H-furo[3,2-c] pyran-2(6H)-one) was originally isolated as claviformin due to its isolation from *Penicillium claviforme*, later it was renamed patulin according to the patulin-producing mold, *Penicillium patulum* (later called *P. urticae*, now *P. griseofulvum*). Later, patulin was isolated under various names; clavicin, clavitin, expansin, gigantic acid, leucopin, mycoin, penicidin and tercinin (Jimenez et al. 1988).

Patulin is produced by certain fungal species of *Penicillium*, *Aspergillus* and *Byssochlamys* growing on fruits, including apples, pears, grapes and other fruits (Hopkins 1993). Patulin in apple and pear juice results from the growth of *Penicillium expansum*, a pathogen of the fruit. The risk arises when unsound fruit is used for the production of juices and other products. The early findings that patulin had antibiotic activity led to its testing against the common cold in humans (Devaraj et al. 1982). It causes edema and bleeding in the lungs and the brain, damage to kidneys, paralysis of motor nerves and also induces cancer in higher organisms (Sydenham et al. 1995). It is commonly found to occur naturally in foodstuffs such as fruit or juices made with fruits partly infected with *Penicillium*, in naturally molded bread and bakery products and in most commercial apple products (Buchanan et al. 1974). Thus, patulin may constitute a serious health hazard for humans as well as animals (Bürger et al. 1988; Paucod et al. 1990; WHO 1996; Sant'Ana et al. 2008). Patulin causes several human health effects, including convulsions, nausea, ulceration, lung congestion, epithelial cell degeneration, cancer, genotoxicity, immunotoxicity, immunosuppression and teratogenicity (Rosett et al. 1957).

9.2.2.4 Tremorgenic Toxins

Fungi belonging to several fungal species of both *Aspergillus* and *Penicillium* produce tremorgenic mycotoxins that elicit either intermittent or sustained tremors in vertebrate species. Over 20 mycotoxins containing a tryptophan-derived indole moiety have demonstrated tremorgenic potential in animals and humans. Tremorgenic mycotoxins are produced by fungi infecting a wide variety of foodstuffs, including dairy or grain-containing products intended for human consumption, stored grains and nuts, and a number of forages consumed by livestock species, as well as food- or beverage-manufacturing byproducts, garbage and compost piles. Tremorgenic toxins cause marked body tremors and excessive discharge of urine, followed by convulsive seizures that often end in death.

9.2.3 Mycotoxins Produced by Dematiaceous Hyphomycetes

9.2.3.1 Tenuazonic Acid

The tetramic acid derivative L-tenuazonic acid was first isolated by Rosett et al. (1957) from *Alternaria tenuis*, a dematiaceous plant pathogen invading a series of plants involved in the post-harvest decay of fruits, grains and vegetables (Shigeura and Gordon 1963). Its structure was established by Stickings (1959) and later investigated (Nolte et al. 1980) by applying NMR spectroscopy and X-ray crystallography, which also described different tautomers of this 3-acetyltetramic acid.

Biological activity of tenuazonic acid was first described by Aver'yanov et al. (2007). Later, it was isolated from *Magnaporthe grisea* (the blast fungus, former name is *Pyricularia oryzae*) as a phytotoxin (Kinoshita et al. 1972) and from *Alternaria longiceps*, *A. kikuchiana*, *A. mali*, *A. alternata*, *A. tenuissima* and *Phoma sorghina* as a mycotoxin (Rothweiler and Tamm 1970; Umetsu et al. 1972, 1974; Steyn and Rabie 1976; Davis et al. 1977; Binder et al. 1973). Additionally, tenuazonic acid has been made responsible for the outbreak of "onyalai," a human hematologic disorder disease occurring in Africa after consumption of sorghum (Rothweiler and Tamm 1966).

9.2.3.2 Cytochalasins

The metabolites isolated from a *Phoma* species have been called phomins. Those from *Zygosporium masonii* were named as zygosporins. The metabolites from *Helminthosporium* species were described as cytochalasins (cytos: cell; chalasis: relaxation). The three groups proposed a systemic nomenclature based on the generic name cytochalasan for the cytochalasins, phomins and zygosporins (Rothweiler and Tamm 1966; Ismaiel and Papenbrock 2015). Phomin (phomine, cytochalasin B), the first cytochalasin reported, was described in 1966 as a macrolide antibiotic with cytostatic activity produced by a *Phoma* species. Later, in a more detailed investigation of the *Phoma* metabolites, a closely related compound with similar activity, dehydrophomin (cytochalasin A), was isolated (Chao and Liu 2006). Aldridge et al. (1967) isolated cytochalasin A and cytochalasin B from *Helminthosporium dematioideum* (*Drechslera dematioidea*).

Over 80 different cytochalasans have been isolated from a number of the fungal genera, including *Aspergillus*, *Phomopsis*, *Penicillium*, *Zygosporium*, *Chaetomium*, *Rosellinia*, *Metarhizium* and so on (Chao and Liu 2006). Of these, cytochalasin B is the most widely studied and has been extensively used for cytological investigations (Joffe 1978).

9.2.4 Mycotoxins Produced by *Fusarium* and Other Species

9.2.4.1 Fusarium Toxins

The importance of *Fusarium* species in the present perspective is that infection may sometimes take place in developing seeds, especially in cereals, and also in maturing fruits and vegetables. An immediate potential for toxin production in foods is obvious.

The crucial role of *Fusarium* strains as mycotoxin producers appears to have remained basically unsuspected until the 1970s. But research has now revealed its role as a causing agent of alimentary toxic aleukia (ATA). This was the responsible agent for the previously reported human mycotoxicosis epidemic in the USSR which killed about 100,000 people between 1942 and 1948. Matossian (1981) has argued convincingly that ATA occurred in other countries, including England, in at least the 16th to 18th centuries.

It is a well-known fact that *Fusarium* species are capable of producing a variety of mycotoxins. Foremost among these are the trichothecenes, of which at least 50 are known: the majority are produced by Fusaria. The most notorious is T-2 toxin, which was responsible for ATA (Trapp et al. 1998). Other *Fusarium* mycotoxins are known to be highly toxic to animals and are also alleged to be accountable for acute and chronic human diseases.

Marasas et al. (2004) intensively studied more than 200 toxigenic *Fusarium* isolates and provided accurate information on species identifications and the corresponding toxins produced. They listed 24 *Fusarium* species with confirmed toxigenicity. The four species judged to be most important from the viewpoint of human health are *F. sporotrichioides*, *F. equiseti*, *F. graminearum* and *F. moniliforme.*

Three groups of toxins, trichothecenes, zearalenones and fumonisins, are produced by several species of *Fusarium*, primarily in moldy corn.

Trichothecenes belong to a major class of mycotoxins produced by a range of fungi from the order Hypocreales, including those of the genera *Fusarium*, *Myrothecium*, *Verticimonosporium*, *Stachybotrys*, *Trichoderma*, *Trichothecium*, *Cephalosporium* and *Cylindrocarpon* (Wilkins et al. 2003; Chao and Liu 2006). More than 120 trichothecenes have been isolated in the past 30 years. The first trichothecene to be isolated was trichothecin from *Trichothecium roseum*, in 1948 by Freeman and Morrison (1948). Diacetoxyscirpenol (DAS) from *Fusarium equiseti* was preliminarily characterized by Brian et al. (1961) and was later followed by nivalenol (NIV) (Tatsuno et al. 1968; Vesonder and Hesseltine 1980) and T-2 toxin (Cole and Cox 1981), both from *F. sporotrichioides*, although they were mis-identified as *F. nivale* and *F. tricinctum*, respectively, in the original articles (Wilkins et al. 2003). However, it was the discovery of 4-deoxynivalenol (DON) from wheat in eastern North America in 1980 (Ueno 1983), which truly sparked the research into the *Fusarium* species and led to the discovery of trichothecenes from other genera.

Trichothecenes (or *Trichothecins*), the principal toxins produced by Fusaria, are sesquiterpenes with a basic 12,13-epoxytrichothec-9-ene ring system. Trichothecenes are produced by species of *Fusarium* and by several other fungi. Trichothecenes are often produced in mixtures even under pure culture conditions, and are very difficult to separate, so the toxicity of many of these compounds remains uncertain. Noncompetitive inhibition of protein synthesis is the biochemical basis of trichothecene toxicity (Bando et al. 2007). They are extremely toxic when fed to swine, where they cause, among other symptoms, listlessness or inactivity, degeneration of the cells of the bone marrow, lymph nodes and intestines, diarrhea, bleeding and death. Other animals, such as cows, chicks and lambs, are also affected. Trichothecenes are responsible for feed refusal, emesis and poor growth in swine and are associated with hemorrhagic syndrome in poultry and with a variety of symptoms and lesions in other animals. Some of the trichothecenes are extremely toxic when consumed or even when in contact with the skin. This toxin was produced by the fungus *Trichothecium roseum* and various *Fusarium* species.

Deoxynivalenol (DON), a trichothecene mycotoxin, also known as vomitoxin or DON is produced by the fungus *Gibberella zeae* (anamorph *Fusarium graminearum*), the cause of *Gibberella* ear rot of corn and of head blight (scab) of wheat. DON is found in wheat and barley, and rarely elsewhere, because *F. graminearum* (and a couple of closely related species) are only pathogens on those crops. The mycotoxin at first causes reduced feeding by the animals and, thereby, slower gain or loss of weight. DON is also known as vomitoxin, because it causes vomiting, especially when consumed by pigs. At higher concentrations of the mycotoxin, the animals are induced to vomit and totally refuse to eat (Roll et al. 1990; Frisvad et al. 2005). The World Health Organization (WHO) considers DON as a neurotoxin with a teratogenic nature and immunosuppressive characteristics, and, as with trichothecenes in general, it has been associated with chronic and fatal intoxication of humans and animals through consumption of contaminated food (Ouanes et al. 2003).

Zearalenones seem to be most toxic to swine, in which they cause abnormalities and degeneration of the reproductive system, the so-called estrogenic syndrome (Munkvold and Desjardins 1997; Lioi et al. 2004).

Fumonisins are produced by *Fusarium moniliforme*, which causes Fusarium ear rot of corn that affects as much as 90% of corn fields (Bolger et al. 2001). Fumonisins are found in maize and sorghum, and not in other crops, because *F. verticillioides* (and close relatives) are found only associated with those crops. Fumonisins are the cause of blind staggers (equine leukoencephalomalacia) in horses, donkeys and mules, pulmonary edema in swine and possibly cancer and neural tube defect in humans (IARC 2002; Marasas et al. 2004; Pal and McSpadden Gardener 2006; Bryła et al. 2013).

Fescue toxicosis affects cattle and horses feeding on plants of the perennial grass tall fescue infected systemically with an endophytic fungus *Acremonium* growing internally through the plant without invading its cells. The fungus actually seems to make the infected plants more resistant to stress, particularly drought. Horses eating tall fescue plants infected with the fungus show only reproductive disorders. Cattle feeding on such plants, in addition to reduced calving and lower milk production, show reduced weight gain, elevated body temperature and rough hair coat; moreover, as in ergotism, feet or other body extremities may develop gangrene and drop off ("fescue foot").

9.3 MANAGEMENT OF PREHARVEST AND POSTHARVEST WORSENING AND SPOILAGE BY PREVENTING MYCOTOXIN CONTAMINATION

Food safety and economic issues employ both preharvest and postharvest management as a strategy to reduce the risk of mycotoxin contamination in food and feed. Preharvest management practices include good cultural practices, biocontrol and development of resistant varieties of crops through new biotechnological techniques. Postharvest management practices, such as adequate storage, detection and decontamination or disposal, as well as continuous monitoring of potential contamination during processing and marketing of agricultural commodities, have proved to be critical and indispensable in ensuring food and feed safety. However, postharvest contamination is usually the result of the preharvest presence of fungal contamination.

Selected aspects of integrated mycotoxin management should involve different phases for preharvest and postharvest management such as the aspects outlined below.

9.3.1 Preharvest Management

Significant levels of mycotoxins can occur in food crops in the field. Some of the management practices for prevention of mycotoxins in the field are:

- Reduction in plant stress through irrigation, mineral nutrition and protection from insect damage.
- Avoidance of environmental conditions that favor infection in the field, for example, drought, insect infestation, primary inoculum, delayed harvesting.

- Good cultural practices *viz.*, crop rotation, cropping pattern, irrigation, timely planting and harvesting and the use of biopesticides have protective actions that reduce mycotoxin contamination of field crops.
- Breeding of cultivars resistant to fungal infection.
- Use of crop protection chemicals that are antifungal agents.
- Identification of plant constituents that disrupt aflatoxin biosynthesis or fungal growth and their use in new biochemical marker-based breeding strategies to enhance resistance in crops.
- Development of transgenic plants resistant to fungal infection.
- Development of transgenic cultivars capable of catabolism/interference with toxin production.
- Development of crops genetically engineered to resist insect damage.
- Development of crop seeds containing endophytic bacteria that excludes toxigenic fungi.
- Exclusion of toxigenic fungi by pre-infection of plants with biocompetitive non-toxigenic fungal strains.
- The fungal genome of *A. flavus* has been sequenced to understand the regulation of aflatoxin formation by environmental factors. This information can be used in the development of host resistance against aflatoxin contamination by studying the effects of various physiological parameters, for example, drought stress on gene expression in toxigenic fungi.

9.3.2 Harvesting Procedure

Mechanical damage to seeds may occur during harvesting. When damage is kept to a minimum during this phase, subsequent contamination is significantly reduced. Field crops should be harvested in a timely manner to reduce moisture or water activity (Aw) level to a point where mycotoxin formation will not occur (Abdallah et al. 2018).

9.3.3 Postharvest Management

Plant diseases caused by fungal contaminants can be managed by the implementation of several strategies, for instance, the promotion of resistant cultivars, the use of well-planned crop rotation and the use of chemical control. The detrimental blow of plant fortification on the surrounding environment and the health of humans and animals has prompted the European Union (EU Directive 2009/128/EC) to promote research on alternative and eco-friendly methods like integrated pest management (IPM) and the use of biological control agents (BCAs). Biological control, henceforth called biocontrol, in plant pathology, aims at utilizing microorganisms to prevent the colonization and/or suppress the spread of harmful plant pathogens (Brooker et al. 1974; Jard et al. 2011; FSSAI 2015).

The management of postharvest worsening and spoilage by fungi in grains, legumes, fodder and commercial feeds depends on certain safety measures and circumstances that must be met before and during harvest and then during storage. If the product is to be used for food and feed purposes, even if the contamination occurs or persists after the preharvest phase, then the hazards associated with toxins must be managed through postharvest procedures. Storage and processing are the major areas where contamination can be prevented. There are various strategies through which postharvest contamination can be prevented without harming the food and feed used for consumption. These postharvest strategies for the prevention of contamination are as follows:

1. *Matured grains devoid of damaged grains.* The stored grain should not be unripe or too old, should be clean, have good germination ability and be free of mechanical damage and broken seeds. The amount of broken grains and foreign materials present is minimized so they cannot be an easy target for pathogens.

2. *Drying of grain to the minimal moisture level.* Field drying of most grains has been an accepted practice since commercial farming began. However, sun and wind are the primary drying agents and may not be available when most needed.

   The simplest and most common solution to maintaining the grain free of storage fungi is through quick air drying and through the use of aeration systems in storage bins in which air moves through the grain at relatively low rates of flow. The airflow removes excess moisture and heat. It can be regulated so that it brings the moisture content of the grain mass to the desired level and reduces the temperature to 8–10 °C, at which insects and mites are dormant and storage fungi are almost dormant. The moisture content is kept at levels below the minimum required level for the growth of storage fungi. Some hardy *Aspergillus* species will grow and cause spoilage of starchy cereal seeds with a moisture content as low as 13.0–13.2% and soybeans with a moisture content of about 11.5–11.8%. Others require a minimum moisture content of 14% or more to cause spoilage. After harvest, the moisture content of the produce must be 12–14% wb for safe storage with minimal deterioration (Zychowski et al. 2013).
3. *Maintenance of storage temperature.* The temperature of stored grain should be kept as low as achievable since most storage fungi grow most rapidly at temperatures between 30 °C and 55 °C; they grow very slowly at 12 °C–15 °C and their growth almost ceases at 5 °C–8 °C. Low temperature also slows down respiration of grain and prevents the augmentation of moisture in the grains (Karabulut and Baykal 2002).
4. *Control of insect, mites and rodents.* Infestation of stored products by insects and mites is kept to a minimum through the use of fumigants. This helps keep the storage fungi from getting started and growing rapidly.
5. *Controlled relative humidity.* Relative humidity is kept at levels below the minimum required for the growth of the common storage fungi.
6. *Appropriate packaging is often a successful way of excluding insects and molds.* The degree to which the grain already has been invaded by fungi before it arrives at a given site should be minimized to control the growth of pathogens.
7. *Length of storage.* Length of time in storage should be minimized because as storage duration increases, the greater the chance there is of infestation by fungal pathogens.
8. *Frequent cleaning of food/feed delivery systems.* Cleaning is almost indispensable to ensure regular and frequent cleaning of all delivery systems.
9. *Short-term storage.* For short-term storage, the evaporative cooler is an option to use for short-term preservation of vegetables and fruits soon after harvest. Likewise, a zero energy cooling system could be used effectively for short-duration storage of fruits and vegetables. It not only reduces the storage temperature but also the required humidity of the storage, which is essential for maintaining the freshness of the commodities (Trooger and Burtler 1977).
10. *Use of antifungal agents.* Antifungal agents such as propionic acid and acetic acid can be used.
11. *Thermal inactivation.* Thermal inactivation is also one of the alternatives for products that are usually heat processed. Fumonisins and ochratoxin levels have been shown to be lower in thermally processed maize and wheat products (Trooger and Burtler 1977; Karabulut and Baykal 2002).

Processed food cannot be safe if prevention, control, good manufacturing practices and quality control are not used at all stages of production. In India, laws are made for the quality control of various articles of food products against contamination and the permissible limits of contaminants are notified. As per the notification of the Government of India, Ministry of Health and Family Welfare (Food Safety and Standards Authority of India) Notification New Delhi, 4th November, 2015, for Food Safety and Standards (Contaminants, Toxins and Residues) (Amendment) Regulations, 2015; no article of food specified in column (3) of Table 9.1 shall contain any crop contaminant specified

TABLE 9.1
Crop Contaminant in Various Articles of Food and Their Permissible Limit as per Government of India Notification

No.	Name of contaminant	Article of food	Limit (µg kg^{-1})
1.	Aflatoxin	Cereal and cereal products	15
		Pulses	15
		Nuts	
		Nuts for further processing	15
		Ready to eat	10
		Dried figs	10
		Oilseeds or oil	
		Oilseeds for further processing	15
		Ready to eat	10
		Spices	30
2.	Aflatoxin M1	Milk	0.5
3.	Ochratoxin A	Wheat, barley and rye	20
4.	Patulin	Apple juice and apple juice ingredients in other beverages	50
5.	Deoxynivalenol	Wheat	1000

in the corresponding entry in column (2) thereof in excess of quantities specified in the corresponding entry in column (4) of Table 9.1.

9.4 CONCLUSION

Prevention of mycotoxicosis in food and fodder by mycotoxins is necessary and it includes both pre- and postharvest strategies. The best way is the avoidance of mycotoxin formation in the field; however, this is frequently inadequate and other strategies are required. To clean or detoxify mycotoxin-contaminated food and feed, the most widespread strategy that came forward in the feed industry was the addition of sorbent materials in the feed to get the desired elimination of toxins by adsorption through the gastrointestinal tract, or to add enzymes or selective microorganisms that are capable of detoxifying definite mycotoxins. A new functional group of feed additives was defined by the Commission Regulation (EC) No 386/2009 of 12 May 2009 as "substances for reduction of the contamination of feed by mycotoxins: substances that can suppress or reduce the absorption, promote the emission of mycotoxins or modify their mode of action."

Depending on their mode of action, these feed additives may act by reducing the bioavailability of the mycotoxins or by degrading or transforming them into less toxic metabolites.

ACKNOWLEDGMENTS

The authors acknowledge the Bioinformatics Center, Banasthali Vidyapith, supported by DBT for providing computation support, and DST for providing networking support through the FIST program at the Department of Bioscience and Biotechnology.

REFERENCES

Abbas, H. K., K. R. N. Reddy, B. Salleh et al. 2010. An overview of mycotoxin contamination in foods and its implications for human health. *Toxin Rev* 29: 3–26.

Abdallah, M. F., M. Ameye, D. S. Sarah et al. 2018. Control of mycotoxigenic fungi and their toxins: An update for the pre-harvest approach. In N. B. Patrick and F. Stepman (Eds.), *Mycotoxins – Impact and Management Strategies* (pp. 59–89). Rijeka: IntechOpen.

AFSSA. 2009. *Évaluation des risques liés à la présence de mycotoxines dans les chaînes alimentaires humaine et animale. Rapport final.* Fougéres: Agence Française de Sé curité Sanitaire Alimentaire.
Agriopoulou, S., A. Koliadima, G. Karaiskakis et al. 2016. Kinetic study of aflatoxins' degradation in the presence of ozone. *Food Control* 61: 221–226.
Aït Mimoune, N., A. Riba, C. Verheecke-Vaessen et al. 2016. Fungal contamination and mycotoxin production by Aspergillus spp. isolated from nuts and sesame seeds. *J Microbiol Biotechnol Food Sci* 5: 301–305.
Akande, K. E., M. M. Abubakar, T. A. Adegbola et al. 2006. Nutritional and health implications of mycotoxin in animal feed. *Pak J Nutr* 5: 398–403.
Aldridge, D. C., J. J. Armstrong and R. N. Speake et al. 1967. The structure of cytochalasins A and B. *J Chem Soc C* 1: 26–27.
Alsberg, G. L. and O. F. Black. 1913. Contributions to the study of maize deterioration: Biochemical and toxicological investigations of *Penicillium puberulum* and *P. stoloniferum*. *US Dep Agric Bureau Plant Ind Bull* 270: 1–47.
Astoreca, A., C. Magnoli, C. Barberis et al. 2007a. Ochratoxin A production in relation to ecophysiological factors by *Aspergillus* section Nigri strains isolated from different substrates in Argentina. *Sci Total Environ* 388: 16–23.
Astoreca, A., C. Magnoli, M. L. Ramirez et al. 2007b. Water activity and temperature effects on growth of *Aspergillus niger, A. awamori* and *A. Carbonarius* isolated from different substrates in Argentina. *Int J Food Microbiol* 119: 314–318.
Aver'yanov, A. A., V. P. Lapikova and M. H. Lebrun. 2007. Tenuazonic acid, toxin of rice blast fungus, induces disease resistance and reactive oxygen production in plants. *Russ J Plant Pathol* 54: 749–754.
Baig, M. and S. Fatima. 2017. Evaluation of mycoflora and mycotoxins contamination of mustard seeds (*Brassica juncea* L.). *Epitome Int J Multidisciplinary Res* 3(1): 11–15.
Bando, E., L. Gonçales, N. K. Tamura et al. 2007. Biomarcadores para avaliação da exposição humana às micotoxinas. *J Brasil de Patologia e Medicina Labora* 43(3): 175–180.
Beardall, J. and J. D. Miller. 1994. Natural occurrence of mycotoxins other than aflatoxin in Africa, Asia and South America. *Mycotox Res* 10: 21–40.
Bennett, J. W. and M. Klich. 2003. Mycotoxins. *Clin Microbiol Rev* 16: 497–516.
Berek, L., I. B. Petri, A. Mesterhazy et al. 2001. Effects of mycotoxins on human immune functions in vitro. *Toxicol In Vitro* 15: 25–30.
Bhatt, R. V. and K. A. V. R. Krishnamachari. 1978. Food toxins and disease outbreaks in India. *Arogya J Health Sci* 4: 92–100.
Bilgrami, K. S. 1984. Mycotoxin in food. *J Indian Bot Soc* 63: 109–120.
Binder, M., C. Tamm, W. B. Turner et al. 1973. Nomenclature of a class of biologically active mould metabolites: The cytochalasins, phomins, and zygosporins. *J Chem Soc Perkin Trans. 1* 11: 1146–1147.
Birkinshaw, J. H., A. E. Oxford and H. Raistrick. 1936. Studies in the biochemistry of microorganisms. XL VIII. Penicillic acid, a metabolite product of *Penicillium puberulum* Bainier and *P. Cyclopium* Westling. *Biochem J* 30: 394–411.
Blount, W. P. 1961. Turkey "X" disease. *J Br Turk Fed* 1(52): 55–61.
Bogs, C., P. Battilani and R. Geisen. 2006. Development of a molecular detection and differentiation system for ochratoxin A producing *Penicillium* species and its application to analyse the occurrence of *Penicillium nordicum* in cured meats. *Int J Food Microbiol* 107: 39–47.
Bolger, M., R. M. Coker, M. DiNovi et al. 2001. Safety evaluation of certain mycotoxins in food: Fumonisins. *Prepared by the 56th Meeting of the joint FAO/WHO Expert Committee on Food Additives (JECFA), WHO Food Additives Series No 47* (pp. 104–275). Geneva: World Health Organization, International Programme on Chemical Safety.
Bragulat, M. R., E. Martinez, G. Castella et al. 2008. Ochratoxin A and citrinin producing species of the genus *Penicillium* from feedstuffs. *Int J Food Microbiol* 126: 43–48.
Bren, U., F. P. Guengerich and J Mavri. 2007. Guanine alkylation by the potent carcinogen aflatoxin B1: Quantum chemical calculations. *Chem Res Toxicol* 20: 1134–1140.
Brian, P. W., A. W. Dawkins, J. F. Grove et al. 1961. Phytotoxic compounds produced by *Fusarium equiseti*. *J Exp Bot* 12: 1–12.
Brooker, D. B., F. W. Bakker-Arkema and C. W. Hall. 1974. *Drying Cereal Grains*. Westport, CT: AV1 Publishing Co.
Bryła, M., R. Jędrzejczak, M. Roszko et al. 2013. Application of molecularly imprinted polymers to determine B1, B2, and B3 fumonisins in cereal products. *J Separation Sci* 36: 578–584.
Buchanan, J. R., N. F. Sommer, R. J. Fortlage et al. 1974. Patulin from *Penicillium expansum* in store fruits and pears. *J Amer Soc Hort Sci* 99: 262–265.

Bueno, D. J., C. H. Casale, R. P. Pizzolitto et al. 2007. Physical adsorption of aflatoxin B1 by lactic acid bacteria and Saccharomyces cerevisiae: A theoretical model. *J Food Prot* 70(9): 2148–2154.

Bui-Klimke, T. R. and F. Wu. 2015. Ochratoxin A and human health risk: A review of the evidence. *Crit Rev Food Sci Nutr* 55(13): 1860–1869.

Bullerman, L. B., D. Ryu and L. S. Jackson. 2002. Stability of fumonisins in food processing. *Adv Exp Med Biol* 504: 195–204.

Bürger, M. G., A. A. Brakhage, E. E. Creppy et al. 1988. Toxicity and mutagenicity of patulin in different test systems. *Arch Toxicol* 12: 347–351.

CAST. 2003. *Mycotoxins: Risks in Plant, Animal and Human Systems. Report No. 139.* Ames, IA: Council for Agricultural Science and Technology.

Chao, J. I. and H. F. Liu. 2006. The blockage of survivin and securin expression increases the cytochalasin B-induced cell death and growth inhibition in human cancer cells. *Mol Pharmacol* 69: 154–164.

Ciegler, A., D. J. Fennell, H. J. Mintzlaff et al. 1972. Ochratoxin synthesis by *Penicillium* species. *Naturwissenschaften* 59: 365–366.

Ciegler, A., R. F. Vesonder and L. K. Jackson. 1977. Production and biological activity of patulin and citrinin from *Penicillium expansum. Appl Environ Microbiol* 33: 1004–1006.

Cole, R. A. and R. H. Cox. 1981. *Handbook of Toxic Fungal Metabolites.* New York: Academic Press.

Creppy, E. E. 1999. Human ochratoxicosis. *J Toxicol Toxin Rev* 18: 277–293.

Davis, N. D., U. L. Diener and G. Morgan-Jones. 1977. Tenuazonic acid production *Alternaria alternata* and *Alternaria tenuissima* isolated from cotton. *Appl Environ Microbiol* 34: 155–159.

Devaraj, H., K. Radha-Shanmugasundaram and E. R. Shanmugasundaram. 1982. Neurotoxic effect of patulin. *Ind J Exp Biol* 20: 230–231.

European Food Safety Authority (EFSA). 2006. Opinion of the scientific panel on the contaminants in the food chain on a request from the commission related to ochratoxin A in food. *EFSA J* 365: 1–56.

Ezzat, S. M., A. E. El-Sayed, M. I. Abou El-Hawa et al. 2007. Morphological and ultrastructural studies for the biological action of penicillic acid on some bacterial species. *Res J Microbiol* 2: 303–314.

Firsvad, J. C. and U. Thrane. 1996. Mycotoxin production by food-borne fungi. In R. A. Samson, E. S. Hoekstra, J. C. Frisvad and O. Filtenborg (Eds.), *Introduction to Food-Borne Fungi* (5th ed., pp. 251–260). Baarn: Centraalbureau voor Schimmelcultures.

Halver, J. E. 1965. Aflatoxicosis and rainbow trout hepatoma. In G. N. Wogan (Ed.), *Mycotoxins in Foodstuffs* pp. 209–234). Cambridge, MA: MIT Press.

Fliege, R. and M. Metzler. 1999. The mycotoxin patulin induces intra- and intermolecular protein cross links *in vitro* involving cysteine, lysine, and histidine side chains, and α-amino groups. *Chem Biol Interact* 123: 85–103.

Food Safety and Standards (Contaminants, Toxins and Residues) (Amendment) Regulations. 2015. Ministry of Health and Family Welfare (Food Safety and Standards Authority of India) Notification New Delhi, 4th November, 2015, F.No P. 15025/264/13-PA/FSSAI dated the 9th April, 2015 in the Gazette of India, Extraordinary, Part III, Section 4.

Freeman, G. G. and R. I. Morrison. 1948. Trichothecin: An antifungal metabolic product of *Trichothecium roseum* Link. *Nature* 162: 30.

Frisvad, J. C., P. Skouboe and R. A. Samson. 2005. Taxonomic comparison of three different groups of aflatoxin producers and a new efficient producer of aflatoxin B1, sterigmatocystin and 3-O-methylsterigmatocystin, *Aspergillus rambellii* sp. nov. *System Appl Microbiol* 28: 442–453.

Gaumy, J. L., J. D. Bailly, V. Burgat et al. 2001. Zéaralénone: Propriétés et toxicité expérimentale. *Rev Méd Vét* 152: 219–234.

Gil-Serna, J., A. González-Salgado, M. A. T. González-Jaén et al. 2009. ITS-based detection and quantification of *Aspergillus ochraceus* and *Aspergillus westerdijkiae* in grapes and green coffee beans by real-time quantitative PCR. *Int J Food Microbiol* 131: 162–167.

Gopal, T., S. Zaki, M. Narayanaswamy et al. 1969. Aflatoxicosis in fowls. *Indian Vet J* 46: 348–349.

Guan, S., C. Ji, T. Zhou et al. 2008. Aflatoxin B(1) degradation by *Stenotrophomonas maltophilia* and other microbes selected using coumarin medium. *Int J Mol Sci* 9(8): 1489–1503.

Gupta, R. C., A. Srivastava and R. Lall. 2018. *Ochratoxin and Citrinin. Veterinary Toxicology: Basic and Clinical Principles* (3rd ed., pp. 1019–1027). Cambridge, MA: Academic Press.

He, J., E. M. K. Wijeratne, B. P. Bashyal et al. 2004. Cytotoxic and other metabolites of *Aspergillus* inhibiting the rhizosphere of Sonoran desert plants. *J. Nat Prod* 67: 1985–1991.

Hendrickse, R. G. 1997. Of sick turkeys, kwashiorkor, malaria, perinatal mortality, heroin addicts and food poisoning: Research on the influence of aflatoxins on child health in the tropics. *Ann Trop Med Parasitol* 91: 787–793.

Hopkins, J. 1993. The toxicological hazards of patulin. *Food Cosmet Toxicol* 31: 455–459.

Horie, Y. 1995. Productivity of ochratoxin A of *Aspergillus carbonarius* in *Aspergillus* section Nigri. *Nippon Kingakukai Kaiho* 36: 73–76.

IARC (International Agency for Research on Cancer Fumonisin B1). 2002. IARC monographs on the evaluation of carcinogenic risks to humans. In *Some Traditional Herbal Medicines, Some Mycotoxins, Naphthalene and Styrene* (Vol. 82, pp. 301–366). Lyon: IARC Press.

Iqbal, S. Z., M. R. Asi and S. Jinap. 2013. Natural occurrence of aflatoxin B1 and aflatoxin M1 in "Halva" and its ingredients. *Food Control* 34(2): 404–407.

Iqbal, S. Z., S. Nisar, M. R. Asi et al. 2014a. Natural incidence of aflatoxins, ochratoxin A and zearalenone in chicken meat and eggs. *Food Control* 43c: 98–103.

Iqbal, S. Z., T. Rabbani, M. R. Asi et al. 2014b. Assessment of aflatoxins, ochratoxin A and zearalenone in breakfast cereals. *Food Chem.* 157c: 257–262.

Ismaiel, A. A. and J. Papenbrock. 2015. Mycotoxins: Producing fungi and mechanisms of phytotoxicity. *Agriculture* 5: 492–537.

Ismaiel, A. A. and N. A. Tharwat. 2014. Antifungal activity of silver ion on ultrastructure and production aflatoxin B1 and patulin by two mycotoxigenic strains, *Aspergillus flavus* OC1 and *Penicillium vulpinum* CM1. *J Mycol Méd* 24: 193–204.

Jard, G., T. Liboz, F. Mathieu et al. 2011. Review of mycotoxin reduction in food and feed: From prevention in the field to detoxification by adsorption or transformation. *Food Add Contam Part A* 28(11): 1590–1609.

Jarvis, B. B. 2002. Chemistry and toxicology of molds isolated from water-damaged buildings. Mycotoxins and food safety. *Adv Exp Med Biol* 504: 43–52.

Jimenez, M., V. Sanchis, R. Mateo et al. 1988. Detection and quantification of patulin and griseofulvin by high-pressure liquid chromatography in different strains of *Penicillium griseofulvin* Dierckx. *Mycotoxin Res* 4: 59–66.

Joffe, A. Z. 1978. *Fusarium poae* and *F. sporotrichioides* as principal causal agents of alimentary toxic aleukia. In T. D. Wyllie and L. G. Morehouse (Eds.), *Mycotoxic Fungi, Mycotoxins, Mycotoxicoses: An Encyclopaedic Handbook* (Vol. 3, pp. 21–86). New York: Marcel Dekker.

Joint FAO/WHO Expert Committee of Food Additives (JECFA). 2001. In Ochratoxin A, in "Safety evaluations of specific mycotoxins." Prepared by the Fifty-Sixth Meeting of the Joint FAO/WHO Expert Committee on Food Additives, 2001, 6–15 February, Geneva.

Karabulut, O. A. and N. Baykal. 2002. Evaluation of the use of microwave power for the control of postharvest diseases of peaches. *Postharvest Biol Technol* 26: 237–240.

Kendra, D. F. and R. B. Dyer. 2007. Opportunities for biotechnology and policy regarding mycotoxin issues in international trade. *Int J Food Microbiol* 119: 147–151.

Khan, S. N. and P. K. Singh. 2000. Mycotoxin producing potential of seed mycoflora of some forest trees. *Indian Forester* 126: 1231–1233.

Kinoshita, T., Y. Renbutsu, I. D. Khan et al. 1972. Distribution of tenuazonic acid production in the genus *Alternaria* and its pathological evaluation. *Ann Phytopathol Soc* 38: 397–404.

Kralj, C. I. and H. Prosen. 2009. An overview of conventional and emerging analytical methods for the determination of mycotoxins. *Int J Mol Sci* 10(1): 62–115.

Kumar, P., D. K. Mahato, M. Kamle et al. 2016. Aflatoxins: A global concern for food safety, human health and their management. *Food Microbiol* 7: 2170.

Lioi, M. B., A. Santoro, R. Barbieri et al. 2004. Ochratoxin A and zearalenone: A comparative study on genotoxic effects and cell death induced in bovine lymphocytes. *Mutat Res Genet Toxicol Environ* 557: 19–27.

Lund, F. and J. C. Frisvad. 2003. *Penicillium verrucosum* in wheat and barley indicates presence of ochratoxin A. *J Appl Microbiol* 95: 1117–1123.

Macri, A., Z. Dancea and A. I. Baba. 2002. Mycotoxin Involvement in Oncogenesis: Bibliographical Investigations. *Bulletin UASVM CN, Romania* 58: 640–644.

Marasas, W. F., R. T. Riley, K. A. Hendricks et al. 2004. Fumonisins disrupt sphingolipid metabolism, folate transport, and neural tube development in embryo culture and in vivo: A potential risk factor for human neural tube defects among populations consuming fumonisin-contaminated maize. *J Nutrition* 134: 711–716.

Matossian, M. K. 1981. Mold poisoning: An unrecognized English health problem, 1550–1800. *Med Hist* 25: 73–84.

Moake, M. M., O. I. Padilla-Zakour and R. W. Worobo. 2005. Comprehensive review of patulin control methods in foods. *Compr Rev Food Sci* 4: 8–21.

Morgavi, D. P. and R. T. Riley. 2007. An historical overview of field disease outbreaks known or suspected to be caused by consumption of feeds contaminated with Fusarium toxins. *Anim Feed Sci Technol* 137(3–4): 201–212.

Munkvold, G. P. and A. E. Desjardins. 1997. Fumonisins in maize. Can we reduce their occurrence? *Plant Dis* 81: 556–564.

Nolte, M. J., Steyn, P. S. and P. L. Wessels. 1980. Structural investigations of 3-acylpyrrolidine-2,4-diones by nuclear magnetic resonance spectroscopy and X-ray crystallography. *J Chem Soc Perkin Trans 1* 5: 1057–1065.

Nonaka, K., T. Chiba, T. Suga et al. 2015. Coculnol, a new penicillic acid produced by a coculture of *Fusarium solani* FKI-6853 and *Talaromyces* sp. FKA-65. *J Antibiot* 68: 530–532.

Ouanes, Z., S. Abid, I. Ayed et al. 2003. Induction of micronuclei by Zearalenone in Vero monkey kidney cells and in bone marrow cells of mice: Protective effect of Vitamin E. *Mutat Res Genet Toxicol Environ* 538: 63–70.

Oxford, A. E. and H. Raistrick. 1935. Studies in the biochemistry of micro-organisms. I- Erythritol, a metabolic product of *Penicillium brevi-compactum* Dierckx and *P. cyclopium* Westling. *Biochem J* 29: 1599.

Oxford, A. E., H. Raistrick and G. Smith. 1942. Anti-bacterial substances from moulds. Part II. Penicillic acid, a metabolic product of *Penicillium puberulum* Bainier and *Penicillium cyclopium* Westling. *Chem Ind* 10: 22–24.

Pagliuca, G., E. Zironi, A. Ceccolini et al. 2005. Simple method for the simultaneous isolation and determination of fumonisin B1 and its metabolite aminopentol-1 in swine liver by liquid chromatography fluorescence detection. *J Chromatogr B* 819: 97–103.

Pal, K. K. and B. McSpadden Gardener. Biological control of plant pathogens. *Plant Heal Instr* 2006: 1–25.

Parent-Massin, D. 2004. Haematotoxicity of trichothecenes. *Toxicol Lett* 153: 75–81.

Paucod, J. C., S. Krivobok and D. Vidal. 1990. Immunotoxicity testing of mycotoxins T-2 and Patulin on Balb/C mice. *Acta Microbiol Hungarica* 37: 331–339.

Pestka, J. J. 2007. Deoxynivalenol: Toxicity, mechanisms and animal health risks. *Anim Feed Sci Technol* 137(3–4): 283–298.

Pfohl-Leszkowicz, A. and A. R. Manderville. 2007. Review Ochratoxin A: an overview on toxicity and carcinogenicity in animals and humans. *Mol Nut Food Res* 51: 61–99.

Pinotti, L., M. Ottoboni, C. Giromini et al. 2016. Mycotoxin contamination in the EU feed supply chain: A focus on Cereal Byproducts. *Toxins* 8: 45.

Reddy, S. M., S. Girisham, V. K. Reddy et al. 2010. *Mycotoxin Problems and Its Management* pp. 1–514). Jodhpur: Scientific Publishers.

Reddy, B. N. and C. R. Raghvender. 2007. Outbreaks of aflatoxicoses in India. *Afr J Food Agric Nutr Dev* 7(5): 1–15.

Reif, K and W. Metzger. 1995. Determination of aflatoxin in medicinal herbs and plant extracts. *J Chromatogr A* 692: 131–136.

Richard, J. L. 2007. Some major mycotoxins and their mycotoxicoses – An overview. *Int J Food Microbiol* 119: 3–10.

Rodrigues, I., J. Handl, E. M. Binder. 2011. Mycotoxin occurrence in commodities, feeds and feed ingredients sourced in the Middle East and Africa. *Food Addit Contam Part B Surveill* 4(3): 168–179.

Roll, R., G. Matthiaschk and A. Korte. 1990. Embryotoxicity and mutagenicity of mycotoxins. *J Environ Pathol Toxicol* 10: 1–7.

Rosett, T., R. H. Sankhala, C. E. Stickings et al. 1957. Biochemistry of microorganisms. CIII. Metabolites of *Alternaria tenuis auct*: Culture filtrate products. *Biochem J* 67: 390–400.

Rothweiler, W. and C. Tamm. 1966. Isolation and structure of phomin. *Expertientia* 22: 750–752.

Rothweiler, W. and C. Tamm. 1970. Isolierung and struktur der Antibiotica phomin und 5-Dehydrophomin. *Helv Chim Acta* 53: 696–724.

Roy, A. K. and H. K. Chourasia. 1990. Inhibition of aflatoxin production by microbial interaction. *J Gen App Microbiol* 38: 59–62.

Roy, A and V. Kumari. 2008. Aflatoxin and citrinin in seeds of some medicinal plants under storage. *Pharm Biol* 29: 62–65.

Roy, A. K., K. Sinha and H. Chourasia. 1988. Aflatoxin contamination of some common drug plants. *Appl Environ Microbiol* 54: 842–843.

Sant'Ana, A. S., A. Rosenthal and P. R. Massaguer. 2008. The fate of patulin in apple juice processing: A review. *Food Res Int* 41: 441–453.

Shigeura, H. T. and C. N. Gordon. 1963. The biological activity of tenuazonic acid. *Biochem* 2: 1132–1137.

Singh, P. K. 2003. Mycotoxin elaboration in Triphala and its constituents. *Ind Phytopath* 56(4): 380–383.

Singh, P. K., S. N. Khan, N. S. Harsh et al. 2001. Incidence of mycoflora and mycotoxins in some edible fruits and seeds of forest origin. *Mycotoxin Res* 17(2): 46–58.

Soriano, J. M., L. Gonzalez and A. I. Catala. 2005. Mechanism of action of sphingolipids and their metabolites in the toxicity of fumonisin B1. *Prog Lipid Res* 44: 345–356.
Steyn, P. S. and C. J. Rabie. 1976. Characterization of magnesium and calcium tenuazonate from *Phoma sorghina. Phytochemistry* 15: 1977–1979.
Stickings, C. E. 1959. Studies in the biochemistry of micro-organisms. Metabolites of *Alternaria tenius auct*: The structure of tenuazonic acid. *Biochem J* 72: 332–340.
Sundstøl Eriksen, G., H. Pettersson and T. Lundh. 2004. Comparative cytotoxicity of deoxynivalenol, nivalenol, their acetylated derivatives and de-epoxy metabolites. *Food Chem Toxicol* 42(4): 619–624.
Sydenham, E. W., H. F. Vismer, W. F. O. Marasas et al. 1995. Reduction of patulin in apple juice samples influence of initial processing. *Food Control* 6: 195–200.
Tanaka, K., Y. Sago, Y. Zheng et al. 2007. Mycotoxins in rice. *Int J Food Microbiol* 119: 59–66.
Tatsuno, T. M., M. Saito, M. Enomoto et al. 1968. Nivalenol, a toxic principle of *Fusarium nivale. Chem Pharm Bull* 16: 2519–2520.
Trapp, S. C., T. M. Hohn, S. McCormick et al. 1998. Characterization of the gene cluster for biosynthesis of macrocyclic trichothecenes in *Myrothecium roridum. Mol Gen Genet* 257: 421–432.
Trooger, J. M. and J. L. Butler. 1977. *Solar Drying of Peanuts in Georgia, Proceedings Solar Grain drying Conferences, Weaver Laboratory* pp. (32–43). Raleigh: North Carolina University.
Ueno, Y. 1983. *Trichothecenes-Chemical, Biological and Toxicological Aspects.* Amsterdam: Elsevier.
Umetsu, N., J. Kaji, K. Aoyama et al. 1974. Toxins in blast diseased rice plants. *Agric Biol Chem* 38: 1867–1874.
Umetsu, N., J. Kaji and K. Tamari. 1972. Investigation on the toxin production by several blast fungus strains and isolation of tenuazonic acid as a novel toxin. *Agric Biol Chem* 36: 859–866.
Van Der Merwe, K. J., P. S. Steyne, L. Fourie et al. 1965. Ochratoxin A, a toxic metabolite produced by *Aspergillus ochraceus. Nature* 205: 1112–1113.
Vashisth, A., R. Singh, D. K. Joshi. 2013. Effect of static magnetic field on germination and seedling attributes in tomato (*Solanum lycopersicum*). *J Agric Phy* 13(2): 182–185.
Vella, F., K. Kpodo, A. K. Sorensen et al. 1995. The occurrence of mycotoxins in fermented maize products. *Food Chem* 56: 147–153.
Vesonder, R. F. and C. W. Hesseltine. 1980. Vomitoxin: Natural occurrence on cereal grains and significance as a refusal and emetic factor to swine. *Process Biochem* 44: 12–14.
Voss, K. A., L. B. Bullerman, A. Bianchini et al. 2008. Reduced toxicity of fumonisin B1 in corn grits by single-screw extrusion. *J Food Prot* 71(10): 2036–2041.
Wilkins, K., K. Nielsen and S. Din. 2003. Patterns of volatile metabolites and nonvolatile trichothecenes produced by isolates of *Stachybotrys, Fusarium, Trichoderma, Trichothecium* and *Memnoniella. Environ Sci Pollut Res* 10: 162–166.
World Health Organization. 1996. Patulin. In WHO food additives series. 35.
Wu, Q., A. Jezkova, Z. Yuan et al. 2009. Biological degradation of aflatoxins. *Drug Metab Rev* 41(1): 1–7.
Xiao, H., S. Madhyastha, R. R. Marquardt et al. 1996. Toxicity of ochratoxin A, its opened lactone form and several of its analogs: Structure-activity relationships. *Toxicol Appl Pharmacol* 137(2): 182–192.
Zychowski, K. E., A. R. Hoffmann, H. J. Ly et al. 2013. The effect of aflatoxin-B1 on red drum (Sciaenops ocellatus) and assessment of dietary supplementation of NovaSil for the prevention of aflatoxicosis. *Toxins* 5: 1555–1573.

10 Biodecontamination of Mycotoxin Patulin

Hongyin Zhang, Gustav Komla Mahunu and Qiya Yang

CONTENTS

10.1 INTRODUCTION

In recent years, global discussions on mycotoxins have been high as there is growing pressure on limited food resources to cater for the increasing human population. Approximately 25% of global crops produced are contaminated with mycotoxins and result in significant losses of about 1 billion metric tons of raw produce and other food products (Lukwago et al., 2019). Mycotoxins occur naturally, while the type of fungi, extent of growth and consequent contamination depend on internal and external environmental factors (Fernández-Cruz et al., 2010). The major groupings of mycotoxins most relevant to food industries are aflatoxins (AFs), ochratoxins-A (OTA), zearalenone (ZEA), moniliformin, deoxynivalenol (DON), fumonisins and patulin (PAT). Although, a specific mycotoxin is known to contaminate particular food items, other previous reports acknowledge the possible occurrence of more than one mycotoxin in food at the same time (Speijers and Speijers, 2004). Such occurring phenomenon though less reported may require further studies to establish clear mechanisms of action and respective control strategies. According to Ianiri et al. (2016), food and feed contaminated with mycotoxins are harmful to human and animal health, together with significant economic losses, especially to exporting countries, where contaminated produce is damaged and rejected. The presence of mycotoxins in these food and feed constituents is attributed to fungal infection of host plants at preharvest or postharvest stages.

Mycotoxins are known to be low molecular weight fungal metabolites capable of causing harm after an extended period of consuming contaminated food (Marroquín-Cardona et al., 2014). Due to their toxicological properties, maximum levels (MLs) for mycotoxins have been established for some food and feed to safeguard animal and human health and assure consumer safety. In 2003, a minimum of 99 countries around the world established regulations on mycotoxin for food and/or feed (FAO and FOODS, 2004).

The purpose of this chapter is to present the available information on PAT in apple fruits and their processed products and the implications for human health. The effect of fungicides on postharvest fungi control and patulin level will be discussed. Additionally, various biological control methods using yeast antagonistic microbes with promising applications in postharvest disease control and PAT reduction, as well as their corresponding modes of action, will be highlighted.

10.2 DISTRIBUTION AND ECONOMIC IMPLICATIONS OF PAT

Most of the fruits that PAT contaminates are seasonal and suitable for consumption for a short time, therefore, they are usually processed or treated in order to be commercially available throughout the year. Since PAT can also be generated in food during storage, while maintaining its stability in food processing techniques, the constant monitoring of different fruit-based products must be conducted to provide a proper appraisal of human exposure to this toxin. Table 10.1 provides a list of countries in which PAT was detected in food and related products.

Consumers not only derive satisfaction from eating fresh apples, but they also provide important nutrients to the human body. Several literature sources have reported on the health benefits of apples (Matsuoka, 2019; Yahia et al., 2019). Processing of fresh apple fruits into various products contributes to value addition and improves all-year-round availability. However, PAT is one of the important mycotoxins produced by a certain number of fungi associated with fruits and vegetable-based products of which apple is highly susceptible. However, apart from pome fruits like apples, loquats and pears, PAT has also been found toxic to other higher plants including cucumber, wheat, peas, corn and flax. *Penicillium expansum* alone was projected to cause postharvest decay between 70% and 80% with the possibility of PAT contamination (Barkai-Golan and Paster, 2011). Possible PAT contamination intensifies the need for food-processing methods that can control and remove the toxic substance from the fruit products, which also increases costs to the food industry (Moake et al., 2005). At refrigeration temperatures, a number of penicillia can produce PAT because it is stable in contaminated products (Stott and Bullerman, 1975), and this occurrence is of particular concern to producers. Common rotting molds found on refrigerated apple products, for instance, are frequent patulin producers during cold storage. Refrigerated storage of foods will not necessarily avoid PAT as many molds are capable of producing PAT at low temperatures. However, the absence of PAT is a direct and very good quality index of apple products.

10.3 TOXICOLOGY OF PATULIN

Investigators including Birkinshaw in the early 1943, undertook the first PAT isolation from *Penicillium expansum* searching for new fungal molecules containing antibiotic properties after the discovery of penicillin by Fleming. Later, the interest in PAT as a potential antibiotic declined after it was found to be toxic to both humans and animals. In recent decades, PAT has become an important chemical contaminant among the various mycotoxins in food (Puel et al., 2010) due to its potential to cause harm and pollution (Spring and Fegan, 2005). Patulin is a secondary metabolite

TABLE 10.1
Countries Around the World Where PAT Was Detected in Food and Other Products

Continent	Country
Asia	India (Saxena et al., 2008), Malaysia (Lee et al., 2014), Japan (Watanabe and Shimizu, 2005), China (Guo et al., 2013)
Africa	South Africa (Leggott and Shephard, 2001; Shephard et al., 2010), Tunisia (Zaied et al., 2012);
Americas	USA Argentina (Funes and Resnik, 2009), Brazil (de Sylos and Rodriguez-Amaya, 1999)
Europe	UK (Atkins, 1994), Turkey (Gökmen and Acar, 1998), Spain (Marín et al., 2011; Murillo-Arbizu et al., 2009), Sweden (Josefsson and Andersson, 1977), Italy (Spadaro et al., 2007)
Australia	Australia (Cressey, 2009)

with a molar mass of 154.12 g mol^{-1} and its molecular formula is 4-hydroxy-4H-furo [3, 2-c] pyran-2(6H)-one (Figure 10.1). During PAT synthesis, one acetyl-CoA molecule in addition to three malonyl-CoA molecules undergo reduction, decarboxylation and oxidation. The 6-MS molecule is converted to m-OH-benzaldehyde, which then undergoes a rearrangement leading to one molecule of PAT (Figure 10.2).

However, several countries worldwide have set regulations in order to decrease PAT in products to the lowest level to protect the health and safety of consumers. For European countries, the regulation 1425/3003 was set in 2003, being one of the first to set up such a regulation (Table 10.2). Patulin regulation was fixed at 50 μg/L as the maximum level for fruit juice and derived products, solid apple products (25 μg/L) and juices and foods (10 μg/L) as raw materials for babies and young infants. Similarly, the US Food and Drug Administration (FDA) adopted 50 μg/L for a PAT concentration limit in fruits and their products (Van Egmond et al., 2008). In the case of China, the Gulf Cooperation Council (GCC) adopted same limits for PAT regulations in apples, fruits and derived products (Table 10.3) (the highest PAT intake of 1.04 ng kg^{-1} body weight [bw]/day for apple juice consumption was reported by the World Health Organization's Global Environment Monitoring System/Food Contamination Monitoring and Assessment Programme [GEMS/Food] (Organization, 2010). The estimated PAT intakes through apple juice among these groups of the population were adults (28.1 ng kg^{-1} bw/day), children (67.5 ng kg^{-1} bw/day) and babies (110 ng kg^{-1} bw/day). The Joint Expert Committee on Food Additives (JECFA) operating under the World Health Organization (WHO) and the Food and Agriculture Organization (FAO) (2005) of the United Nations (UN) presented reports that offered in-depth data on PAT toxicity and other mycotoxins (Van Egmond et al., 2008). In addition, there has been adequate documentation of the adverse effects of PAT on the health of consumers after long exposures to contaminated foods.

According to the International Agency for Research on Cancer (IARC), PAT is classified as a category 3 carcinogen (Zoghi et al., 2017). Patulin has been found to cause different acute effects (gastrointestinal tract distension, intestinal hemorrhage and epithelial cell degeneration), chronic

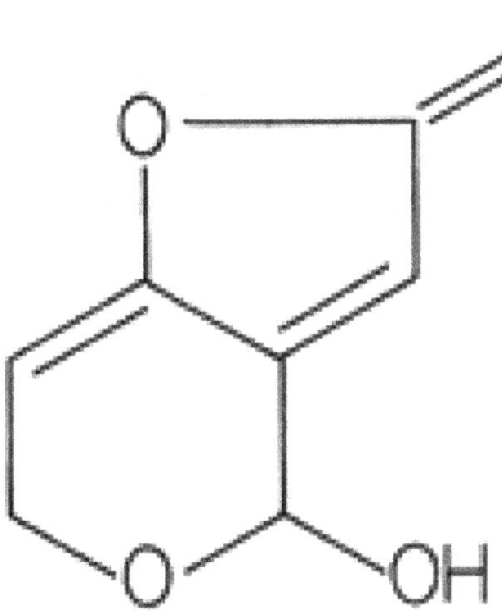

FIGURE 10.1 The structure of patulin (Stott and Bullerman, 1975).

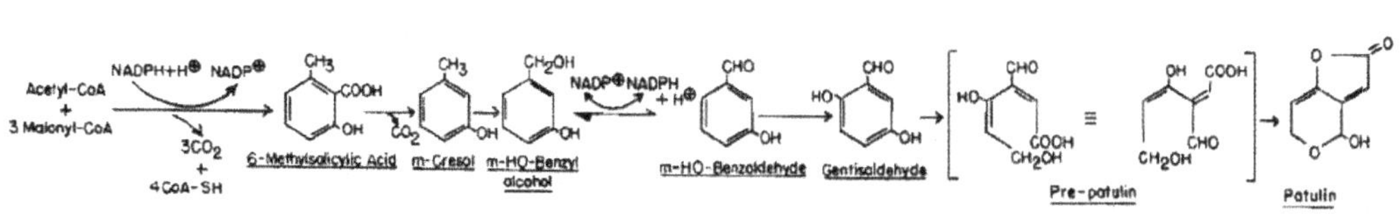

FIGURE 10.2 The main pathway for patulin biosynthesis in *Penicillium* spp. (Forrester and Gaucher, 1972).

TABLE 10.2
EU Maximum Levels for PAT in Fruits and Their Processed Products

Patulin/product	Maximum level (µg/L)
Fruit juice, concentrate fruit juice as reconstituted and fruit nectars	50.0
Spirit drinks, cider and other fermented drinks derived from apples or containing apple juice	50.0
Solid apple products, including apple compote and apple puree intended for direct consumption	25.5
Apple juice and solid apple products, including apple compote and apple puree for infants and young children and labeled and sold as such	10.0
Baby foods other than processed cereal-based foods for infants and young children	10.0

effects (genotoxic, neurotoxic, teratogenic, immunotoxic, cytotoxic), inhibition of protein synthesis and inhibition of DNA and RNA synthesis (Abrunhosa et al., 2016; Glaser and Stopper, 2012; Moake et al., 2005).

As a food contaminant, PAT has the potential to cause oxidative damage to cells (Liu et al., 2006; Speijers and Speijers, 2004), while the interaction between PAT and hormone-production systems can be destructive and alter the immune system (Marin et al., 2013). According to Magan and Olsen (2004), PAT has the ability to react with sulfhydryl groups, which explains cytotoxic and certain genotoxic effects. Generally, PAT cytotoxicity has a typical feature of causing rapid induction of calcium influx and total lactic dehydrogenase release causing the damage of structural integrity of the plasma membrane in mammalian cells (Riley and Showker, 1991) and a reduction on gap junction mediated intercellular communications (Burghardt et al., 1992). It was also reported that after penetrating the gastric wall, PAT degrades rapidly, and its toxicity is not systemic. However, the reaction with glutathione, and perhaps proteins, could contribute partially to PAT degradation, whereas a major diminution of glutathione in gastric tissue is likely to influence PAT to produce localized toxic effects (Rychlik et al., 2004). It was reported that foods high in protein and low in carbohydrates seem unfavorable for excessive PAT production. The combination of this phenomenon with PAT reactivity with sulfhydryl groups appears to minimize the risk of PAT incidence in these foods.

Early investigations indicated that patulin can be removed with active fermentation through yeasts or by an aqueous solution with cysteine or glutathione, and patulin can disappear in 4–5 days (Halász et al., 2009; Stinson et al., 1979; Valletrisco et al., 1988).

TABLE 10.3
Recommended Maximum Levels for PAT in Apple Fruits and Their Processed Products According to Countries/Agencies Worldwide (FAO and FOODS, 2004)

Country	Products	PAT (µg kg^{-1})
EU	Fruit juices, concentrated fruit juice as reconstituted and fruit nectars	50
China	Fruit products containing apple or hawthorn (excluding Guo Dan Pi, a Chinese-style fruit snack)	50
	Fruit or vegetable juice containing apple or hawthorn juice	50
	Alcoholic beverages containing apple or hawthorn	50
Codex, GCC, Kenya, Nigeria	Apple juice	50
South Africa	Apple juice, apple juice ingredients in other juices	50
USA	Apple juice, apple juice concentrates and products	50
India	Apple juice, apple juice ingredients in other beverages	50
Japan	Apple juice, food made using only apple juice as a raw material	50

GCC, Gulf Cooperation Council.

The main PAT-producing organism in apples is known to be ubiquitous (Coulombe Jr, 1993), and this characteristic leads to the development of various symptoms in mycotoxicoses. Patulin is involved in three major mechanisms of action: (1) alteration in the content, absorption and metabolism of nutrients; (2) damage to organs and their functions; and (3) suppression of the immune system. However, the effect of PAT on the immune system predisposes the organism to disease infection and consequently production loss occurs. The exposure of organisms to the toxin suppresses the immune system which is complicated by opportunistic diseases.

10.4 OCCURRENCE OF PATULIN IN APPLE FRUIT AND ITS DERIVED PRODUCTS

Different species of fungi produce mold (*Penicillium*, *Aspergillus*, *Paecilomyces* and *Byssochlamys*) (Varga et al., 2010). Three *Aspergillus* species of the Clavati group produce PAT: *A. clavatus*, *A. giganteus*, and *A. longivesica*. Patulin is produced by 13 species of *Penicillium*: *P. carneum*, *P. clavigerum*, *P. concentricum*, *P. coprobium*, *P. dipodomyicola*, *P. expansum*, *P. gladioli*, *P. glandicola*, *P. griseofulvum*, *P. marinum*, *P. paneum*, *P. sclerotigenum* and *P. vulpinum* (Frisvad et al., 2004). Out of the 13 *Penicillium* species, *P. expansum* commonly known as blue mold is the main cause of rot in pome fruits, characterized by rapid soft rot and formation of the blue mold. Figure 10.3 shows blue mold rot on apple fruit and is adopted from Janisiewicz (1999).

Various parameters influence PAT, which include the amount of free water (a_w), temperature, oxygen, the condition of the substrate and pH (Yiannikouris et al., 2007). The inter-relation of these factors affects PAT production. To be able to develop effective techniques for controlling PAT accumulation in foods, the mechanisms by which the various parameters influence PAT production need to be well understood (Barkai-Golan and Paster, 2011). Most *Penicillium* species are saprophytic based on their ecological classification, thus surviving on dead and decaying matter or soil (Pitt, 2000). PAT is a colorless, crystalline compound with a melting point of 110 °C. It is stable in water solution up to 105–125 °C and pH 3.5–5.5 (González-Osnaya et al., 2007). Similarly, a pH range of 3.2–3.8 supports active production of PAT by the fungus (Moss, 1991).

In order to establish the link between pH of the flesh of apple fruit and *P. expansum* colonization, sections of apple fruit were studied. From these studies, it was observed that pH was between 4 and 6 in non-decayed tissues, whereas in decayed tissues, the pH declined from 3.6 to 4.1 within 4 to 6 days after inoculation (Prusky et al., 2004, 2014). As a response to modification of pH (upwards

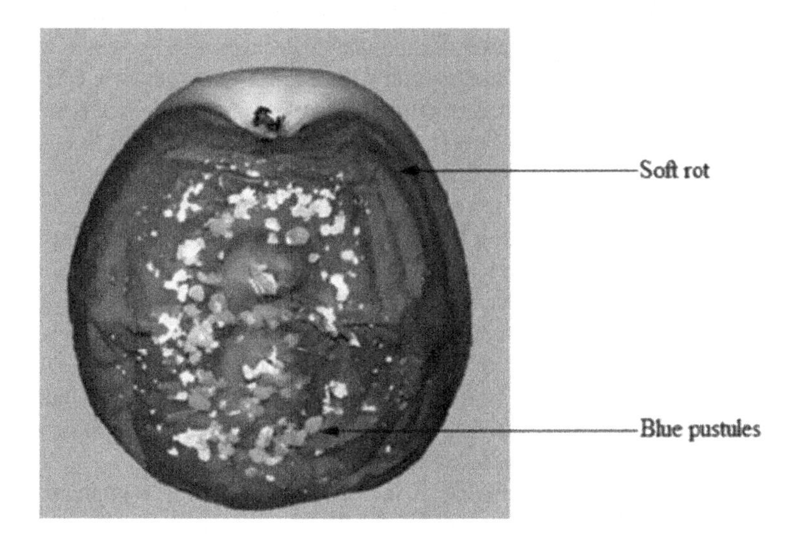

FIGURE 10.3 Blue mold rot on apple fruit (Janisiewicz, 1999)

or downwards) in the host, the pH activity could decline, which could also enhance pathogen virulence in response to host signals such as alkalization by ammonification of the host tissue in *Colletotrichum* and *Alternaria* or acidification by secretion of organic acids in *Penicillium*, *Botrytis* and *Sclerotinia* (Prusky et al., 2014). The rise in pH of a solution can increase the breakdown of PAT content; thus, when pH was 6 at 100 °C, PAT declined by 50% (Collin et al., 2008). PAT is not easily affected by heat (it is stable and survives the pasteurization process); for this reason, PAT will require higher temperature activity to breakdown (Salas et al., 2012). PAT production is temperature dependent and during incubation, high heat applied alone or in combination with other factors can kill or reduce the fungi spores and mycelia, even though it may not be able to totally destroy the toxin. According to an earlier report, *Penicillium expansum* is a psychrophilic fungus that is capable of growing in low temperatures (Deming, 2002). The storage of fruit in cold temperatures may not prevent decay caused by blue mold (Morales et al., 2007a) and subsequent buildup of PAT. Within the range of 20–25 °C is considered the optimum temperature for PAT production, but as low as 5 °C can also elicit its production. Some varieties of apples (such as Golden Delicious and Fuji) with higher amounts of organic acids reported more PAT accumulation in storage at room temperature. Similarly, Golden Delicious and Red Delicious apple varieties exhibited significantly higher PAT accumulation as a result of their lower acidity (Konstantinou et al., 2011). This implies that cultivars of apples must be considered a critical factor influencing the biocontrol of *P. expansum* and subsequent accumulation of PAT in the fruit (Spadaro et al., 2013). It was reported that the pH value of the apple variety regulated the accumulation of PAT during storage; as such, at 0 °C to 1 °C storage, fresh fruits containing a lower pH were reported to be more predisposed to PAT accumulation (Morales et al., 2008b). Clearly, fruit cultivars have differences in their vulnerability to blue mold decay and possible PAT accumulation (Neri et al., 2010).

Controlled atmosphere (CA) storage functions effectively in mold control and PAT production. Low O_2 content (1.5–2.5%) and high CO_2 (up to 3%) provide favorable conditions for extended storage of apple fruit (Juhneviča et al., 2011). Irrespective of the positive impact of CA storage on prolonged shelf life, it also has the potential to compromise fruit quality since mold growth and PAT pollution increases with extended storage time (de Souza Sant'Ana et al., 2008). CA-stored apples appeared to intensify PAT concentration in juice as compared with juice extracted directly from freshly harvested/picked apples and the extended floor storage time prior to processing caused a significant increase in PAT accumulation in apple juice (Baert et al., 2012).

The distribution of PAT in rotten apple fruit showed PAT content present in juice extracted from different sections of decay (Bandoh et al., 2009). In Figure 10.4, PAT concentration varied from the outer layer toward the inner part of apple pulp where "section a" represents complete decay with PAT content of 40 μg kg^{-1}; "section b" (after the completely decayed area and extending up to 5 mm) with PAT 0.14 μg kg^{-1}; and "section c" (the section within 5–10 mm) with only PAT 0.003 μg kg^{-1}. Welke et al. (2009) reported that PAT can penetrate up to 10 mm of the surrounding healthy looking pulp. There is evidence that PAT pollution is not restricted to the decayed tissues alone. This is supported by other reports (Laidou et al., 2001), where diffusion of PAT concentration through the pulp declines with increasing distance from the completely rotten region to the central part of the fruit that seems healthy. Similarly, small infected sections on apples were found to contain PAT and this did not correlate with lesion diameter (Beretta et al., 2000).

Although, the rational is that patulin is commonly detected in *P. expansum*-infected apples, however, PAT could be absent in infected apple fruits (Neri et al., 2010). The absence of PAT in *P. expansum*-infected apples could be attributed to the incompetence of the fungus to produce PAT (Barad et al., 2014; Sanzani et al., 2012). Such incidences may be controlled by a number of factors such as fruit variety and degree of ripening, the fungal strain, the presence of other microbes or conditions of postharvest storage (Ballester et al., 2015).

A large proportion of sugar in apple fruits is sucrose; changes of this component during ripening of fruit may be linked to fungal metabolism and PAT synthesis. The presence of other sugars like glucose or maltose also leads to a high production of PAT by *P. expansum*. The utilization of

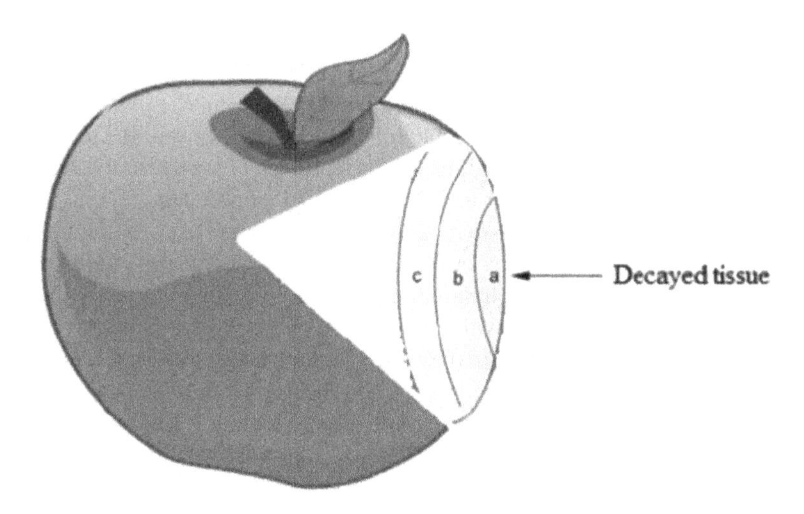

FIGURE 10.4 Distribution of patulin through a rotten apple fruit. The concentration of PAT decreases from "a" (the visibly decayed outer area) toward "c" (the inner cortex). Diagram adopted from Mahunu et al. (2015).

sugars is for carbon sources for the growth of pathogens in apples and the subsequent production of PAT and induced expression of pH modulators (Barad et al., 2016a). The decline in the growth of *P. expansum* may correspond with the decrease in the available sugars or other nutrients, indicating that PAT production may be connected to nutritional stress (McCallum et al., 2002).

Pathogenicity demonstrated by *P. expansum* has a direct effect on patulin content by means of either gene-disruption or RNAi mutants (Barad et al., 2016b; Sanzani et al., 2012). For instance, the mutants still produce some patulin to explain the lack of correlation between the degree of pathogenicity expressed by the *Penicillium* strain and its ability to produce PAT (Ballester et al., 2015).

Estimations of PAT in both raw and processed apple fruits in different countries worldwide have been well described. Murillo-Arbizu et al. (2009) reported on test results from 100 apple juice samples collected from separate markets in Spain, where PAT concentrations were between 0.7 and 118.7 μg L^{-1} (average 19.4 μg l^{-1}) compared with the level across Europe. However, 11% of the samples were beyond the highest permissible limits in the EU (50 μg L^{-1}). Spadaro et al. (2007) also indicated that 34.8% of 135 apple products collected in Italy were contaminated with PAT at concentrations of 1.58–55.41 μg kg^{-1}. Further, out of 45 apple product samples obtained in Argentina, ten were contaminated with a PAT concentration between 17 and 221 μg kg^{-1}, with a mean concentration of 61.7 μg kg^{-1} (Funes and Resnik, 2009). They found that 50% of the apple fruit jam samples studied were PAT contaminated with an average concentration of 123 μg kg^{-1}. The following PAT levels were detected in samples collected from various countries such as Iran (50–285 μg L^{-1}) (Cheraghali et al., 2005), Saudi Arabia (5–152.5 μg L^{-1}) (Al-Hazmi, 2010) and Romania (0.7–101.9 μg L^{-1}) (Oroian et al., 2014).

Also, varied PAT concentrations have been reported in China. The Chinese Academy of Preventive Medicine noticed that 76.9% of fruit juice, fruit jam and other semi-manufactured products were found contaminated with PAT with concentrations between 18 and 953 ppm. In this survey, rate of PAT occurrence in fresh fruit samples was lower (19.6%) with concentrations of 4–262 μg kg^{-1} (Zhang et al., 2009). In another study, 1987 apple juice concentrates were collected from Shanxi Province in China on four sequential processing seasons (2006 to 2010) to study the dietary exposure to PAT content. In the processing season of 2007 and 2008, only four product samples out of the total mentioned above were PAT (78.0 μg kg^{-1}) contaminated. Samples collected in 2006–2007, 2008–2009 and 2009–2010 processing seasons recorded PAT concentrations of 29.0 μg kg^{-1}, 22.7 μg kg^{-1} and 20.0 μg kg^{-1}, respectively. From the results, PAT was detected in more than 90% of the samples collected with lower than 30 μg kg^{-1} concentration, which is important for the apple juice industry.

Over the years, apples have been treated with fungicides to prevent *P. expansum* infection during storage. It was reported that neither selection of fruits that appear healthy will prevent PAT in final products (Chen et al., 2012). It implies that the initial occurrence of postharvest fungal pathogens on harvested fruits must be checked in order to guarantee PAT-free products.

Similarly, growth studies of *P. expansum* on the surface of apple jam for the period of 28 days at 15 °C was conducted to determine the production and distribution of fungal metabolites in the sample (approximately 6 cm high divided into three equal layers). First, the study indicated that the growth rate of the pathogen increased with storage time, which corresponded with a decrease in temperature. Second, PAT was detected in the entire 2 cm layers of the apple jam. The concentrations of PAT in the upper two layers of the jar matched the exposures above the health-based guidance value (HBGV) for a normal serving size. Although reduction of PAT content is often based on the removal of the moldy fragments of the food, the study finally confirmed that this practice is inadequate to avert unhealthy contact (Olsen et al., 2019).

10.5 EFFECT OF POSTHARVEST FUNGICIDE APPLICATION ON DECAY AND PATULIN LEVELS IN APPLES

Traditional methods including sorting, washing, density segregation and thermal treatments have been used for control of PAT-producing pathogens (Hojnik et al., 2017). These methods can produce substantial prevention and decontamination of PAT, whereas thermal treatment, for instance, can cause thermal degradation of toxins. Usually, the processing period is very long with a high energy demand that eventually increases the cost and the treated food products lose quality significantly after heat treatment (Jouany, 2007). As research in the food industry advances in the search for effective decontamination efficiency in mycotoxins, novel non-thermal methods (UV- and gamma-irradiation, pulsed-light treatments and control atmospheric pressure) are being developed. The change in chemical structure of PAT resulting in its degradation is influenced by non-thermal methods. The presence of water in treated products, the extent of PAT contamination and intensity of exposure determine the decontamination efficiency (REF: prevention and control of animal feed contamination by mycotoxins and reduction of their adverse effects in livestock). Some of the non-thermal methods such as gamma irradiation (dosage between 1 and 20 kGy) have achieved a mycotoxin removal rate of about 90% in solution. In effect, the degradation of the mycotoxin in the solution was perhaps attributed to the formation of free radicals produced by the radiolysis of water in the product (Hojnik et al., 2017).

In the case of contaminated solid and dry food products (containing low moisture content), the use of gamma irradiation for decontamination of the mycotoxin has proven less effective (Calado et al., 2014). Nevertheless, the attempt to introduce high doses of gamma irradiation could negatively affect food product quality (Kottapalli et al., 2003). UV light irradiation has been used directly for decontaminating PAT in apple juice. Application of UV light irradiation at 222 nm produced 99% PAT content decrease, although some photosensitive substances in the juice (healthy ascorbic acid) were lowered by the treatment (Zhu et al., 2014).

For many years, chemical agents including many synthetic compounds have produced significant decontamination effects, probably as cost-effective treatment methods. Notwithstanding these outcomes, synthetic fungicides are very expensive with high environmental and human health risks and prolonged time of treatment is not appropriate for high quality food preservation.

Globally, each year about 23 million kg fungicides are used to reduce the incidence of diseases (Karabulut et al., 2005). Among the schemes employed, the application of fungicides such as thiabendazole has been relied upon for many years, since it is difficult to cull rotten apples unless the decayed areas are visible on the surface. Synthetic fungicides, therefore, are conventionally used for the control of postharvest diseases and mycotoxin decontamination of food. Irrespective of the role fungicides have played in postharvest treatment, chemical residues remain a potential risk coupled with the development of fungicide-resistant species. Given the fact that continuous mishandling of fungicides is also characterized as an environmental hazard, most countries have gradually restricted or completely banned their applications. Fungal pathogens such as *P. italicum* and *P. digitatum* in citrus fruits have developed resistance to fungicides (Fogliata et al., 2000), whereas 77% of the *P. expansum* isolates were identified as resistant to thiabendazole (Morales et al., 2008a). Also, 70% of *P. expansum* isolates from apples, pears, and grapes demonstrated resistance to thiabendazole (Cabañas et al., 2009). According to Baraldi et al. (2003), the magnitude of *P. expansum* and increased severity of blue mold on fruits could be attributed to the presence of thiabendazole. Following this, it was also revealed that captan was accountable for enhancing PAT production (Paterson, 2007). Fungicides have been reported as stress drivers for mycotoxin-producing fungi, for which reason mycotoxin content can increase even though *P. expansum* growth is suppressed (Kazi et al., 1997).

In order to determine the ability to decrease the resistance of fungal strains to fungicides, several new "low-risk fungicides" (LRFs) have been studied. Some of the LRFs include fludioxonil (Calvert et al., 2008), iprodione (Ochiai et al., 2002), azoxystrobin and pyrimethanil (Kanetis et al., 2007). The filtrates of the LRFs have shown the capacity to reduce fungal growth (Kuiper-Goodman, 2004). The inhibitory properties of iprodione and fludioxonil against *Candida albicans* growth were investigated, and the fungicide toxicities were capable of suppressing hyphal formation (Ochiai et al., 2002). In as much as fruit producers are likely to continue to use these LRFs in excess, it may eventually render their potency weak (Morales et al., 2010). The next section focuses on the application and manipulation of biocontrol microbes as antagonistic agents and their associated potential to decontaminate PAT in fruit.

10.6 BIOCONTROL MICROBES AS ANTAGONISTIC AGENTS AND POTENTIAL PATULIN DECONTAMINANTS

Biological methods fall into the category of mycotoxin decontamination processes, which are based on the ability of representative microorganisms including yeast, bacteria, molds, actinomycetes and algae to either eliminate or reduce mycotoxins present in foods and other products. The obvious benefit of biocontrol decontamination is the involvement of zero chemicals and the quality of treated foods not being compromised. Biological control agents (BCAs) represent less than 1% of the annual market share of the chemicals used for crop protection, estimated to cost US$ 16,000 million. More commercially viable microbial antagonists (such as bacteria and yeast) have the tendency to biodegrade or detoxify PAT in order to protect commodities in storage without compromising produce quality. Generally, yeasts respond favorably to different extreme stressful environmental conditions (low and high temperature, desiccation, wide range of relative humidity, law oxygen levels, pH changes, UV radiation) that exist before and after harvest (Droby et al., 2016). The authors described yeasts as possessing special qualities to adapt to the micro-environment such as high sugar concentration, high osmotic pressure and low pH existing in the wounded tissues of fruit. Yeasts are easy to multiply in large quantities and most of the species are able to grow speedily on inexpensive substrates in fermenters (Spadaro et al., 2010). Compared with filamentous fungi, yeasts exhibit simple nutritional needs that permit them to colonize dry surfaces for extended durations (Droby et al., 2016).

In apples, blue mold decay has been controlled using *Cryptococcus albidus* (Fan and Tian, 2001), *Candida sake* and *Pantoea agglomerans* (Morales et al., 2008b), *Rhodotorula mucilaginosa* (Li et al., 2011, 2019), *Yarrowia lipolytica* (Yang et al., 2017; Zhang et al., 2017a), *Sporidiobolus pararoseus* (Abdelhai et al., 2019; Li et al., 2017a), *Candida guilliermondii* (Chen et al., 2017; Coelho et al., 2009), *Cryptococcus podzolicus* (Wang et al., 2018), *Pichia caribbica* (Mahunu et al., 2018) and *Hanseniaspora uvarum* (Apaliya et al., 2018).

According to Castoria et al. (2005), *Rhodotorula glutinis* LS11 cells demonstrated ability to resist PAT concentration and further degraded it *in vitro*. Thus, *Rhodotorula glutinis* LS11 cells were able to metabolize PAT and/or reduce its accumulation or synthesis. A recent study conducted by Li et al. (2019) also validated *R. mucilaginosa* JM19 stain as a promising candidate for control of PAT contamination in food and raw materials. This stain (*R. mucilaginosa* JM19 1 × 10^8 cells/L) was able to significantly reduce PAT by 90% after 21 h in MES buffer at 35 °C. With at initial patulin concentration (100 μg/mL), the *R. mucilaginosa* JM19 was able to detoxify above 50%. By this process of detoxification, *R. mucilaginosa* JM19 degraded PAT into desoxypatulinic acid, a lesser non-toxic compound.

Similarly, Coelho et al. (2007) indicated that initial PAT concentration (223 μg) declined by >83% after co-incubation with *Pichia ohmeri* 158 for 2 d at 25 °C, followed by 99% reduction after 5 d, and eventually untraceable after 15 d. Also, PAT incubated with *Pichia caribbica* yeast for 15 d at 20 °C, decreased contamination in apples than in the control and the same response by *P. caribbica* against PAT was noticed in the *in vitro* tests (Cao et al., 2013). Despite the performance of antagonistic yeasts, recent studies have reported a significant interest in substances capable of augmenting their biocontrol efficacy against postharvest storage pathogens (Droby et al., 2016; Wisniewski et al., 2016). Tolaini et al. (2010) in their study revealed the application of *Lentinula edodes* (LS28) against *P. expansum* in apples. This study also showed that *L. edodes* (LS28) reduced approximately 50% of apple decay caused by *P. expansum*. It also enhanced the activity of antioxidant enzymes catalase (CAT) and glutathione peroxidase as host defense response to the presence of disease infection in the biocontrol system. Other studies on the efficacy of antagonistic yeast indicated that *Cryptococcus laurentii* is correlated with its level of superoxide dismutase (SOD) and CAT production, and is responsible for the yeast resistance to oxidative stress (Castoria et al., 2005). The application of antioxidant quercetin significantly reduced the incidence of apple blue mold. It is evident that quercetin may induce resistance to *P. expansum* in apples through acting on gene transcription levels involved in several distinct metabolic processes (Romanazzi et al., 2016; Sanzani et al., 2012).

10.7 IMPROVING THE PERFORMANCE AND STABILITY OF ANTAGONISTIC YEASTS IN PATULIN REDUCTION

The satisfactory and stable approaches for PAT detoxification are extremely important in order to achieve success with any biocontrol representative. Several investigations have reported on various techniques to boost the consistency and effectiveness of postharvest biocontrol agents. The agents are categorized as salts and organic acids, food additives, glucose analogs and various physical (thermal and non-thermal) treatments. Each antagonistic yeast within the biocontrol system has its specific unique characteristic response to the amending agent and therefore needs precise protocol to undertake commercial assessment. Indeed, addition of substances at non-toxic but acceptable concentrations have enriched the bio-efficiency of yeasts, where the outcome of the interaction between two or more agents essentially yielded a higher yeast population or improved the use of yeast biomass at a lower concentration (Apaliya et al., 2018). Some of the substances include salicylic acid (Babalar et al., 2007; Zhang et al., 2008, 2010), brassinosteroids (Zhu et al., 2010), sodium bicarbonate (Cerioni et al., 2012; Youssef et al., 2014; Zhu et al., 2012), chitosan (Meng et al., 2010; Romanazzi and Feliziani, 2016), phytic acid (Mahunu et al., 2016; Yang et al., 2015), indole-3-acetic acid (Yu et al., 2008), boron (Cao et al., 2012; Qin et al., 2010), harpin (Tang et al., 2015; Zhu and Zhang, 2016), phosphatidylcholine (Li et al., 2016), glycine betaine (Zhang et al., 2017b), jasmonic acid (Li et al., 2017b), potassium phosphite (Lai et al., 2017), trehalose (Apaliya et al., 2017, 2018), bamboo (Mahunu et al., 2018), β-glucan (Wang et al., 2018), methyl thujate (Ji et al., 2018), $CaCl_2$ (Tournas and Katsoudas, 2019) and baobab (Abdelhai et al., 2019).

Given the possibility of enhancing yeast performance, it is pivotal to search for substances that exhibit the ability to counter mycotoxin activity as well as be relevant in PAT detoxification. Recently, a biocompatibility study by Mahunu et al. (2018) showed *Pichia caribbica* supplemented with bamboo leaf flavonoid (0.01% w/v) reduced or inhibited PAT accumulation concentration both *in vitro* and *in vivo* after a 20 d incubation period at 20 °C. Also, an examination showed that *R. mucilaginosa* augmented with phytic acid (4 μmol/mL) was able to decrease the initial patulin concentration of 20 μg/mL to 1.561 μg/mL, and the degradation rate was 92.2% after 24 h *in vitro* incubation at 28 °C (Yang et al., 2015).

Furthermore, the distinct and multifaceted modes of action exhibited by biocontrol yeasts reduced PAT contamination on the bases of their competition for space, nutrients and their wounding competence in apples (Spadaro et al., 2013). Likewise, the improvement of biocontrol efficacy is influenced by yeast resistance to the oxidative stress that is caused by the reactive oxygen species (ROS) generated in fruit wounds (Castoria et al., 2003) and lytic enzyme production (Castoria et al., 2001). Findings of Lima et al. (2011) indicated that the fungal pathogen cannot readily develop resistance against biocontrol methods due to the multifaceted mode of action of BCAs.

Similarly, the direct application of physical treatment such as ultraviolet-C (UV-C) to control fungi has been well investigated besides other methods for postharvest fruit disease prevention. UV-C was found to stimulate responses in fruits (Stevens et al., 2005), which confirmed earlier studies on the relationship between UV-C dosage, the number of fungal-spore germination and the number of infected wound lesions on fruit surfaces (Stevens et al., 1998). Indeed, the assimilation of low doses (200 and 400 Gy) of γ-irradiation together with a BCA (*P. fluorescens*) was described as suitable for extension of physicochemical qualities and to reduce postharvest losses of apples (Mostafavi et al., 2013). Application of UV-C at a low-dose, which reduced pathogen inoculum and induced host resistance was considered as doubly effective on fungi but did not show evidence of PAT degradation in apples. As a result, UV-C inhibited *P. expansum* infection and growth but it was unable to prevent PAT distribution in the apple once inoculation had occurred. After γ-irradiation treatment, phenolic compounds rapidly accumulated in the fruit tissues commonly attributed to the ability of UV light to stimulate the activity of PAL, which is the key enzyme in the biosynthesis of phenols in plant tissues (Charles et al., 2008). Generally, there is a connection between phenolic content and antioxidant activity, and some phenolic compounds are known for their antioxidant activities (Zhang et al., 2013). Similarly, antioxidant enzymes and related enzymes were shown to play a key role in eradicating reactive oxygen species (Wang et al., 2011). It was also reported that oxidant activities can elicit and regulate the biosynthesis of mycotoxins produced by various *Aspergillus* species (*Aspergillus flavus*, *A. parasiticus*, and *A. ochraceus*) (Reverberi et al., 2008). Previous studies also showed that natural antioxidants from compounds have been recognized for postharvest fungal control and inhibition of other mycotoxins including aflatoxin, ochratoxin and PAT (Reverberi et al., 2005; Ricelli et al., 2007).

On the other hand, the exhibition of microbial activity by PAT appears to play a direct role in microbial competition and survival, thus protecting the fungal pathogen and functions partly as a defense mechanism against yeast antipathy (Castoria et al., 2008). From investigations, PAT inhibited the growth of 75 species of Gram-positive and Gram-negative bacteria and also expressed antiviral and anti-protozoal actions (Moake et al., 2005). This also implies that PAT activity has the capacity to enhance fungal growth and improve their competing ability in an environment with other microorganisms (Tikekar, 2009).

In general, the function of mycotoxin-producing fungi is to activate pathogenesis. However, fungal mutants unable to produce toxins might demonstrate substantially less virulence than their toxin-producing counterparts. As initially described by Desjardins and Hohn (1997), *Penicillium* produces PAT-mycotoxicity which plays a role in pathogenesis. This observation was confirmed by Harris et al. (1999), who indicated that trichothecene mycotoxin produced by *Fusarium graminearum* plays a crucial role in pathogenicity in maize; thus, the activity undoubtedly weakens the defense mechanism of the host cell, causing the host to become vulnerable to fungal infection. The presence of trichothecene might modify the penetrability of the host cell membrane, whereas PAT generates excessive oxidative stress resulting in cell death (Speijers and Speijers, 2004). Tikekar (2009) indicated that fungal toxins are able to increase the competitiveness and pathogenicity of fungi and subsequently support the fungi to survive proficiently in the environment they occupy.

Since characteristically PAT is unstable, the BCAs screened and selected to control patulin-producing fungi must have the capacity to defend the antimicrobial activity of PAT. The assumption is that the antimicrobial activity produced by PAT can persist on fruit surfaces all through the preharvest and postharvest stages, therefore when the bioefficacy of agents is enhanced, it will be able to destroy the toxicity and compete with the pathogen (Paster and Barkai-Golan, 2008). In other words, the yeasts, for instance, are able to colonize, survive and persist on dry fruit surfaces for longer periods of time during storage. Interestingly, the majority of reports on postharvest BCAs co-incubated with yeast cells showed a superior efficacy for eradicating PAT or inhibited fungal infection. This approach based on the use of mediating technologies at critical points of the disease control process also offers possible energy sources for microbes or adds to oxidative stress against fungi in the biocontrol system.

10.8 MECHANISMS FOR ANTAGONISTIC YEAST ACTION IN PATULIN REDUCTION

So far, studies have shown advances in the mechanisms of PAT degradation through the use of BCAs. Expounding the mechanisms of degradation as well as identifying key functional factors in yeasts will promote the development of systems that will assure efficient removal of PAT in foods. Biocontrol agents have convincingly proven to carry the efficacy to overcome PAT, yet the mechanisms of action of degradation are repeatedly poorly understood. The most important mechanisms of biological removal or degradation of PAT suggested include microbial cell biosorption of PAT (Guo et al., 2012; Wang et al., 2015; Yue et al., 2013), PAT degradation by microbial enzymes (Zhu et al., 2015), PAT-glutathione adducts or reaction of PAT with the thiol group present in protein extracts and for that reason terminating its functional properties (Luz et al., 2017). It has also been observed that the presence of PAT in growth media induces the synthesis of PAT degrading enzymes in PAT resistant/degrading yeasts (Zheng et al., 2017). Also, earlier report by Park and Troxell (2002) indicated that the capacity of BCAs to reduce mycotoxin levels is governed by the following factors: (i) the stability of the mycotoxin; (ii) the nature of the degradation process; (iii) interactions between mycotoxin and food or host; and (iv) mycotoxin–mycotoxin interactions. Similarly, the control of PAT contamination in fruits is achieved by any of these following three categories: (i) through prevention of initial PAT pollution during harvesting, processing, and storage; (ii) PAT removal during processing; and (iii) use of post-production treatment agents for PAT removal or detoxification.

The use of supplementary agents has shown the potency to enrich the antifungal activity of yeast as an important mechanism of action against PAT, although some yeast such as *Pichia ohmeri* strain 158 alone in broth cultures was able to decrease PAT production (Coelho et al., 2007). *Lentinula edodes* yeast develops poorly in a highly oxidative environment (Tolaini et al., 2010), whereas *Rhodotorula glutinis* LS11 strain reinforces its enzymatic antioxidant potential under similar conditions in order to prevent PAT biosynthesis (Castoria et al., 2008). From the literature, *R. glutinis* LS11 strain is capable of resisting extremely high concentrations of PAT (500 μg mL^{-1}) and metabolizes it under aerobic conditions. Using thin-layer chromatography (TLC)-based detection, two products were formed and their retention factors (Rf) were 0.46 and 0.38 (Castoria et al., 2005). From the study, the formation of the compound(s) with Rf of 0.46 seemed to be more stable than that of the other compound. The compound with Rf of 0.46 was the main product detected at the end of the degradation process, whereas with Rf of 0.38, it could not be detected anymore. The study by Ricelli et al. (2007) showed that in apple juice approximately 96% of PAT was degraded after incubation with *Gluconobacter oxydans*. The use of *G. oxydans* converted PAT to ascladiol, a less-toxic compound. *Sporobolomyces* sp. strain IAM 13481 was also found to transform PAT into two different end-products (desoxypatulinic acid and (Z)-ascladiol) (Ianiri et al., 2013). Again, *Candida guilliermondii* was found to be a major yeast that degraded PAT to an E-ascladiol toxic compound (Figure 10.5) (Chen et al., 2017). Essentially, rapid biodegradation of PAT depends on the viable/

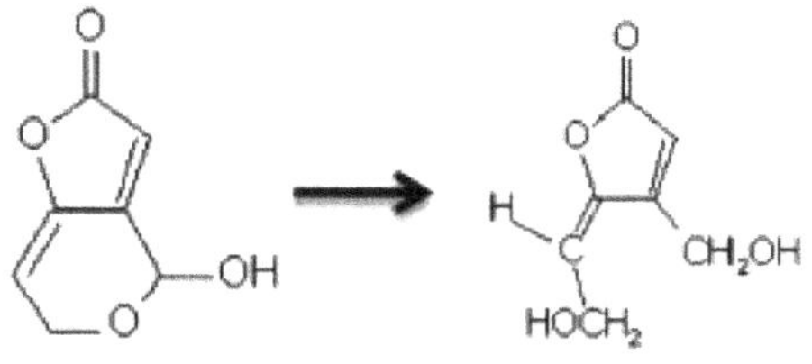

Patulin → E-ascladiol

FIGURE 10.5 Biodegradation by *Candida guilliermondii* to convert patulin to E-ascladiol (Chen et al., 2017).

living yeast cells with greater efficacy to eliminate PAT than dead cells. According to Zhu et al. (2015), supported by Chen et al. (2017), intercellular protein showed the ability to degrade PAT, which implies that the biodegradation of PAT is an intercellular occurrence in yeast.

According to Coelho et al. (2007), the presence of antagonistic yeast cells *in vitro* decreased PAT contamination through two main mechanisms: (i) by PAT adsorption on the yeast cell wall; and (ii) PAT absorption into yeast cells. With respect to the mode of action, the microbes and additives (such as zeolites, bentonites, clays and activated carbons) function together to bind toxin to their surface through adsorption or transform it into less toxic metabolites known as biotransformation (Kolosova and Stroka, 2011). The studies performed by Yiannikouris et al. (2004) reported that yeast strains with a high cell wall glucan (HCWG) content are tightly bound to toxins compared with strains with low cell wall glucan (LCWG) content. This occurrence was the case when their total cell wall (TCW) fraction and alkali-insoluble glucan (AIG) fraction were exposed to the toxins. The authors observed that the adsorption efficacy of AIG fraction was the highest when its chitin content was lower than that of the TCW fraction, which increased glucan flexibility and accessibility of the toxin to the glucan network. In similar studies, Moran (2004) described the yeast cell wall as a natural source of two oligosaccharides (β-glucan and mannan) most potent in inducing defense. Mostly, β-D-glucans are known to assist in inducing defense against fungal infections (Novak and Vetvicka, 2008) and Moran (2004) had previously demonstrated the mechanism of action involved in mycotoxin–glucan interaction.

Tian et al. (2007) reported on the enzyme β-1,3-glucanase as one of the well categorized pathogenesis-related (PR) proteins, acting directly in pathogen cell wall degradation or indirectly releasing oligosaccharides to elicit defense reactions. Also, Janisiewicz and Korsten (2002) stated that the yeast extracellular β-glucanases and chitinases depolymerize fungal cell walls, which symbolizes an essential biocontrol mechanism of fungi in postharvest. Following reports of previous authors (Reverberi et al., 2005; Ricelli et al., 2007), Zhang et al. (2013) also confirmed the importance of enzymes in pathogen cell wall degradation and antioxidant function in toxigenic fungi.

A non-specific phytotoxic substance, oxalic acid (OA) is also able to elicit plant defense reactions through suppression of the oxidative burst of the host plant (Kim et al., 2008). It plays a key role in pathogenesis and fungal growth. The presence of postharvest treatment with BCAs induces host resistance of fruits (apples and citrus) and stimulation of enzymatic activities including phenylalanine ammonia lyase (PAL) (Droby et al., 2003; El Ghaouth et al., 2003).

Extracellular compounds produced by yeast strains have the potential to prevent and possibly kill mycotoxin-producing strains (Coelho et al., 2007). To achieve this, the yeasts secrete a low-molecular-mass protein or glycoprotein toxin that can kill sensitive cells belonging to the same or related yeast genera, but this does not involve direct cell-cell contact. Furthermore, the biocontrol activities of such yeasts do not depend on the production of harmful antibiotics, rather they are influenced by their ability to compete for space and nutrients on fruit surfaces or in wounded tissue (Paster and Barkai-Golan, 2008). The findings of Castoria et al. (2005) also identified lower PAT accumulation in infected apples that were pretreated with low concentrations of *Rhodotorula glutinis* strain LS11, as compared with the level in control (non-treated) infected fruits. These authors also indicated that

PAT degradation was hastened after the yeast cells survived and increased their population in the infected apples. A similar trend of results was detected in an *in vitro* test, where surviving yeast cells in decayed apples may possibly metabolize PAT and/or reduce its accumulation. PAT production was estimated to decrease in accordance with the incubation time (20 days at 15 °C), which was possibly due to organic acids present in the culture that cause PAT to leach from the vacuole and mycelium of *P. expansum* itself (Morales et al., 2007b). In the same study, the growth of *P. expansum* was detected to be slower at the early stages of incubation while a sufficient energy source was present in the medium. However, growth of *P. expansum* increased in the late phase after the energy source was almost exhausted in addition to the presence of many intermediaries.

10.9 CONCLUSIONS

Fungal infections contaminate raw apple fruit prior to processing with the possibility of PAT synthesis. Therefore, there are well-established standards and regulations to prevent PAT crossing over to humans, especially infants. Various processing methods have been used satisfactorily, but still cannot guarantee complete removal or degradation of PAT from contaminated commodities. Synthetic fungicides have been found feasible for PAT control, but by themselves they are capable of acting as a potential stress driver for mycotoxin-producing fungi. Recent searches point to potential biocontrol agents being used after significant progress has been made in their commercialization. Other research directions point to enhanced biocontrol efficacy with two or more bioactive substances or compounds without compromising consumer safety. It is expected that biocontrol agents or products will gain increasing recognition with a wide acceptability in the coming years as a component of postharvest disease management. Lastly, molecular studies, when conducted, could further explain the complex mechanisms involved in mycotoxin elimination. Thus, it can be concluded that PAT can be degraded in several ways, by biotransformation, biodegradation or removal of PAT through microbial cell biosorption of PAT, microbial enzymes degrading PAT and PAT-glutathione adducting or PAT reacting with the thiol group existing in protein extracts. Finally, the presence of PAT in growth media induces the production of degrading enzymes in patulin resistant/degrading yeasts.

ACKNOWLEDGMENTS

This work was supported by the National Natural Science Foundation of China (31772369, 31571899), the National Key Research Project (subproject) of China (2016YFD0400902-04) and the Agricultural Independent Innovation Fund in Jiangsu Province [CX(18)2028]. The authors thank some of the experts in this field for sharing their valuable knowledge over the years.

REFERENCES

Abdelhai, M.H., H.E.O.J. Tahir, Q. Zhang et al. 2019. Effects of the combination of Baobab (*Adansonia digitata* L.) and *Sporidiobolus pararoseus* Y16 on blue mold of apples caused by *Penicillium expansum*. *Biological Control* 134: 87–94. doi: 10.1016/j.biocontrol.2019.04.009

Abrunhosa, L., H. Morales, C. Soares et al. 2016. A review of mycotoxins in food and feed products in Portugal and estimation of probable daily intakes. *Critical Reviews in Food Science and Nutrition* 56: 249–265.

Al-Hazmi, N.A. 2010. Determination of Patulin and Ochratoxin A using HPLC in apple juice samples in Saudi Arabia. *Saudi Journal of Biological Sciences* 17: 353–359.

Apaliya, M.T., H. Zhang, Q. Yang et al. 2017. *Hanseniaspora uvarum* enhanced with trehalose induced defense-related enzyme activities and relative genes expression levels against *Aspergillus tubingensis* in table grapes. *Postharvest Biology and Technology* 132: 162–170.

Apaliya, M.T., H. Zhang, X. Zheng et al. 2018. Exogenous trehalose enhanced the biocontrol efficacy of *Hanseniaspora uvarum* against grape berry rots caused by *Aspergillus tubingensis* and *Penicillium commune*. *Journal of the Science of Food and Agriculture* 98(12): 4665–4672. doi: 10.1002/jsfa.8998

Atkins, D. 1994. The UK's Food Chemical Surveillance Programme. *British Food Journal* 96: 24–29.

Babalar, M., M. Asghari, A. Talaei et al. 2007. Effect of pre- and postharvest salicylic acid treatment on ethylene production, fungal decay and overall quality of Selva strawberry fruit. *Food Chemistry* 105: 449–453.

Baert, K., F. Devlieghere, A. Amiri et al. 2012. Evaluation of strategies for reducing patulin contamination of apple juice using a farm to fork risk assessment model. *International Journal of Food Microbiology* 154: 119–129.

Ballester, A.-R., M. Marcet-Houben, E. Levin et al. 2015. Genome, transcriptome, and functional analyses of *Penicillium expansum* provide new insights into secondary metabolism and pathogenicity. *Molecular Plant-Microbe Interactions* 28: 232–248.

Bandoh, S., M. Takeuchi, K. Ohsawa et al. 2009. Patulin distribution in decayed apple and its reduction. *International Biodeterioration and Biodegradation* 63: 379–382.

Barad, S., E. Sionov and D. Prusky. 2016b. Role of patulin in post-harvest diseases. *Fungal Biology Reviews* 30(1): 24–32.

Barad, S., E.A. Espeso, A. Sherman et al. 2016a. Ammonia activates pacC and patulin accumulation in an acidic environment during apple colonization by *Penicillium expansum*. *Molecular Plant Pathology* 17(5): 727–740.

Barad, S., S.B. Horowitz, I. Kobiler et al. 2014. Accumulation of the mycotoxin patulin in the presence of gluconic acid contributes to pathogenicity of *Penicillium expansum*. *Molecular Plant-Microbe Interactions* 27: 66–77.

Baraldi, E., M. Mari, E. Chierici et al. 2003. Studies on thiabendazole resistance of *Penicillium expansum* of pears: Pathogenic fitness and genetic characterization. *Plant Pathology* 52: 362–370.

Barkai-Golan, R. and N. Paster. 2011. *Mycotoxins in Fruits and Vegetables*. Cambridge, MA: Academic Press.

Beretta, B., A. Gaiaschi, C.L. Galli et al. 2000. Patulin in apple-based foods: Occurrence and safety evaluation. *Food Additives & Contaminants* 17: 399–406.

Burghardt, R.C., R. Barhoumi, E.H. Lewis et al. 1992. Patulin-induced cellular toxicity: A vital fluorescence study. *Toxicology and Applied Pharmacology* 112: 235–244.

Cabañas, R., M. Abarca, M. Bragulat et al. 2009. Comparison of methods to detect resistance of *Penicillium expansum* to thiabendazole. *Letters in Applied Microbiology* 48: 241–246.

Calado, T., A. Venâncio and L. Abrunhosa. 2014. Irradiation for mold and mycotoxin control: A review. *Comprehensive Reviews in Food Science and Food Safety* 13: 1049–1061.

Calvert, G.M., J. Karnik, L. Mehler et al. 2008. Acute pesticide poisoning among agricultural workers in the United States, 1998–2005. *American Journal of Industrial Medicine* 51: 883–898.

Cao, B., H. Li, S. Tian et al. 2012. Boron improves the biocontrol activity of *Cryptococcus laurentii* against *Penicillium expansum* in jujube fruit. *Postharvest Biology and Technology* 68: 16–21.

Cao, J., H. Zhang, Q. Yang et al. 2013. Efficacy of *Pichia caribbica* in controlling blue mold rot and patulin degradation in apples. *International Journal of Food Microbiology* 162: 167–173.

Castoria, R., L. Caputo, F. De Curtis et al. 2003. Resistance of postharvest biocontrol yeasts to oxidative stress: A possible new mechanism of action. *Phytopathology* 93: 564–572.

Castoria, R., F. De Curtis, G. Lima et al. 2001. *Aureobasidium pullulans* (LS-30) an antagonist of postharvest pathogens of fruits: Study on its modes of action. *Postharvest Biology and Technology* 22: 7–17.

Castoria, R., V. Morena, L. Caputo et al. 2005. Effect of the biocontrol yeast *Rhodotorula glutinis* strain LS11 on patulin accumulation in stored apples. *Phytopathology* 95: 1271–1278.

Castoria, R., S.A. Wright and S. Drobyy. 2008. Biological control of mycotoxigenic fungi in fruits. In R. Bargain Golan and N. Paster (Eds.), *Mycotoxins in Fruits and Vegetables* (pp. 311–333). Cambridge, MA: Academic Press.

Cerioni, L., L. Rodríguez-Montelongo, J. Ramallo et al. 2012. Control of lemon green mold by a sequential oxidative treatment and sodium bicarbonate. *Postharvest Biology and Technology* 63: 33–39.

Charles, M.T., N. Benhamou and J Arul. 2008. Physiological basis of UV-C induced resistance to *Botrytis cinerea* in tomato fruit: III. Ultrastructural modifications and their impact on fungal colonization. *Postharvest Biology and Technology* 47: 27–40.

Chen, X., J. Li, L. Zhang et al. 2012. Control of postharvest radish decay using a *Cryptococcus albidus* yeast coating formulation. *Crop Protection* 41: 88–95.

Chen, Y., H.M. Peng, X. Wang et al. 2017. Biodegradation Mechanisms of Patulin in *Candida guilliermondii*: An iTRAQ-Based Proteomic Analysis. *Toxins* 9: 48.

Cheraghali, A.M., H.R. Mohammadi, M. Amirahmadi et al. 2005. Incidence of patulin contamination in apple juice produced in Iran. *Food Control* 16: 165–167.

Coelho, A.R., M.G. Celli E.Y.S. Ono et al. 2007. *Penicillium expansum* versus antagonist yeasts and patulin degradation in vitro. *Brazilian Archives of Biology and Technology* 50: 725–733.

Coelho, A.R., M. Tachi, F.C. Pagnocca et al. 2009. Purification of *Candida guilliermondii* and *Pichia ohmeri* killer toxin as an active agent against *Penicillium expansum*. *Food Additives and Contaminants* 26: 73–81.
Collin, S., E. Bodart, C. Badot et al. 2008. Identification of the main degradation products of patulin generated through heat detoxication treatments. *Journal of the Institute of Brewing* 114: 167–171.
Coulombe Jr, R.A. 1993. Biological action of mycotoxins. *Journal of Dairy Science* 76: 880–891.
Cressey, P.J. 2009. Mycotoxin risk management in New Zealand and Australian food. *World Mycotoxin Journal* 2: 113–118.
de Sylos, C.M. and D.B. Rodriguez-Amaya. 1999. Incidence of patulin in fruits and fruit juices marketed in Campinas, Brazil. *Food Additives and Contaminants* 16: 71–74.
de Souza Sant'Ana, A., A. Rosenthal and P.R. de Massaguer. 2008. The fate of patulin in apple juice processing: A review. *Food Research International* 41: 441–453.
Deming, J.W. 2002. Psychrophiles and polar regions. *Current Opinion in Microbiology* 5: 301–309.
Desjardins, A.E. and T.M. Hohn. 1997. Mycotoxins in plant pathogenesis. *Molecular Plant-Microbe Interactions* 10: 147–152.
Droby, S., M. Wisniewski, A. El Ghaouth et al. 2003. Influence of food additives on the control of postharvest rots of apple and peach and efficacy of the yeast-based biocontrol product Aspire. *Postharvest Biology and Technology* 27: 127–135.
Droby, S., M. Wisniewski, N. Teixidó. 2016. The science, development, and commercialization of postharvest biocontrol products. *Postharvest Biology and Technology* 122: 22–29.
El Ghaouth, A., C.L. Wilson and M. Wisniewski. 2003. Control of postharvest decay of apple fruit with *Candida saitoana* and induction of defense responses. *Phytopathology* 93: 344–348.
Fan, Q. and S. Tian. 2001. Postharvest biological control of grey mold and blue mold on apple by *Cryptococcus albidus* (Saito) Skinner. *Postharvest Biology and Technology* 21: 341–350.
FAO, J. and M.H.I. FOODS. 2004. *Food and Agriculture Organization of the United Nations*. Rome: FAO.
FAO/ WHO Expert Committee on Food Additives. 2005. Evaluation of certain food additives. *Sixty-third report of the Joint FAO/WHO expert Committee on Food Additives*. WHO Technical Report Series, No. 928.
Fernández-Cruz, M.L., M.L. Mansilla and J.L. Tadeo. 2010. Mycotoxins in fruits and their processed products: Analysis, occurrence and health implications. *Journal of Advanced Research* 1: 113–122.
Fogliata, G., L. Torres, L. Ploper. 2000. Detection of imazalil-resistant strains of *Penicillium digitatum* Sacc. in citrus packinghouses of Tucumán Province (Argentina) and their behavior against currently employed and alternative fungicides. *Revista industrial y agrícola de Tucumán* 77: 71–75.
Forrester, P. and G. Gaucher. 1972. m-Hydroxybenzyl alcohol dehydrogenase from *Penicillium urticae*. *Biochemistry* 11: 1108–1114.
Frisvad, J.C., J.M. Frank, J. Houbraken et al. 2004. New ochratoxin A producing species of Aspergillus section Circumdati. *Studies in Mycology* 50: 23–43.
Funes, G.J. and S.L. Resnik. 2009. Determination of patulin in solid and semisolid apple and pear products marketed in Argentina. *Food Control* 20: 277–280.
Glaser, N. and H. Stopper. 2012. Patulin: Mechanism of genotoxicity. *Food and Chemical Toxicology* 50: 1796–1801.
Gökmen, V. and J. Acar. 1998. Incidence of patulin in apple juice concentrates produced in Turkey. *Journal of Chromatography A* 815: 99–102.
González-Osnaya, L., J.M. Soriano, J.C. Moltó et al. 2007. Exposure to patulin from consumption of apple-based products. *Food Additives and Contaminants* 24: 1268–1274.
Guo, C., T. Yue, Y. Yuan et al. 2012. Biosorption of patulin from apple juice by caustic treated waste cider yeast biomass. *Food Control* 32: 99–104.
Guo, Y., Z. Zhou, Y. Yuan et al. 2013. Survey of patulin in apple juice concentrates in Shaanxi (China) and its dietary intake. *Food Control* 32: 570–573.
Halász, A., R. Lásztity, T. Abonyi et al. 2009. Decontamination of mycotoxin-containing food and feed by biodegradation. *Food Reviews International* 25: 284–298.
Harris, L., A.E. Desjardins, R. Plattner et al. 1999. Possible role of trichothecene mycotoxins in virulence of *Fusarium graminearum* on maize. *Plant Disease* 83: 954–960.
Hojnik, N., U. Cvelbar, G. Tavčar-Kalcher et al. 2017. Mycotoxin decontamination of food: Cold atmospheric pressure plasma versus "classic" decontamination. *Toxins* 9: 151.
Ianiri, G., A. Idnurm and R. Castoria. 2016. Transcriptomic responses of the basidiomycete yeast *Sporobolomyces* sp. to the mycotoxin patulin. *BMC Genomics* 17: 210. doi: 10.1186/s12864-016-2550-4

Ianiri, G., A. Idnurm, S.A. Wright et al. 2013. Searching for genes responsible for patulin degradation in a biocontrol yeast provides insight into the basis for resistance to this mycotoxin. *Applied and Environmental Microbiology* 73: 3101–3115.

Janisiewicz, W. 1999. *Blue Mold, Penicillium spp. Fruit Disease Focus.* Kearneysville, WA: USDA Appalachian Fruit Research Station.

Janisiewicz, W.J. and L. Korsten. 2002. Biological control of postharvest diseases of fruits. *Annual Review of Phytopathology* 40: 411–441.

Ji, D., T. Chen, D. Ma et al. 2018. Inhibitory effects of methyl thujate on mycelial growth of *Botrytis cinerea* and possible mechanisms. *Postharvest Biology and Technology* 142(4): 46–54.

Josefsson, E. and A. Andersson. 1977. Analysis of patulin in apple beverages sold in Sweden. *Archives de l'Institut Pasteur de Tunis* 54: 189–196.

Jouany, J.P. 2007. Methods for preventing, decontaminating and minimizing the toxicity of mycotoxins in feeds. *Animal Feed Science and Technology* 137: 342–362.

Juhneviča, K., G. Skudra and L. Skudra. 2011. Evaluation of microbiological contamination of apple fruit stored in a modified atmosphere. *Environmental and Experimental Biology* 9: 53–59.

Kanetis, L., H. Förster and J.E. Adaskaveg. 2007. Comparative efficacy of the new postharvest fungicides azoxystrobin, fludioxonil, and pyrimethanil for managing citrus green mold. *Plant Disease* 91: 1502–1511.

Karabulut, O.A., U. Arslan, K. Ilhan et al. 2005. Integrated control of postharvest diseases of sweet cherry with yeast antagonists and sodium bicarbonate applications within a hydrocooler. *Postharvest Biology and Technology* 37: 135–141.

Kazi, S., R. Paterson and N. Abo-Dahab. 1997. Effect of 2-deoxy-D-glucose on mycotoxins from apples inoculated with *Penicillium expansum*. *Mycopathologia* 138: 43–46.

Kim, K.S., J.-Y. Min and M.B. Dickman. 2008. Oxalic acid is an elicitor of plant programmed cell death during *Sclerotinia sclerotiorum* disease development. *Molecular Plant-Microbe Interactions* 21: 605–612.

Kolosova, A. and J. Stroka. 2011. Substances for reduction of the contamination of feed by mycotoxins: A review. *World Mycotoxin Journal* 4: 225–256.

Konstantinou, S., G. Karaoglanidis, G. Bardas et al. 2011. Postharvest fruit rots of apple in Greece: Pathogen incidence and relationships between fruit quality parameters, cultivar susceptibility, and patulin production. *Plant Disease* 95: 666–672.

Kottapalli, B., C.E. Wolf-Hall, P. Schwarz et al. 2003. Evaluation of hot water and electron beam irradiation for reducing Fusarium infection in malting barley. *Journal of Food Protection* 66: 1241–1246.

Kuiper-Goodman, T. 2004. Risk assessment and risk management of mycotoxins in food. In N. Magan and M. Olesan, *Mycotoxins in Food.* Cambridge, UK: Woodhead Publishing, pp. 3–31.

Lai, T., Y. Wang, Y. Fan et al. 2017. The response of growth and patulin production of postharvest pathogen *Penicillium expansum* to exogenous potassium phosphite treatment. *International Journal of Food Microbiology* 244: 1–10.

Laidou, I., C. Thanassoulopoulos and M. Liakopoulou-Kyriakides. 2001. Diffusion of patulin in the flesh of pears inoculated with four post-harvest pathogens. *Journal of Phytopathology* 149: 457–461.

Lee, T.P., R. Sakai, N.A. Munaf et al. 2014. High performance liquid chromatography method for the determination of patulin and 5-hydroxymethylefurfural in fruit juices marketed in Malaysia. *Food Control* 38: 142–149.

Leggott, N.L. and G.S. Shephard. 2001. Patulin in South African commercial apple products. *Food Control* 12: 73–76.

Li, Q., C. Li, P. Li et al. 2017a. The biocontrol effect of *Sporidiobolus pararoseus* Y16 against postharvest diseases in table grapes caused by *Aspergillus niger* and the possible mechanisms involved. *Biological Control: Theory and Applications in Pest Management* 113: 18–25. doi: 10.1016/j.biocontrol.2017.06.009

Li, X., H. Tang, C. Yang et al. 2019. Detoxification of mycotoxin patulin by the yeast *Rhodotorula mucilaginosa*. *Food Control* 96: 47–52.

Li, T., Y. Xu, L. Zhang et al. 2017b. The Jasmonate-activated transcription factor MdMYC2 regulates Ethylene Response Factor and ethylene biosynthetic genes to promote ethylene biosynthesis during apple fruit ripening. *The Plant Cell* 29: 1316–1334.

Li, W., H. Zhang, P. Li et al. 2016. Biocontrol of postharvest green mold of oranges by *Hanseniaspora uvarum* Y3 in combination with phosphatidylcholine. *Biological Control* 103: 30–38.

Li, R., H. Zhang, W. Liu et al. 2011. Biocontrol of postharvest gray and blue mold decay of apples with *Rhodotorula mucilaginosa* and possible mechanisms of action. *International Journal of Food Microbiology* 146: 151–156.

Lima, G., R. Castoria, F. De Curtis et al. 2011. Integrated control of blue mould using new fungicides and biocontrol yeasts lowers levels of fungicide residues and patulin contamination in apples. *Postharvest Biology and Technology* 60: 164–172.
Liu, B.-H., T.-S. Wu, F.-Y. Yu et al. 2006. Induction of oxidative stress response by the mycotoxin patulin in mammalian cells. *Toxicological Sciences* 95: 340–347.
Lukwago, F. B., I. M. Mukisa, A. Atukwase et al. 2019. Mycotoxins contamination in foods consumed in Uganda: A 12-year review (2006–2018). *Scientific African* 3: e00054.
Luz, C., F. Saladino, F. Luciano et al. 2017. In vitro antifungal activity of bioactive peptides produced by *Lactobacillus plantarum* against *Aspergillus parasiticus* and *Penicillium expansum*. *LWT-Food Science and Technology* 81: 128–135.
Magan, N. and M. Olsen. 2004. *Mycotoxins in Food: Detection and Control*. Sawston: Woodhead Publishing.
Mahunu, G.K., H. Zhang, M.T. Apaliya et al. 2018. Bamboo leaf flavonoid enhances the control effect of *Pichia caribbica* against *Penicillium expansum* growth and patulin accumulation in apples. *Postharvest Biology and Technology* 141: 1–7.
Mahunu, G.K., H. Zhang, Q. Yang et al. 2015. Biological control of patulin by antagonistic yeast: A case study and possible model. *Critical Reviews in Microbiology*: 41(1): 1–13.
Mahunu, G.K., H. Zhang, Q. Yang et al. 2016. Improving the biocontrol efficacy of *Pichia caribbica* with phytic acid against postharvest blue mold and natural decay in apples. *Biological Control* 92: 172–180.
Marin S., A. Ramos, C. Cano-Sancho et al. 2011. Patulin contamination in fruit derivatives, including baby food from Spanish market. *Food Chemistry* 124: 563–568.
Marin, S., A. Ramos, G. Cano-Sancho et al. 2013. Mycotoxins: Occurrence, toxicology, and exposure assessment. *Food and Chemical Toxicology* 60: 218–237.
Marroquín-Cardona, A., N. Johnson, T. Phillips et al. 2014. Mycotoxins in a changing global environment–A review. *Food and Chemical Toxicology* 69: 220–230.
Matsuoka, K. 2019. Anthocyanins in apple fruit and their regulation for health benefits. In *Anthocyanins-Novel Antioxidants in Human Health and Diseases Prevention*. Rejika: IntechOpen.
McCallum, J., R. Tsao and T. Zhou. 2002. Factors affecting patulin production by *Penicillium expansum*. *Journal of Food Protection* 65: 1937–1942.
Meng, X.-H., G.-Z. Qin and S.P. Tian. 2010. Influences of preharvest spraying *Cryptococcus laurentii* combined with postharvest chitosan coating on postharvest diseases and quality of table grapes in storage. *LWT - Food Science and Technology* 43: 596–601.
Moake, M.M., O.I. Padilla-Zakour and R.W. Worobo. 2005. Comprehensive review of patulin control methods in foods. *Comprehensive Reviews in Food Science and Food Safety* 4: 8–21.
Morales, H., S. Marín, A. Rovira et al. 2007b. Patulin accumulation in apples by *Penicillium expansum* during postharvest stages. *Letters in Applied Microbiology* 44: 30–35.
Morales, H., S. Marín, A.J. Ramos et al. 2010. Influence of post-harvest technologies applied during cold storage of apples in *Penicillium expansum* growth and patulin accumulation: A review. *Food Control* 21: 953–962.
Morales, H., S. Marín, X. Centelles et al. 2007a. Cold and ambient deck storage prior to processing as a critical control point for patulin accumulation. *International Journal of Food Microbiology* 116: 260–265.
Morales, H., S. Marín, L. Obea et al. 2008a. Ecophysiological characterization of *Penicillium expansum* population in lleida (Spain). *International Journal of Food Microbiology* 122: 243–252.
Morales, H., V. Sanchis, J. Usall et al. 2008b. Effect of biocontrol agents *Candida sake* and *Pantoea agglomerans* on *Penicillium expansum* growth and patulin accumulation in apples. *International Journal of Food Microbiology* 122: 61–67.
Moran, C.A. 2004. Functional components of the cell wall of *Saccharomyces cerevisiae*: Applications for yeast glucan and mannan. In *ALLTECHS Annual Symposium*, pp. 283–296.
Moss, M. 1991. The environmental factors controlling mycotoxin formation. In Smith J. E. and R. A. Anderson (Eds.), *Mycotoxins and Animal Foods*. Boca Raton: CRC Press, pp. 37–56.
Mostafavi, H.A., S.M. Mirmajlessi, H. Fathollahi et al. 2013. Integrated effect of gamma radiation and biocontrol agent on quality parameters of apple fruit: An innovative commercial preservation method. *Radiation Physics and Chemistry* 91: 193–199.
Murillo-Arbizu, M., S. Amézqueta, E. González-Peñas et al. 2009. Occurrence of patulin and its dietary intake through apple juice consumption by the Spanish population. *Food Chemistry* 113: 420–423.
Neri, F., I. Donati, F. Veronesi et al. 2010. Evaluation of *Penicillium expansum* isolates for aggressiveness, growth and patulin accumulation in usual and less common fruit hosts. Internat. *Journal of Food Microbiology* 143: 109–117.

Novak, M. and V. Vetvicka. 2008. β-glucans, history, and the present: Immunomodulatory aspects and mechanisms of action. *Journal of immunotoxicology* 5: 47–57.
Ochiai, N., M. Fujimura, M. Oshima et al. 2002. Effects of iprodione and fludioxonil on glycerol synthesis and hyphal development in *Candida albicans*. *Bioscience, Biotechnology, and Biochemistry* 66: 2209–2215.
Olsen, M., R. Lindqvist, A. Bakeeva et al. 2019. Distribution of mycotoxins produced by *Penicillium* spp. inoculated in apple jam and crème fraiche during chilled storage. *International Journal of Food Microbiology* 292: 13–20.
Organization, W.H. 2010. *World Health Organization Working to Overcome the Global Impact of Neglected Tropical Disease. First WHO Report on Neglected Disease*. Geneva, Switzerland: World Health Organization.
Oroian, M., S. Amariei and G. Gutt. 2014. Patulin in apple juices from the Romanian market. *Food Additives & Contaminants: Part B* 7: 147–150.
Park, D.L. and T.C. Troxell. 2002. US perspective on mycotoxin regulatory issues. In J.W. DeVries, M.W. Trucksess and L.S. Jackson (Eds.), *Mycotoxins and Food Safety*. Boston, MA: Springer, pp. 277–285.
Paster, N. and R. Barkai-Golan. 2008. Mouldy fruits and vegetables as a source of mycotoxins: Part 2. *World Mycotoxin Journal* 1: 385–396.
Paterson, R.R.M. 2007. Some fungicides and growth inhibitor/biocontrol-enhancer 2-deoxy-d-glucose increase patulin from *Penicillium expansum* strains in vitro. *Crop Protection* 26: 543–548.
Pitt, J. 2000. Toxigenic fungi and mycotoxins. *British Medical Bulletin* 56: 184–192.
Prusky, D., S. Barad, N. Luria et al. 2014. pH Modulation of host environment, a mechanism modulating fungal attack in postharvest pathogen interactions. In D. Prusky and M.L. Gullino (Eds.), *Post-Harvest Pathology*. Boston, MA: Springer, pp. 11–25.
Prusky, D., J.L. McEvoy, R. Saftner et al. 2004. Relationship between host acidification and virulence of *Penicillium* spp. on apple and citrus fruit. *Phytopathology* 94: 44–51.
Puel, O., P. Galtier and I.P. Oswald. 2010. Biosynthesis and toxicological effects of patulin. *Toxins* 2: 613–631.
Qin, G., Y. Zong, Q. Chen et al. 2010. Inhibitory effect of boron against *Botrytis cinerea* on table grapes and its possible mechanisms of action. *International Journal of Food Microbiology* 138: 145–150.
Reverberi, M., A. Fabbri, S. Zjalic et al. 2005. Antioxidant enzymes stimulation in *Aspergillus parasiticus* by *Lentinula edodes* inhibits aflatoxin production. *Applied Microbiology and Biotechnology* 69: 207–215.
Reverberi, M., S. Zjalic, A. Ricelli et al. 2008. Modulation of antioxidant defense in *Aspergillus parasiticus* is involved in aflatoxin biosynthesis: A role for the ApyapA gene. *Eukaryotic Cell* 7: 988–1000.
Ricelli, A., F. Baruzzi, M. Solfrizzo et al. 2007. Biotransformation of patulin by *Gluconobacter oxydans*. *Applied and Environmental Microbiology* 73: 785–792.
Riley, R.T. and J.L. Showker. 1991. The mechanism of patulin's cytotoxicity and the antioxidant activity of indole tetramic acids. *Toxicology and Applied Pharmacology* 109: 108–126.
Romanazzi, G. and E. Feliziani. 2016. Use of chitosan to control postharvest decay of temperate fruit: Effectiveness and mechanisms of action. In S.Bautista-Banos, G. Romanazzi and A.J. Aparicio (Eds.), *Chitosan in the Preservation of Agricultural Commodities*. Academic Press, pp. 155–177.
Romanazzi, G., S.M. Sanzani, Y. Bi et al. 2016. Induced resistance to control postharvest decay of fruit and vegetables. *Postharvest Biology and Technology* 122: 82–94.
Rychlik, M., F. Kircher, V. Schusdziarra et al. 2004. Absorption of the mycotoxin patulin from the rat stomach. *Food and Chemical Toxicology* 42: 729–735.
Salas, M.P., C.M. Reynoso, G. Céliz et al. 2012. Efficacy of flavanones obtained from citrus residues to prevent patulin contamination. *Food Research International* 48: 930–934.
Sanzani, S.M., M. Reverberi, M. Punelli et al. 2012. Study on the role of patulin on pathogenicity and virulence of *Penicillium expansum*. *International Journal of Food Microbiology* 153: 323–331.
Saxena, N., P.D. Dwivedi, K.M. Ansari et al. 2008. Patulin in apple juices: Incidence and likely intake in an Indian population. *Food Additives and Contaminants* 1: 140–146.
Shephard, G.S., L. van der Westhuizen, D.R. Katerere et al. 2010. Preliminary exposure assessment of deoxynivalenol and patulin in South Africa. *Mycotoxin Research* 26: 181–185.
Spadaro, D., A. Ciavorella, Z. Dianpeng et al. 2010. Effect of culture media and pH on the biomass production and biocontrol efficacy of a *Metschnikowia pulcherrima* strain to be used as a biofungicide for postharvest disease control. *Canadian Journal of Microbiology* 56: 128–137.
Spadaro, D., A. Ciavorella, S. Frati et al. 2007. Incidence and level of patulin contamination in pure and mixed apple juices marketed in Italy. *Food Control* 18: 1098–1102.

Spadaro, D., A. Lorè, A. Garibaldi et al. 2013. A new strain of *Metschnikowia fructicola* for postharvest control of *Penicillium expansum* and patulin accumulation on four cultivars of apple. *Postharvest Biology and Technology* 75: 1–8.

Speijers, G.J.A. and M.H.M. Speijers. 2004. Combined toxic effects of mycotoxins. *Toxicology Letters* 153: 91– 98.

Spring, P and D.F. Fegan. 2005. Mycotoxins–a rising threat to aquaculture. *Feedmix* 13: 5–9.

Stevens, C., V. Khan, J. Lu et al. 1998. The germicidal and hormetic effects of UV-C light on reducing brown rot disease and yeast microflora of peaches. *Crop Protection* 17: 75–84.

Stevens, C., V.A. Khan, C.L. Wilson et al. 2005. The effect of fruit orientation of postharvest commodities following low dose ultraviolet light-C treatment on host induced resistance to decay. *Crop Protection* 24: 756–759.

Stinson, E., S. Osman and D. Bills. 1979. Water-soluble products from patulin during alcoholic fermentation of apple juice. *Journal of Food Science* 44: 788–789.

Stott, W. and L. Bullerman. 1975. Patulin: A mycotoxin of potential concern in foods. *Journal of Milk and Food Technology* 38: 695–705.

Tang, J., Y. Liu, H. Li et al. 2015. Combining an antagonistic yeast with harpin treatment to control postharvest decay of kiwifruit. *Biological Control* 89: 61–67.

Tian, S.P., H.J. Yao, X. Deng et al. 2007. Characterization and expression of β-1, 3-glucanase genes in jujube fruit induced by the microbial biocontrol agent *Cryptococcus laurentii*. *Phytopathology* 97: 260–268.

Tikekar, R.V. 2009. *Ultraviolet Light Induced Degradation of Patulin and Ascorbic Acid in Apple Juice*. (PhD thesis).

Tolaini, V., S. Zjalic, M. Reverberi et al. 2010. *Lentinula edodes* enhances the biocontrol activity of *Cryptococcus laurentii* against *Penicillium expansum* contamination and patulin production in apple fruits. *International Journal Food Microbiology* 138: 243–249.

Tournas, V. and E. Katsoudas. 2019. Effect of $CaCl_2$ and various wild yeasts from plant origin on controlling *Penicillium expansum* postharvest decays in golden delicious apples. *Microbiology Insights* 12: 1178636119837643.

Valletrisco, M., I. Niola and C. Stefanelli. 1988. Decontamination tests on fruit juices contaminated with patulin. In R. Walker and E. Quattrucci (Eds.), *Nutritional and Toxicological Aspects of Food Processing: Proceedings of an Interntional Symposium held at the Istituto Superiore di Sanita, Rome, Italy, 14–16 April* 1987. London, UK: Taylor & Francis.

Van Egmond, H., M. Jonker and B. Rivka. 2008. Regulations and limits for mycotoxins in fruits and vegetables. *Mycotoxins in Fruits and Vegetables* 1: 45–74.

Varga, J., S. Kocsubé, K. Suri et al. 2010. Fumonisin contamination and fumonisin producing black Aspergilli in dried vine fruits of different origin. *International Journal of Food Microbiology* 143: 143–149.

Wang, Y., Y. Li, W. Xu et al. 2018. Exploring the effect of β-glucan on the biocontrol activity of *Cryptococcus podzolicus* against postharvest decay of apples and the possible mechanisms involved. *Biological Control* 121: 14–22. doi: 10.1016/j.biocontrol.2018.02.001

Wang, Y., F. Tang, J. Xia. 2011. A combination of marine yeast and food additive enhances preventive effects on postharvest decay of jujubes (*Zizyphus jujuba*). *Food Chemistry* 125: 835–840.

Wang, L., Z. Wang, Y. Yuan et al. 2015. Identification of key factors involved in the biosorption of patulin by inactivated lactic acid bacteria (LAB) cells. *PloS One* 10: e0143431.

Watanabe, M. and H. Shimizu. 2005. Detection of patulin in apple juices marketed in the Tohuku District, Japan. *Journal of Food Protection* 68: 610–612.

Welke, J.E., M. Hoeltz, H.A. Dottori. 2009. Effect of processing stages of apple juice concentrate on patulin levels. *Food Control* 20: 48–52.

Wisniewski, M., S. Droby, J. Norelli et al. 2016. Alternative management technologies for postharvest disease control: The journey from simplicity to complexity. *Postharvest Biology and Technology* 122: 3–10.

Yahia, E.M., P. García-Solís and M.E.M. Celis. 2019. Contribution of Fruits and Vegetables to Human Nutrition and Health. *Postharvest Physiology and Biochemistry of Fruits and Vegetables*. Elsevier, pp. 19–45.

Yang, Q., H. Wang, H. Zhang et al. 2017. Effect of *Yarrowia lipolytica* on postharvest decay of grapes caused by Talaromyces rugulosus and the protein expression profile of *T. rugulosus*. *Postharvest Biology and Technology* 126: 15–22.

Yang, Q., H. Zhang, X. Zhang et al. 2015. Phytic acid enhances biocontrol activity of *Rhodotorula mucilaginosa* against *Penicillium expansum* contamination and patulin production in apples. *Frontiers in Microbiology* 6: 1296. doi: 10.3389/fmicb.2015.01296

Yiannikouris, A., J. Francois, L. Poughon et al. 2004. Adsorption of zearalenone by beta-D-glucans in the *Saccharomyces cerevisiae* cell wall. *Journal of Food Protection* 67: 1195–1200.

Yiannikouris, A., J.-P. Jouany, G. Bertin et al. 2007. Counteracting mycotoxin contamination: The effectiveness of Saccharomyces cerevisiae cell wall glucans in Mycosorb® for sequestering mycotoxins. Nutritional biotechnology in the feed and food industries: In *Proceedings of Alltech's 23rd Annual Symposium. The New Energy Crisis: Food, Feed or Fuel*, pp. 11–19.

Youssef, K., S.M. Sanzani, A. Ligorio et al. 2014. Sodium carbonate and bicarbonate treatments induce resistance to postharvest green mould on citrus fruit. *Postharvest Biology and Technology* 87: 61–69.

Yu, T., H. Zhang, X. Li et al. 2008. Biocontrol of *Botrytis cinerea* in apple fruit by *Cryptococcus laurentii* and indole-3-acetic acid. *Biological Control* 46: 171–177.

Yue, T., C. Guo, Y. Yuan et al. 2013. Adsorptive removal of patulin from apple juice using Ca-alginate-activated carbon beads. *Journal of Food Science* 78: T1629–T1635.

Zaied, C., S. Abid, W. Hlel et al. 2012. Occurrence of patulin in apple-based foods largely consumed in Tunisia. *Food Control* 31: 263–267.

Zhang, H., L. Chen, Y. Sun et al. 2017a. Investigating proteome and transcriptome defense response of apples induced by *Yarrowia lipolytica. Molecular Plant-Microbe Interactions* 30(4): 301–311. doi: 10.1094/MPMI-09-16-0189-R

Zhang, H., L. Ma, M. Turner et al. 2010. Salicylic acid enhances biocontrol efficacy of *Rhodotorula glutinis* against postharvest Rhizopus rot of strawberries and the possible mechanisms involved. *Food Chemistry* 122: 577–583.

Zhang, H., L. Ma, L. Wang et al. 2008. Biocontrol of gray mold decay in peach fruit by integration of antagonistic yeast with salicylic acid and their effects on postharvest quality parameters. *Biological Control* 47: 60–65.

Zhang, Y., Y. Liu and F. Xing. 2009. Domestic production status of concentrated apple juice and harm of patulin on its quality. *Food Science and Technology* 34: 54–57.

Zhang, H., Z. Liu, B. Xu et al. 2013. Burdock fructooligosaccharide enhances biocontrol of *Rhodotorula mucilaginosa* to postharvest decay of peaches. *Carbohydrate Polymers* 98: 366–371.

Zhang, X., G. Zhang, P. Li et al. 2017b. Mechanisms of glycine betaine enhancing oxidative stress tolerance and biocontrol efficacy of *Pichia caribbica* against blue mold on apples. *Biological Control* 108: 55–63.

Zheng, X., Q. Yang, X. Zhang et al. 2017. Biocontrol agents increase the specific rate of patulin production by *Penicillium expansum* but decrease the disease and total patulin contamination of apples. *Frontiers in Microbiology* 8: 1240.

Zhu, R., K. Feussner, T. Wu et al. 2015. Detoxification of mycotoxin patulin by the yeast *Rhodosporidium paludigenum. Food Chemistry* 179: 1–5.

Zhu, Y., T. Koutchma, K. Warriner et al. 2014. Reduction of patulin in apple juice products by UV light of different wavelengths in the UVC range. *Journal of Food Protection* 77: 963–971.

Zhu, R., L. Lu, J. Guo et al. 2012. Postharvest control of green mold decay of citrus fruit using combined treatment with sodium bicarbonate and *Rhodosporidium paludigenum. Food and Bioprocess Technology* 6, 1–6.

Zhu, Z. and X. Zhang. 2016. Effect of harpin on control of postharvest decay and resistant responses of tomato fruit. *Postharvest Biology and Technology* 112: 241–246.

Zhu, Z., Z. Zhang, G. Qin et al. 2010. Effects of brassinosteroids on postharvest disease and senescence of jujube fruit in storage. *Postharvest Biology and Technology* 56: 50–55.

Zoghi, A., K. Khosravi-Darani, S. Sohrabvandi et al. 2017. Effect of probiotics on patulin removal from synbiotic apple juice. *Journal of the Science of Food and Agriculture* 97: 2601–2609.

11 Decontamination of Mycotoxigenic Fungi by Phytochemicals

Avantina S. Bhandari and Madhu Prakash Srivastava

CONTENTS

11.1 INTRODUCTION

Food decay by spoilage fungi causes considerable economic losses and constitutes a health risk for consumers due to the potential for fungi to produce mycotoxins. Mycotoxins are metabolic intermediates or products, found as a differentiation product in restricted taxonomic groups, not essential to growth and life of the producing organism and biosynthesized from one or more general metabolites by a wider variety of pathways than is available in general metabolism.

Mycotoxins occurring in food commodities are secondary metabolites of filamentous fungi, which frequently occur in major food crops in the field and continue to contaminate cereals, oilseeds nuts and spices during storage (Reddy et al., 2010b). Traditionally, toxigenic fungi contaminating agricultural grains have been conventionally divided into two groups: those that invade seed crops have been described as "field" fungi (e.g., *Cladosporium*, *Fusarium*, *Alternaria* spp.), which reputedly gain access to seeds during plant development; and "storage" fungi (e.g., *Aspergillus*, *Penicillium* spp.), which proliferate during storage. Currently, four types of toxigenic fungi can be distinguished: (1) plant pathogens such as *Fusarium graminearum* and *Alternaria alternata*; (2) fungi that grow and produce mycotoxins on senescent or stressed plants, for example, *F. moniliforme* and *Aspergillus flavus*; (3) fungi that initially colonize the plant and increase the feedstock's susceptibility to contamination after harvesting, for example, *A. flavus*; and (4) fungi that are found on the soil or decaying plant material that occur on the developing kernels in the field and later proliferate in storage if conditions permit, for example, *P. verrucosum* and *A. ochraceus*.

A vast number of fungal metabolites have the potential to contaminate food and feed (Table 11.1), however, few are classified as relevant for public food and feed safety. These are aflatoxins, ochratoxin A, zearalenone, trichothecenes like deoxynivalenol and T-2 toxin, as well as the fumonisins, along with the emerging mycotoxins such as enniatins and *Alternaria* toxins (Stoev, 2015). Among

TABLE 11.1
Mycotoxigenic Fungi and Mycotoxins

Name	Abbreviation	Species/genus/group	Chemical group
Citreoviridin	CIV	*Penicillium citreo-viride*	Pyranone derivative
Citrinin	CTN	*Penicillium*, *Monascus*, *Aspergillus terreus*	Benzopyran compound
Cyclopiazonic acid	CPA	*Penicillium*, *Aspergillus*	Indole tetramic acid
Frequentin	FRE	*Penicillium frequentans*	Carbocyclic compound
Gliotoxin	GT	*A. fumigatus*	Epipolythiodioxopiperazine
Mycophenolic acid	MPA	*Penicillium*	Meroterpenoid compound
Ochratoxin A	OTA	*Aspergillus*, *Penicillium*	Benzopyran compound
Palitantin	PAL	*Penicillium*	Cyclohexane derivative
Patulin	PAT	*Penicillium*, *Aspergillus*	Benzopyran compound
Penicillic acid	PA	*Penicillium*, *Aspergillus*	Isopropylidene tetronic acid
Penitrem A	PNT	*Penicillium*	Indole diterpene alkaloid
Roquefortine	RQF	*Penicillium*	Diketopiperazine compound
Rubratoxin B	RB	*Penicillium rubrum*	α,β-unsaturated lactone
Viomellein	VIM	*Penicillium*	Benzopyran compound
3-Acetyldeoxynivalenol	3-ADON	*Fusarium*	Trichothecenes, sesquiterpenoid
4-Deoxynivalenol	DON	*Fusarium*	Trichothecenes, sesquiterpenoid
Beauvericin	BEA	*Fusarium*	Hexadepsipeptide compound
Diacetoxyscirpenol	DAS	*Fusarium*	trichothecenes, sesquiterpenoid
Enniatin	ENN	*Fusarium*	Cyclic depsipeptide
Fumonisins	FB1	*Fusarium*	Monoterpenes
Fusaric acid	FA	*Fusarium*	Picolinic acid derivative, carboxylic acid
HT-2 toxin	-	*Fusarium*	Trichothecenes, sesquiterpenoid
Moniliformin	MON	*Fusarium*	Cyclobutanedione compound
Nivalenol	NIV	*Fusarium*	Trichothecenes, sesquiterpenoid
T-2 toxin	–	*Fusarium*	Trichothecenes, sesquiterpenoid
Zearalenone/F-2 toxin	ZEA	*Fusarium*	Estrogenic compound
Aflatoxin	AFB1	*Aspergillus*	Difuranocoumarin derivative
Tenuazonic acid	TA	*Alternaria tenuis*	3-acetyl tetramic acid
Cytochalasin	CB	*Phoma*	Polyketide-amino acid hybrid

these mycotoxins, aflatoxin B1 (AFB1), fumonisin B1 (FB1) and ochratoxin A (OTA) are the most toxic to mammals, causing a variety of toxic effects including hepatotoxicity, teratogenicity and mutagenicity, resulting in diseases such as toxic hepatitis, hemorrhage, edema, immunosuppression, hepatic carcinoma, equine leukoencephalomalacia (LEM), esophageal cancer and kidney failure (Adegoke and Letuma, 2013). AFB1 has been classified as a class I human carcinogen, while FB1 and OTA have been classified as class 2B (probable human) carcinogens by the International Agency for Research on Cancer (IARC, 1993). Several outbreaks of mycotoxicoses, diseases in humans and animals caused by various mycotoxins, have been reported after the consumption of mycotoxin-contaminated food and feed.

11.2 MYCOTOXIN-INDUCED PHYTOTOXICITY

These mycotoxin-producing fungi colonize plant tissues and products both externally and internally, leading to accumulation of mycotoxins within the tissues. The presence of these toxins is associated with disease symptoms and at times bringing about subtle biological changes in the physiology of the infected plant or product (Balendres et al., 2019). Some of these mycotoxins are phytotoxic and others are non-phytotoxic. Several reports highlight the toxic effect of mycotoxins on animals and cell lines, but little information is available about the mode of action of these harmful metabolites on plant cells (Ismaiel and Papenbrock, 2015).

11.2.1 *Aspergillus* and Aflatoxins

The discovery of aflatoxin has revolutionized research on molds and mold metabolites because of its potency as a toxin and carcinogen. Aflatoxins are difuranocoumarin derivatives produced in a polyketide pathway by many strains of *Aspergillus flavus* and *A. parasiticus*. Among 18 different types of aflatoxins produced by *A. flavus* strains, AFB1 is the most toxic, mutagenic and carcinogenic type (Ismaiel and Papenbrock, 2015). The phytotoxic effects of the aflatoxins have been investigated on the basis of the remarkable inhibitory effect on chlorophyll and carotenoid synthesis and reduction of seed germination, seedling growth, inhibition of root and hypocotyl elongation.

Aflatoxin has been reported to occur within apparently healthy, intact seeds, which suggest that the toxin can be transported from contaminated soil to the fruit. AFB1 can be translocated from the roots to the stems and leaves. The increased inhibition of root and shoot extension as AFB1 concentration increased was suggested to be correlated with the increasing disruption of the organelles. At higher toxin concentrations, inhibition of root and shoot elongation was accompanied ultrastructurally by derangement of cytoplasmic constituents, dissolution of membranes, particularly the tonoplast, loss of ribosomes, organellar disruption and disappearance of the endoplasmic reticulum. Aflatoxin inhibits chlorophyll a, chlorophyll b and protochlorophyllide biosynthesis that result in the virescence or albinism in the affected plants. Suppression of protein and nucleic acid levels by aflatoxin has also been observed in germinating seeds and it also inhibits chromatin bound DNA-dependent polymerase activity. Inhibition of protein synthesis was attributed to the non-availability of m-RNA, whereas inhibition of DNA synthesis was due to the binding of aflatoxin to DNA during replication or due to the inhibition of DNA polymerase. Electron microscopic studies also revealed the inhibition of grana formation in chloroplasts of leaves treated with aflatoxins (Samuel and Valentine, 2014).

Reduction in the number of tillers in plants treated with aflatoxin may be due to the accumulation of DNA damage in cells, which leads to apoptosis. Aflatoxin was reported to arrest cell cycle and induce apoptosis in cultured cells. The apoptotic pathway is the only option for a cell when DNA repair systems are overburdened due to too much damage. It has been demonstrated that among its diverse functions, the p53 gene normally prevents DNA replication in cells that have DNA damage by maintaining the cell in G2/M phase allowing more opportunity for DNA repair. Cells with inactivated p53 might therefore survive abnormally and allow further DNA damage to accumulate, a situation which favors carcinogenesis. AFB1 exhibited a genotoxic effect and several types of chromosomal aberrations have been detected during meiosis such as chromosome stickiness, outside bivalents, bridges, laggards, unequal division and micronuclei (Wang et al., 2019).

11.2.2 *Penicillium* and Toxins

The *Penicillium* genus dominated fungal flora, with mycotoxigenic species such as *P. aurantiogriseum, P. baarnense, P. canescens, P. chrysogenum, P. citrinum, P. commune P. crustosum, P. cyclopium, P. expansum, P. fenneliae, P. frequentans, P. hirsutum, P. madriti, P. palitans, P. roqueforti, P. thomii, P. verrucosum* and P. *viridicatum.* These species have been reported to produce a number of toxins such as citrinin (CTN), cyclopiazonic acid (CPA), ochratoxin A (OTA), patulin (PAT), penicillic acid (PA), penitrem A (PNT), roquefortine (RQF), frequentin (FRE), palitantin (PAL), mycophenolic acid (MPA), viomellein (VIM), gliotoxin (GT), citreoviridin (CIV) and rubratoxin B (RB). Phytotoxic effect due to different toxins synthesized by different species of *Penicillium* have been summarized below (Ismaiel and Papenbrock, 2015).

11.2.2.1 Patulin

Patulin (4-hydroxy-4H-furo [3,2-c] pyran-2 (6H)-one), PAT isolated from *Penicillium claviforme* and named as claviformin, later renamed as PAT according to the PAT-producing mold, *Penicillium patulum* (known as *P. urticae*, now *P. griseofulvum*). It has been isolated from different fungal genera such as *Aspergillus clavatus, A. giganteus, A. longivesica, A. terreus*; *Byssochlamys fulva, B. nivea*; *Penicillium aurantiogriseum, P. canescens, P. carneum, P. clavigerum, P. concentricum, P. coprobium, P. claviforme, P. chrysogenum, P. cyclopium, P. dipodomyicola, P. divergens, P. expansum, P. glandicola, P. gladioli, P. granulatum, P. griseofulvum, P. lapidosum, P. leucopus, P. marinum, P. melinii, P. patulum, P. paneum, P. roqueforti, P. sclerotigenum, P. variabile, P. vulpinum, P. purpurogenum* and *Paecilomyces saturatus.* Later, PAT was isolated under various names as clavicin, clavitin, expansin, gigantic acid, leucopin, mycoin, penicidin and tercinin.

Mainly its effects cause several human health issues including convulsions, nausea, ulceration, lung congestion, epithelial cell degeneration, carcinogenicity, genotoxicity, immunotoxicity, immunosuppression and teratogenicity. PAT inhibited various enzymes containing thiol groups, and alcohol and lactic dehydrogenases and muscle aldolase.

The phytotoxic toxic effect has been observed on germinating seeds, young seedlings and isolated plant tissue of plants, which had continuous applications of PAT until maturity. Inhibition of plant elongation phases due to PAT effect accompanied a reduction in seed number, seed weight and number of flowers; gains in biomass were also observed.

Due to the phytotoxic action of PAT inhibition of protein, RNA and DNA synthesis occurs. This genotoxic effect might be related to its ability to react with sulfhydryl groups and induce oxidative damage. Due to its electrophilic character, PAT has been reported to react in vitro with cellular nucleophiles such as proteins and GSH. Respiratory inhibition in both germinating apple pollen and soybean suspension cultures seems to be inhibited by PAT at concentrations below toxic levels. In a recent study (Ismaiel and Papenbrock, 2015), the mechanism of PAT phytotoxic action through its effect on the endogenous glutathione (GSH) concentration of maize seedlings was reported.

11.2.2.2 Penicillic Acid

The mycotoxin penicillic acid (3-methoxy-5-methyl-4-oxo-2, 5-hexadienoic acid), a secondary metabolite of *Penicillium puberulum,* was first isolated and named during the investigation of the possible connection between the incidence of pellagra and mold deterioration of maize. PA has also been isolated from the following *Penicillium* species: *P. aurantiovirens, P. aurantiogriseum, P. canescens, P. cyclopium, P. corneum, P. frequentans, P. fenelliae, P. freii, P. hirsutum, P. madriti, P. martensii, P. melanocladium, P. polonicum, P. roqueforti, P. radicicola, P. stoloniferum, P. thomii, P. tricolor, P. tupliae, P. viridicatum* and *P. verrucosum.* Different *Eupenicillium* spp. such as *E. bovifimosum* and *E. baarnese* were also found to produce PA. *Aspergillus* species such as *A. auricomus, A. alliaceus, A. cervinus, A. melleus, A. ochraceus A. ostianus A. sclerotiorum* and *A. wentii* also synthesized this toxin.

The phytotoxic effect caused by PA inhibited the growth of young plant roots, caused anatomical changes as browning of roots and collapse in the root structure and reduced seedling respiration (Ismaiel and Papenbrock, 2015).

The mode of action of this mycotoxin is due to its interaction with the SH-residues in enzymes. It has been observed that the PA inhibited the thiol enzymes alcohol dehydrogenase and lactic dehydrogenase. The phytotoxicity of PA increases due to its accumulation at the low temperatures of typical storage conditions.

11.2.2.3 Ochratoxin A

Ochratoxin A (L-phenylalanine-N-[(5-chloro-3,4-dihydro-8-hydroxy-3-methyl-1-oxo-1H-2-ben zopyrane-7-yl)carbonyl]-(R)-isocoumarin) is a secondary metabolite synthesized by *Aspergillus ochraceus*. Subsequent studies revealed that a variety of mold fungus species, including *Aspergillus carbonarius, A. niger* and *Penicillium verrucosum* also synthesized ochratoxins. Recently, *A. westerdijkiae* and *A. steynii*, two new species from the *Aspergillus* section Circumdati are reported to be stronger OTA producers than *A. ochraceus. Penicillium nordicum* and *P. verrucosum* frequently isolated from cereal crops, meat products and cheese varieties are known to produce OTA. Because of its widespread occurrence on a large variety of agricultural commodities and the potential health risks, mainly toward humans, OTA has been classified as a possible human carcinogen (Aarane et al., 2018).

OTA exposure can cause significant inhibition in the growth of plants and induce necrotic lesions in leaves. Moreover, preferential inhibition of root growth in seedlings by OTA was observed. OTA induces cell death in plants due to the occurrence of an oxidative burst and the deposition of callose and phenolic compounds (autofluorescence).

The toxicity due to OTA is caused due to interference with metabolic systems involving phenylalanine, promotion of membrane lipid peroxidation, disruption of calcium homeostasis, inhibition of mitochondrial respiration, inhibition of protein synthesis and DNA damage. OTA rapidly causes hypersensitive responses and significantly accelerates the increase of reactive oxygen species (ROS) and malondialdehyde, and enhances antioxidant enzyme defense responses and xenobiotic detoxification. Second, OTA stimulation causes dynamic changes in the expression of transcription factors and activates the membrane transport system. Lastly, there is a metabolic shift from a highly active to a weak state due to concomitant persistence of compromised photosynthesis and photorespiration. The process of toxicity caused by OTA is mediated through ethylene, salicylic acid, jasmonic acid, and mitogen-activated protein kinase signaling molecules.

11.2.3 Mycotoxins Produced by Dematiaceous Hyphomycetes

Alternaria, Fusarium Helminthosporium, Phoma, Pyrenophora (sexual state: *Drechslera*) and *Zygosporium* commonly associated with leaves, wood, cereals and other grasses are genera of saprobic and plant pathogenic dematiaceous fungi with a worldwide distribution. The species of these fungal genera are known to produce tenuazonic acid (TA) and cytochalasins and produce phytotoxicity (Ismail and Papenbrock, 2015).

11.2.3.1 Tenuazonic Acid

The tetramic acid derivative L-tenuazonic acid (5S, 8S)-3-acetyl-5-sec-butyl-pyrrolidine-2, 4-dione) was isolated from *Alternaria tenuis*, a plant pathogen involved in the postharvest decay of fruits, grains and vegetables. Later, as a mycotoxin, it was isolated from *Magnaporthe grisea* (the blast fungus, former name was *Pyricularia oryzae*) and from other species of *Alternaria* as *Alternaria longiceps, A. kikuchiana, A. mali, A. alternata, A. tenuissima* and *Phoma sorghina*. TA is said to be responsible for the outbreak of "onyalai," a human hematologic disorder disease occurring in Africa after the consumption of sorghum.

TA, a non-specific phytotoxin exhibits significant phytotoxic effects on both monocotyledonous and dicotyledonous plants by inhibiting seed germination and seedling growth. TA induces leaf necrosis as well as causing the browning of edges of dead areas on susceptible varieties.

The stunted plant cell growth induced by TA is due to the inhibition of protein synthesis at the ribosome level. It has been demonstrated that TA has a ribosome-binding site as it inhibits the binding of radioactive protein synthesis inhibitors anisomycin and trichodermin on the ribosome. It was also reported that TA forms complexes with ions, including iron or copper. The ultrastructure of parenchyma cells of Japanese pear leaves treated with TA supported the toxic effect of TA on the basis of degeneration of chloroplast and endoplasmic reticulum in leaf tissue near the treated area with toxin.

11.2.3.2 Cytochalasins

Over 80 different cytochalasins, a group of polyketide-amino acid hybrid compounds of fungal secondary metabolites, have been isolated from *Aspergillus*, *Phomopsis*, *Penicillium*, *Zygosporium*, *Chaetomium*, *Rosellinia*, *Metarhizium* and so on. Cytochalasins have significant commercial and research values due to their diverse arrays of biological activities and complex molecular structures. Phomin (phomine, cytochalasin B, CB), a macrolide antibiotic with cytostatic activity is produced by a *Phoma* species. Later, during more detailed studies of the Phoma metabolites, a closely related compound with similar activity, dehydrophomin (cytochalasin A, CA), was isolated. CA and CB are also isolated from *Helminthosporium dematioideum* (*Drechslera dematioidea*) and CD and CD from *Metarhizium anisopliae*. These compounds are intraconvertible by oxidation-reduction reactions and had the empirical formulas of $C_{29}H_{35}NO_5$ and $C_{29}H_{37}NO_5$.

Cytochalasin reacts in the following manner to cause deleterious effect on plants

(Cimmino et al., 2015). It attacks the cytoskeleton leading to disruption of endocytosis and cell deregulation and ends in cell death. CA blocked the accumulation of phytoalexin and phenylalanine ammonia-lyase in pea tissues. CD altered the actinomycin system of the onion epidermal cells, disintegrated the actin filaments and caused formation of large flat-sheet-like cytoplasmic reticulum sacs in the epidermal cells of onion. The mycotoxin resulted in dislocated mitotic spindles, disrupted phragmoplasts, symmetric divisions and finally embryogenesis.

Pyrenophora semeniperda, an ascomycete seed pathogen was reported to produce phytotoxins CB, as well as CA, CF, deoxaphomin and the three novel cytochalasins, Z1, Z2 and Z3. *P. semeniperda* was also able to produce phytotoxic sesquiterpenoid penta-2, 4-dienoic acid, named pyrenophoric acid that showed strong phytotoxicity by reducing coleoptile length. Pyrichalasin H isolated from a *Digitaria* isolate (IFO 7287) of the blast fungus inhibited the growth of rice seed and induced characteristic curling of the shoot.

Zygosporin D and two new cytochalasins were isolated from *Metarhizium anisopliae*. Of these cytochalasins, only zygosporin D was an effective inhibitor of shoot elongation of rice seedlings (Fujii et al., 2000).

11.2.3.3 Trichothecenes

More than 120 trichothecenes are known to be produced by a range of fungi from the genera *Fusarium*, *Myrothecium*, *Verticimonosporium*, *Stachybotrys*, *Trichoderma*, *Trichothecium*, *Cephalosporium* and *Cylindrocarpon*. Members of the genus *Fusarium* produce a range of chemically different phytotoxic compounds, such as fusaric acid (FA), fumonisins (fumonisin B1, FB1), beauvericin (BEA), enniatin (ENN), moniliformin (MON) and trichothecenes. These possess a variety of biological activities and cause morphological, physiological and metabolic effects including necrosis, chlorosis, growth inhibition, wilting, inhibition of seed germination and effects on calli leading to mortality and enabling them to mediate a wide variety of plant diseases, including wilts, stalk rot, root rot and leaf rot in many important crop and ornamental plants.

11.3 STRATEGIES TO MANAGE FUNGAL GROWTH AND MYCOTOXIN PRODUCTION

A large share of annual crops can suffer colossal losses due to the ravages of insects and diseases, thus causing a serious threat to agricultural production. The congenial climatic conditions and conducive environment play an important role in causing severe loss of agricultural commodities in tropical and subtropical countries. The deterioration of stored food products due to fungi and mycotoxins in humid and warm areas of the world has added a new dimension to the gravity of the problem. Mycotoxigenic fungi contamination leads to reduced yield and poor quality of products and mycotoxicosis among humans and livestock.

Among several approaches employed to check fungal growth and mycotoxin biosynthesis, pre-harvest management is considered the most important mitigating strategy. Breeding of resistance cultivars provides an effective and environmentally sound strategy but their incorporation into commercial cultivars is, however, slow, time consuming and complex. There are other methods in use to prevent mycotoxin contamination such as drying methods, storage in proper conditions, irradiation and chemical agents (Adegoke and Letuma, 2013).

The use of synthetic pesticides is the most effective and widespread way of chemical control for eliminating harmful microorganisms and diseases from raw plant materials and foods. These compounds possess inherent toxicity that endangers the health of farm operators, consumers and the environment. Upon entering the food chain, they cause ecological imbalance and destroy microbial diversity as they are persistent in nature. Use of these field fungicides and food additives has led to the development of resistant strains that have necessitated utilization of higher concentrations, with the consequent increase in toxic residues in food products.

11.4 RATIONALE TO CONSIDER NATURAL RESOURCES

The increasing public demand to make a shift and experience "green consumerism," along with the restrictions imposed by food industries and regulatory agencies on the use of synthetic food protectants and additives have led to renewed interest in searching for alternative strategies that are practical, economical, economical, eco-friendly and protective for both public health and the environment.

The use of plant-derived natural products in agriculture is not a new concept and dates back to solving agricultural problems ever since humanity took to farming. The control of various pests and pathogens through the use of natural plant extracts from neem *Pyrethrum*, tobacco and so on is well documented in early scriptures.

In most traditional farming systems in many developing countries, this approach has long been used. Farmers possess substantial knowledge of identifying and utilizing those plants with antimicrobial properties against insects and pathogens. Numerous studies have demonstrated that plant extracts contain diverse bioactive components that can control mold growth. The metabolites produced by plants are a promising alternative because plants generate a wide variety of compounds, either as part of their development or in response to stress or pathogen attack.

The potential beneficial attributes of utilizing natural phytochemicals present in plants against mycotoxigenic fungi and mycotoxins provide an opportunity to avoid chemical preservatives; they may help to overcome the rising incidence of drug resistance among pathogenic microorganisms and their mode of action could be different; they may alleviate the side effects associated with synthetic pesticides; and be less toxic toward humans and animals. These phytochemicals are usually associated with multiple beneficial effects on the plants on which they are applied. They are often growth promoting, more systemic and beyond the symptomatic treatments of disease. It has also been conceived, as a possibility, that crude plant extracts might be more affordable to subsistence farmers as they are readily available and are probably cheaper to produce. Hence, efforts to develop plant-derived natural products as a viable option and consideration of their potential in disease management systems in both developed and developing countries does not seem to be out of line (Pretorious and Watt, 2011).

11.5 ANTIMICROBIALS FROM PLANT ORIGINS

Agricultural microbiologists are continuously evaluating the healing attributes in plants, mostly from indigenous origin. Over the years, efforts have also been directed to search for new antifungal materials from natural resources for food preservation (Raveesha, 2011). Plants, herbs, essential oils and spices in powdered or extract forms were frequently evaluated for their use to detoxify microbes and toxins due to the presence of betalain, flavonoids, phenolics, phytoalexins and thiosulfonates. The botanicals can be utilized in managing the deleterious effects of mycotoxigenic fungi and mycotoxins in foods by using plant extracts and essential oils and to develop effective anti-mycotoxigenic natural products for eco-friendly management.

Fumigation is one of the best methods to protect commodities against contamination during storage. Plant extracts for fumigation have already been considered an eco-friendly alternative to prevent fungal growth and the production of mycotoxins (Prakash et al., 2015). The vapor phases of purified essential oils or plant extracts preserve their bioactivity, hence making them highly effective fumigants for protecting stored products (da Cruz Cabral et al., 2013).

11.5.1 Plant Extract

Several edible botanical extracts have been reported to possess antimicrobial and antioxidant activities due to their phenolic alignments. The inhibitory effects of these extracts have been attributed to phytophenols, the secondary metabolites existing in nearly 8000 structures. These structures resemble with tannin and phenolic acid.

The ethanolic extract of various plant genera viz., *Areca catechu, Allium cepa, Cassia bakeriana, Catharanthus roseus, Centella, asiatica, Chromolaena odorata, Citrus reticulata, Curcuma longa, Eryngium foetidum, Eucalyptus globulus, Garcinia mangostana, Hibiscus sabdariffa, Jussiaea repens, Lantana camara, Lycopersicon esculentum, Momordica charantia, Morinda citrifolia, Piper betle, P. nigrum, P. sarmentosum, Psidium guajava, Punica granatum, Raphanus sativus, Syzygium aromaticum, Stemona tuberosa* and *Thunbergia laurifolia* exhibited their ability to inhibit toxin-producing fungi (Saleem et al., 2017). Similarly, Pundir (2010) studied the efficacy of 22 plant extracts against food associated fungi and found clove and ginger are more effective than other plant extracts.

11.5.1.1 Aflatoxigenic Fungi and Aflatoxins

Health hazards from exposure to toxic chemicals and economic considerations make natural plant extracts an ideal alternative to protect food and feed from fungal contamination (Sinha, 2018). Many strategies have been developed to manage aflatoxins in crops, among them, control by natural products appears to be the most promising approach for managing aflatoxins in postharvested crops.

Olive callus is reported to contain caffeic acid, catechin, coumarin and o-, p- or m-coumaric acid. It was observed that crude ethanolic extract of olive callus facilitated the reduction by 90% of aflatoxin synthesis but did not show any inhibitory effect on the growth of *Aspergillus* (Shialy et al., 2015).

In another study, aflatoxin production by *A. parasiticus* was suppressed depending on the concentration of the plant aqueous extract added to the culture media at the time of spore inoculation. Aflatoxin production in fungal mycelia grown for 96 h in culture media containing 50% neem leaf and seed extracts was inhibited by 90 and 65%, respectively (Razzaghi-Abyaneh et al., 2005). It was found that growth was inhibited significantly and controlled with both alcoholic and water extracts of all ages and of the concentrations used. Kumar et al. (2016) reported inhibitory effect of neem extracts on biosynthesis of aflatoxins (groups B and G) in fungal mycelia.

Ageratum conyzoides (baume, chick weed), a popular medicinal plant having therapeutic properties that have been attributed to volatile terpenes and coumarins, and non-volatile phenolic acids and flavonoids. Nogueira et al. (2010) found nearly 48% inhibition of *A. flavus* growth in Sabouraud agar with 5.0 µL in a 6 mm diameter of filter paper disk of the *A. conyzoides* extract. The component found in the highest concentration in the extract was precocene II (46.3%).

Rosmarinus officinalis (rosemary), which also has a large amount of precocene II, demonstrated important changes in the mitochondria and endomembrane system of fungal cells (Rasooli et al., 2008). The antifungal action of these plants is attributed to the presence of carvacrol, in the case of rosemary, as well as its synergistic action with other compounds.

Carum carvi (caraway) is an important member of the Apiaceae family, which includes carrot and fennel. Studies demonstrated that *C. carvi* lacks inhibitory activity against *Aspergillus parasiticus* growth, although it can affect AFB1 and AFG1 production. The main components of this plant are myristicin and dillapiole (two phenylpropanoids) that are responsible for the inhibition of aflatoxin production (Razzaghi-Abyaneh et al., 2009).

Foeniculum vulgare (fennel) has been widely known and used as a medicinal plant for many years (Alinezhad et al., 2011). The roots of this plant contain approximately 90% dillapiol, which may be responsible for the inhibitory action of AFG1 production. However, it has been reported that high concentrations of *F. vulgare* extracts are required for growth inhibition of *A. flavus* and *A. parasiticus* (Razzaghi-Abyaneh et al., 2007).

Chenopodium ambrosioides (Mexican tea, wormseed), belongs to the Amaranthaceae family and is used in the treatment of respiratory and digestive problems in Central and South America. Kumar et al. (2007) demonstrated that the essential oil from its leaves (100 µg/mL) caused the lysis of mycelia and spores, thereby reducing aflatoxin production of two aflatoxigenic strains of *A. flavus*, Navjot 4NSt and Saktiman 3NSt.

Alseeni et al. (2019) compared the antifungal activity of the aqueous extracts of some herbals against *A. flavus* and *A. parasiticus*. The results demonstrated that clove and garlic at 10% inhibited growth and also reduced the level of aflatoxin production in rice. There have been a number of reports citing the inhibitory effects of onion extracts on *A. flavus* growth, with an ether extract of onions, thio-propanol-S-oxide, being demonstrated to inhibit growth. In another study, onion extract ceased about 60.44% aflatoxin synthesis in maize. In addition, Welsh onion ethanol extracts suppressed the fungal growth and aflatoxin production of some strains of aflatoxin-producing fungi. Efficacy of various concentrations of four plant extracts prepared from garlic, neem leaf, ginger and onion bulb were studied on growth reduction of *A. flavus* on mustard. It was also found that betel leaf extract at 10,000 ppm completely inhibited the growth of *A. flavus*.

Syzygium aromaticum (clove) effectively inhibited the mycelial growth of *A. flavus* and aflatoxin production (Reddy et al., 2009).

The crude extracts from spices as mint, sage, bay, anise and ground red pepper caused inhibitory effect on the growth of *A. parasiticus* NRRL 2999 and its aflatoxin production in vitro. Antifungal activity of selected Turkish spices (black cumin, coriander, cumin, dill, laurel, oregano, parsley, spearmint, white mustard) was observed on some food-borne fungi and it was found that ground oregano showed an inhibitory effect on *A. flavus* and *A. niger*. Saxena and Mathela (1996) found antifungal activity of new compounds from *Nepeta leucophylla* and *N. clarkei* against *Aspergillus* sp. Terpenoids from natural sources were found to be effective against growth of *Aspergillus* species (Pizzolitto et al., 2015).

Some traditionally useful plants *Ocimum gratissimum*, *Cymbopogon citratus*, *Xylopia aethiopica*, *Monodora myristica*, *Syzygium aromaticum*, *Cinnamomum verum* and *Piper nigrum* are effective in inhibiting formation of non-sorbic acid, a precursor in the aflatoxin synthesis pathway. Powdered leaf of *Ocimum* has been successfully used in inhibiting mold development on stored soybean for 9 months. The powder extracts of *Cymbopogon citratus* inhibited the growth of fungi including toxigenic species such as *A. flavus* and *A. fumigatus*. Powdered leaves of basil (*O. gratissimum*) and cloves (*S. aromaticum*) in combination with some packaging materials protected groundnut kernels artificially inoculated with *A. parasiticus* (Tofa Begam et al., 2019).

Khaya senegalensis bark protected maize against insect-initiated risk of aflatoxin development. Pepper extracts have been shown to reduce aflatoxin production in *A. parasiticus* IFO 30179 and *A. flavus var columnaris* S46. Aqueous extracts of *Lupinus albus*, *Ammi visnaga* and *Xanthium pungens* were found to arrest the growth of *Aspergillus* and also affected aflatoxin synthesis.

11.5.1.2 Ochratoxigenic Fungi and Ochratoxins

Ochratoxin A, a nephrotoxic and carcinogenic mycotoxin, contaminates agricultural products, including cereals, coffee, dried fruits, wine and pork (Reddy et al., 2010a). Various studies have been conducted to reduce ochratoxigenic fungi and ochratoxin contamination using plant extracts (Zeinoddin and Khalesi, 2019; Perczak et al., 2016). In an investigations on the effect of neem extracts on mycelial growth, sporulation, morphology and OTA production by *P. verrucosum* and *P. brevicompactum*, only the inhibition of fungal growth was observed without any effect on OTA production (Mossini et al., 2009). Reddy et al. (2007a) reported the efficacy of certain plant extracts on mycelial growth of *A. ochraceus* and OTA biosynthesis.

The antitoxigenic potential of spices was tested against an OTA-producing strain of *Aspergillus*. Curcumin, a constituent of turmeric, has been identified as an efficient photosensitizer for inactivation of *Aspergillus flavus* conidia in maize (Temba et al., 2019). It has been demonstrated that curcumin completely inhibited mycelial growth of *Aspergillus alliaceus* isolate 791 at 0.1% (w/v) and decreased OTA production by 70% at 0.01% (w/v). Clove completely inhibited the vegetative growth of the fungi *A. ochraceus*, while garlic and laurel were found to completely inhibit OTA production.

Cinnamon and anis inhibited the synthesis of OTA starting from the concentration of 3% and mint starting from 4% (Pereira et al., 2006). The effects of four alkaloids on the biosynthesis of OTA and ochratoxin B (OTB) were examined on four OTA-producing *Aspergilli*: *A. auricomus*, *A. sclerotiorum* and two isolates of *A. alliaceus*. Natural alkaloids of *Piper longum* piperine and piperlongumine significantly inhibited OTA production at 0.001% (w/v) for all the species of *Aspergillus* investigated.

11.5.1.3 Fumonisin-producing Fungi and Fumonisins

Several species of the genus *Fusarium* are economically relevant because, apart from their ability to infect and cause tissue destruction on important crops such as corn, wheat and other small grains on the field, they produce mycotoxins in the crops in the field and in storage grains. Fumonisins are mycotoxins produced mainly by the fungi *F. verticillioides* and *F. proliferatum* (Dambolena et al., 2010). Different plant extracts from *Azadirachta indica*, *Artemessia annua*, *Eucalyptus globulus*, *Ocimum sanctum* and *Rheum emodi* were tested to control *F. solani*. All plant extracts showed significant reduction of pathogens (Joseph et al., 2008). Recently, Anjorin et al. (2008) reported the effect of neem extract on control of *F. verticillioides* in Maize. In another study, Amin et al. (2009) reported the efficacy of garlic tablet against *Fusarium* sp. associated with cucumber and found that garlic tablet effectively inhibited all the fungi tested. Still today, there are no reports on the effect of plant extracts on fumonisin biosynthesis.

11.5.2 Essential Oils

Plant products, especially essential oils from higher plants, are most likely to be biodegradable, renewable in nature and perhaps safer to human health, and are recognized as one of the most promising groups of natural compounds for the development of safer antifungal agents (Varma and Dubey, 2001). Essential oils from plant origin had been widely used for antiparasitic, bactericidal, fungicidal, virucidal, medicinal and cosmetic applications. Nowadays, their use in the pharmaceutical, sanitary, cosmetics, agricultural and food industries has come to the forefront.

Several essential plant oils are involved in the mechanisms of plant protection involving a range of antimicrobial effects (Saleem et al., 2017) on different species of pathogenic bacteria and fungi and in vivo (Prakash et al., 2012). Essential oils have low risk in the development of antimicrobial resistance and are non-toxic, being classified as GRAS (generally recognized as safe) by the United States Food and Drug Administration (Cardile et al., 2009).

Although, the exact picture of the mechanisms associated with antifungal activities of essential oils is not very clear; still it is worth mentioning that terpenoids and phenolics, the major constituents present in essential oils, play a crucial role in antimicrobial activity. These compounds are able to cause structural and functional damage due to their low molecular weight and lipophilic nature. Essential oils disrupt the membrane permeability and the osmotic balance of the cell (Grata, 2016) and damage the enzymatic system of the microbial cells including mitochondrial enzymes as lactate, malate and succinate dehydrogenases. These enzymes are involved in ATP biosynthesis (Nazzaro et al., 2017) as well as H^+-ATPase activity; hindrance of these processes leads to intracellular acidification and cell degeneration. Some compounds, such as monoterpenes and limonene, have been identified as potential inhibitors of pectin methylesterase, which is responsible for building the main components of the cell wall in fungi (Marei et al., 2012). The concentration of essential oils also governs the inhibition of mycelial growth and sporulation (Perczak et al., 2016). The mechanism of action is by denaturing enzymes responsible for spore germination and interfering with amino acids involved in germination.

Studies attempting to understand the mechanisms of action and toxicological safety of the use of plants are still scarce, preventing more effective use of these compounds in agriculture, livestock production and industry. Essential oil antimicrobial actions involve several chemical compounds found in plants, and this activity cannot be attributed to a single cell mechanism, but a set of them (Kitic et al., 2005).

11.5.2.1 Control of Aflatoxigenic Fungi and Aflatoxins

Numerous essential oils of plant origin have been tested for their antimicrobial and aflatoxin inhibiting ability. Essential oils extracted from aloe, cinnamon, coriander, eucalyptus, ginger, lavender, marigold, mangosteen, pepper, pomelo, rosemary, safflower, thyme and whitewood were tested for their potential to inhibit *Aspergillus*. It was found that whitewood oil exhibited highest inhibition of the test fungus followed by oil from cinnamon and lavender respectively (Thanaboripat et al., 2007).

Large-scale application of different higher plant products—azadirachtin from *Azadirachta indica*, eugenol from *Syzygium aromaticum*, carvone from *Carum carvi* and allyl isothiocyanate from mustard and horseradish oil—have attracted the attention of microbiologists to other plant chemicals for use as antimicrobials (Reddy et al., 2007b; Singh et al., 2008).

Plant essential oils from *Azadirachta indica* and *Morinda lucida* were found to inhibit the growth of *A. flavus* and significantly reduced aflatoxin synthesis in inoculated maize grains. Zeringue et al. (2001) observed the increase of 11–31% of dry mycelial mass along with a slight decrease (5–10%) in AFB1 production in 5-day-old aflatoxigenic *Aspergillus* sp., submerged cultures containing either 0.5 or 1.0 mL clarified neem oil (CNO) in 0.1%. Recently, essential oils extracted from the seeds of neem, mustard, black cumin and asafoetida were evaluated for their antifungal activity against seed borne fungi viz., *A. niger* and *A. flavus* (Sitara et al., 2008). All extracted oils except mustard, showed fungicidal activity of varying degrees against test species.

Sindhu et al. (2011) observed the bioactivity of α-phellandrene, terpinolene and p-cymene present as an active compound extracted from turmeric leave oil against aflatoxin-producing fungi.

Several reports are available on the inhibitory effect of clove oil on *A. flavus* and *A. parasiticus* as well as several other fungi. Clove oil and its major component, eugenol, have been extensively used to control rice infested with fungus treated at 2.4 mg eugenol/g of grains; the inoculum of *A. flavus* failed to grow and thus AFB1 biosynthesis on rice was prevented (Reddy et al., 2007b).

Juglal et al. (2002) studied the effectiveness of nine essential oils to control the growth of mycotoxin-producing molds and noted that clove, cinnamon and oregano were able to prevent the growth of *A. parasiticus* while clove (ground and essential oil) markedly reduced the aflatoxin synthesis in infected grains. Complete inhibition of mycelial growth of *A. flavus* and *A. versicolor* was observed at 250 μg eugenol/ml in yeast-sucrose broth, but the biosynthesis of AFB1 in vitro was arrested only at 125 μg mL^{-1} of eugenol. The effect of cinnamon oil, clove oil, cinnamic aldehyde and eugenol on growth and aflatoxin production by *A. parasiticus* showed promising results as they inhibited mold growth and subsequent toxin production effects. Complete suppression of *A. flavus* growth was observed by using essential oils namely citronellol, cinnamaldehyde geraniol, nerol and thymol.

Kumar et al. (2010) studied the efficacy of *O. sanctum* essential oil and its major component, eugenol against the fungi causing biodeterioration of foodstuffs during storage. *O. sanctum* and eugenol were found efficacious in checking growth of *A. flavus* and also inhibited the AFB1 production completely at 0.2 and 0.1 μg mL^{-1}, respectively. The essential oil from *Ocimum basilicum* (basil) at a concentration of 3000 μg/mL had fungistatic and fungicidal activity against *A. parasiticus* and *A. flavus*. These effects are attributed to the two substances found in the highest concentration in the oil, estragole (50%) and ocimene (11.2%), which are known for their antifungal activity.

Apart from neem and clove oils, various essential oils from plant origin have been used for the reduction of mycotoxins. Recently, Singh et al. (2008) extracted essential oils from different parts of 12 plants belonging to eight angiospermic families and tested for activity against two toxigenic strains of *A. flavus*. The oil of the spice plant *Amomum subulatum* was found to be effective against two strains of aflatoxigenic fungi, completely inhibiting their mycelial growth at 750 μg mL^{-1} and AFB1 production at 500 μg mL^{-1}.

Kumar et al. (2007) extracted oil from the leaves of *Chenopodium ambrosioides* and tested against the aflatoxigenic strain of test fungus *A. flavus*. The oil completely inhibited the mycelial growth at 100 μg mL^{-1} and significantly reduced AFB1.

In another study, Jardim et al. (2008) studied antifungal activity of essential oil from the Brazilian epazote by the poison food assay at the concentrations of 0.3, 0.1 and 0.05% against postharvest deteriorating fungi (*A. flavus, A. glaucus, A. niger and A. ochraceus*). Mycelial growth of all the test fungi was completely arrested at 0.3% concentration and by 90 to 100% at 0.1% concentration.

Vilela et al. (2009) analyzed the bioactivity of *Eucalyptus globulus* essential oil and its main component (1,8-cineol) singly against *A. flavus* and *A. parasiticus*. They observed that 1,8-cineol has lower antifungal activity than the essential oil. These findings suggest that other components found in lower levels in the oil may be critical for promoting synergism and enhancing the effects. The effect of eucalyptus oil on growth and aflatoxin production by *A. flavus* was tested at three levels, viz., 0.05, 0.1 and 0.2 mL/50 mL SMKY medium. After 6 days of incubation on 0.05 and 0.1 mL supplemented SMKY medium, growth and toxin production were inhibited, while at 0.2 mL concentration, there was no growth. Thanaboripat et al. (2007) studied the effects of 16 essential oils from aromatic plants against mycelia growth of *A. flavus* IMI 242684. The results showed that the essential oil of *Melaleuca cajuputi* gave the highest inhibition followed by the essential oils of *Cinnamomum cassia* and lavender *Lavandula officinalis*, respectively.

Adegoke et al. (2000) worked on finding the inhibitory concentrations of the essential oil of the spice *Aframomum danielli* for the aflatoxigenic mold, *A. parasiticus*. Kumar et al. (2009) studied the efficacy of essential oil from *Mentha arvensis* to control storage molds of chickpea. The oil effectively reduced mycelia growth of *A. flavus*. During screening of essential oils for their antifungal activity against *A. flavus*, the essential oil of *Cymbopogon citratus* was found to exhibit fungitoxicity. Essential oil, containing mainly garcinol, from the tropical shrub/tree *Garcinia indica* at 3000 ppm inhibited both *A. flavus* growth and AFB1 production (Tamil Selvi et al., 2003).

Essential oils have two different modes of action: reduced fungal growth or aflatoxin biosynthesis and secretion. Therefore, inhibition of AFB1 production may not be completely attributed to reduction of fungal growth. Tian et al. (2011a) demonstrated that 5 μL/mL and 4 μL/mL of essential oils from *Cicuta virosa* var. *latisecta* were sufficient for antifungal and anti-aflatoxigenic activities of *A. flavus*, respectively. However, in another study by Razzaghi-Abyanehet al. (2009), they reported that the oil extract from *Carum carvi* was only effective in reducing aflatoxin inhibition, with no effect on the vegetative growth of *A. parasiticus*.

Essential oils prepared from flowers, herbs, seeds, leaves and roots have variable inhibitory activities against the growth of toxigenic fungi (*A. flavus and A. parasiticus*), and aflatoxin production

(Youssef et al., 2016). Inhibition of growth and mycotoxin decontamination reported in those studies varied from 48 to 100%, with a huge variation among the extract's concentrations used in the assays (0.25 to 3000 µg/mL). The variation in the results may partially be attributable to different processes used for extraction in each experiment, leading to different concentrations of essential oils in the extracts. However, strong antifungal activities at lower extract concentrations (0.25 to 5.0 µL/mL) were reported for extracts from *Ageratum conyzoides*, *Carum carvi*, *Origanum vulgare* and *Rosmarinus officinalis* (de Sousa et al., 2013; Nogueira et al., 2010; Azizkhani et al., 2015).

Plants in the family Lamiaceae have two types of phenols, carvacrol and thymol (Pinto et al., 2006). Studies have demonstrated that these substances are able to inhibit the growth of a wide range of microorganisms, including fungi and bacteria (Kedia et al., 2014). These two classes of compounds affect the integrity of the cell membrane in microorganisms, thus changing the pH homeostasis and the inorganic ion balance (Lambert et al., 2001). Carvacrol is found in the highest percentage in *Origanum* spp., ranging from 50 to 86% depending on the species (Kulisic et al., 2004). This plant has been widely employed for a long time as a therapeutic agent and as an antioxidant in the food industry. However, its antimicrobial and anti-aflatoxin characteristics are still issues that should be studied due to conflicting results described in different studies (Carmo et al., 2008).

11.5.2.2 Control of Ochratoxigenic Fungi and Ochratoxins

Essential oils from *Origanum vulgare* (oregano), *Menta arvensis* (mint), *Ocimum basilicum* (basil), *Salvia officinalis* (sage) and *Coriandrum sativum (coriander)* have been shown to be effective against ochratoxin-producing fungi, with oregano and mint oils completely inhibiting the growth of *A. ochraceus* NRRL 3174 and OTA production after 21 days at the concentration of 1000 ppm. Recently, Mossini et al. (2009) conducted in vitro trials to evaluate the effect of *Azadirachta indica* (neem) oil on mycelial growth, sporulation, morphology and OTA production by *P. verrucosum* and *P. brevicompactum*. Oil extracts exhibited significant reduction of growth and sporulation of the fungi. No inhibition of OTA production was observed. Essential oils of 12 medicinal plants were tested for inhibitory activity against *A. ochraceus* and OTA production (Soliman and Badea, 2002). The oils of thyme and cinnamon completely inhibit all the test fungi and OTA at 3000 ppm.

11.5.2.3 Control of Fumonisin-producing Fungi and Fumonisin

Several reports are available on the use of plant essential oils against fumonisin-producing fungi and fumonisin biosynthesis. Recently, Sitara et al. (2008) evaluated essential oils extracted from the seeds of neem (*Azadirachta indica*), mustard (*Brassica campestris*), black cumin (*Nigella sativa*) and asafoetida (*Ferula asafoetida*) against seed borne fungi viz., *F. oxysporum*, *F. moniliforme*, *F. nivale* and *F. semitectum*. All the oils extracted except from mustard, showed fungicidal activity of varying degrees against test species.

Essential oil extracted from the leaves of *Chenopodium ambrosioides* demonstrated antimicrobial activity against the postharvest deteriorating fungi *Fusarium oxysporum* and *F. semitectum* (Kumar et al., 2007; Jardim et al., 2008). Vegetative growth of both fungi was completely inhibited at 0.3% concentration.

Dambolena et al. (2010) investigated the constituents and efficacy of essential oils from *O. basilicum* and *O. gratissimum* against *F. verticillioides* infection and fumonisin production from samples collected from different locations in Kenya. There were inhibitory effects on growth of the test fungi, however, the extent of inhibition was widely dependent upon the composition and concentration of oils. When maize was treated with *O. basilicum* oil, no effects were observed on the FB1 biosynthesis but *O. gratissimum* essential oil induced a significant inhibitory effect on FB1 production with respect to control. Juglal et al. (2002) reported spice oils of eugenol, cinnamon, oregano, mace, nutmeg and turmeric, and aniseed displayed antifungal activity against *F. moniliforme* and 78% reductions in FB1 formation by this fungus when treated with 2 $\mu L\ mL^{-1}$ clove oil.

Among 75 different essential oils tested against *F. oxysporum* f. sp. *ciceris*, the most active essential oils found were those of lemongrass, clove, cinnamon bark, cinnamon leaf, cassia, fennel, basil and evening primrose (Pawar and Thanker, 2007). The effect of cinnamon, clove, oregano, palmarosa and lemongrass oils on FB1 accumulation by one isolate each of *F. verticillioides* and *F. proliferatum* in non-sterilized naturally contaminated maize grain at 0.995 and 0.950 a_w and at 20 and 30 °C was evaluated. The concentration used was 500 mg kg^{-1} maize. Under these conditions, it was shown that antimycotoxigenic ability only took place at the higher water availabilities and mostly at 20 °C. Only cinnamon, lemongrass and palmarosa oils were somewhat effective. Moreover, it was suggested that competing mycoflora play an important role in FB1 accumulation. It was concluded that the efficacy of essential oils in real substrates, such as cereals, may be much lower than in synthetic media; different essential oils may be found to be useful at different concentrations. Their effectiveness is highly dependent on both abiotic and biotic factors involved (Marín et al., 2003).

11.6 PERSPECTIVES FOR USING PLANT PRODUCTS IN ACTIVE PACKAGING

Active packaging is an advanced, innovative and original food-packaging concept that combines food technology, food safety packaging and material sciences in an effort to meet consumer demand for fresh-like, safe products (Han et al., 2018). The application of preservatives with food packaging material offers many advantages as they not only prevent microbial contamination, but also only low levels of the agents come into contact with food. These preservation compounds can be directly introduced into the bulk mass of the products or applied to their surfaces, which subsequently get migrated into the food or headspaces surrounding the food. This type of antimicrobial packaging is suitable for vacuum-packed products. In other non-migrating approaches, preservatives are applied to packaging surfaces that inhibit target microorganisms when they come into contact with them. The application of plant extracts and generally of essential oils as antimicrobial and antioxidant agents in food and food packaging has no safety restrictions because they are considered GRAS by the Food and Drug Administration (2016).

Usually, various kinds of wax-coated papers and boards are utilized as packaging material to increase the shelf life and provide water resistance to packed products. However, incorporation of natural active ingredients along with the packaging material not only provides shelf-life extension but also helps in protection from unavoidable microbial spoilage (Begam et al., 2019). There are several success stories of *Cananga odorata*, *Hedychium spicatum*, *Origanum majorana*, *Coriandrum sativum* and *Commiphora myrrha* essential oil-based packaging used as preservatives of chickpea (Prakash et al., 2012).

Tian et al. (2011b) evaluated the essential oil from *Anethum graveolens* (dill) as a potential food preservative on cherry tomatoes infected with *A. flavus* and healthy tomatoes within in vivo models. As a result, fungal development was completely inhibited at 120 μg/mL of essential oil for the contaminated tomatoes and 100 μg/mL for healthy tomatoes. Feng and Zeng (2017) demonstrated the efficacy of the *Cassia fistula* essential oil as an alternative to synthetic chemicals to control *Alternaria alternata* postharvest contamination, this essential oil demonstrated results with concentrations of 300–500 μg/mL. Anžlovar et al. (2017) showed another example of using essential oil with a fumigation technique. Essential oil of thyme was tested for the protection of wheat grain from *Alternaria alternata*, *Alternaria infectoria*, *Aspergillus flavus*, *Epicoccum nigrum* and *Fusarium poae*. The fungitoxicity potential was good against all the strains tested (Anžlovar et al., 2017). These results differ from those with the fumigation of *Boswellia carterii* essential oil on black pepper. The authors found an increased protection against *A. flavus* during a 6 month period of storage

Despite promising results shown by using plant extracts and essential oils in antimicrobial active packaging, several limitations have also been identified in their application (Jayasena and Jo, 2013). The antimicrobial potency of essential oil constituents depends on the pH, temperature and level of microbial contamination present in the food, therefore, much higher concentrations of essential oils will be required to achieve desirable effects (Malhotra et al., 2015). In addition, their use may cause negative sensory effects because of their intense aroma, which can partially limit their use as preservatives in food. The major obstacle to the use of essential oils in foodstuffs is the reproducibility of their activity caused by variations in the bioactive components and their strong aroma which may restrict their applications (Ribeiro et al., 2017).

To alleviate these problems, some studies have demonstrated that combining lower levels of essential oils with other antimicrobial compounds and other preservation technologies such as MAP, high hydrostatic pressure or low-dose irradiation synergistic effect can be obtained (Ozogul et al., 2017). Furthermore, incorporating the volatile component of the essential oils into films and edible coatings, encapsulation of essential oils in polymers with biodegradable and edible coatings, sachets or nanoemulsions can provide better results (Noshirvani et al., 2017; Rattanapitigorn, 2018).

There are considerable differences in the ways that antimicrobials are used at the laboratory scale and in real-time applications. Regulatory issues and technical constraints together are some of the main limiting factors for proper commercialization of active antimicrobial packaging (Realini and Marcos, 2014). In order to achieve good results and for successful implementation of active packaging solutions in the market, it is essential to select the right package for the antimicrobial agent and environmental conditions faced by a particular food product.

11.7 CONCLUSION

The development of new reliable techniques to eliminate the mycotoxin contamination of foods and commodities is a very important task for the food industry, as well as for food production. In this context, considerable experimental research has demonstrated that essential oils and aqueous plant extracts inhibit the fungal development and/or the biosynthesis of toxins, hence demonstrating a potential for their use in food products. However, there is limited information available on the use of essential oils or aqueous plant extracts directly on food commodities to prevent mycotoxigenic fungal growth or aflatoxin production. Therefore, more studies are necessary to evaluate the potential application of plant extracts under field conditions, particularly on stored cereals and their manufactured products. As a first step, the procedures for preparation of plant extracts and/or essential oils need standardization. Additionally, further studies are necessary to identify the main active compounds of plant extracts and understand their mechanisms of action, as well as to determine the safety levels for their use by the food or feed industry. Some practical aspects of using essential oils and aqueous plant extracts in food products also need to be investigated, especially their potential effects on the sensory characteristics of foods and their shelf life for the maintenance of antifungal properties under different environmental conditions.

A multidisciplinary approach involving researchers from the fields of biotechnology, food technology engineering and material science is required to join hands to find and create new active antimicrobial packages with a promising future (Fang et al., 2017).

REFERENCES

Aarane, M., M.B.S. Ratnaseelan, I. Tsilioni et al. 2018. Effects of mycotoxins on neuropsychiatric symptoms and immune process. *Clin. Ther.* 40(6): 903–917.

Alinezhad, S., A. Kamalzadeh, M. Shams-Ghahfarokhi *et al.* 2011. Search for novel antifungals from 49 indigenous medicinal plants: Foeniculum vulgare and Platycladus orientalis as strong inhibitors of aflatoxin production by Aspergillus parasiticus. *Ann. Microbiol.* 61: 673–681.

Alseeni, M.N., E.M. Allheani, S.Y. Qusti et al. 2019. Antimicrobial activities of some plant extracts against some fungal pathogens. *J. Pharm. Biol. Sci.* 14(2): 1–10. doi: 10.1016/j.ijfoodmicro.2006.10.017

Adegoke, G.O., S.B. Fasoyiro and B. Skura. 2000. Control of microbial growth, browning and lipid oxidation by the spice *Aframomum danielli*. *Eur. Food Res. Technol.* 211: 342–345.

Adegoke, O.G. and P. Letuma. 2013. Strategies for prevention and reduction of mycotoxins in developing countries. In H.A. Makun (ed.), *Mycotoxin and Food Safety in Developing Countries* (pp. 123–136). Rijeka, Croatia: IntechOpen.

Amin, A.B.M.R., M.M. Rashid and M.B. Meah. 2009. Efficacy of garlic tablet to control seed-borne fungal pathogens of cucumber. *J. Agric. Rural Dev.* 7: 135–138.

Anjorin, S.T., H.A. Makun, T. Adesina et al. 2008. Effects of *Fusarium verticillioides*, its metabolites and neem leaf extract on germination and vigour indices of maize (*Zea mays* L.). *Afr. J. Biotechnol.* 7: 2402–2406.

Anžlovar, S., M. Likar and J.D. Koce. 2017. Antifungal potential of thyme essential oil as a preservative for storage of wheat seeds. *Acta Bot. Croat.* 76: 64–71. doi: 10.1515/botcro-2016-0044

Awuah, R.T. and W.O. Ellis. 2002. Effects of some groundnut packaging methods and protection with *Ocimum* and *Syzygium* powders on kernel infection by fungi. *Mycopathologia* 154: 29–36.

Azizkhani, M. and F. Tooryan. 2015. Antioxidant and antimicrobial activities of rosemary extract, mint extract and a mixture of tocopherols in beef sausage during storage at 4°C. *J. Food Saf.* 35: 128–136. doi: 10.1111/jfs.12166

Balendres, M.A.O., P. Karlovsky and C.J.R. Cumagun. 2019. Mycotoxigenic fungi and mycotoxins in agricultural crop commodities in the Philippines: A review. *Foods* 8: 249. doi: 10.3390/foods8070249

Begam, T., J. Mahmud, Md. N. Islam et al. 2019. Essential oils and biodegradable Packaging materials: Application on food preservations. *Sci. Rev.* 5(1): 1–7.

Cardile, V., A. Russo, C. Formisano *et al.* 2009. Essential oils of Salvia bracteata and Salvia rubifolia from Lebanon: Chemical composition, antimicrobial activity and inhibitory effect on human melanoma cells. *J. Ethnopharmacol.* 126(2): 265–272. doi: 10.1016/j.jep.2009.08.034

Cimmino, A., M. Masi, M. Evidente et al. 2015. Fungal phytoxins with potential herbicidal activity: Chemical and biological characterization. *Nat. Prod. Rep.* 32: 1629–1653. doi: 10.1039/C5NP00081E

Carmo, E.S., E. de Oliveira Lima and E.L. de Souza. 2008. The potential of Origanum vulgare L. (Lamiaceae) essential oil in inhibiting the growth of some food-related Aspergillus species. *Braz. J Microbiol.* 39(2): 362–367. doi: 10.1590/S1517-83822008000200030

Dambolena, J.S., M.P. Zunino, A.G. Lopez et al. 2010. Essential oils composition of *Ocimum basilicum* L. and *Ocimum gratissimum* L. from Kenya and their inhibitory effects on growth and fumonisin production by Fusarium verticillioides. *Innov. Food Sci. Emerg. Technol.* 11: 410–414.

da Cruz Cabral, L., V. Fernández Pinto and A. Patriarca. 2013. Application of plant derived compounds to control fungal spoilage and mycotoxin production in foods. *Int. J. Food Microbiol.* 166(1): 1–14.

de Sousa, L.L., S.C.A. de Andrade, A.J.A.A. Athayde et al. 2013. Efficacy of Origanum vulgare L. and Rosmarinus officinalis L. essential oils in combination to control postharvest pathogenic Aspergilli and autochthonous mycoflora in Vitis labrusca L. (table grapes). *Int. J. Food Microbiol.* 165(3): 312–318. doi: 10.1016/j.ijfoodmicro.2013.06.001

Fang, Z., Y. Zhao, R.D. Warner et al. 2017. Active and intelligent packaging in meat industry. *Trends Food Sci. Technol.* 61: 60–71.

Feng, W. and X. Zheng. 2017. Essential oils to control Alternaria alternata in vitro and in vivo. *Food Control* 18: 1126–1130. doi: 10.1016/j.foodcont.2006.05.017

Food and Drug Administration. 2016. *Code of Federal Regulations (CFR). Title 21: Food and Drugs. Chapter I - Food and Drug Administration, Department of Health and Human Services, Subchapter B - Food for Human Consumption (Continued), Part 182 - Substances Generally Recognized as Safe (GRAS), Subpart a - General Provisions, Subpart 182.20 - Essential Oils, Oleoresins, and Natural Extractives.* Washington, DC, USA: Office of the Federal Register.

Fujii, Y., H. Tani, M. Ichinoe et al. 2000. Zygosporin D and two new cytochalasins produced by the *fungus Metarrhizium anisopliae. J. Nat. Prod.* 63(1): 132–135.

Grata, K. 2016. Sensitivity of *Fusarium solani* isolated from Asparagus on essential oils. *Ecol. Chem. Eng. A* 23(4): 453–464.

Han, J-W., L-R. Garcia, J-P. Qian et al. 2018. Food packaging: A comprehensive review and future trends. *Compr. Rev. Food Sci. Food Saf.* 17: 860–877.

IARC. 1993. *Some Naturally Occurring Substances: Food Items and Constituents, Heterocyclic Aromatic Amines and Mycotoxins* (Vol. 56, pp: 489–521). Geneva, Switzerland: International Agency for Research on Cancer.

Ismaiel, A.A. and J. Papenbrock. 2015. Mycotoxins: Producing fungi and mechanisms of phytotoxicity. Agriculture 5: 492–537. doi: 10.3390/agriculture5030492

Jardim, C.M., G.N. Jham, O.D. Dhingra et al. 2008. Composition and antifungal activity of the essential oil of the Brazilian *Chenopodium ambrosioides* L. *J. Chem. Ecol.* 34: 1213–1218.

Jayasena, D.D. and C. Jo. 2013. Essential oils as potential antimicrobial agents in meat and meat products: A review. *Trends Food Sci. Technol.* 34(2): 96–108.

Joseph, B., M.A. Dar and V. Kumar. 2008. Bioefficacy of plant extracts to control *fusarium solani* f. sp. melongenae incitant of brinjal wilt. *Global J. Biotechnol. Biochem.* 3: 56–59.

Juglal, S., R. Govinden and B. Odhav. 2002. Spice oils for the control of co-occurring mycotoxin producing fungi. *J. Food Prot.* 65: 683–687.

Kedia, A., B. Prakash, P.K. Mishra et al. 2014. Antifungal and antiaflatoxigenic properties of Cuminum cyminum (L.) seed essential oil and its efficacy as a preservative in stored commodities. *Int. J. Food Microbiol.* 168–169: 1–7. doi: 10.1016/j.ijfoodmicro.2013.10.008

Kitic, D., G. Stojanovic, R. Palic et al. 2005. Chemical composition and microbial activity of the essential oil of Calamintha nepeta (L.) Savi ssp. nepeta var. subisodonda (Borb.) Hayek from Serbia. *J Essent. Oil Res.* 17: 701–703.

Kulisic, T., A. Radonic, V.Katalinic et al. 2004. Use of different methods for testing antioxidative activity of oregano essential oil. *Food Chem.* 85: 633–640.

Kumar, P., D.K. Mahato, M. Kamle et al. 2016. Aflatoxins: A global concern for food safety, human health and their management. *Front. Microbiol.* 7: 2170. doi: 10.3389/fmicb.2016.02170

Kumar, R., A.K. Mishra, N.K. Dubey et al. 2007. Evaluation of Chenopodium ambrosioides oil as a potential source of antifungal, antiaflatoxigenic and antioxidant activity. *Int. J Food Microbiol.* 115(2): 159–164. doi: 10.1016/j.ijfoodmicro.2006.10.017

Kumar, A., R. Shukla, P. Singh et al. 2009. Use of essential oil from *Mentha arvensis* L. to control storage moulds and insects in stored chickpea. *J. Sci. Food Agric.* 89: 2643–2649.

Kumar, A., R. Shukla, P. Singh et al. 2010. Chemical composition, antifungal and anti-aflatoxigenic activities of *Ocimum sanctum* L. essential oil and its safety assessment as plant based antimicrobial. *Food Chem. Toxicol.* 48: 539–543.

Lambert, R.J., P.N. Skandamis, P.J. Coote et al. 2001. A study of the minimum inhibitory concentration and mode of action of oregano essential oil, thymol and carvacrol. *J. Appl. Microbiol.* 91(3): 453–462. doi: 10.1046/j.1365-2672.2001.01428.x

Marin, S., A. Velluti, A. Munoz et al. 2003. Control of fumonisin B1 accumulation in naturally contaminated maize inoculated with *Fusarium verticillioides* and *Fusarium proliferatum*, by cinnamon, clove, lemongrass, oregano and palmarosa essential oils. *Eur. Food Res. Technol.* 217: 332–337.

Malhotra, B., A. Keshwani and H. Kharkwal. 2015. Antimicrobial food packaging: Potential and pitfalls. *Front. Microbiol.* 6: 611. doi: 10.3389/fmicb.2015.00611

Marei, G.I.K., M.A. Abdel Rasoul and S.A.M. Abdelgaleil. 2012. Comparative antifungal activities and biochemical effects of monoterpenes on plant pathogenic fungi. *Pestic. Biochem. Physiol.* 103: 56–61. doi: 10.1016/j.pestbp.2012.03.004

Mossini, S.A.G., C.C. Arroteia and C. Kemmelmeier. 2009. Effect of neem leaf extract and Neem oil on *Penicillium* growth, sporulation, morphology and ochratoxin A production. *Toxins* 1: 3–13.

Nazzaro, F., F. Fratianni, R. Coppola et al. 2017. Essential oils and antifungal activity. *Pharamaceuticals* 10(4): 86. doi: 10.3390/ph10040086

Nogueira, J.H.C., E. Gonçalez, S.R. Galleti et al. 2010. Ageratum conyzoides essential oil as aflatoxin suppressor of Aspergillus flavus. *Int. J. Food Microbiol.* 137(1): 55–60. doi: 10.1016/j.ijfoodmicro.2009.10.017

Noshirvani, N., B. Ghanbarzadeh, C. Gardrat et al. 2017. Cinnamon and ginger essential oils to improve antifungal, physical and mechanical properties of chitosan-carboxymethyl cellulose films. *Food Hydrocolloids* 70: 36–45.

Ozogul, Y., I. Yuvka, Y. Ucar et al. 2017. Evaluation of effects of nanoemulsion based on herb essential oils (rosemary, laurel, thyme and sage) on sensory, chemical and microbiological quality of rainbow trout fillets during ice storage. *LWT Food Sci. Technol.* 75: 677–684.

Pawar, V.C. and V.S. Thaker. 2007. Evaluation of the anti-*Fusarium oxysporum* f. sp. *cicer* and anti-*Alternaria porri* effects of some essential oils. *World J. Microbiol. Biotechnol.* 23: 1099–1106.

Perczak, A., D. Gwiazdowska, M. Katarzyna et al. 2016. Antifungal activity of selected essential oils against *Fusarium culmorum* and *F.graminearum* and their secondary metabolites in wheat seeds. *Arch. Microbiol.* 201: 1085–1097. doi: 10.1007/s00203-019-01673-5

Pereira, M.C., S.M. Chalfoun, C.J. Pimenta et al. 2006. Spices, fungi mycelial development and ochratoxin A production. *Sci. Res. Essays* 1: 38–42.

Pinto, E., C. Pina-Vaz, L. Salgueiro et al. 2006. Antifungal activity of the essential oil of Thymus pulegioides on Candida, Aspergillus and dermatophyte species. *J. Med. Microbiol.* 55(Pt10): 1367–1373. doi: 10.1099/jmm.0.46443-0

Pizzolitto, R.P., C.L. Barberis, J.S. Dambolena et al. 2015. Inhibitory effect of natural phenolic compounds on *Aspergillus parasiticus* growth. *J. Chem.* 2015(19): 1–7. doi: 10.1155/2015/547925

Prakash, B., P.K. Mishra, A. Kedia et al. 2014. Antifungal, antiaflatoxin and antioxidant potential of chemically characterized Boswellia carterii Birdw essential oil and its in vivo practical applicability in preservation of Piper nigrum L. fruits. *Lebensm. Wiss. Technol.* 56: 240–247. doi: 10.1016/j.lwt.2013.12.023

Prakash, B., P. Singh, A. Kedia et al. 2012. Assessment of some essential oils as food preservatives based on antifungal, antiaflatoxin, antioxidant activities and in vivo efficacy in food system. *Food Res. Int.* 49: 201–208. doi: 10.1016/j.foodres.2012.08.020

Prakash, B., A. Kedia, P.K. Kumar et al. 2015. Plant essential oils as food preservatives to control moulds, mucotoxin contamination and oxidative deterioration of agri-food commodities- Potentials and challenges. *Food Control* 47: 381–391.

Pretorious, J.C. and E. van der Watt. 2011. Natural products from plants: Commercial prospects in terms of antimicrobial, herbicidal and Bio-stimulatory activities in an integrated pest management system. In N.K. Dubey (Ed.), *Natural Products in Plant Pest Management* (pp. 42–90). Wallingford, UK: CAB International.

Pundir, R. 2010. Antifungal activity of twenty-two ethanolic plant extracts against food-associated fungi. *J. Pharm. Res.* 3(1): 506–510.

Rasooli, I., M.H. Fakoor, D. Yadegarinia et al. 2008. Antimycotoxigenic characteristics of Rosmarinus officinalis and Trachyspermum copticum L. essential oils. *Int. J. Food Microbiol.* 122(1-2): 135–139. doi: 10.1016/j.ijfoodmicro.2007.11.048

Rattanapitigorn, P. 2018. Essential oils from Plant extracts and their application as antimicrobial agents in food products. *J. Food Technol., Siam Univ.* 13(2): 1–10.

Raveesha, K.A. 2011. Antimicrobials of plant origin to prevent the biodeterioration of Grains. In N.K. Dubey (Ed.), *Natural Products in Plant Pest Management* (pp. 91–108). Wallingford, UK: CAB International.

Razzaghi-Abyaneh, M., A. Allameh, T. Tiraihi et al. 2005. Morphological alterations in toxigenic *Aspergillus parasiticus* exposed to neem (*Azadirachta indica*) leaf and seed aqueos extracts. *Mycopathologia* 159: 565–570.

Razzaghi-Abyaneh, M., M. Shams-Ghahfarokhi, M.-B., Rezaee et al. 2009. Chemical composition and antiaflatoxigenic activity of Carum carvi L., Thymus vulgaris and Citrus aurantifolia essential oils. *Food Control* 20: 1018–10124. doi: 10.1016/j.foodcont.2008.12.007

Razzaghi-Abyaneh, M., T. Yoshinari, M. Shams-Ghahfarokhi et al. 2007. Dillapiol and Apiol as specific inhibitors of the biosynthesis of aflatoxin G1 in Aspergillus parasiticus. *Biosci. Biotechnol. Biochem.* 71(9): 2329–2332.

Realini, C.E. and B. Marcos. 2014. Active and intelligent packaging systems for a modern society. *Meat Sci.* 98: 404–419.

Reddy, K.R.N., H.K. Abbas, C.A. Abel et al. 2010a. Mycotoxin contamination of beverages: Occurrence of patulin in apple juice and ochratoxin A in coffee, beer and wine and their control methods. *Toxins* 2: 229–261.

Reddy, K.R.N., S.B. Nurdijati and B. Salleh. 2010b. An overview of plant-derived products on control of mycotoxigenic fungi and mycotoxins. *Asian J. Plant Sci.* 9(3): 126–133. doi: 10.3923/ajps.2010.126.133

Reddy, K.R.N., C.S. Reddy, H.K. Abbas et al. 2008. Mycotoxigenic fungi, mycotoxins and management of rice grains. *J. Toxicol. Toxin Rev.* 27: 287–317.

Reddy, K.R.N., C.S. Reddy and K. Muralidharan. 2009. Potential of botanicals and biocontrol agents on growth and aflatoxin production by Aspergillus flavus infecting rice grains. *Food Control* 20: 173–178.

Reddy, K.R.N., C.S. Reddy and K. Muralidharan. 2007a. Exploration of ochratoxin a contamination and its management in rice. *Am. J. Plant Physiol.* 2: 206–213.

Relddy, C.S., K.R.N. Reddy, M. Prameela et al. 2007b. Identification of antifungal component in clove that inhibits *Aspergillus* spp. colonizing rice grains. *J. Mycol. Plant Pathol.* 37: 87–94.

Reddy, K.R.N., B. Salleh, B. Saad et al. 2010. An overview of mycotoxin contamination in foods and its implications for human health. *Toxin Rev.* 29: 3–26.

Ribeiro-Santos, R., M. Andrade and N. Melo. 2017. Use of essential oils in active food packing: Recent advances and future trends. *Trends Food Sci. Technol.* 61: 132–140. doi: 10.1016/j.tifs.2016.11.021

Saleem, F., B. Sadia and F. Saeed Awan. 2017. Control of Aflatoxin production using Herbal plant extract. In L. Abdulra'Uf (Ed.), *Aflatoxin-Control, Analysis, Detection and Health Risks.* Rijeka, Croatia: IntechOpen.

Samuel, A.T. and I.T. Valentine. 2014. Effect of total aflatoxin on the growth characteristics and chlorophyll level of sesame (*Sesamum indicum* L.). *N. Y. Sci. J.* 7(4): 8–13.

Saxena, J. and C.S. Mathela. 1996. Antifungal activity of new compounds from *Nepeta leucophylla* and *Nepeta clarkei. Appl. Environ. Microbiol.* 62: 702–704.

Shialy, Z., M. Zarrin, B.S. Nejad et al. 2015. In vitro antifungal properties ofvPistacia atlantica and olive oil extracts on different fungal species. *Curr. Med. Mycol.* 1(4): 40–45 doi: 10.18869/acadpub.cmm.1.4.40

Sindhu, S., B. Chempakam, N.K. Leela et al. 2011. Chemoprevention by essential oil of turmeric leaves (*Curcuma longa* L.) on the growth of Aspergillus flavus and aflatoxin production. *Food Chem. Toxin* 49(5): 1188–1192. doi: 10.1016/j.fct.2011.02.014

Singh, P., B. Srivastava, K. Ashok et al. 2008. Efficacy of essential oil of *Amomum subulatum* as a novel aflatoxin B_1 suppressor. *J. Herbs Spices Med. Plants* 14: 208–218.

Sinha, B.K. 2018. Studies on the inhibitory effects of some plant extracts on mycoflora and Aflatoxin Production. *Acta Sci. Microbiol.* 1(6): 23–25.

Sitara, U., I. Niaz, J. Naseem et al. 2008. Antifungal effect of essential oils on *in vitro* growth of pathogenic fungi. *Pak. J. Bot.* 40: 409–414.

Soliman, K.M. and R.I. Badeaa. 2002. Effect of oil extracted from some medicinal plants on different mycotoxigenic fungi. *Food Chem. Toxicol.* 40(11): 1669–1675. doi: 10.1016/S0278-6915(02)00120-5

Stoev, S.D. 2015. Feedborne mycotoxicosis, risk assessment and underestimated hazard of masked mycotoxins and joint mycotoxin effects or interaction. *Environ. Toxicol. Pharmacol.* 39: 794–809.

Tamil Selvi, A., G.S. Joseph and G.K. Jayaprakasha. 2003. Inhibition of growth and aflatoxin production in *Aspergillus flavus* by *Garcinia indica* extract and its antioxidant activity. *Food Microbiol.* 20: 455–460.

Temba, B.A., M.T. Fletcher, G.P. Fox et al. 2019. Curcumin-based photosensitization inactivates Aspergillus flavus and reduces aflatoxin B1 in maize kernels. *Food Microbiol.* 82: 82–88.

Thanaboripat, D., Y. Suvathi, P. Srilohasin et al. 2007. Inhibitory effect of essential oils on the growth of *Aspergillus flavus*. *KMITL Sci. Technol. J.* 7: 1–7.

Tian, J., X. Ban, H. Zeng et al. 2011a. Chemical composition and antifungal activity of essential oil from Cicuta virosa L. var. latisecta Celak. *Int. J. Food Microbiol.* 145(2-3): 464–470. doi: 10.1016/j.ijfoodmicro.2011.01.023

Tian, J., X. Ban, H. Zeng et al. 2011b. In vitro and in vivo activity of essential oil from dill (Anethum graveolens L.) against fungal spoilage of cherry tomatoes. *Food Control* 22: 1992–1999. doi: 10.1016/j.foodcont.2011.05.018

Varma, J. and N.K. Dubey. 2001. Efficacy of essential oils of *Caesulia axillaris* and *Mentha arvensis* against some storage pests causing biodeterioration of food commodities. *Int. J. Food Microbiol.* 68: 207–210.

Vilela, G.R., G.S. de Almeida, M.A.B.R. D'Arce et al. 2009. Activity of essential oil and its major compound, 1,8-cineole, from Eucalyptus globulus Labill., against the storage fungi Aspergillus flavus Link and Aspergillus parasiticus Speare. *J. Stored Prod. Res.* 45: 108–111. doi: 10.1016/j.jspr.2008.10.006

Wang, H., P. Liao, S.X. Zeng et al. 2019. It takes a team: A gain-of-function story of p53-R249S. *J. Mol. Cell Biol.* 11(4): 277–283.

Youseef, M.M., Q. Pham, P.N. Achar et al. 2016. Antifungal activity of essential oils on Aspergillus parasiticus isolated from peanuts. *J. Plant Prot. Res.* 56: 139–142. doi: 10.1515/jppr-2016-0021

Zeinoddin, M. and M. Khalesi. 2019. Biological detoxification of ochratoxin A in plants and plant products. *Toxin Rev.* 38(3): 187–199. doi: 10.1080/15569543.2018.1452264

Zeringue, H.J., B.Y. Shih and D. Bhatnagar. 2001. Effects of clarified neem oil on growth and aflatoxin B1 formation in submerged and plate cultures of aflatoxigenic *Aspergillus* spp. *Phytoparasitica* 29: 361–366.

12 Prevention and Control Strategies of Aflatoxin Contamination

Pramila Pandey, Narendra Shankar Pandey, and Rachna Chaturvedi

CONTENTS

12.1 INTRODUCTION

Postharvest diseases are diseases that develop on the harvested parts of plants such as seeds and fruits and also on vegetables. They cause decay on a particular part or deteriorate the whole product. The plant product may be infected by microorganisms at any time during handling, harvesting and consumption (Singh et al. 2017). Postharvest losses are estimated to be to the tune of 10 to 35% per year despite the use of modern storage facilities and techniques. The reduction in product quality and quantity render the produce unfit for human consumption. Aside from direct economic considerations, diseased produce poses a potential health risk as they produce toxic secondary metabolites of unrelated chemical structures and biological properties that are known as mycotoxins. Among numerous fungal genera, predominant producers of these toxic elements are *Alternaria*, *Aspergillus*, *Fusarium* and *Penicillium*.

Mycotoxins, the natural contaminants of food and agricultural products all over the world, have emerged as an issue of serious concern as they are associated with a wide array of negative effects and other complications (Omatayo et al. 2019). Lack of awareness of the occurrence and risks of mycotoxins, poor agricultural practices and undiversified diets predispose populations to dietary mycotoxin exposure in developing and underdeveloped countries (Misihairabgwi et al. 2019). It is estimated that close to US$5 billion are lost every year due to the association with fungal infestations and product contamination by toxigenic plant pathogens. Generally, mycotoxins represent a significant threat to human health as they can be carcinogenic, neurotoxic and toxic to the endocrine or immune systems (Marin et al. 2013).

Nearly 400 types of mycotoxins have been discovered since the 1960s. Dominant among these are the *Aflatoxins*, known as the "hidden poison" due to its concealed and adverse effects on various biological pathways in humans, particularly among children, in whom it leads to delayed development, stunted growth, liver damage and liver cancer (Pandey et al. 2019). However, aflatoxin contamination occurs more during postharvest than preharvest conditions, and it is a major risk factor for hepatocellular carcinoma, which has been identified as chronic infection with hepatitis B (HBV) and hepatitis C (HCV) viruses and dietary exposure to aflatoxins (Wild and Hall 2000). In developing countries, about 4.5 billion individuals may be vulnerable to the harmful effects of aflatoxins (Horn and Green 1995). Intake of polluted foodstuffs causes aflatoxicosis in humans and animals. Aflatoxicosis can be powerful and persistent and sometimes acute conditions may be responsible for increasing mortality or death rates. Long-lasting conditions have unassailable suppression and malignant growth. In human beings, it is categorized by an unsettled stomach in the form of vomiting and nausea, stomach annoyance, pneumonic edema, pulmonary edema, tiredness and may result in death with cerebral edema and the contribution of fatty liver disease, kidney disorder, heart failure and so on (Diener et al. 1987). In animals, aflatoxicosis is diagnosed by duct dysfunction, less intake of feed, anemia, jaundice, liver harm, declining milk and egg production and occasionally by immunity suppression (Payne 1998). In plants, aflatoxins are the cause of poor seed germination, underdeveloped seed plant growth and root elongation. It conjointly suppresses synthesis of chlorophyll, carotenoid and some enzymes (Pettit 1986).

Aflatoxins are highly toxic, mutagenic, teratogenic and carcinogenic compounds synthesized by fungi belonging to several *Aspergillus* species, mainly *Aspergillus bombycis*, *A. flavus*, *A. fumonisins*, *A. nomius*, *A. ochraceous*, *A. parasiticus* and *A. pseudotamari* (IARC 1993; Bhat et al. 2010). The production of aflatoxins is associated with spore production by species of *Aspergillus* (Calvo et al. 2002). Aflatoxins are fairly thermostable and are robust to the degradation process. Among the 13 compounds so far identified, aflatoxins B1, B2, G1 and G2, and the aflatoxin metabolic byproducts M1 and M2, are considered to be the chief toxic metabolites. They are produced by the polyketide pathway and their chemical structure incorporates dihydrofuran and tetrahydrofuran moieties coupled to a substituted coumarin. Most of the toxigenic strains of *Aspergillus flavus* characteristically harvest only two types of aflatoxins such as B1 and B2 (Ferron and Deguine 2005), but the maximum number of strains of *Aspergillus parasiticus* could harvest all four types of toxins (Dorner 2004). The main types of aflatoxins B1 and B2 show a bluish color, while G1 and G2 affords a somewhat yellowish greenish color beneath ultraviolet light. *Aspergillus flavus* yields exclusively AFB1 and AFB2, however, it also produces cyclopiazonic acid and *Aspergillus parasiticus* synthesizes AFBI, AFB2, AFG1 and AFG2. (Janisiewicz and Korsten 2002).

12.2 DIFFERENCES BETWEEN STRAINS OF *ASPERGILLUS*

Though *Aspergillus parasiticus* and *Aspergillus flavus* have many similarities, they are completely dissimilar from one another on the basis of their color and dimensions in the form of length of hypha. Phialides are the most important characteristic that distinguishes the species of the *Aspergillus* genus. The phialides of *Aspergillus flavus* are biseriate types compared with *Aspergillus parasiticus* that have uniseriate types (Upadhyaya et al. 2002). The additional most important feature that differentiates the two fungal species is their proliferation in different environments; *Aspergillus flavus* is mostly adapted to the aerial and foliar environment, and principally found in corn, oilseed and tree nuts, while *Aspergillus parasiticus* is mostly confined to soil and is found in peanuts (Waliyar and Adomou 2000). However, the following criteria appears to be more appropriate. *Aspergillus flavus* mainly has two forms: one is the S type and the other is the L type based on genetic features, morphological characters and physiological types. The S type of strain produces more aflatoxins than the isolates of L type strains (Waliyar et al. 2003). The S type of strain is able to frequently develop very minute sclerotia that are less than 400 μm and produce smaller conidia when compared with isolates of the L type of strain. They have sclerotia that are often greater than 400 μm (Beattie et al. 1989).

12.3 STAGES AT WHICH AFLATOXIN IS SYNTHESIZED IN AGRICULTURAL PRODUCTS

Aspergillus is found to proliferate profusely in tropical and subtropical regions of the world as the environmental conditions favor their growth and sporulation. There are chances of fungal contamination at numerous phases, for example, it may be present in the pre- and postharvest stages, during storage and transportation of the products from one place to another. In the postharvesting phase, infestation might occur because of unsuitable dry conditions, storing in polythene packets, damage during shelling or storage in a warm ventilated type of environment. While in the preharvesting phase, the field fungi can gain entry to growing crops due to environmental pressure (hot and dry conditions and soil wetness, mechanical damage, pathogenic injury made by arthropods, fowls, rodents, nematodes and so on) or suspended accumulation (Pandey et al. 2019). Adulteration due to aflatoxin contamination depends on dampness, temperature range, storing process and different factors of soil condition (Chand-Goyal and Spotts 1997). The perfect state for the development of fungi in grain is moisture content about 18% (that is equivalent to 85% of relative humidity) and a temperature range around 12–42 °C with an ultimate point at 27–30 °C in different tropical and subtropical areas (Cooke et al. 1995). An important point is the time of incubation, which has to be considered due to its impact on the creation of poison by *Aspergillus* species. The aflatoxin is obtained after 14 days of incubation at 30 °C. Expanded length of incubation time may be responsible for a decrease in aflatoxin level due to the re-adsorption or degradation by the parasite. The contagious expansion is synthetic with 20% CO_2 and 10% O_2 levels. Some metals, for example, zinc and manganese, are important for aflatoxin development. The combination of iron and cadmium declines in the development and synthesis of aflatoxin (Wills et al. 1989; Sharma et al. 2009). Under the most suitable conditions (i.e., high temperature and moisture level) in the summer season, the dormant fungus either produces hyphae or conidia (abiogenetic spores) which are disseminated through the air and may be responsible for infection on different agricultural products (Hua et al. 1999).

12.4 CONTROL MEASURES FOR AFLATOXIN CONTAMINATION

Increasing knowledge and awareness of the deleterious effects caused by the consumption of food and feeds contaminated with mycotoxins have turned the focus toward developing procedures for the inactivation of these toxic metabolites (Lavkor and Var 2017). Although good agricultural practices are being tried to minimize aflatoxin contamination at field level, various other methodologies have been adopted to eliminate these toxins when present in harvested products.

12.4.1 Physical Strategies

12.4.1.1 General Physical Methods

- Rodent-proof rooms are useful.
- Cold storage of feeds should have a moisture level approximately less than 100 g/kg.

- Use of rapid drying may be common.
- Use of hand picking or photoelectric detecting machines for the removal of fungi-contaminated seeds.

12.4.1.2 Heat

- Heating and cooking under pressure can destroy nearly 70% of aflatoxin in rice (Coomes et al. 1966).
- Dry and oil roasting can reduce the amount of aflatoxin B1 by about 50–70% (Feuell 1966).

12.4.1.3 Radiation

Ionizing radiation such as gamma rays can stop the growth of fungi. It has been reported that gamma irradiation (5–10 M-rad) caused a reduction of aflatoxin (Sommer and Fortlage 1969). The irradiation, however, could not completely destroy the toxin and its mutagenicity (Chipley and Uraih 1980). The treatment combination of gamma irradiation and ammonization should, therefore, be attempted for more aflatoxin decontamination.

12.4.2 Chemical Strategies

12.4.2.1 Chemicals

Chemical treatment is suggested with the aim of reducing unavoidable and unpredictable aflatoxin contamination. There are several chemical agents used for decontamination, which can be divided into categories such as alkaline, acids, reducing agents, oxidizing reagents and many others, for example, chlorinating agents and salts (Karlovsky et al. 2016). The important chemicals are acetic acid, benzoic acid, propionic acid, citric acid, ammonia gas hydrogen peroxide, copper sulfate, ammonium hydroxide, formaldehyde methylamine ozone gas, phosphoric acid, phosphine gas, calcium hydroxide, sodium bisulfide, sodium bicarbonate, sodium bisulfite and sodium hypochlorite, which are utilized to inhibit the growth of fungi and in turn reduce the production of aflatoxin But there are some drawbacks associated with this method as the formation of toxic residue by treatment with these chemicals may be responsible for the risk of potent health problems.

12.4.2.2 Fungicides and Fumigants

Fungicides are utilized widely for postharvest disease control in foods and fodder grown in the field (Hauser Hahn et al. 2008). The timing of the use of fungicide essentially depends on the target pathogen and at what time the contamination occurs. For postharvest pathogens that contaminate agricultural products before harvest, field use of fungicides is frequently essential. This may include the repeated use of protectant fungicides during the developing season and strategic systematic fungicides to inactivate contaminations previously settled and to check on the further development of diseases. In postharvest conditions, fungicides are frequently applied to control contaminations previously settled on the surface tissues of products or to safeguard against diseases which may happen during storage and handling. On account of quiescent field diseases, systematic fungicides are commonly utilized so that during harvest, fungicides must have the option to enter the site of contamination to be successful. On account of diseases which happen during and after harvest, fungicides can be utilized to interfere with pathogens. The success of fruitful fungicides depends to a great extent on the degree to which contamination has spread at the hour of fungicide application and how adequately the fungicide enters the host tissue. The fungicides for the control of wound-attacking pathogens should be applied at the earliest opportunity after harvest because if the infection has progressed, control will be hard to accomplish. The typical methodology with controlling injury pathogens is to keep up a specific convergence of the fungicide at the damage site, which will check pathogen advancement until the injury has mended. In this sense, a large portion of the "fungicides" that are utilized as postharvest treatments are, in a real sense, fungistatic as opposed to fungicidal in their activity under ordinary use. Disinfectants, for example, sodium hypochlorite, can be utilized to kill pathogen propagules on the surface of organic products, but cannot control pathogens once they have entered the tissue. Postharvest fungicides can be applied as dips, sprays, fumigants, treated wraps and box liners or in waxes and coatings. Dips and sprays are usually utilized and, relying on the compound, can appear as aqueous solutions, suspensions or emulsions. Fungicides ordinarily applied as dips or sprays can incorporate the benzimidazoles (e.g., benzoyl and thiabendazole) and the triazoles (e.g., prochlor and imazalil). Benzimidazole fungicides are valuable for the control of numerous significant postharvest pathogens, for example, *Penicillium* and *Colletotrichum* (Bowen et al. 1997). However, due to the development of pesticide-resistant strains, they are not in use now.

An example of a fumigant is sulfur dioxide for the control of different postharvest diseases. Different fumigants used in specific circumstances are carbon dioxide, ozone and smelling salts. Natural product wraps or box liners impregnated with the fungicide biphenyl are utilized in certain nations for the control of *Aspergillus* and *Penicillium*. Fungicides to control postharvest ailments of citrus and some other natural products are regularly applied to organic products in wax on the pressing line (Steen et al. 2001).

12.4.2.3 Feed Additives

Feed additives or supplements as mineral clays, microorganisms and yeast cell walls are mixed with contaminated diets to minimize the effect of mycotoxins on consumers prior to intake or during ingestion. These supplements adsorb or detoxify aflatoxins in the digestive tracts. Potential absorbent materials include activated carbon, aluminosilicates, complex indigestible carbohydrates and synthetic polymers (Thomas et al. 2018).

12.4.3 Biological Detoxification Strategies

The use of both physical and chemical processes outlined earlier to decontaminate food and feed is restricted by very high expense and some nutrient quality loss. The concept of detoxifying mycotoxins through biological conversion is now gaining popularity. The application of microorganisms or enzymes to contaminated products can detoxify aflatoxins by metabolism or degradation; the process is an irreversible and environmentally friendly method as it does not leave toxic residues or unwanted byproducts (Hassan and Zhou 2018). Biological control will include different types of microbes such as yeast, bacteria and most of the non-toxigenic strains of *Aspergillus flavus* and *Aspergillus parasiticus*, which might be utilized for the detoxification of aflatoxins by the process of microbial binding and biotransformation (Hua et al. 1999). These procedures may be difficult and expensive. One of the most important methods to evade this potential hazard of aflatoxin contamination is the application of renewable, harmless and biodegradable natural extracts of plants to remove aflatoxin adulteration (World Health Organization 1988, 2018).

12.4.3.1 Bacteria

Some bacteria such as *Lactobacilli* spp., *Bacillus subtilis*, *Ralstonia* spp. *Pseudomonas* spp. and *Burkholderia* species may play important roles in the inhibition of fungal growth and the production of aflatoxin by *Aspergillus* species. Palumbo et al. (2006) reported that a number of *Ralstonia Pseudomonas*, *Bacillus* and *Burkholderia* strains isolated from the samples of California almond could completely inhibit growth of *Aspergillus flavus*. Several other strains of *Pseudomonas solanacearum* and *Bacillus subtilis* isolated from the non-rhizosphere region of maize soil have inhibition in aflatoxin gathering (Nesci et al. 2005). Application of the soil bacterium *Nocardia corynebacterioides* is supposed to remove aflatoxin B, G and M from a variety of products.

12.4.3.2 Yeasts

Some of the most common saprophytic yeast species (such as *Pichia anomala and Candida krusei*) have been proven as the best agents of biocontrol against *Aspergillus flavus*. As similar to bacteria, these saprophytic yeast strains were able to greatly suppress growth of *Aspergillus* in favorable *in vitro* conditions (Masoud and Kaltoft 2006). Although they were also considered to be probable biocontrol agents for the management of the production of aflatoxins, some field experiments are required to test their abilities in minimizing aflatoxin contamination under *in vivo* circumstances.

12.4.3.3 Non-toxigenic *Aspergillus* Strains

Some of the most important reasonable non-toxigenic strains of *Aspergillus flavus* and/or *Aspergillus parasiticus* have a great level of success in the control of aflatoxin contamination (Pitt and Hocking 2006; Mahato et al. 2019). A significant decrease in aflatoxin contamination in the range of 70–90% has been detected steadily by the use of these non-toxigenic strains of *Aspergillus* (Dorner 2004; Pitt and Hocking 2006; Dorner 2008). This strategy is completely based on the use of non-toxigenic strains in competition to eliminate naturally toxigenic strains in a similar position and to contest for different crop substrates. In the late 1980s, Cotty (1990) verified non-toxigenic *Aspergillus flavus* strains for their capability in the reduction of aflatoxin contamination of cottonseed crops. Recently, a marketable biopesticide product (called Afla-Guard) has been produced and marketed based on the *Aspergillus flavus* strain NRRL21882. Additionally, the non-toxigenic *Aspergillus flavus* strains CT3 and K49 have been verified in the United States and exhibited good efficacies in reduction of aflatoxin contamination in corn crops (Abbas et al. 2006). In Africa, non-toxigenic strain BN30 had a good effect in dropping the quantity of toxin produced in maize when co-inoculated with the highly toxigenic S type strain (Cardwell and Henry 2004). In Australia, it was observed that using non-toxigenic strains could reduce aflatoxin contamination in peanut crops by approximately 95%.

12.4.3.3.1 Factors for the Reduction of Aflatoxin By Non-toxigenic Aspergillus *Strains*

12.4.3.3.1.1 Formulation The formulation is the type of combination that deals with the competitive strain and the carrier/substrate. Initially, it was observed that there was a direct application of suspension of homogenized culture of non-toxigenic *Aspergillus parasiticus* to the soil surface before plantation was done, which was found to be very effective in a significant reduction of aflatoxin contamination, but it was too expensive to apply this type of formulation for large-scale protection of fields (Dorner et al. 1992). Another method employed to protect from toxin contamination was to apply it on the grains. This type of method is achieved by formulation by inoculation, incubation with agitation and then dried at about 50 °C. The grains are then stored at approximately 5 °C until they are utilized for further processing (Dorner 2004). When the treated grains are sown in the field, the non-toxigenic strains resume their growth and produce numerous conidia on the surface of the grains. These conidia are then dispersed in the soil and compete with naturally occurring toxigenic strains.

12.4.3.3.1.2 Inoculum Rate Inoculum rate is considered as one of the most important factors influencing the efficiency of different biocontrol agents. There must be a strong relationship between efficacy and inoculum rate of different available biocontrol agents for the decrease of contamination. Furthermore, a higher degree of control measures might be achieved when plots or fields are treated and retreated with different biocontrol agents in subsequent years and so on (Pitt and Hocking 2006).

12.4.3.3.1.3 Optimum Time and Temperature Soil temperature may be considered as the essential factor affecting growth and sporulation of non-toxigenic fungi. It is observed that *Aspergillus flavus* propagates at temperatures below 10 °C on the growth medium in *in vitro* environments, but *in vivo* circumstances, it exhibited the development of different biocontrol strains not presented easily when the soil temperature is found to be below 20 °C (Pitt and Hocking 2006). So application of non-toxigenic strains to soil must be delayed until the soil temperature reaches at least 20 °C. In Arizona, USA, the first half of June and the second half of April are known as the most suitable times for the application of these non-toxigenic agents. In Georgia, USA, the biocontrol agent NRRL21882 was applied between about 50 and 70 days after planting of the peanut crop (Dorner 2004).

12.4.3.3.1.4 Application of Herbicide Herbicides are sometimes regularly applied in agricultural fields where most non-toxigenic strains are used. The outcome of the use of herbicides commonly depends on the incubation period of 9 days, herbicides like trifluralin and paraquat did not inhibit growth of *Aspergillus* until the concentration of those herbicides were five times more than the recommended dose. After 16 days, the fungus developed in specific media, such as in the Czapek Yeast Extract agar medium adjusted with suitable herbicide concentration with up to ten times the suggested dose, but the herbicides did not show noticeable inhibition in *Aspergillus* growth (Pitt and Hocking 2006).

12.5 TRADITIONAL STRATEGIES FOR THE PREVENTION OF AFLATOXIN CONTAMINATION

It is very important for the decrease or removal of mycotoxins that there must be some knowledge about their fungal sources. The growth of fungi in crop yields and different agricultural products is the main source of toxin production and is related to the concentration of the different toxic substances. Many factors are important such as plant susceptibility to fungi, fittingness of fungal substrate, climate, moisture content and physical damage of seeds are all involved in increasing the production of mycotoxins. Toxin-producing fungi may attack at preharvesting periods, harvest time, during postharvest handling and in storage. According to the site of infestation, toxicogenic fungi can be divided into three groups: (a) field fungi, *Fusarium*, for example, *Fusarium moniliforme*, *Fusarium roseus*, *Fusarium tricinctum and Fusarium nivale*; (b) storage fungi, *Aspergillus* and *Penicillium*, for example, *Aspergillus flavus* and *Aspergillus parasiticus*; and (c) advanced deterioration fungi, for example, *Aspergillus clavatus*, *Aspergillus fumigatus*, *Chaetomium*, *Scopulariopsis*, *Rhizopus*, *Mucor* and *Absidia*. The prevention of mycotoxins in our atmosphere is a big task.

12.6 METHODS FOR THE PREVENTION OF AFLATOXIN CONTAMINATION

In general, prevention of the contamination of fungi and their mycotoxins in agricultural commodities can be divided into the three levels in Table 12.1.

12.7 EFFECT OF MEDICINAL PLANTS ON AFLATOXIN-PRODUCING FUNGI

It is evident from modern research that phytophenols, which are considered as secondary plant metabolites, exist with over 8000 structures (Begossi et al. 2002). These structures usually look like the structure of phenolic acid and tannin. It has been established that phytophenols have cell reinforcement, antiatherogenic, anti-inflammatory, antiallergenic, antimicrobial and antithrombotic actions. These plant compounds exhibited significant biological action in the degradation of numerous microbes. Most of the plants, herbs, essential oils and flavors in powder form or in the form of concentrates are utilized in the detoxification of microorganisms because of the presence of phenolic, betalain, phytoalexin and thiosulfate flavonoids. Sinapic acid, syringaldehyde and acetosyringone were the plant phenolics that mostly repressed the production of aflatoxin type B1. Nevertheless, vanillyl $(CH_3)_2CO$, thymol, salicylic acids, cinnamic acid and vanillin are different phenolic aggravates that are responsible for checking the development of *Aspergillus flavus* by concentrating on oxidative mitochondrial apprehension as the fortification outline (Cole et al. 1995). Several medicinal plants exhibited different specific properties such as antimicrobial, antifungal, anthelmintic, antibiotic antiarthritic, antiviral, anti-inflammatory, antirheumatic and antihemorrhoidal. The different types of restorative plants, which are local to the area of Southeast Asia, include unfriendly Asiatic pennywort (*Centella asiatica*), cucumber (*Momordica*

TABLE 12.1
Primary, Secondary, and Tertiary Methods for the Prevention of Aflatoxin Contamination

Method	Purpose
Primary prevention: minimize fungal infestation and aflatoxin contamination	
Cultivation of resistant varieties of fungi	Limit fungal invasion and toxin production during crop growth
Control field infection by following appropriate phytosanitary measures	Limit fungal inoculum in the field
Seed treatment and fungicide application	Limit fungal invasion during crop growth
Appropriate scheduling of planting, preharvest, harvest and postharvest	Avoid drought and other abiotic stresses
Application of soil amendments, e.g., gypsum, farmyard manure	Increase soil nutrients (especially calcium) and water-holding capacity, promoting growth of antagonistic native soil microflora
Lower moisture content of seeds after postharvest and during storage	Limit fungal invasion and growth during storage
Add preservatives to prevent insect infestation and fungal contamination during storage	Limit fungal invasion during storage
Secondary prevention: eliminate or limit fungal contamination	
Sorting of contaminated products	Reduce aflatoxin contamination in the final product
Re-drying products	Limit fungal growth and toxin synthesis
Appropriate storage conditions	Limit fungal growth and toxin synthesis
Detoxification of contaminated products	Chemical inactivation or binding of aflatoxins through the use of clay dietary supplements or ammonization
Tertiary prevention	
Complete destruction of contaminated products	No chance of any growth of fungi
Detoxification or devastation of mycotoxins to the negligible level	No toxin production

charantia), betel vine (*flautist betel*), betel nut (*Areca catechu*), bogus coriander (*Eryngium foetidum*), Cha Phlu (*Piper sarmentosum*), clove (*Syzygium aromaticum*), Chinese radish (*Raphanus sativus*), Tasmanian blue gum (*Eucalyptus globulus*), Indian mulberry (*Morinda citrifolia*), Madagascar periwinkle (*Catharanthus roseus*), bmangosteen (*Garcinia mangostana*), mandarin (*Citrus reticulate*), Non Taai Yak (*Stemona tuberosa*), Raang Chuet (*Thunbergia laurifolia*), Saab Sue (*Chromolaena odorata*), turmeric (*Curcuma longa*), water primrose (*Jussiaeda repens*), onion (*Allium cepa*), pepper (*Piper nigrum*), pomegranate (*Punica granatum*), tomato (*Lycopersicon esculentum*), support bloom (*Lantana camara*), roselle (*Hibiscus sabdariffa*) and wishing tree (*Cassia bakeriana*). They were tried for their capacity to control aflatoxins creating microbes and hence aflatoxin contamination.

12.8 BIOTECHNOLOGICAL APPROACHES

Biotechnological approaches to increase host plant resistance through the use of antifungal and antimycotoxin genes have also begun (Ojiambo et al. 2018). This approach received a major boost with the successful establishment of peanut regeneration and transformation protocols (Anjaiah et al. 2006) and led to the transformation of popular peanut cv. JL24 with a rice chitinase gene to help prevent invasion by fungal pathogens. These transgenic events have now advanced to the T3 generation, with three events showing good resistance to *A. flavus* infection (<10% infection) in *in vitro* seed inoculation tests.

12.9 CONCLUSIONS

Aflatoxins, the "silent killers," have been considered to be fatal when food and feed are contaminated and used by consumers. There are many techniques that are occasionally employed to mitigate their toxicity. Given the adverse impact on health because of the age-old problem of aflatoxin contamination, no stable and sustainable solution has been observed so far. It is essential to minimize aflatoxin contamination in the entire food chain by adopting modern pre- and postharvest management practices in the processing, packaging, transportation and storage stages. The recent interest in genetic resistance will provide a much needed defense from invasion of mycotoxigenic fungi in the field and efficient steps during postharvest management will ensure fewer aflatoxins in produce (Pandey et al. 2019).

REFERENCES

Abbas, H.K., R.M. Zablotowicz, H.A. Bruns et al. 2006. Bio control of Aflatoxin in corn by inoculation with non-aflatoxigenic Aspergillus flavus isolates. *Biocontrol Sci Technol* 16(5): 437–449.

Anjaiah, V., R.P. Thakur and N. Koedam. 2006. Evaluation of bacteria and Trichoderma for biocontrol of preharvest seed infection by *Aspergillus flavus* in groundnut. *Biocontrol Sci Technol* 16(5): 431–436.

Beattie, B., W.B. McGlasson and N.L. Wade. 1989. *Postharvest Diseases of Horticultural Produce, Volume 7: Temperate Fluit.* Melbourne, Australia: CSIRO.

Begossi, A., N. Hanazaki and J. Tamashiro. 2002. Medicinal plants in the Atlantic forest (Brazil): Knowledge, use and conservation. *Hum Ecol* 30: 19.

Bhat, R., R.V. Rai and A.A. Karim. 2010. Mycotoxins in food and feed: Present status and future concerns. *CRFSFS* 9: 57–81.

Bowen, K. L., A.K. Hagan and J.R. Weeks. 1997. Number of tebuconazole applications for maximizing disease control and yield of peanut in growers' fields in Alabama. *Plant Dis* 81: 927–931.

Calvo, A.M., R.A. Wilson, J.W. Bok et al. 2002. Relationship between secondary metabolism and fungal development. *Microbiol Mol Biol Rev* 66: 447–459. doi: 10.1128/MMBR.66.3.447-459

Cardwell, K.F. and S.H. Henry. 2004. Risk of exposure to and mitigation of effect of aflatoxin on human health: A West African example. *J Toxicol Toxin Rev* 23(2&3): 217–247.

Chand-Goyal, T. and R.A. Spotts. 1997. Biological control of postharvest diseases of apple and pear under semi-commercial and commercial conditions using three saprophytic yeasts. *Biol Control* 10: 199–206.

Chipley, J.R. and N. Uraih. 1980. Inhibition of Aspergillus growth and aflatoxin release by derivatives of benzoic acid. *Appl Environ Microbiol* 40: 352–357.

Cole, R.J., J.W. Dorner and C.C. Holbroook. 1995. Advances in mycotoxin elimination and resistance. In H.E. Pattee and H.T. Stalker (Eds.), *Advances in Peanut Science* (pp. 456–474). Stillwater, OK: American Peanut Research and Education Society.

Cooke, A., D. Persley, B. Beattie et al. 1995. *Postharvest Diseases of Horticultural Produce, Volume 2: Tropical Fruit.* Queensland, Australia: DPI.

Coomes, T.J., P.C. Crowther, A.J. Feuell et al. 1966. Experimental detoxification of groundnut meals containing aflatoxin. *Nature* 290: 406.

Cotty, P.J. 1990. Effect of atoxigenic strains of Aspergillus flavus on aflatoxin contamination of developing cottonseed. *Plant Dis* 74(3): 233–235.

Diener, U.L., R.J. Cole, T.H. Sanders et al. 1987. Epidemiology of aflatoxin formation by *Aspergillus flavus. Ann Rev Phytopath* 25: 249–270.

Dorner, J.W. 2004. Biological control of aflatoxin contamination of crops. *J Toxicol Toxin Rev* 23(2&3): 425–450.

Dorner, JW. 2008. Management and prevention of mycotoxins in peanuts. *Food Addit Contam* 25(2): 203–208.

Dorner, J.W., R.J. Cole, P.D. Blankenship et al. 1992. Use of a bio competitive agent to control preharvest aflatoxin in drought stressed peanuts. *J Food Prot* 55(11): 888–892.

Ferron, P. and J.P. Deguine. 2005. Aflatoxin formation by *Aspergillus flavus. Ann Rev Phytopath* 25: 249–270.

Feuell, A.J. 1966. Aflatoxin in groundnuts IX, Problems of detoxification. *Trop Sci* 8: 61.

Hassan, Y.I. and T. Zhou. 2018. Promising detoxification stategies to mitigate Mycotoxins in food and feed. *Toxins* 10: 116.

Hauser Hahn, I.S., R. Dutzmann, R. Freissleben et al. 2008. Prosaro–A new fungicide for control of Fusarium and myco-toxins in cereals. *Cereal Res. Commun* 36(Suppl. B): 711–712.

Horn, B.W. and R.L. Green. 1995. Vegetative compatibility within populations of *Aspergillus flavus*, *A. parasiticus*, and *A. tamari* from a peanut field. *Mycologia* 87: 324–332.

Hua, S.S.T., J.L. Baker and M.E. Spiritu. 1999. Interaction of saprophytic yeast, with a non-mutant of *Aspergillus nidulance*. *Appl Environ Microbiol* 65(5): 2738–2740.

International Agency for Research on Cancer (IARC). 1993. Toxins derived from *Fusarium moniliforme*: Fumonisins B1, B2 and Fusarin C. In *Monograph on the Evaluation of Carcinogenic Risk to Humans No. 56* (pp. 445–466). Lyon, France: International Agency for Research on Cancer (IARC).

Janisiewicz, W. and L. Korsten. 2002. Biological control of postharvest diseases of fruits. *Ann Rev Phytopathol* 40: 411–441.

Karlovsky, P., M. Suman, M. Berthiller et al. 2016. Impact of food processing and detoxification treatments on mycotoxin contamination. *Mycotoxin Res* 32: 179–205.

Lavkor, I. and I. Var. 2017. The control of Aflatoxin contamination at harvest, drying, pre-storage and storage periods in peanuts: The new approach. Chapter 3. In L. Abdulra'Uf (Ed.), *Aflatoxin-Control, Analysis, Detection and Health Risks*. Rijeka, Croatia: IntechOpen.

Mahato, D.K., K. EunLee, M. Kamble et al. 2019. Aflatoxins in food and feed: An overview on prevalence, detection and control strategies. *Front Microbiol* 10: 2266. doi: 10.3389/fmicb.2019.02.266

Marin, S., A.J. Ramos, G. Cano-Sancho et al. 2013. Mycotoxins: Occurrence, toxicology, and exposure assessment. *Food Chem Toxicol* 60: 218–237.

Masoud, W. and C.H. Kaltoft. 2006. The effects of yeasts involved in the fermentation of coffea arabica in East Africa on growth and ochratoxin A (OTA) production by Aspergillus ochraceus. *Int J Food Microbiol* 106(2): 229–234.

Misihairabgwi, J.M., C.N. Ezekiel, M. Sulyok et al. 2019. Mycotoxin contamination of foods in Southern Africa: A review (2007–2016). *Crit. Rev. Food Sci. Nutr.* 59(1): 43–58. doi: 10.1080/10408398.2017.1357003.

Nesci, A.V., R.V. Bluma and M.G. Etcheverry. 2005. In vitro selection of maize rhizobacteria to study potential biological control of Aspergillus section Flavi and aflatoxin production. *Eur J Plant Pathol* 113(2): 159–171.

Ojiambo, P.S., P. Battilani, W.C. Jeffrey et al. 2018. Cultural and genetic approaches to manage Aflatoxin contamination:Recent insights provide oppurtunities for improved control. *Phytopathol Rev* 108: 1024–1037. doi: 10.1094/PHYTO-04-18-0134-RVW.

Omatayo, O.P., A.O. Omatayo, M. Mwanza et al. 2019. Prevalence of mycotoxin and their consequences on human health. *Toxicol. Res.* 35: 1–7.

Palumbo, J.D., J.L. Baker and N.E. Mahoney. 2006. Isolation of bacterial antagonists of Aspergillus flavus from almonds. *Microbial Ecol* 52(1): 45–52.

Pandey, M.K., R. Kumar, A.K. Pandey et al. 2019. Mitigating Aflatoxin contamination in groundnut through a combination of genetic resistance and post-harvest management practices. *Toxins* 11: 315. doi: 10.3390/toxins11060315.

Payne, G.A. 1998. Process of contamination by aflatoxin producing fungi and their impacts on crops. In K.K. Sinha and D. Bhatnagar (Eds.), *Mycotoxins in Agriculture and Food Safety* (pp. 279–300). New York: Marcel Dekker, Inc.

Pettit, R.E. 1986. Incidence of aflatoxin in groundnuts as influenced by seasonal changes in environmental conditions – A review. In M.V.K. Sivakumar, S.M. Virmani and S.R. Beckerman (Eds.), *Agrometeorology of Groundnut. Proceedings of an International Symposium (21–26 August* 1985, *ICRISAT Sahelian Center, Niamey, Niger)* (pp. 163–174). Patancheru, India: ICRISAT.

Pitt, J.I. and A.D. Hocking. 2006. Mycotoxins in Australia: Biocontrol of aflatoxin in peanuts. *Mycopathologia* 162(3): 233–243. doi: 10.1007/s11046-006-0059-0

Sharma, R.R., D. Singh and R. Singh. 2009. Biological control of postharvest diseases of fruits and vegetables by microbial antagonists: A review. *Biol Control* 50(3): 205–221.

Sommer, N.F. and R.J. Fortlage. 1969. Ionizing radiation for control of postharvest diseases of fruits and vegetables. *Adv Food Res* 15: 147.

Steen Van der, C., L. Jacxsens, F. Devlieghere et al. 2001. Combining high oxygen atmospheres with low oxygen modified atmosphere packaging to improve the keeping quality of strawberries and raspberries. *Postharvest Bio Technol* 26: 49.

Thomas, M., W.H. Hendricks and A.F.B. vander Poel. 2018. Size distribution analysis of wheat, maize and soybeans and energy efficiency using different methods for coarse grinding. *Anim Feed Sci Technol* 240: 11–21.

Upadhyaya, H.D., S.N. Nigam and R.P. Thakur. 2002. *Genetic enhancement for resistance to Aflatoxin contamination in groundnut*. In *Summary Proceedings of the Seventh ICRISAT Regional Groundnut Meeting for Western and Central Africa 6–8 Dec.* 2000, *Cotonu, Benin.*

Waliyar, F. and M. Adomou. 2000. *Summary Proceedings of the 7th ICRISAT Regional Groundnut Meeting for Western and Central Africa, (6–8 December* 2000, *Cotonou, Benin*) (pp. 29–36). Patancheru, India: ICRISAT.

Waliyar, F., S.V. Reddy, K. Subramanyam et al. 2003b. Importance of mycotoxins in food and feed in India. *Aspects Appl Biol* 68: 147–154.

Wild, C.P. and A.J. Hall. 2000. Primary prevention of hepatocellular carcinoma in developing countries *Mutat Res* 462(2–3): 381–393.

Wills, R.B.H., W.E. McGlasson, D. Graham et al. 1989. *Postharvest: An Introduction to the Postharvest Handling of Fruit and Vegetables* (3rd ed.). Sydney, Australia: University of New South Wales Press.

World Health Organization. 1988. *Food Irradiation. A Technique for Preserving and Improving the Safety of Food.* Geneva, Switzerland: World Health Organization.

World Health Organization. 2018. *Aflatoxins. Food Safety Digest.* Ref.No.: WHO/NHM/FOS/RAM/18.1. Geneva, Switzerland: World Health Organization.

13 Nano Magic Bullets
An Ecofriendly Approach to Managing Plant Diseases

Neeta Sharma and Avantina S. Bhandari

CONTENTS

13.1 INTRODUCTION

Over the past few decades, focus and resources have been directed and concentrated in finding significant solutions to increase food production and reduce food losses. It is projected that the current scenario of losses and constraints due to climatic conditions, soil fertility and availability of water, arable land and low-cost energy suggests that production increase could fall to 0.87% toward 2030 and to 0.5% toward 2030–2050. Alexendratos and Bruinsma (2012) reported that food supplies would need to increase by 60% in order to meet food shortages by 2050. According to the data presented in a report by the UN in March 2013, the current world population is expected to reach 10.5 billion by 2050, further adding to global food security concerns.

This alarming rise in populations combined with improved nutritional standards and shifting dietary preferences will exert pressure for increases in the global food supply; this increase will translate into 33% more human mouths to feed, with greatest demand growth in the poor communities of the world. The world will have to struggle hard to produce more food in the face of rising populations, limited energy supplies and degradation of our fresh water and soil. In addition, a production short of demand, greater geographical inequity in production and demand combined with possibly more challenging weather and subsequent speculation in food markets, could leave us in an even worse condition than that seen in the current crisis if appropriate options for increasing food supply and security are not considered and implemented in the near future (Sharma, 2014).

Globally, more than 800 million people do not have adequate food supply; 1.3 billion live on less than US$1 a day and at least 10% of all global food production is lost to various abiotic and biotic factors. Diseases caused by biotic factors, a general phenomenon at the field level well after harvesting is a matter of grave concern. It has been estimated that a minimum of 47,000,000 metric tons of durables and 60,000,000 metric tons of perishables fall victim to various pathogens. Postharvest pathogens cause enormous losses, nearly to the tune of several million dollars to commodities that are meant for consumption. A rough estimate suggests that up to 40% of the produce is lost during postharvest handling and diseases.

Considerable attention has been drawn toward the enormity of food shortages due to postharvest diseases and causal pathogens. Promoting food security through loss reduction is the most feasible and sustainable method of increasing food production and ensuring adequate food supply to billions of mouths.

13.2 A WAY OUT OF THIS CRISIS

So far, the management of postharvest diseases has heavily relied upon the use of synthetic chemicals. The world became so obsessed with the idea of increasing crop productivity that it completely forgot that the produce should be safe for human consumption and the environment. These hazardous chemicals contaminate the environment, enter the food chain, destroy native insect pests, build residual toxicity in animal and plant tissues and introduce health issues in humans leading to hormonal imbalances, respiratory troubles and even cancer; targeting non-target microorganisms and increasing the incidence of pest resurgence due to the development of resistance in pests toward these chemicals.

The farming community the world over has realized that indiscriminate use of pesticides has raised its ugly head, leading to pollution of all kinds, which has increased the momentum to say "No" to hazardous chemical use. The worldwide interdependence of markets for agricultural products has increasingly brought to the forefront the need to develop strategies that can mitigate the adverse effects on the environment and will result in products that are safe for human consumption.

Several postharvest technologies have been developed to improve quality, shelf life and manage postharvest loss due to pathogens, thereby ensuring produce quality and consumer safety. Understandably, alternatives to chemical products/pesticides that are cost-effective, less harmful to the environment and allow the minimum use of chemicals in terms of fewer or reduced rates and doses of application have started gaining popularity in the market (Vermeulen et al., 2012).

13.3 NANOTECHNOLOGY: A NEW HOPE

These challenges of sustainability, food security and climate change are engaging scientists in exploring the field of nanotechnology as a new source of key improvements in different fields (Figure 13.1). The development of nanotechnology has provided an exciting and novel frontier to nearly all fields of industrial applications with a profound impact on human life and welfare (Linkov et al., 2011). Amid this management turmoil, all hopes in the present scenario have settled upon the use of the latest technology of applying nanoparticles in the field of agriculture, including horticulture. It is anticipated that due to the worldwide popularity and expansion of this technology, its market value will reach US$75.8 billion by 2020 (Research and Markets, 2015).

The subject of nanotechnology deals with fabrication, study and manipulation of matter at an atomic level in the size range of 1–100 nm, which is designated as "nanoparticles". These particles express entirely new properties which enable them to demonstrate extended plasticity at a high temperature; hardiness, breaking strength and toughness at a low temperature; elevated chemical reactivity and surface energy; and more mobility in the body of organisms, including cellular entry (Banik and Sharma, 2011). The application of nanotechnology for detection and managing plant pathogens, host–pathogen interactions, development of nanoparticles through microbes and other uses in agriculture has also been explored in plant hormone delivery, seed germination, transfer of target gene nanobarcoding, nanosensors and controlled release of agrichemicals (Elmer and White, 2018a).

13.4 SYNTHESIS OF NANOPARTICLES

Several conventional physical and chemical protocols such as melt mixing, laser pyrolysis or ablation, vapor deposition, sputtering, thermolysis, photoreduction, micro-emulsion and sol-gel, ultrasonic fields and ultraviolet irradiation have been successfully employed for nanoparticle synthesis for decades (Singh et al., 2018a). Nanoparticles can be designed with unique biological, chemical and physical properties to enhance all the properties of the parent metal at the nano scale. Transformation of bulk elements to the nano level not only reduces its size, but also leads to the formation of different shapes such as spherical, triangular, truncated triangular, octahedral, rod and flower shaped. The variation in geometric shapes is advantageous in the application of these particles as antimicrobials since the bioactivity of nanoparticles is directly proportional to the surface

FIGURE 13.1 Applications of nanoparticles.

area available for interaction with biological components. These particles have the potential to inhibit diseases of seeds, roots, seedlings, foliage and fruits, providing protection against pathogens like bacteria, viruses, fungi and insects.

Current research in biosynthesis of nanometals using plant extracts has opened a new era of fast and non-toxic methods for production of nanoparticles. The biological synthesis of metal nanoparticles (especially gold and silver nanoparticles) using plants (inactivated plant tissue, plant extracts and living plants) has received more attention as a suitable alternative to chemical procedures and physical methods (Hussain et al., 2016).

Synthesis of metal nanoparticles using plant extracts is very cost-effective, and therefore, can be used as an economic and valuable alternative for the large-scale production of metal nanoparticles. Several metal nanoparticles have been fabricated using a biosynthetic approach (Kharissova et al., 2013). Plant extracts may act both as reducing and capping agents in nanoparticle synthesis. The bio-reduction of metal nanoparticles by combinations of biomolecules found in plant extracts (e.g., enzymes, proteins, amino acids, vitamins, polysaccharides and organic acids such as citrates) is environmentally benign, yet chemically complex.

13.5 TYPES OF NANOPARTICLES

Nano-sized particles of different metals, pesticides and growth promoters have been explored for plant disease management either as nanoparticles alone, acting as protectants; or as nanocarriers for fungicides, herbicides, nematicides, insecticides and RNA-interference molecules (Abdellatif et al., 2016; Worrall et al., 2018). Nanoparticles are commonly used as carriers to absorb or attach, encapsulate and entrap the active molecules in developing effective agricultural formulations. Nanotubes are biologically active and more environmentally friendly. Commonly used nanoparticles (Figure 13.2) as carriers for fungicides, insecticides, herbicides and RNAi including molecules are summarized as follows.

13.5.1 CARBON NANOTUBES

Carbon is considered to be a brick molecule for simple as well as complex architectural designs of almost all molecules. Although many types have been synthesized, most commonly in use are carbon nanotubes, graphene oxides and fullerenes. More recently, the antimicrobial activity of carbon nanomaterials has been demonstrated against bacteria (Iravani, 2011; Chen et al., 2013) and fungi (Sarlak et al., 2014; Wang et al., 2014), along with a positive effect on plant growth (Tripathi et al.,

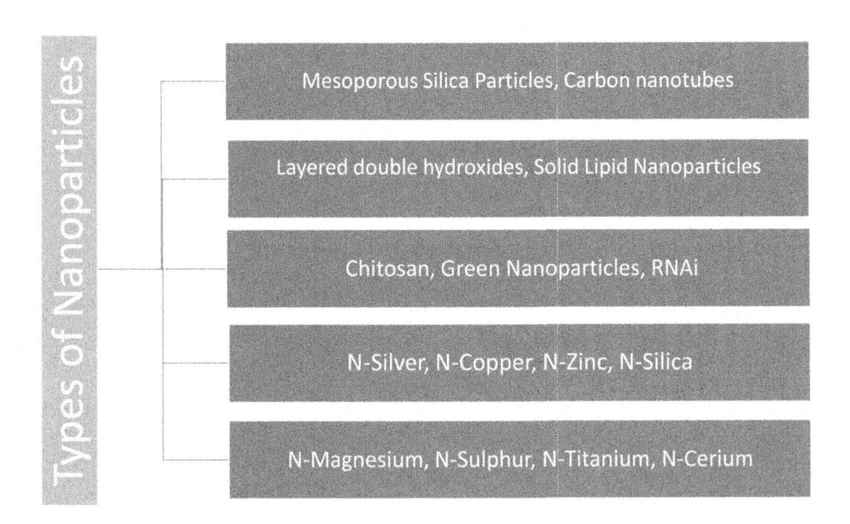

FIGURE 13.2 Types of nanoparticles.

2011). The area of focus at present includes research for producing carbon nanofibers to strengthen natural fibers, for example, those from coconut and sisal and synthesizing nanoparticles that contain pesticides and control their release.

Mainly, researchers are concentrating on carbon nanotubes (CNTs), which are synthesized from graphene oxide sheets to have a single cylindrical wall or multiple walls at the nanoscale. Significant reduction in the radial measurements was observed of *Aspergillus niger*, *A. oryzae* and *Fusarium oxysporum* when culture on agar was amended with a reduced form of graphene oxide in different concentrations; 50% inhibitory concentrations for *F. oxysporum* were 50µg/mL while both the species of Aspergillus were inhibited at 100 µg/mL concentrations.

In the studies conducted by Wang et al. (2016b), it was found that single wall carbon nanotubes were the most toxic to *F. graminearum* and *F. poae* followed by multiwalled carbon nanotubes, graphene oxide (oxidized) and graphene oxide (reduced). Fullerene at 50 µg/mL inhibited spores of Botrytis cinerea (Hao et al., 2017). Graphene oxides at 250 µg/mL killed 95% of the cells of *Xanthomonas oryzae pv. oryzae* compared with 13.3% mortality when bismerthiazol was used as a bactericide (Cota et al., 2013).

Reports have claimed that when tomato seeds were sown in soil impregnated with CNTs, these nanoparticles penetrated the hard integument of the germinating seed and they exerted growth-enhancing effects. They envisaged that this enhanced growth was due to increased water uptake caused by penetration of CNTs

(Khodakovsky et al., 2000). Since these CNTs are growth promoting, they will not have any adverse effects on the plants and will also serve as a vehicle to deliver desired molecules to the plants/plant parts and thus can help in protecting them from any disease

13.5.2 Nano-copper

The antimicrobial properties of copper are well established, thus choosing nano Cu is a logical choice for plant disease management. It has been proved through various studies that nano Cu products were either equal or superior to the commercial Cu-based products, delivered less Cu/ha and did not cause phytotoxicity.

In experimental studies, Giannousi et al. (2013) examined foliar application of nano- CuO, Cu_2O and Cu/Cu_2O composites and compared them with the registered commercial Cu-based fungicides, including Cuprofix Dispress^R, Kocide 2000 35 WG, Kocide Opti 30WG and Ridomil Gold Plus, for their ability to suppress *Phytophthora infestans* in tomato; they observed that the most effective product for suppressing the leaf lesions was nano CuO at the concentration of 150–340 µg/mL, followed by the nano Cu/Cu_2 O composite. Nanoparticle composites possessing a core shell Cu, multivalent Cu and fixed quaternary ammonium copper significantly reduced the severity of *Xanthomonas perforans* causing bacterial spot disease in greenhouse and field when compared with Cu fungicides (Strayer-Scherer et al., 2018).

Copper nanoparticles have also been explored by examining their efficacy as nano-fertilizers/supplements to enhance disease resistance (Elmer and White, 2016). When nano CuO particles were applied as foliar spray to the leaves of young seedlings of tomato and eggplants at 500 µg/mL and then planted into potting mix infested with *F. oxysporum f.sp. lycopersici* or *Verticillium dahliae*, disease severity ratings were consistently lower than the controls. Eggplants and tomato treated with nanoparticles of copper oxide produced 24% more yield than controls.

Elmer et al. (2018b) were successful in managing wilt disease when they extended the foliar application technique by spraying CuO nanoparticles on watermelon seedlings and transplanting the seedlings into potting mix infested with *F. oxysporum f.sp, niveum.* Plants treated with nano CuO particles showed elevated polyphenol oxidase activity suggesting the reason for disease suppression. Several reports on the antifungal activity of CuO nanoparticles have been reported against *Alternaria alternata*, *Botrytis cinerea*, *Curvularia lunata*, *F.oxysporum* and *Phoma destructiva* (Bramhanwade et al., 2016; Kanhed et al., 2014).

With the increasing number of Cu-tolerant bacterial strains that that have emerged and the high amount of Cu currently being applied in agriculture, it could be argued that any strategy using Cu would further erode the use of copper as a management tool (Elmer and White, 2018b). Arguments against these claims are that the amount of copper being supplied in nano form is much less than the conventional Cu-mancozeb. Furthermore, it was found that foliar spray of nanoparticles to young seedlings resulted in only 1–2 mg of CuO/seedling (roughly equivalent to 7.5–15.0 g/ha, assuming 7500 plant/ha). Arguably, these studies have demonstrated that nano Cu delivers a more active load of copper at significantly smaller nominal rates, thus supporting the growing consensus that nano Cu and its composites offer multiple benefits in plant disease management.

13.5.3 Nano-Mg Particles

It has been demonstrated that MgO nanoparticles increase systemic disease resistance against *Ralstonia solanacearum* if applied preventatively on tomatoes. Simultaneous application with nano-MgO (500 or 1000 μg/mL) and the pathogen had only a marginal effect, whereas, pre-exposed roots to 350 μg/mL of nano-MgO and then inoculated with *R. solanacearum* after 4–8 days showed a significant reduction in disease resistance. It is reported that ß-1,3-glucanase and tyloses appeared in vascular tissues of the hypocotyls of plant exposed to nanoparticles of MgO as seen during histological studies. Further nano-MgO treated roots produced rapid generation of reactive oxygen species along with upregulation of PR1, jasmonic acid, ethylene and systemic resistance related genes (Imada et al., 2016). Liao et al. (2017) reported suppression of bacterial spot at 200 μg/mL of MgO nanoparticle on tomato plants. Wani and Shag (2012) demonstrated that the conidial germination of *Alternaria alternata*, *Fusarium oxysporum*, *Rhizopus stolonifer* and *Mucor plumbeus* was inhibited by MnO nanoparticle as compared with nano ZnO particles

13.5.4 Nanosilica Particles

Metalloid silica nanoparticles have found multiple uses in biosensors, biocatalysts, immunochemistry and so on. The first mention of nano-Si was the use of an Ag-Si composite against powdery mildew at 0.3 μg/mL (Park et al., 2006). These studies were a result of an Ag-Si composite, and as the elemental effect of the two metals was not analyzed separately, it is not clear which one of these metals could be more effective in suppressing the disease. However, maize grown in a loamy soil amended with nano-Si at 5, 10 or 15 kg/ha and compared with untreated control or bulked silica at 15 kg/ha had the lowest amount of disease caused by *Aspergillus niger* and *Fusarium oxysporum* at higher rates of nano-Si particles (Koch et al., 2016).

13.5.5 Mesoporous Silica Nanoparticles

One such effort is the use of aluminum silicate nanotube with active ingredients; these can be easily synthesized with controlled shape, size and structure, making them highly advantageous delivery (Datnoff et al., 2007). Porous hollow silica nanoparticles (PHSNs) or mesoporous silica nanoparticles (MSNs) are loaded with pesticides into the inner core to protect the active molecules inside the nanoparticles against degradation, and therefore, provide a sustained release. The advantage is that aluminum silicate nanotubes sprayed on plant surface are easily picked up in insect hair and so they consume pesticide filled nanotubes. It is a documented fact that silica can enhance stress resistance to plants, including plant diseases through promotion of plant physiological activity and growth, but it has no direct antimicrobial effect and also mesoporous SiNP can deliver DNA and chemicals into plants, thus creating a powerful tool for targeted delivery into plant cells. Spherical silica nanoparticles, with an array of honeycomb-like channels, can be filled with chemicals or molecules. These nanoparticles have a unique "capping" strategy that seals the chemicals inside.

13.5.6 Nanosilver

Since 1000 BC, silver vessels were used to serve food and drinks as it was thought that they possessed antiseptic and antimicrobial properties. Silver in an ionic state exhibits high antimicrobial activity, however, due to its high reactivity, it is unstable and easily gets oxidized or reduced into a metal depending on the surrounding media and so cannot exert antimicrobial activity. Silver nanoparticles have a high surface area and high fraction of surface atoms, and have greater antimicrobial effect as compared with bulk silver.

Commendable efforts have been made to explore the antimicrobial properties of silver nanoparticles against human pathogens, but studies to observe its effect against plant pathogens are still in their infancy (Lamsel et al., 2011). Colloidal nanosilver is a well-dispersed and stabilized solution and is more adhesive on bacteria and fungus, hence it is better than fungicides. In unicellular microorganisms, the antimicrobial effect of silver is by inactivating enzymes having metabolic functions in the microorganisms by oligodynamic action. Double capsulated nanosilver was prepared by chemical reaction of silver ion with the aid of physical methods, reducing agents and stabilizers, which were stable and dispersive in aqueous solution.

There are many reports on the in vitro fungicidal effects of nanosilver particles but very few studies have been conducted to elucidate the application of AgPNs as antifungal compounds in managing plant diseases in the field and in turn promote overall growth.

Several plant pathogenic microorganisms belonging to ascomycetes and basidiomycetes develop sclerotia as primary surviving structures to avoid unfavorable conditions as they exhibit resistance against abiotic factors such as extreme temperatures, drought and fungicides and they can survive for a very long period in this dormant stage. Sclerotium forming pathogens including *Sarracenia minor*, *Sclerotinia sclerotiorum* and *Rhizoctonia solani* are widespread in the world and cause many important diseases in a wide range of host plants. Min et al. (2009) found that nano-sized silver particles strongly inhibited fungal growth and sclerotial germination. Patel et al. (2014) during the course of study found that the growth of *Sclerotium rolfsii* was strongly inhibited by the fungicidal effect of silver nano-emulsion.

When the colloidal solution of silver nanoparticle (5000 ppm) diluted in 10 ppm of 500 kg was sprayed on rose plants infected with *Sphaerotheca pannosa var. rosae* and observations made after 2 days revealed that nearly 95% of rose powdery mildew faded out and did not recur for a week (Sharon et al., 2010). Antifungal activity of nanosilver colloids was also studied against *Bipolaris sorokiniana* and *Magnaporthe grisea* (Agarwal and Rathore, 2014). Spherical and nano-sized AgNPs prepared by ultrasound wave induced a reaction between silver nitrate solution and leaf extract of *Centella asiatica*. These AgNPs exhibited antifungal activity against *Macrophomina theicola* B1 fungi, which were isolated from mandarin peels (Boa et al., 2018).

When silver nanoparticles were sprayed on bean leaves, complete suppression of sun-hemp rosette virus was observed. Elbeshehy et al. (2015) reported that when faba bean plants were challenged with bean yellow mosaic virus, and sprayed with silver nanoparticles 24 h post-infection, the results were positive as compared with spray application before infection, or simultaneously at the time of inoculation.

The effective concentration of nano-sized silica silver on suppression of growth of many fungi were studied and it was observed that *Botrytis cinerea*, *Colletotrichum gloeosporioides*, *Curvularia lunata*, *Macrophomina phaseolina*, *Magnaporthe grisea*, and *Rhizoctonia solani* showed 100% growth inhibition at 10 ppm of the nanoparticles of silver-silica by well diffusion assay (Park et al., 2006; Anguilar-Mendez et al., 2011; Krishnaraj et al., 2012), whereas *Bacillus subtilis*, *Azobacter chroococum*, *Rhizobium tropici*, *Pseudomonas syringae* and *Xanthomonas campestris pv. vesicatoria* showed complete inhibition in growth at 100 ppm. These SNPs penetrated and damaged the hyphal wall resulting in plasmolysis of hyphae. Higher concentration (3200 ppm) of silica silver nanoparticles caused chemical injuries on cucumber and pansy plants.

13.5.7 Nano Zinc

Studies to evaluate the potential for combining nano Zn formulations with other existing strategies for improving crop health and managing diseases gave positive and encouraging results (Duffy, 2007).

The antimicrobial activity of Zn nanoparticles to plant pathogens has been explored and reported from several experimental laboratories (Indumathy and Mala, 2013; Rajiv et al., 2013; Hafez et al., 2014), however, maximum reports of inhibition *in vitro* have been observed in the case of bacterial pathogens. Zinkcide™, a zinc-based nano product is used for controlling citrus canker (Young et al., 2017). Grahem et al. (2016) compared two formulations on citrus and found that both Zinkcide™ SG4 and particulate Zinkcide™ SG6 were equally effective in suppressing canker lesions caused by *Xanthomonas citri subsp. citri*. Two other important diseases, citrus scab caused by *Elsinoe fawcettii* and melanose by *Diaporthe citri* on grapefruit were minimized by this Zn-based formulation.

A range of fungal pathogens including *Alternaria alternata*, *Botrytis cinerea*, *Fusarium oxysporum*, *Mucor plumbeus*, *Penicillium expansum*, *Rhizoctonia solani*, *Rhizopus stolonifer* and *Sclerotinia sclerotiorum* (He et al., 2011), as well as root knot nematode (Kaushik and Dutta, 2017), were inhibited by nano Zn particles.

Derbalah et al. (2013) made studies, for two years, and compared foliar applications of ZnO nanoparticles at 500 μg/mL with the conventional fungicide tetraconazole along with additional treatments using silica nanoparticles, diatomaceous earth and six bacterial bioagents to suppress *Cercospora* leaf blight of beet. Nano Zn particle was second only to tetraconazole at increasing leaf dry weight, root yield and sugar content as it significantly reduced disease severity. Similarly, Dimkpa et al. (2013) also confirmed these findings in the interaction of the Zn nanoparticle with the biocontrol agent *Pseudomonas chlororaphis* O6 and *Fusarium graminearum in vitro.*

The ZnO nanoparticle having a concentration ranging from 25 to 200 ppm was added to 95% chitosan solution. The sterilized fresh fig fruits of uniform size were dipped in 100 ml of ZnO solution. Application of Zinc oxide nanoparticles to fig fruits delayed ripening, increased the firmness, slowed the color change and reduced the growth of microbes compared with uncoated fruits (Sagili et al., 2018). These findings demonstrate the potential for combining nano Zn-based particles as viable alternative to conventional strategies (Sardella et al., 2017).

13.5.8 Other Nanoparticles

There is an ardent desire to explore the role of other metal nanoparticles in plant disease management as well. Despite the use of sulfur as a fungicide in the past, only a few reports are available describing the effect of S nanoparticles. In vitro experiments were conducted to compare and assess the antimicrobial activity of elemental S, sulfur 80WP and S nanoparticle, using detached leaf assay technique (Gogoi et al., 2013). The results demonstrated that nano-S performed better than the fungicides and inhibited conidial germination and cleistothecia became sterile with distorted appendages. Similar results were observed by Rao and Paria (2013) as it was found that nano-S could inhibit *A. niger*, *F. solani*, *F. oxysporum* and *Venturia inaequalis*.

Adisa et al. (2017) investigated the effect of nano-Ce on diseased plants of tomato infected with *F. oxysporum* f.sp. *lycopersici*. They found that nano-Ce reduced disease severity by 48% and the total biomass of the plant also increased.

Nano-Ti is another metal particle that is getting the attention of researchers due to its photocatalytic activity (Paret et al., 2013a). Paret et al. (2013b) developed a light-activated nano-TiO_2/Zn nanoparticle composite to suppress bacterial leaf spot on rose caused by *Xanthomonas* sp. Corredor et al. (2009) applied iron nanoparticles coated with carbon to pumpkin plants for treating specific plants that were infected. Boxi et al. (2016) also supported disease suppressive effect of nano-TiO_2.

13.5.9 Chitosan Nanoparticles

Chitosan is another popular nanoparticle that has favorable biological properties such as biocompatibility, biodegradability and non-allergenicity. It is antimicrobial with low toxicity to animals and humans (Cota et al., 2013). Chitosan is hydrophobic in nature having low solubility in aqueous media; therefore, it is commonly mixed with co-polymer, inorganic or organic to improve its solubility (Kashyap et al., 2015). Chitosan with reactive amine and hydroxyl groups allows modification, graft reaction and ionic interactions enabling improvement of chitosan properties. Chitosan adheres well to the epidermis of stem and leaves, prolonging the contact time and facilitating the uptake of the bioactive molecules (Malerba and Cerana, 2016). Chitosan particles have shown antifungal properties against *Fusarium* crown, root rot in tomato, *Botrytis* bunch rot in grapes and *Pyricularia grisea* in rice (Kashyap et al., 2015). Chitosan nanoparticles induce viral resistance in plant tissues by protecting them against infections caused by mosaic virus of alfalfa, snuff, peanut, potato and cucumber (Chirkov, 2002). Chitosan has also shown effectiveness against *Aphis nerii* (oleander aphid), *Spodoptera littoralis* (cotton leaf worm), root knot nematode (*Meloidogyne javanicum*) and nymphs of pear psylla (*Cacopsylla pyricola*).

Chitosan-silver nanoparticle composite demonstrated antifungal properties against mango anthracnose (Chowdappa et al., 2014). When low water-soluble fungicides such as pyraclostrobin and kaempferol, were loaded onto chitosan-lactide and lecithin/chitosan co-polymer nanoparticles, respectively, effective disease suppression of *Fusarium oxysporum* was observed (Ilk et al., 2017).

13.5.10 Solid Lipid Nanoparticles

These nanoparticles are composed of lipids that are solid at room temperature and provide a matrix to entrap lipophilic active molecules without the use of organic solvents; due to decreased mobility of the active in the solid matrix, solid lipid nanoparticles (SLNs) can provide controlled release of various lipophilic components (Ekambaram et al., 2012; Borel and Sabliov, 2014). The major drawback being is their low loading efficiency and the leakage of active ingredients during storage (Tamjidi et al., 2013).

13.5.11 Layered Double Hydroxides

These are the clays that form hexagonal sheets entrapping layers of active molecules in the interlayer spaces (Derbalah et al., 2013). Positively charged delaminated layered double hydroxides (LDHs) lactate nanoparticles have been shown to facilitate the transport of biologically active materials across plant cell wall barriers (Boa et al., 2016) and these nano-LDH particle breakdowns under acidic conditions.

13.5.12 Nanoparticles and RNAi

The discovery of the RNAi pathway has heralded a new and innovating approach for managing pests and pathogens, thus research into the topical application of dsRNA has emerged as a highly appealing alternative (Worrall et al., 2018).

RNAi, a conserved eukaryotic mechanism, is involved in the growth, development and host defense against viruses and transposons that can also be hijacked to target insects, fungi, viruses and weeds (Borges and Martienssen, 2015). In plants, RNAi is triggered by dsRNA, which is processed into small-interfering RNA (siRNA) by Dicer-like (DCL) enzymes. These siRNAs are incorporated into an RNA-induced silencing complex (RISC); siRNAs then direct the RISCs through base pairing to degrade the pathogen RNA by preventing it from being used as a translation template (Baulcombe et al., 2004)

To date, nanoparticles acting as carriers of RNAi-inducing molecules have been targeted against viruses and aphids (Thiaru et al., 2017). Mitter et al. (2017a) successfully afforded viral protection against cucumber mosaic virus (CMV) and pepper mild mottle virus (PMMoV) using plants sprayed with dsRNA loaded onto LDH nanoparticles, called BioClay. Significantly, a single spray of BioClay protected plants for 20 days after application on sprayed, and on newly emerged, unsprayed leaves, while plants sprayed with naked dsRNA succumbed to viral infection. Li-Byarlay et al. (2013) used siRNA loaded onto perfluorocarbon nanoparticles to target three different aphid species viz., *Acyrthosiphon pisum*, *Aphis glycines* and *Schizaphis graminum*.

It was observed that the nanoparticle-loaded siRNA had a significantly higher gene knock down than siRNA alone. Each report shows the potential of nanoparticle-facilitated, topical application of RNAi, due to the potential to target a wide range of fungal pathogens (Wang et al., 2016), insect pests (Ghosh et al., 2017; San Miguel and Scott, 2016) and plant viruses (Mitter et al., 2017b) with RNAi.

Although exogenous application of RNAi-inducing molecules for crop protection holds a strong appeal over pesticides, but due to its reduced toxicity, topical application of RNAi still faces its own hurdles such as short longevity due to environmental degradation and difficulties with site-specific uptake by the targeted pest. Since its discovery, the RNAi pathway has emerged as a powerful tool to combat plant pests and pathogens by genetic modification, however, as genetically modified organisms are controversial and highly regulated in most countries, it requires more detailed studies.

13.5.13 Green Nanoparticles

Green synthesis is defined as the use of environmentally compatible materials such as bacteria, fungi and plants in the synthesis of nanoparticles (Jain and Kothari, 2014). Conventional methods used in preparing metal nanoparticles are associated with environmental risks of toxic and hazardous chemicals used for synthesis purposes, instability of nanoparticles and production of hazardous byproducts besides high-energy input and costly downstream processing. Therefore, opting for ecofriendly, non-toxic and sustainable methods for fabricating myriads of nanoparticles are the current areas of research and interest. Novel methods of ideally synthesizing NPs are thus thought as formed at ambient temperatures, neutral pH, low costs and in an environmentally friendly fashion.

The superiority of biological methods for nanoparticle synthesis is gaining impetus because of several advantages: the whole process of synthesis is rapid, stable and requires a wide range of non-toxic biomolecules of low cost; no hazardous byproducts are formed; additional reduction or stabilizing agents are not required; and a biocompatible coating on the nanoparticle's surface which provides additional active surface area for interaction in the biological environment. Moreover, the shape, size and disparity of nanoparticles can be regulated by modifying and optimizing the conditions such as pH, temperature and salt concentration of the reaction mixture during biosynthesis (Iravani et al., 2014). Biosynthesized nanoparticles (Figure 13.3) offer a viable, efficient and

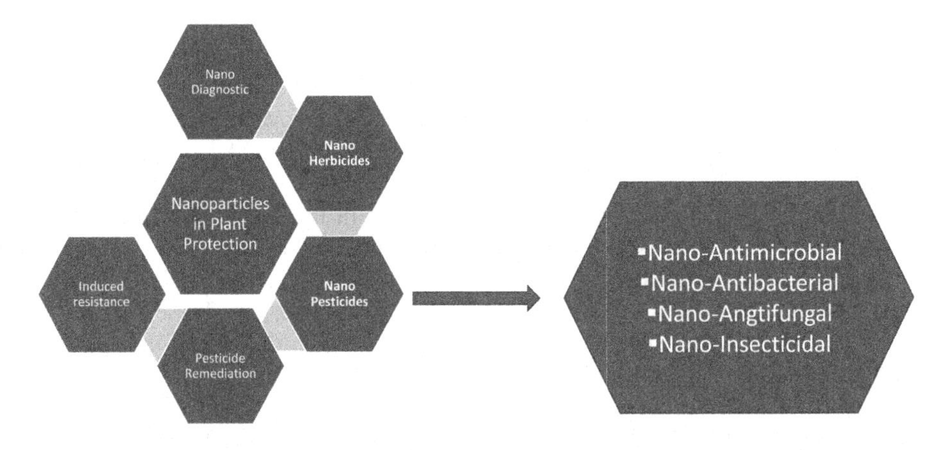

FIGURE 13.3 Nanoparticles in plant protection.

environmentally friendly option for plant growth promotion, stress tolerance and plant disease management (Hussain et al., 2016).

Biosynthesis routes employ biological materials such as plant extracts, sugars, polyphenols, vitamins and microorganisms which are used as reducing and capping agents in the synthesis process leading to more stabilized and biocompatible nanoparticles with higher longevity (Kharissova et al., 2013). Most importantly, the bio-fabricated nanoparticles exhibit relatively lower toxicity compared with chemically produced nanoparticles (Órdenes-Aenishanslins et al., 2014).

Microorganisms usually exhibit a process called bio-reduction, which involves the accumulation of metallic ions in order to reduce their toxicity (Mishra et al., 2016). Microorganisms reduce intracellularly with the help of various reducing ions present either inside the cell and on the cell wall, or extracellularly by different metabolites. Nanoparticles formed from microorganisms through nucleation and surface growth could be entrapped by the additional surface (capping layer), often exhibiting excellent stability (Singh et al., 2018b). Plants also possess the reducing capability because of various flavonoids, proteins and water-soluble biomolecules. The stability and biocompatibility of green nanoparticles correspond to their capping layer, which usually forms during the synthesis of biogenic nanoparticles, and originates from the corresponding biological extracts used for synthesis. This layer affects the biological activity of nanoparticles and is useful in long-term stability.

Hence, nanoparticle synthesis from biological agents such as actinomycetes, algae, bacteria, fungi, plant extracts and oils has been exploited (Singh et al., 2016). Silver nanoparticles synthesized extracellularly by *Alternaria alternata* were found to cause significant enhancement in the antifungal action of triazole fungicide fluconazole against *Candida albicans*, *Phoma glomerata* and *Trichoderma* sp. (Gajbhiye et al., 2009).

Raliya et al. (2014) demonstrated positive effect of biosynthesized MgO nanoparticles using *Aspergillus flavus* on cluster bean (*Cyamopsis tetragonoloba*). Biosynthesized silver nanoparticles (AgNP) showed strong antifungal activity against spot blight of wheat caused by *Bipolaris sorokiniana* (Mishra et al., 2014b). Apart from this, many earlier studies have confirmed the antimicrobial activity of biosynthesized AgNPs against a broad range of phytopathogens pointing toward their exciting possibilities in agriculture (Krishnaraj et al., 2012; Gopinath and Velusamy, 2013; Lee et al., 2013; Paulkumar et al., 2014). Although a large number of microorganisms are used to synthesize green NPs, fungi, mainly *Verticillium* sp., *Aspergillus flavus*, *Aspergillus fumigatus*, *Phanerochaete chrysosporium* and *Fusarium oxysporum* are considered to be the most efficient systems for the biosynthesis of metal and metal sulfide containing NPs (Panpatte et al., 2016; Kitching et al., 2015). Similarly, AgNPs synthesized from *Serratia* sp. BHU-S4 were also found to inhibit melanin biosynthesis genes in *B. sorokiniana* (Mishra and Singh, 2015b). Additionally, ZnO nanoparticles synthesized from *Aspergillus fumigatus* showed significant improvement in overall plant health along with enhancement in rhizosphere microbial population, acid phosphatase, alkaline phosphatase and phytase activity in cluster bean rhizosphere (Raliya and Tarafdar, 2013). Furthermore, Raliya et al. (2015) found stimulating impact of biosynthesized TiO_2 nanoparticles using *Aspergillus flavus* on plant growth of *Vigna radiata* and the rhizosphere microbial population.

Microorganisms, such as diatoms, *Pseudomonas stutzeri*, *Desulfovibrio desulfuricans* NCIMB 8307, *Clostridium thermoaceticum* and *Klebsiella aerogens* are used to synthesize silicon, gold, zinc sulfide and cadmium sulfide nanoparticles, respectively.

Biogenic AgNPs obtained from *Brevibacterium frigoritolerans* DC2, *Sporosarcina koreensis* DC4 and *Bhargavaea indica* DC1 showed antimicrobial activity against *Vibrio parahaemolyticus*, *Salmonella enterica*, *Bacillus anthracis*, *Bacillus cereus*, *Escherichia coli* and *Candida albicans* (Singh et al., 2015).

In addition, these nanoparticles showed enhancement in the antimicrobial efficacy of conventional antibiotics such as lincomycin, oleandomycin, vancomycin, novobiocin, penicillin G and rifampicin when applied together. Thus, the findings suggest that combining the current antibiotics with green metallic nanoparticles can be further helpful for enhancing their antimicrobial activity. Moreover, a comparative study between biological and chemical nanoparticles demonstrated that biological nanoparticles exert a higher antimicrobial effect than the chemically synthesized nanoparticles. The biological synthesized nickel nanoparticles from *Desmodium gangeticum* are more monodispersed and have higher antioxidant, antibacterial and biocompatible activities in LLC PK1 (epithelial cell lines) than chemically synthesized nanoparticles. Specifically, in terms of antibacterial activity, they tested both the nanoparticles against *S. aureus*, *Klebsiella pneumonia*, *P. aeruginosa*, *Vibrio cholerae* and *Proteus vulgaris*, and found that chemically synthesized nickel nanoparticles were not at all active against *K. pneumonia*, *P. aeruginosa* and *P. vulgaris*, whereas biological nanoparticles showed antimicrobial activity against these microorganisms. For *S. aureus*, chemical nanoparticles were less active than the biological ones. It was also observed that biologically synthesized zinc nanoparticles have more antimicrobial potential against *Salmonella typhimurium* ATCC 14028, *B. subtilis* ATCC 6633 and *Micrococcus luteus* ATCC 9341 compared with chemically synthesized zinc nanoparticles (Singh et al., 2018b).

Sabir et al. (2014) also proposed that application of ZnO NPs can revolutionize the agricultural sector and could solve the current problem of food demand due to their antimicrobial and fertilizer action, especially if biogenic synthesis of these nanoparticles is considered.

Plants are said to be nature's "chemical factories". Plant-mediated synthesis of nanoparticle is the green chemistry approach that connects nanotechnology with plants. Among biological alternatives, plants and their extracts seem to be the best option. They are cost efficient and require low maintenance. For example, *Medicago sativa* and *Sesbania* plant species are used to formulate gold nanoparticles. Likewise, inorganic nanomaterials made of silver, nickel, cobalt, zinc and copper can be synthesized inside live plants, such as *Artemisia*, *Brassica juncea*, *Cannabis sativa*, *Helianthus annuus* and *Medicago sativa* (Ghormade et al, 2011; Irvani, 2011; Ali et al., 2015; Singh et al., 2018b).

Numerous studies elaborate on the use of plant extracts to synthesize AgNPs with significant antimicrobial activities. Leaf extracts of *Acalypha*, *Murraya*, *Ocimum*, *Solidago* and *Xanthium*; seed extracts of *Acacia* and *Macrotyloma*; root extracts of *Trianthema*; stem extract of *Ocimum sanctum* and fruit extracts of banana (peel) and *Carica papaya* were shown to have the potential of silver nanoparticle synthesis. A concentration of 6.25 μg/mL inhibited the growth of target pathogens (Moodley et al., 2017). Copper nanoparticles (CuNPs) obtained from *Sida acuta* showed antimicrobial activity against *Escherichia coli*, *Proteus vulgaris*, and *Staphylococcus aureus* (van Viet et al. 2016).

Jagana et al. (2017) carried out investigations on the management of banana anthracnose by using green nanoparticles. They used different metal nanoparticles synthesized from leaf extracts of *Azadirachta indica* and *Trachyspermum ammi* against *Colletotrichum musae*. In all the treatments, significant inhibition of spores could be observed. Zinc oxide also revealed its antibacterial activity against *S. aureus*, *E. coli*, and *P. aeruginosa*.

Essential oils with fungicidal properties (Solgi et al., 2009) evaporate too quickly for large-scale commercial use. Janatova et al. (2015) successfully incorporated five individual essential oil components into MSN to show higher antifungal activity against *Aspergillus niger* when compared with the bulk ES component. Similar results were obtained using SLNs to stabilize *Zataria multiflora* essential oil providing protection against six fungi. Spherical and nano-sized AgNPs were synthesized by inducing a reaction between silver nitrate solution and the leaf extract of *Centella asiatica*. These synthesized nanoparticles of silver exhibited antifungal activity against *Macrophomina theicola* B1 fungi (Boa et al., 2018).

The above-mentioned studies confirmed that agricultural applications of biosynthesized nanoparticles provide new insight on precision farming technology. As agricultural nanotechnology develops, the potential to provide a new generation of pesticides and other active green particles for plant disease management will greatly increase. Despite the fact that biogenic metallic nanoparticles are biocompatible in nature with high stability and amenable for biomedical applications, a balance between price, process and scalability is still a considerable challenge. Especially for microorganisms involved in biogenic nanoparticles production, there is a requirement of sophisticated instruments throughout the process for the maintenance, production and purification of nanoparticles. These heavy instruments required for the complete process of nanoparticle production and purification make the methodology comparatively expensive. In the case of plants, the requirement of natural resource management, which includes plant culturing and maintenance, is an important issue that needs to be addressed. However, the advantages of biogenic metallic nanoparticles over physiochemically obtained nanoparticles cannot be overlooked for future research and commercialization in the field of antimicrobial applications. Moreover, there is a growing interest in studying the fate, transport and toxicity of biosynthesized nanoparticles and hence more attention should be given in this direction (Khadija et al., 2016).

13.6 HOW NANOPARTICLES WORK

In today's world, nanomaterials are being used in various sectors like agriculture, biomedical, cosmetics, electronics, energy production, pharmaceuticals and telecommunications. Such an increase in the production of nanomaterial-based products and their enormous use in various fields further raises the possibility of their interaction with living organisms and the environment (Rawat et al., 2018). Various studies have reported the mode of action of NPs in plants, animals and microbial systems (Du et al., 2017). However, the effect of NPs is governed by various factors like size, shape, dose, concentration and type of NPs, the developmental stages and the target plant/animal species and the duration of exposure to NPs. Higher concentrations of NPs generated cellular toxicity while optimized use of NPs augmented the growth of plants. Moreover, use of NPs is evident as a potential antimicrobial agent as metals act against pathogens through several different methods (Figure 13.4), such as direct attachment to the cell surface, disruption of cell membrane and membrane potential, damage of cellular components, generation of oxidative stress and inhibition of proteins/enzymes (Irvani et al., 2014).

13.6.1 Action of Nanoparticles Mediated ROS Toxicity

Plants control reactive oxygen species (ROS) effectively and use it beneficially as a signaling molecule to control specialized processes such as plant growth, defense, hormonal signaling and development (Ahmad et al., 2010). These partially reduced or activated derivatives of oxygen are highly reactive and toxic, and can lead to the oxidative destruction of cell components.

Recently, it has been observed that NPs have great potential to elevate ROS levels in plants, which in turn contributes significantly to their cytotoxic effects. The induction of ROS levels by NPs leads to the activation of a defense pathway to combat the oxidative stress. Hence the enhanced levels and activity of the components in the anti-oxidative system is considered as a reliable indicator of the elevated ROS levels/oxidative stress caused by NPs. H_2O_2 generation, MDA content and

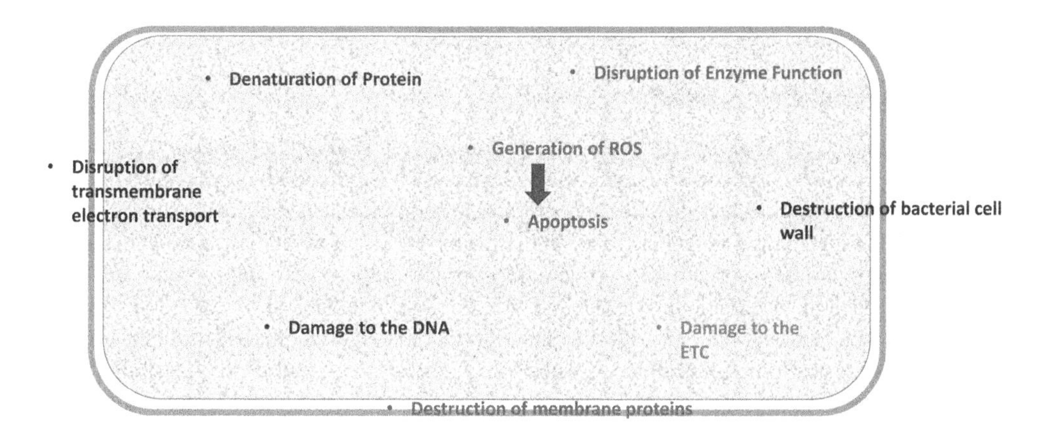

FIGURE 13.4 Mode of action of nanoparticles.

activities of CAT, SOD and APX are the most frequently detected indicators of NP-induced oxidative cytotoxicity. Whether ROS acts as damaging, protective or signaling factors depends on the delicate equilibrium between ROS production and scavenging at the proper site and time. When the rate of ROS generation overweighs the rate of ROS scavenging, it leads to a harmful imbalance in the anti-oxidative system of the plant. This imbalance (often called as oxidative burst) results in widespread damage in cells including oxidative damage to vital biomolecules such as nucleic acids and proteins, per oxidation of membrane lipids, and also the activation of the programmed cell death (PCD) pathway. The consequential disintegration of membranes can facilitate the entrance of metals and NPs. A further increase in membrane permeability is caused by the action of thiobarbituric acid reactive species (TBARS), induced by the NPs (Arruda et al., 2015). A destabilized membrane causes reduced integrity of the cellular contents and can lead to electrolyte leakage; finding their way into the cell, NPs redefine the cellular environment, causing major alterations in the persisting physiological and biochemical conditions.

The imbalance in the cellular redox system leads to multiple impacts on the physiology of the plant. Two predominant effects are an increase in the intracellular ROS load and reduction in photosynthetic capacity of the cell. The maintenance of optimal cellular ROS level is closely related to the proper functioning of photosynthetic apparatus, electron transport chain and photorespiration. Hence, the increased ROS load due to NP exposure can have indirect but significant implications on these inevitable processes, imposing disadvantageous effects on the photosynthetic capacity of the plants (Deng et al., 2014; Khan et al., 2017). Mitochondrial dysfunction and nuclear and organellar DNA damage can be considered as direct effects.

The most studied NPs for their ability to induce oxidative stress on plants are metal NPs such as CeO_2, CuO, TiO_2, ZnO, Al_2O_3 and Fe_2O_3 (Du et al., 2017). Various studies with CeO_2 on different species suggest that CeO_2 exposure can cause an increase in the activities of CAT, SOD and APX (Rico et al., 2013; Morales et al., 2013; Du et al., 2015). At the same time, very high concentrations of CeO_2 can be inhibitory to the activities of these enzymes (Gui et al., 2015). In a study, dose-dependent increase and decrease in H_2O_2 content were respectively observed in rice seedlings when exposed to higher and lower concentrations of CeO_2 (Rico et al., 2013). Several studies indicate that an elevated level of protectants and increased activity of enzymes such as APX, CAT, POD, SOD, GR and so on are caused by CuO NPs on various plants (Da Costa and Sharma, 2016; Dimkpa et al., 2012; Nair and Chung, 2015). On the other hand, altered activity/levels of enzymatic and non-enzymatic antioxidative agents were observed in many other studies (Wang et al., 2016b; Rajeshwari et al., 2015). These reports are a general reflection of the potential of different metal NPs to induce remarkable changes in the defense system of the plants against oxidative stress. Although all the studies generally point out an enhanced level of ROS and antioxidative components, most of the studies with high concentrations have suggested toxic effects on plants.

13.6.2 Genotoxic Effects of NPs

Exposure of plants to NPs induces different kind of damages on DNA right from the sequence level to the chromosome level. Exposure of cells to AgNPs causes a significant increase in chromosomal aberrations, micronucleus induction and a reduction in mitotic index (Patlolla et al., 2012). Presence of micronuclei (at interphase) is a manifestation of chromosome fragmentation that occurred in the previous cell cycle (Ghosh et al., 2010). A decrease in mitotic index (MI) and increase in chromosomal aberrations in meristematic regions are considered as reliable indications of cytotoxic effects. Aberrations such as chromatid breaks, isochromatid breaks, acentric fragments, minutes, translocations and gaps were predominantly induced when *Vicia faba* was exposed to AgNPs (Patlolla et al., 2012). AgNPs are also reported to cause reduction of MI and increase in chromosomal aberrations viz., chromatin bridge, stickiness, disturbed metaphase, multiple chromosomal breaks and cell disintegration in *Allium cepa* (Kumari et al., 2009). Exposure to single-walled carbon nanotubes (SWCNTs) resulted in chromatin condensation in *Arabidopsis* (Shen et al., 2010).

Lopez-Moreno et al. (2010) reported the genotoxic impact of NPs in edible crops, that showed CeO_2 NPs could induce DNA damage in soybean (*Glycine max*). A study in *Zea mays* and *Vicia narbonensis* suggested that a short-term exposure to TiO_2 NPs delays germination process considerably (Castiglione et al., 2011), justified by a reduced mitotic index observed. A reduction in MI suggests a reduction in cell division, which might be resulting in the delay of germination. This is further supported by a concentration dependent increase in chromosomal aberrations and micronuclei formation. Wang et al. (2011) have suggested that TiO_2 NPs can disrupt microtubular network in *Arabidopsis*. Conversely, a similar brief exposure to nano TiO2 caused a stimulatory effect on the mitotic activity in the root tips of *Allium cepa* (Klancnik et al., 2011). An increase in duration of exposure eliminated this stimulatory effect but did not cause any genotoxicity. However, the delayed germination imposed by TiO_2 (Castiglione et al., 2011) also cannot be generalized as a universal effect of NPs as studies with seven different metal NPs on rice and maize viz. TiO_2, SiO_2, CeO_2, Fe_3O_4, Al_2O_3, ZnO and CuO showed no undesirable effect on germination rate (Yang et al., 2015). This finding was supported by the observations by Lopez-Moreno et al. (2010) on soybean. Yet both these studies have identified a genotoxic effect of the respective NPs. Demir et al. (2014) used root meristem cells of *Allium cepa* to study the genotoxic effects of ZnO and TiO_2 NPs. With the help of modified comet assay, they found a dose-dependent impact of ZnO and a high-concentration effect of TiO2 on the integrity of genetic material. DNA damage caused by TiO_2 in *A. cepa* was correlated with a reduction in its root growth (Ghosh et al., 2010; Pakrashi et al., 2014). It can be inferred from these studies that a strong species-dependency operates in concert with the genotoxic potential of NPs. The morpho-physiological manifestations of these genotoxic effects may also vary with the species exposed.

In addition to the direct damages to the integrity of genomic DNA by NPs, several studies have considered abnormalities in overall gene expression patterns as a part of genotoxic effects of NPs. According to Mustafa and Komatsu (2016), exposure to NPs causes wide alterations in the transcriptome and proteome by changing the levels of genes involved in antioxidant defense and detoxification, related to stress signaling and energy metabolism, levels of phytochelatins and metallothioneins and cell death pathways. According to a microarray-based study conducted by Kaveh et al. (2013), up-regulated genes were primarily associated with the response to metals and oxidative stress (e.g., vacuolar cation/proton exchanger, superoxide dismutase, cytochrome P450-dependent oxidase, and eroxidase), while down-regulated genes were more associated with the response to pathogens and hormonal stimuli, for example, auxin-regulated genes involved in organ size (ARGOS), ethylene signaling pathway and systemic acquired resistance (SAR) against fungi and bacteria.

13.7 APPLICATION OF NANOTECHNOLOGY IN AGRICULTURE

Sustainable agriculture entails a minimum use of agrochemicals that can eventually protect the environment and conserve different species from extinction. To address the increasing challenges of sustainable production and food security, significant technological advancements and innovations have been made in recent years in the field of agriculture (Dwivedi et al., 2016).

Over the last two decades, a significant amount of research has been carried out on nanotechnology, emphasizing its numerous applications in agriculture sectors (Lv et al., 2018). Moreover, the major concern in agricultural production is to enable accelerated adaptation of plants to progressive climate change factors, such as extreme temperatures, water deficiency, salinity, alkalinity and environmental pollution with toxic metals without threatening existing sensitive ecosystems (Shang et al., 2019). Notably, nanomaterials enhance the productivity of crops by increasing the efficiency of agricultural inputs to facilitate site-targeted controlled delivery of nutrients, thereby ensuring the minimal use of agri-inputs.

Nanotechnology represents a new frontier and is anticipated to become a major thrust in modern agriculture by offering potential applications (Figure 13.5). This integrated approach has great potential to cope up with global challenges of food production, sustainability climate changes and security. These magical green particles can find many uses in agricultural science as summarized below.

One such use is as diagnostic probes. These small metal particles achieve certain characteristics, which make them suitable for development as diagnostic probes (Sharon et al., 2010). These properties are structural sturdiness in spite of atomic granularity and enhanced or delayed particle aggregation depending on surface modification, large surface to volume ratio, strong affinity to the target, enhanced photoemission, high electrical and heat conductivity and improved surface catalytic activity (Banik and Sharma, 2011).

Quantum dots have emerged as an important tool for the detection of a specific biological marker in the medicinal field with extreme accuracy. They have been used in DNA detection, cell tracking, cell labeling and *in vivo* imaging. These dots are spherical to rod like, fluorescent, crystalline particles, few nm in diameter of semiconductors whose excisions are confined in all three spatial dimensions (Sharon and Sharon, 2008).

Carbon nanosensors as electrodes have the potential to be developed as an electro chemical sensor to detect pesticide residue in and over the plant surface. Though, not so far observed in plant disease diagnosis, it is applicable to animal diseases (Shastry et al., 2010).

Nanobiosensors for precision farming in early and beforehand detection of infection in non-symptomatic plants followed by targeted delivery of treatment would be an essential component (Fang and Ramasamay, 2015). In addition, the development and exploitation of nanosensors in precision farming to measure and monitor crop growth, soil conditions, diseases, uses and penetration of agrochemicals and environmental pollution have substantially improved human control of soil and plant health, quality control and safety assurance, contributing much to sustainable agriculture and environmental systems (Balaure et al., 2017). Nugaeva et al. (2005) successfully demonstrated the use of micro-mechanical cantilever arrays for detection of spores of *Aspergillus niger* and *Saccharomyces cerevisiae*. This biosensor detected the target fungi in the range of 10^3–10^6 cfu ml^{-1} in the investigations made by them. Proteins like concanavalin A, immunoglobulin G or fibronectin, having different affinities to bind to the molecular structures present on fungal cell surface, were surface grafted on micro-fabricated uncoated as well as gold coated silicon cantilevers. Spore germination and immobilization of the test fungi led to a shift in resonance frequency which was measured by dynamically operated cantilever arrangements that took less time in contrast to several days in conventional methods.

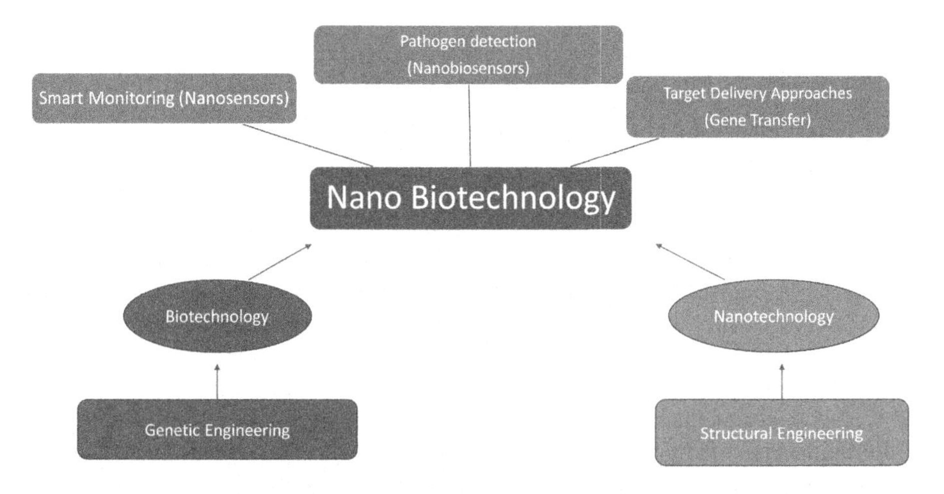

FIGURE 13.5 Applications of nanotechnology in agriculture.

Fluorescent nanosilica impregnated with tris 2,2'bipyridyl dichlororuthenium (II) hexahydrate (Rubpy), an organic dye probe has potential for rapid diagnosis of plant diseases. These nanoparticles conjugated with the secondary antibody of goat anti-rabbit lgG were used for detection of Xanthomonas axonopodis pv. vesicatoria causing bacterial spots on solanaceous plants.

13.7.1 Nanofabrication

Various infection processes and behaviors of pathogens inside the host could well be understood by artificially creating plant parts such as stomata and xylem with the help of nanofabrication techniques. Meng et al. (2005) studied *Uromyces appendiculatus* (rust diseases in beans), *Colletotrichum graminicola* (anthracnose in corn) and *Xylella fastidiosa* (xylem restricted bacterium causing Pierce's disease of grape wine). It would help in understanding information beforehand on the penetration of fungi or vascular characters to obstruct the movement of vascular pathogens like bacteria and fungi. Such studies will benefit plant breeders looking for specific stomatal characteristics to prevent entry inside the host through the stomata or leaf surface to prevent appressorium formation.

13.7.2 Targeted Drug Delivery

Considerable research is being directed toward developing biodegradable polymeric nanoparticles for drug delivery and tissue engineering in view of their applications in controlling the release of drugs, stabilizing labile molecules from degradation and site specific targeting (Shang et al., 2019). Nanocapsules are vesicular systems in which the drug is confined to a cavity surrounded by polymer membranes, whereas nanospheres are matrix systems in which the drug is physically and uniformly dispersed. They can be tailor-made to achieve both controlled drug release and disease specific localization by turning the polymer characteristics and surface chemistry. These systems can provide cellular or tissue delivery of drugs, improve bioavailability, sustain the release of drugs or solubilize drugs for the delivery system. An interesting and fascinating area using nanoparticles having a magnetic core consisting of Fe nanoparticles was used to develop a smart treatment delivery system. Carbon coated Fe nanoparticles were used in the treatment of *Cucurbita pepo* in vitro (González-Melendi et al., 2008).

Liposomes, the spherical vesicles that have at least one phospholipid bilayer, and dendrimers, the branched tree-like nanoparticles composed of a central core with attached multiple functional groups are being utilized as vehicles for transporting materials to tissues previously unreachable. Chemical agents and DNA conjugate to the dendrimer and are delivered to the diseased plant part via enhanced permeability.

Syngenta (world-leading agri-company) uses nanoemulsions in its growth regulator Primo MAXX®, which if applied prior to the onset of stress such as heat, drought, disease or traffic can strengthen the physical structure of turfgrass, which allows it to withstand ongoing stresses throughout the growing season. Another encapsulated product Karate® ZEON from Syngenta delivers a broad control spectrum on insecticide which breaks open on contact with leaves. However, the encapsulated product "gutbuster" only breaks open to release its contents when it comes into contact with alkaline environments, such as the stomach of certain insects (Syngenta's US Patent No. 6,544,540). The ultimate aim is to tailor these products in a controlled release in response to different signals, for example, magnetic fields, heat, ultrasound and moisture. New research also aims to make plants use water, pesticides and fertilizers more efficiently to reduce pollution and to make agriculture more ecofriendly.

13.7.3 Nanoformulations of Pesticides

Chemical pesticides are commonly used in agricultural fields to control various pests and weeds. However, due to the adverse effect on the environment and the threat to living beings upon prolonged exposure to these hazardous chemicals, various alternative techniques have been developed to manage diseases in the past. Recently, bimetallic particles, metal oxides, and carbon and halloysite nanotubes have been used for detection, degradation and removal of pesticides (Ratwani et al., 2014).

Major chemical companies are now trying to make potential pesticides at the nanoscale to increase the effectiveness of pesticides. One of the most efficient nanomaterials is aluminosilicate nanotube; when sprayed on plant surfaces, the pests easily take it up, and consumes the pesticide-filled nanotubes and is killed. Mesoporous silica nanoparticles have the ability to deliver DNA and drugs into plants. Silver and TiO2 particles have an antimicrobial effect (Kumar et al., 2017).

"Nano-5" is projected as a natural mucilage organic solution to control several plant pathogens and pests besides improving crop productivity. Several bacterial pathogens such as *Azotobacter chroococcum, Bacillus subtilis, Rhizobium tropici, Pseudomonas syringae, Xanthomonas campestris pv. vesicatoria* and fungal pathogens viz., *Botrytis cinerea, Colletotrichum gloeosporioides, Magnaporthe grisea, Pythium ultimum* and *Rhizoctonia solani* were effectively managed by nano-sized silica silver particles in vitro.

Fungicide formulation containing nanoparticles as Banner MAXX (active ingredient propiconazole), ApronMAXX (active ingredient fludioxonil) and RFC for seed treatment; Primo MAXX (chlorophyll derivative of cyclohexenone) has been developed as a plant growth promoter but also aids in withstanding abiotic and biotic stresses (Gogoi et al., 2009).

Nano-Gro, launched by Agro Nanotechnology Corporation, Florida, has shown promising results as it has helped in increasing the yield by 20% to a maximum of 50% in the case of grain yield of sunflowers. The product, prepared by mixing several bio-chemicals, is certified to be purely organic and harmless to plants and soil and is able to eliminate blast disease from infected rice plants. Nano-5 has been found to eliminate many bacterial, fungal and viral diseases.

13.7.4 Nanoparticles as Antimicrobials

In the recent past, metal NPs have been utilized as antimicrobial agents against various microorganisms including multidrug resistant bacteria (Durairaj et al., 2012). The activity and effects of metals can be further tuned through manipulating their size, as metal NPs possess higher surface area to volume ratio in comparison with the bulk metals, enabling them to contact and interact with target molecules in a better way.

Proteomic analysis of Ag NPs treated *E. coli* cells revealed that Ag NPs could penetrate and disrupt the bacterial membranes, leading to a considerable loss of intracellular potassium. Besides this, Ag NPs also inhibit the process of ATP synthesis, where the potential targets could be protein thiol groups (key respiratory enzymes). Ag NPs and Ag^+ ions are both effective against bacteria, whereas Ag NPs act in nanomolar concentrations and Ag^+ ions in the micro-molar ranges (Lok et al., 2006). Ag^+ ions react with thiol groups of proteins and cause uncoupling of respiratory electron transport from oxidative phosphorylation and inhibit respiratory enzymes. They interfere with membrane permeability to protons and phosphate. Interaction of Ag NPs with nucleic acids might cause impairment of DNA replication (Duran et al., 2010). Involvement of the three main reasons for antibacterial activity of biogenic Ag NPs have been demonstrated, i.e., release of Ag^+ ions, physical contact and ROS formation. Release of Ag^+ ions is mainly responsible for antibacterial action of Ag NPs. Beside this, the secondary factors are physical contact and ROS formation. Biogenic NPs might be more benign to the environment in comparison with chemically synthesized NPs (Sintubin et al., 2011).

Gene expression analysis has been performed in *E. coli* against antibacterial effect of Au NPs. Among differently expressed genes, chemotaxis related genes are found most significantly up regulated (increased motility), while most down-regulated genes are related with energy coupled transport, electrochemical gradient and ATP biosynthesis. Au NPs treated *E. coli* cells were marked with severely decreased activity of F-type ATP synthase, lowered ATP levels and consequently weakened metabolism. This was found to be correlated with the collapse of membrane potential in Au NPs treated *E. coli* cells. Moreover, a decline was also recorded in the expression of a catalase (alkyl hydroperoxide reductase, Ahp) that is involved in scavenging low levels of H_2O_2. However, no changes are observed in the expression of genes related to oxidative stress response, SOS response and DNA damage.

Antibacterial property of ZnO nano-fluid is not greatly affected in the presence of stabilizing agents such as polyvinylpyrrolidone (PVP) and polyethylene glycol (PEG) (Zhang et al., 2007). On the contrary, a significant inducing impact of an anionic surfactant i.e., SDS has been recorded on antimicrobial activity of Ag NPs against *Pseudomonas aeruginosa* CCM 3955. Higher surface area to volume ratio of ZnO NPs (12nm) is attributed to more antibacterial activity in comparison with large sized (47 nm) NPs. An *E. coli* cell of 2 μM can accommodate a greater number of ZnO NPs of 12 nm in comparison with ZnO NPs of 47 nm (Padmavathi and Vijayaraghvan, 2016). A similar finding has been reported by other researchers that the size of ZnO NPs and antibacterial activity are inversely related; smaller ZnO NPs possess more surface area in contact to the bacterial cell wall, thus higher antibacterial activity as compared with larger NPs (Raghupati et al., 2011; Pal et al., 2007; Lok et al., 2007; Liu and Hurt, 2010). It has also been reported that smaller sized Cu NPs are able to show higher antibacterial activity against both *E. coli* and *S. aureus*. Moreover, *E. coli* was found to be more susceptible than *S. aureus* to CuO NPs. Antibacterial activity of CuO NPs is mediated through generation of ROS, which eventually triggered oxidative stress (Applerot et al., 2012).

Besides the size of NPs, their morphology also contributes to the antibacterial response. The shape of ZnO NPs can affect the way of their internalization into bacterial cell walls; rods and wired ZnO NPs are more easily internalized in comparison with spherical ZnO NPs (Yang et al., 2009). In addition, the higher the number of polar facets, the more oxygen will interact, which may lead to increased generation of reactive oxygen species that will finally affect the photo-catalytic activity of ZnO (Li et al., 2008). Such photo-catalytic activity of various metal oxides is mainly responsible for the generation of reactive oxygen species that have been a reported cause of antibacterial activity (Prasad et al., 2011).

In addition to the size and shape of NPs, their concentration has also been reported as an important factor for their activity. CaO NPs have shown antibacterial activity against *Staphylococcus epidermidis*, *Pseudomonas aeruginosa* and *Candida tropicalis* in decreasing order. A smaller concentration of CaO NPs was found sufficient for their bactericidal effect against *Staphylococcus epidermidis* (Gram-positive) among all the tested bacterial cultures. A thicker layer of peptidoglycans (negatively charged) makes the Gram-positive bacteria more vulnerable against these positively charged NPs. The antibacterial activity of CaO NPs depends on dose, concentration, treatment time and bacterial species (Roy et al., 2013).

13.7.5 Antiviral Activity of Nanoparticles

Most of the findings on antiviral properties of nanoparticles are observed and studied on human viruses. Analysis through transmission electron microscopy revealed that Ag NPs are able to interact directly with viral particles and it is hypothesized that Ag NPs of 10 nm might bind to viral DNA (a template for RNA synthesis), thus inhibiting the transcription of viral RNA. Ag NPs prevent the initial stage of replication, thus acting as a virucidal agent, which inhibits virus entry into host cells. Ag NPs bind to gp120 (a glycoprotein expressed on the surface of the virus envelope, which is essential for viral entry through interaction with host cell surface receptors) and thus restricts its attachment to the host cell and entry into host cells (Lara et al., 2010). Antiviral activity of Ag NPs and chitosan composites depends on the concentration and size of Ag NPs in the composite. Interestingly, chitosan alone is found unable to exert antiviral activity, suggesting an indispensable role of Ag NPs for antiviral activity of the composite. Complete antiviral mechanism is yet to be understood, however, some clues are observed about interaction of virions and composites (Mori et al., 2013).

13.7.6 Antifungal Activity of Nanoparticles

Ag NPs attack the cell membranes, thereby disturbing membrane potential. TEM analysis revealed that AgNPs create pits on the surface of yeast cells; consequently, pore formation occurs followed by a leakage of ions and other cell constituents, which finally leads to cell death. In addition, the physiological alterations taking place in yeast cells in response to Ag NPs treatment blocked the cell cycle at the G2/M phase in *C. albicans* targeting cellular processes which are essential for regular bud growth. Endo et al. (1997) reported that any kind of inhibition in the process of bud growth is correlated with membrane damage. Therefore, antifungal activity of Ag NPs seems to be due to inhibition of the budding process that possibly has caused damage to membrane integrity. Ag NPs have also shown antifungal activity against plant pathogenic fungi. Antifungal activity of NPs is concentration dependent as it may occur because of a high density of NP solution, which can saturate and adhere to fungal hyphae and causes inactivation of plant pathogenic fungi (Kim et al., 2012).

Antifungal activity of ZnO NPs has been recorded against two postharvest pathogenic fungi, i.e., *Botrytis cinerea* and *Penicillium expansum*. It has been found that ZnO NPs may have two distinct mechanisms of action against these two fungi. In *Botrytis cinerea*, ZnO NPs cause distortion in the cellular structure of fungal hyphae, while in *Penicillium expansum*, ZnO NPs target the reproductive system of fungus via inhibiting the development of conidiospores and conidia (He et al., 2011).

13.8 PACKAGING OF PLANT PRODUCTS

A major problem in food science is determining and developing an effective packaging material. Improved food packaging needs materials that have good strength, barrier properties and stability to heat and cold. These are being achieved using nanocomposite materials. Bayer Polymers have produced a nanocomposite "hybrid system" film "Durethan", enriched with silicate nanoparticles that reduce the entrance of oxygen and other gases and preserve moisture, thus preventing food from spoiling. When this plastic is processed into a thin film and wrapped over food, it does a better job than previous plastics of preventing food from going bad on the shelf and it also helps prevent odors from one food mixing with another. Antimicrobial packaging of edible food films made with cinnamon or oregano oil, or nanoparticles of zinc, calcium and other materials that kill bacteria, is being tried. Green packaging using nanofibers made from lobster shells or organic corn (both are antimicrobial and biodegradable) is also a food safety effort.

13.9 EPILOGUE

In the present scenario of drastic climate change, the threat of food scarcity due to increasing population trends and changing dietary habits means that global agricultural systems are facing numerous, unprecedented challenges. In order to achieve food security, advanced nano-engineering is a handy tool for boosting crop production and assuring sustainability. The synthesis of nanoparticles from biological sources has been proposed as an environmentally friendly and cost-effective alternative to chemical and physical methods.

Nanotechnology offers a feasible option and opens the possibility for a wide variety of biological research and its use at molecular and cellular levels.

Nanotechnology protects the environment indirectly through the use of alternative (renewable) energy supplies, and uses filters or catalysts to reduce pollution and clean-up existing pollutants. The integration of biology and nanotechnology into nanosensors has greatly increased their potential to sense and identify environmental conditions or impairments. Therefore, nanotechnology can not only reduce uncertainty, but also coordinate management strategies of agricultural production as an alternative to conventional technologies.

Nanomaterial engineering supports the development of agricultural production by increasing the efficiency of inputs and minimizing relevant losses. Nanomaterials as unique carriers of agrochemicals facilitate the site-targeted controlled delivery of nutrients with increased crop protection. In addition, nanomaterials offer a wider specific surface area to fertilizers and pesticides. Due to their direct and intended applications in the precise management and control of inputs (fertilizers, pesticides, herbicides), nanobiosensors support the development of high-tech agricultural farms. Nanoparticles for the delivery of drugs or nutrients and for therapy of nearly all pathological suffering of plants are underway. Smart sensors and smart delivery systems will help to combat viruses and other crop pathogens. Nanostructured catalysts will be available "on demand" in doses that will increase the efficiency of pesticides.

New tools with nanodevices capable of efficiently replacing many cellular types of machinery are underway. Nanorobotic devices roaming inside the body can give a plethora of information for detecting and mitigating biological hazards.

In many instances, agro-nanotech innovations offer short-term techno-fixes to the problems faced in modern industrial agriculture. Still, the full potential of nanotechnology in the agricultural and food industry is yet to be realized. It is gradually moving from theoretical knowledge toward the application regime.

In the near future, nanoscale devices could be used to make agricultural systems "smart".

Use of nanotechnology could permit rapid advances in agricultural research, such as reproductive science and technology, early detection of stresses and alleviating stress effects, conversion of agricultural and food wastes to energy and other useful byproducts through enzymatic nanobioprocessing, disease prevention and treatment in plants and animals (Mishra et al., 2017).

13.9.1 Points to Ponder

Nanotechnology, a new area of research, offers a viable solution to various problems encountered in the agricultural and medical area. The application of nanotechnologies in the field of agriculture seems quite promising, however, the potential risks involved are no different than those anticipated from any other industry. Consequently, with growing public concern on nanotoxicity and its direct or indirect environmental impact, considerable attention is required for employing biosynthesized nanoparticles for agricultural purposes.

The ability for these materials to infiltrate the human body is well known, however, no information is available that aims to know about associated risk factors, toxicity levels and the environmental impact of biosynthesized nanoparticles. Use of nanopesticides may create a new kind of contamination in soil and water bodies due to enhanced transport, longer persistence and higher reactivity of particles. Through the rapid distribution of nanoparticles to food products, whether it is in the food itself or part of the packaging, it is anticipated that nanoparticles will come in direct contact with virtually everyone. The main concern is that of the unknown, thus, the implications and regulations of nanotechnology used in food are deeply concerning. Since there is no standardization for the use and testing of nanotechnology, products incorporating nanomaterials are being produced without a check. The environmental group ETC (Action Group on Erosion, Technology and Concentration) stated that "the merger of nanotech and biotech has unknown consequences for health, biodiversity and the environment".

In this context, several regulatory bodies, i.e., USFDA (US Food and Drug Administration), OECD (Organization for Economic Cooperation and Development) and ISO (International Standard Organization) are undertaking challenges in this direction. The legislation on soil is being enforced by the USFDA, while ISO and OECD provide guidelines and suggestions to the regulatory bodies. The OECD and non-OECD countries follow different regulatory laws in the application of nanotechnology in agri/food and feed sectors (Amenta et al., 2015).

The EU regulation, Registration, Evaluation, Authorisation and Restriction of Chemicals (REACH) addresses the use of nanomaterials in food additives, supplements and food contact materials, and in plant protection products (European Commission, 2013).

Globally, only the EU and Switzerland have established nano-specific legislation for agriculture and food/feed sectors, whereas other non-EU countries have non-mandatory frameworks binding with non-legal guidance (Mishra et al., 2017).

The differences in opinion and standardized regulatory laws throughout the globe, as well as proper utilization of nanotechnology for agricultural benefits, are facing difficulties and not flourishing as was once thought they would.

Regulatory policies promising applications of nanotechnology in all areas cannot be overlooked. Hence, at this point, it is desirable to consider and assess the risk factors involved on the basis of which regulatory policies to manage risk should be framed to address biosafety issues.

In this context, the following points should be considered as suggested by Mishra et al. (2017).

- In order to understand the accurate, active and non-toxic dose of nanoparticle concentration, dependent studies in the natural soil system are required. A clear overview of the soil physicochemical characteristics of agricultural fields where nanoparticles are to be applied may help in reducing their risk toward plant and soil biota. This will help in exploring and validating the permissible level of nanoparticle dose within safety limits.
- The experimental design must be set in the natural environment (growing the plants in soil) to give a precise depiction of the environmental impact of nanoparticles.
- Altering the soil environment for improving soil conditions could assist in reducing transport, bioavailability and further toxicity of nanoparticles with a significant positive impact in the agroecosystem.
- An understanding of the transgenerational and trophic chain transfer effects of nanoparticle applications on plants must be included to gain comprehensive knowledge of permissible limits of nanoparticles. Combining the selection of permissible levels together with studying the transgenerational and trophic chain transfer effects could provide adequate safety assessments.
- Future research must be directed toward scrutinizing the ways to evade risk factors associated with nanoparticle usage. Studying nanoparticle synthesis with only a few applications limited to laboratory conditions does not contribute to the complete acceptance of nanotechnology in the agricultural sector.
- Redeeming the environmentally friendly approach of green synthesis of nanoparticles. It is believed that biosynthesized nanoparticles may possess relatively little or no toxicity, and hence, future research must precisely focus on their practical utility. Therefore, the scientific community must work together to improve future research based on a more realistic approach.

While there is no direct evidence of harm to people or the environment at this stage, simultaneous uncertainty and negative perception, vis-à-vis nanotechnological interventions in the agricultural sector, must be taken seriously. There is an ardent need to make extensive efforts in forwarding and improving future research based on recognized knowledge gaps.

Furthermore, there is an enormous scope of research in this underexplored, emerging and challenging area. Therefore, considerable efforts must be devoted to in-depth study on the environmental impact of biosynthesized nanoparticles. Keeping this in mind, it is believed that the meticulous application of biosynthesized nanoformulations in the agricultural system can eventually remove its negative perception.

REFERENCES

Abdellatif, K.F., R.H. Abdelfattah and M.S.M. El-Ansary. 2016. Green nanoparticles engineering on root knot nematode infecting eggplants and their effect on plant DNA modification. *Iran. J. Biotechnol.* 14(4): 250–59.

Adisa, I.O., J.A. Hernandez-Viezcas, W.H. Elmer et al. 2017. Evaluating the role of CeO_2 nanoparticle in the suppression of *Fusarium wilt* disease in tomato plant. In *Paper Presented at the 6th Sustainable Nanotechnology Conference, Los Angeles, CA.*

Agrawal, S. and P. Rathore. 2014. Nanotechnology Pros and cons to agriculture: A review. *Int. J. Curr. Microbiol. Appl. Sci.* 3: 43–45.

Ahmad, P., C.A. Jaleel, M.A. Salem et al. 2010. Roles of enzymatic and nonenzymatic antioxidants in plants during abiotic stress. *Critical Rev. Biotechnol.* 30(3): 161–175. doi: 10.3109/07388550903524243

Ali, M., B. Kim, K.D. Belfield et al. 2015. Inhibition of Phytophthora parasitica and P. capsici by silver nanoparticles synthesized using aqueous extract of *Artemisia absinthium*. *Phytopathology* 105(9): 1183–1190.

Alexendratos, N. and J. Bruinsma. 2012. *World Agriculture Towards 2030/2050: The 2012 Revision. 12-03.* Rome, Italy: FAO.

Amenta, V., K. Aschberger, M. Arena et al. 2015.Regulatory aspects of nanotechnology in the agri/feed/food sector in EU and non EU countries. *Regul. Toxicol. Pharmacol.* 73: 463–467. doi: 10.1016/j.yrtph.2015.06.016

Anguilar-Mendez, M.A., E.S. Martin-Martinez, L. Ortega-Arroyo et al. 2011. Synthesis and characterization of silver nanoparticles: Effect on phytopathogen *Colletotrichum gloesporioides*. *J. Nanopart. Res.* 13: 2525–2532.

Applerot, G., J. Lellouche, A. Lipovsky et al. 2012. Understanding the antibacterial mechanism of CuO NPs: Revealing the route of induced oxidative stress. *Small* 8(21): 3326–3337. doi: 10.1002/smll.201200772

Arruda, S. C. C., A.L.D. Silva, R.M. Galazzi et al. 2015. NPs applied to plant science: A review. *Talanta* 131: 693–705. doi: 10.1016/j.talanta.2014.08.050

Banik, S. and P. Sharma. 2011. Plant pathology in the era of nanotechnology. *Indian Phytopath.* 64(2): 120–127.

Balaure, P.C., D. Gudovan and I. Gudovan. 2017. Nanopesticides: A new paradigm in crop protection. In A.H. Grumezescu (Ed.), *New Pesticides and Soil Sensors*, pp. 129–192. Amsterdam, Netherlands: Academic Press, Elsevier.

Boa, V.-V.Q., L.D. Vuong and L.V. Luan. 2018. Biomimetic synthesis of silver nanoparticles for preparing preservative solutions for Mandarins (Citrus Deliciosa Tenore). *Nono LIFE* 8(1): 1–10

Bao, W., J. Wang, Q. Wang et al. 2016. Layered double hydroxide nanotransporter for molecule delivery to intact plant cells. *Sci. Rep.* 6: 1–9.

Baulcombe, D. 2004. RNA silencing in plants. *Nature* 431: 356–363.

Borel, T. and C. Sabliov. 2014. Nanodelivery of bioactive components for food applications: Types of delivery systems, properties, and their effect on ADME profiles and toxicity of nanoparticles. *Annu. Rev. Food Sci. Technol.* 5: 197–213.

Borges, F. and R.A. Martienssen. 2015. The expanding world of small RNAs in plants. *Nat. Rev. Mol. Cell Boil.* 16: 727–741.

Boxi, S.S., K. Mukherjee and S. Paria. 2016. Ag doped hallow TiO_2 nanoparticles as an effective green fungicide against *Fusarium solani* and *Venturia inaequalis* phytopathogens. *Nanotechnology* 27: 085103.

Bramhanwade, K., S. Shende, S. Bonde et al. 2016. Fungicidal activity of Cu nanoparticles against *Fusarium solani and Venturia inaequalis* phytopathogens. *Nanotechnology* 27: 085103.

Castiglione, M.R., L. Giorgetti, C. Geri et al. 2011. The effects of nano-TiO_2 on seed germination, development and mitosis of root tip cells of Vicia narbonensis L. and Zea mays L. *J. Nano Res.* 13(6): 2443–2449. doi: 10.1007/s11051-010-0135-8

Chen, J., X. Wang and H. Han. 2013. A new functionof graphene oxide emerges inactivating phytopathogenic bacterium *Xanthomonas oryzae pv.oryzae*. *J. Nanopart. Res.* 15(5): 1658.

Chowdappa, P., S. Gowda, C.S. Chethana et al. 2014. Antifungal activity of chitosan-silver nanoparticle composite against *Colletotrichum gloeosporoides* associated with mango anthracnose. *African J. Microbiol. Res.* 8: 1803–1812.

Chirkov, S. 2002. The antiviral activity of chitosan (review). *Appl. Biochem. Microbiol.* 38: 1–8.

Cota-Arriola, O., M. Onofre Cortez-Rocha, A. Burgos-Hernández et al. 2013. Controlled release matrices and micro/nanoparticles of chitosan with antimicrobial potential: Development of new strategies for microbial control in agriculture. *J. Sci. Food Agric.* 93: 1525–1536.

Corredor, E., P.S. Testillano, M.J. Coronado et al. 2009. Nanoparticle penetration and transport in living pumpkin plants: In situ subcellular identification. *BMC Plant Biology* 9: 45.

Da Costa, M.V.J. and P.K. Sharma. 2016. Effect of copper oxide NPs on growth, morphology, photosynthesis, and antioxidant response in *Oryza sativa*. *Photosynthetica* 54(1): 110–119. doi: 10.1007/s11099-015-0167-5

Datnoff, L.E., F. A. Rodrigues and K.W. Seebold. 2007. Silicon and plant diseases. In L.E. Datnoff et al. (Eds.), *Mineral Nutrition and Plant Diseases*. St. Paul, MN: APS Press.

Derbalah, A.S., S.M. El-Moghazy and M.I. Godah. 2013. Alternative control methods of sugar beet leaf spot disease caused by the fungus Cercospora beticola (Sacc). *Egypt J. Biol. Pest Control* 23(2): 247–253.

Demir, E., N. Kaya and B. Kaya. 2014. Genotoxic effects of zinc oxide and titanium dioxide NPs on root meristem cells of *Allium cepa* by comet assay. *Turk. J. Biol.* 38: 31–39. doi: 10.3906/ biy-1306-11

Deng, Y.Q., J.C. White and B.S. Xing. 2017. Interactions between engineered nanomaterials and agricultural crops: Implications for food safety. *J. Zhejiang Univ. Sci A* 15(8): 552–572. doi: 10.1631/jzus.A1400165

Dimkpa, C.O., J.E. McLean, D.W. Britt et al. 2013. Antifungal activity of ZnO nanoparticles and their interactive effect with a biocontrol bacterium on growth antagonism of plant pathogen Fusarium graminearum. *Biometals* 26(6): 913–924.

Dimkpa, C. O., J. E. McLean, D. E. Latta et al. 2012. CuO and ZnO NPs: Phytotoxicity, metal speciation, and induction of oxidative stress in sand-grown wheat. *J. Nanopart. Res.* 14(9): 1125. doi: 10.1007/s11051-012-1125-9

Du, W., J. L. Gardea-Torresdey, R. Ji et al. 2015. Physiological and biochemical changes imposed by CeO_2 NPs on wheat: A life cycle field study. *Environ. Sci. Technol.* 49(19): 11884–11893. doi: 10.1021/acs.est.5b03055

Du, W., W. Tan, J.R. Peralta-Videa et al. 2017. Interaction of metal oxide NPs with higher terrestrial plants: Physiological and biochemical aspects. *Plant Physiol. Biochem.* 110: 210–225. doi: 10.1016/j.plaphy.2016.04.024

Duffy, B. 2007. Zinc and plant disease. In L.E. Dantoff, w.H. Elmer and D.M. Huber (Eds.), *Mineral Nutrition and Plant Disease* (pp. 155–176). St. Paul, MN: APS Press.

Durairaj, R., A.N. Amirulhusni, N.K. Palanisamy et al. 2012. Antibacterial effect of Ag NPs on multi drug resistant *Pseudomonas aeruginosa. World Acad. Sci. Eng. Technol.* 6: 210–213.

Duran, N., P.D. Marcato, R.D. Conti et al. 2010. Potential use of Ag NPs on pathogenic bacteria, their toxicity and possible mechanisms of action. *J. Brazilian Chem. Soc.* 21(6): 949–959. doi: 10.1590/S0103-50532010000600002

Dwivedi S., Q. Saquib, A.A. Al-Khedhairy et al. 2016. Understanding the role of nanomaterials in agriculture. In D.P. Singh, H.B. Singh and R. Prabha (Eds.), *Microbial Inoculants in Sustainable Agricultural Productivity* (pp. 271–288). New Delhi, India: Springer.

Ekambaram, P., A.A.H. Sathali and K. Priyanka. 2012. Solid lipid nanoparticles: A review. *Sci. Rev. Chem. Commun.* 2: 80–102.

Elbeshehy, E.K.F., A.M. Elazzazy, G. Aggelis. 2015. Silver nanoparticles synthesis mediated by new isolates of Bacillus spp., nanoparticle characterization and their activity against Bean Yellow Mosaic Virus and human pathogens. *Front. Microbiol.* 6: 453.

Elmer, W.H., R. De-La Torre-Roche, L. Pogano et al. 2018b. Effect of mettaloid and metallic oxide nanoparticles on Fusarium wilt of water melon. *Plant. Dis.* 102(7): 1394–1401.

Elmer, W.H. and J. White. 2016. Nanoparticle of CuO improves growth of eggplant and tomato in disease infested soils. *Environ. Sci. Nano.* 3: 1072–1079.

Elmer, W. and J.C. White. 2018a. The future of nanotechnology in plant pathology. *Ann. Rev. Phytopath.* 56: 111–133. doi: 10.1146/annurevev-phyto-080417-050108

Endo, M., K. Takesako, I. Kato et al. 1997. Fungicidal action of Aureobasidin A, a cyclic depsipeptide antifungal antibiotic, against *Saccharomyces cerevisiae. Antimicrobial Agents and Chemotherapy* 41: 672–676.

European Commission. 2013. Proposal for a Regulation of the European parliament and of the Council on Novel Foods. Available at: http://ec.europa.eu/food.food/biotechnology/novelfood/documents/novel-cloning_com2013-894_final_en.pdf

Fang, Y. and R.P. Ramasamy. 2015. Current and prospective methods for plant disease detection. *Biosensors* 5(3): 537–561.

Gajbhiye, M., J. Kesarwani, A. Ingle et al. 2009. Fungus-mediated synthesis of silver nanoparticles and their activity against pathogenic fungi in combination with fluconazole. *Nanomed. Nanotechnol. Bio. Med.* 5(4): 382–386.

Ghosh, M., M. Bandyopadhyay and A. Mukherjee. 2010. Genotoxicity of titanium dioxide (TiO_2) NPs at two trophic levels: Plant and human lymphocytes. *Chemosphere* 81(10): 1253–1262. doi: 10.1016/j.chemosphere.2010.09.022

Ghosh, S.K.B., W.B. Hunter and A.L. Park. 2017. Double strand RNA delivery system for plant-sap-feeding insects. *PLoS One* 12: e0171861.

Ghormade, V., M.V. Deshpande and K.M. Paknikar. 2011. Perspectives for nano-biotechnology enabled protection and nutrition of plants. *Biotechnol. Adv.* 29: 792–803. doi: 10.1016/j.biotechadv.2011.06.007

Giannousi, K., I. Avramidis, and C. Dendrion-Samara. 2013. Synthesis, characterization and evaluation of copper based nanoparticles as agrochemicals against *Phythophthora infestans. RSC Adv.* 3(44): 21743–21752.

Gogoi, R., P. Dureja and P.K. Singh. 2009. Nanoformulations–A safer and effective option for agrochemicals. *Ind. Farming* 59(8): 7–12.

Gogoi, R., P.K. Singh, R. Kumar et al. 2013. Suitability of nano-sulphur for biorational management of powdery mildew of okra (*Abelmoschus esculentus*) caused by *Erysiphe cichoracearum*. *J. Plant Pathol. Microbiol.* 4(4): 71–75.
González-Melendi, P., R. Fernandez-Pacheco, M.J. Coronado et al. 2008. Nanoparticles as smart treatment delivery systems in plants: Assessment of different techniques of microscopy for their visualization in plant tissues. *Ann. of Bot.* 101: 187–195.
Gopinath, V. and P. Velusamy. 2013. Extracellular biosynthesis of silver nanoparticles using Bacillus sp. GP-23 and evaluation of their antifungal activity towards *Fusarium oxysporum*. *Spectrochiem. Acta A Mol. Biomol. Spectrosc.* 106: 170–174. doi: 10.1016/j.saa.2012.12.087
Graham, J.H., E.G. Johnson, M.E. Myers et al. 2016. Potential of nano-formulated zinc oxide for control of citrus canker on grapefruit trees. *Plant Dis.* 100(12): 2442–2447.
Gui, X., Z. Zhang, S. Liu et al. 2015. Fate and phytotoxicity of CeO_2 NPs on lettuce cultured in the potting soil environment. *PLoS One* 10(8): e0134261. doi: 10.1371/journal.pone.0134261
Hafez, E.F., H.S. Hassan, M.F. Elkady et al. 2014. Assessment of antibacterial activity for synthesized zinc oxide nanorods against plant pathogenic strains. *Int. J. Sci. Technol. Res.* 3(9): 318–324.
Hao, Y., X. Cao, C. Ma et al. 2017. Potential applications and antifungal activities of engineered nanomaterials against gray mold disease agent *Botrytis cinerea* on rose petals. *Front. Plant Sci.* 8: 1332.
He, L., Y. Liu, A. Mustapha et al. 2011. Antifungal activity of zinc oxide nanoparticles against Botrytis cineria and Penicillium expansum. *Microbiol. Res.* 166(3): 207–215.
Hussain, I., N.B. Singh, A. Singh et al. 2016. Green synthesis of nanoparticles and its potential application. *Biotechnol. Lett.* 38: 545–560. doi: 10.1007/s10529-015-2026-7
Ilk, S., N. Saglam and M. Özgen. 2017. Kaempferol loaded lecithin/chitosan nanoparticles: Preparation, characterization, and their potential applications as a sustainable antifungal agent. *Artif. Cells Nanomed. Biotechnol.* 45: 907–916.
Imada, K., S. Sakai, H. Kajihara et al. 2016. Magnesium oxide nanoparticle induce systemic resistance in tomato against bacterial wilt disease. *Plant Pathol.* 65(4): 551–560.
Indhumathy, M. and R. Mala. 2013. Photocatalytic activity of zinc sulphate nano material on phytopathogens. *Int. J. Agric. Environ. Biotechnol.* 6(4s): 737–743.
Iravani, S. 2011. Green synthesis of metal nanoparticles using plants. *Green Chem.* 13: 2638–2650. doi: 10.1039/c1gc15386b
Iravani, S., H. Korbekandi, S. Mirmohammadi et al. 2014. Synthesis of silver nanoparticles: Chemical, physical and biological methods. *Res. Pharm. Sci.* 9: 385–406.
Jagana, D., Y.R. Hegde and R. Lella. 2017. Green nanoparticle: A novel approach for the management of banana anthracnose caused by *Colletotrichum musae*. *Int. J. Curr. Microbiol. Appl. Sci.* 6(10): 2638–2650.
Jain, D. and S. Kothari. 2014. Green synthesis of silver nanoparticles and their application in plant virus inhibition. *J. Mycol. Plant Pathol.* 44: 21.
Janatova, A., A. Bernardos, J. Smid et al. 2015. Long-term antifungal activity of volatile essential oil components released from mesoporous silica materials. *Ind. Crop. Prod.* 67: 216–220.
Kanhad, P., S. Birla, S. Gaikwad et al. 2014. In vitro antifungal efficacy of copper nanoparticles against selected crop pathogenic fungi. *Mater. Lett.* 115: 13–17.
Kashyap, P.L., X. Xiang and P. Heiden. 2015. Chitosan nanoparticle based delivery systems for sustainable agriculture. *Int. J. Boil. Macromol.* 77: 36–51.
Kaushik, H. and P. Dutta. 2017. Chemical synthesis of zinc oxide nanoparticle: Its application for antimicrobial activity and plant health management. In *Paper Presented at the 109th Annual meeting of the American Phytopathological Society, San Antonio, TX, August.*
Kaveh, R., Y.S. Li, S. Ranjbar et al. 2013. Changes in *Arabidopsis thaliana* gene expression in response to Ag NPs and Ag ions. *Environ. Sci. Technol.* 47(18): 10637–10644. doi: 10.1021/es402209w
Khadeeja, P., V. Banes and L. Ledwani. 2016. Green synthesis of nanoparticles: Their advantages and disadvantages. *AIP Conf. Proc.* 1724: 020048. doi: 10.1063/1.4945168
Khan, M. N., M. Mobin, Z. K. Abbas et al. 2017. Role of nanomaterials in plants under challenging environments. *Plant Physiol. Biochem.* 110: 194–209. doi: 10.1016/j.plaphy.2016.05.038
Kharissova, O.V., H.V.R. Dias, B.I. Kharisov et al. 2013. The greener synthesis of nanoparticles. *Trends Biotechnol.* 31: 240–248. doi: 10.1016/j tibtech.2013.01.003
Khodakovsky, A., P. Schroder and W. Sweldens. 2000. Progressive geometry compression. In *Siggraph 2000 Computer Graphics Proceedings* (pp. 271–278).
Kim, S. W., J. H. Jung, K. Lamsal et al. 2012. Antifungal effects of silver nanoparticles (AgNPs) against various plant pathogenic fungi. *Korean Soc. Mycol. Mycobiol.* 40: 53–58.

Kitching, M., M. Ramani and E. Marsili. 2015. Fungal biosynthesis of gold nanoparticles: Mechanism and scale up. *Microb. Biotechnol.* 8: 904–917. doi: 10.1111/1751-7915.12151

Klančnik, K., D. Drobne, J. Valant et al. 2011. Use of a modified Allium test with nanoTiO_2. *Ecotoxic. Environ. Safety* 74(1): 85–92. doi: 10.1016/j.ecoenv.2010.09.001

Koch, A., D. Biedenkopf, A. Furch et al. 2016. An RNAi-based control of *Fusarium graminearum* infections through spraying of long dsRNAs involves a plant passage and is controlled by the fungal silencing machinery. *PLoS Pathog.* 12: e1005901.

Krishnaraj, C., R. Ramachandran, K. Mohan et al. 2012. Optimization for rapid synthesis of silver nanoparticles and its effect on phytopathogenic fungi. *Spectrochim. Acta Part A Mol. Biomol. Spectrosc.* 93: 95–99.

Kumar, M., T.N. Shamsi, R. Parveen et al. 2017. Application of nanotechnology in enhancement of crop productivity and integrated pest management. In R. Prasad et al. (Eds.), *Nanotechnology* (pp. 361–371). Singapore: Springer.

Kumari, M., A. Mukherjee and N. Chandrasekaran. 2009. Genotoxicity of Ag NPs in *Allium cepa*. *Sci. Total Environ.* 407(19): 5243–5246. doi: 10.1016/j.scitotenv.2009.06.024

Lamsal, K., S.W. Kim, J.H. Jung et al. 2011. Application of silver nanoparticles for the control of *Colletotrichum* species in vitro and pepper anthracnose disease in field. *Mycobiology* 39: 194–199.

Lara, H.H., N.V. Ayala-Nuez, L. Ixtepan-Turrent et al. 2010. Mode of antiviral action of Ag NPs against HIV-1. *J. Nanobiotechol.* 8(1): 1. doi: 10.1186/1477-3155-8-1

Lee, K.J., S.H. Park, M. Govarthanan et al. 2013. Synthesis of silver nanoparticles using cow milk and their antifungal activity against phytopathogens. *Mater. Lett.* 105: 128–131. doi: 10.1016/j.matlet.2013.04.076

Li, G., T. Hu, G. Pan et al. 2008. Morphology– function relationship of ZnO: Polar planes, oxygen vacancies, and activity. *J. Phys. Chem. C* 112(31): 11859–11864. doi: 10.1021/jp8038626

Liao, Y.Y., A.L. Strayer, J.C. White et al. 2017. Magnesium oxide nanomaterial, a novel bactericide for control of bacterial spot of tomato without accumulating in fruits. In *Paper presented at 109th Annual Meeting of the American Phytopathological Soc., San Antonio, TX.*

Li-Byarlay, H., Y. Li, H. Stroud et al. 2013. RNA interference knockdown of DNA methyl-transferase 3 affects gene alternative splicing in the honey bee. *Proc. Natl. Acad. Sci. USA* 110: 12750–12755.

Linkov, I., M.E. Bates, L.J. Canis et al. 2011. A decision-directed approach for prioritizing research into the impact of nanomaterials on environment and human health. *Nat. Nanotechnol.* 6: 784–787. doi: 10.1038/nnano.2011.163

Liu, J. and R.H. Hurt. 2010. Ion release kinetics and particle persistence in aqueous nano-Ag colloids. *Environ. Sci. Technol.* 44(6): 2169–2175. doi: 10.1021/es9035557

Lok, C. N., C.M. Ho, R. Chen et al. 2006. Proteomic analysis of the mode of antibacterial action of Ag NPs. *J. Proteome Res.* 5(4): 916–924. doi: 10.1021/pr0504079

Lok, C. N., C.M. Ho, R. Chen et al. 2007. Ag NPs: Partial oxidation and antibacterial activities. *J Bio. Inorg. Chem.* 12: 527–534. doi: 10.1007/s00775-007-0208-z

Lopez-Moreno, M. L., G. de la Rosa, J. Hernandez-Viezcas et al. 2010. Evidence of the differential biotransformation and genotoxicity of ZnO and CeO_2 NPs on soybean (*Glycine max*) plants. *Environ. Sci. Technol.* 44(19): 7315–7320. doi: 10.1021/es903891g

Lv M., Y. Liu, J.H. Geng et al. 2018. Engineering nanomaterials-based biosensors for food safety detection. *Biosens. Bioelectron.* 106: 122–128. doi: 10.1016/j.bios.2018.01.049

Malerba, M. and R. Cerana. 2016. Chitosan effects on plant systems. *Int. J. Mol. Sci.* 17: 996.

Meng, Y., Y. Li, C.D. Galvani et al. 2005. Upstream migration of *Xylella fastidiosa* via pilus driven twitching motility. *J. Bacteriol.* 187(16): 5560–5567.

Min, J.S., K.S. Kim, S.W. Kim et al. 2009. Effects of colloidal silver nanoparticles on sclerotium-forming phytopathogenic fungi. *J. Plant Pathol.* 25: 376–380.

Mishra, S., C. Kesarvani, P.C. Abhilash et al. 2017. Integrated approach of agri-nanotechnology: Challenges and future trends. *Front. Plant Sci.* 8(1896): 471. doi: 10.3389/fpls.2017.00471

Mishra, S., C. Kesarvani, A. Singh et al. 2016. Microbial nanoformulation: Exploring potential for coherent nano farming. In V.K. Gupta et al. (Eds.), *The Handbook of Microbial Bioresources* (pp. 107–120). London, UK: CABI.

Mishra, S. and H.B. Singh. 2015b. Silver nanoparticles mediated altered gene expression of melanin biosynthesis genes in *Bipolaris sorokiniana*. *Microbiol. Res.* 172: 16–18. doi: 10.1016/jj.micres.2015.01.006.

Mishra, S., B.R. Singh, A. Singh et al. 2014b. Biofabricated silver nanoparticles act as strong fungicides against *Bipolaris sorokiniana* causing spot bloch disease in wheat. *PLoS One* 9(5): 97881.

Mitter, N., E.A. Worrall, K.E. Robinson et al. 2017a. Clay nanosheets for topical delivery of RNAi for sustained protection against plant viruses. *Nat. Plants* 3: 16207.

Mitter, N., E.A. Worrall, K.E. Robinson et al. 2017b. Induction of virus resistance by exogenous application of double-stranded RNA. *Curr. Opin. Virol.* 26: 49–55.

Moodley, J.S., S.B.N. Krishna, K. P. Sershen et al. 2018. Green synthesis of silver nano particles from *Moringa oleifera* leaf extracts and its antimicrobial potential. *Adv. Nat. Sci. Nanosci. Nanotechnol.* 9(1): 015011. doi: 10.1088/2043-6254/aaabb2

Morales, M.I., C.M. Rico, J.A. Hernandez-Viezcas et al. 2013. Toxicity assessment of cerium oxide NPs in cilantro (*Coriandrum sativum* L.) plants grown in organic soil. *J. Agric. Food Chem.* 61(26): 6224–6230. doi: 10.1021/jf401628v

Mori, Y., T. Ono, Y. Miyahira et al. 2013. Antiviral activity of Ag nanoparticle/chitosan composites against H1N1 influenza A virus. *Nanoscale Res. Lett.* 8(1): 1–6. doi: 10.1186/1556-276X-8-93

Mustafa, G. and S. Komatsu. 2016. Toxicity of heavy metals and metal-containing NPs on plants. *Biochim. Biophys. Acta Proteins Proteomics* 1864(8): 932–944. doi: 10.1016/j.bbapap.2016.02.020

Nair, P.M.G. and I.M. Chung. 2015. Study on the correlation between copper oxide NPs induced growth suppression and enhanced lignification in Indian mustard (*Brassica juncea* L.). *Ecotoxicol. Environ. Saf.* 113: 302–313. doi: 10.1016/j.ecoenv.2014.12.013

Nugaeva, N., K.Y. Gfeller, N. Backman et al. 2005. Micromechanical cantilever array sensor for selective fungal immobilization and fast growth detection. *Biosens. Bioelectron.* 21(6): 849–856.

Ordenes-Aenishanslins, N.A., L.A. Saona, V.M. Duran et al. 2014. Use of titanium dioxide nanoparticles biosynthesized by *Bacillus mycoides* in quantum dot sensitized solar cells. *Microb. Cell Fact* 13: 90. doi: 10.1186/s12934-014-0090-7

Padmavathy, N. and R. Vijayaraghavan. 2008. Enhanced bioactivity of ZnO NPs—An antimicrobial study. *Sci. Technol. Adv. Mater.* 9(3): 035004. doi: 10.1088/1468-6996/9/3/035004

Pal, S., Y.K. Tak and J.M. Song. 2007. Does the antibacterial activity of Ag NPs depend on the shape of the nanoparticle? A study of the gram-negative bacterium *Escherichia coli*. *Appl. Environ. Microbiol* 73: 1712–1720. doi: 10.1128/AEM.02218-06

Panpatte D.G., Y.K. Jhala, H.N. Shelat et al. 2016. Nanoparticles: The next generation technology for sustainable agriculture. In D.P. Singh, H.B Singh and R. Prabha (Eds.), *Microbial Inoculants in Sustainable Agricultural Productivity* (pp. 289–300). New Delhi, India: Springer.

Patel, N., P. Desai, N. Patel et al. 2014. Agro-nanotechnology for plant fungal disease management: A review. *Int. J. Curr. Microbiol. App. Sci.* 3(10): 71–84.

Paret, M.L., A.J. Palmateer and G.W. Knox. 2013a. Evaluation of light-activated nanoparticle formulation of titanium oxide with zinc for management of bacterial leaf spot on rosa 'Noare'. *Hortscience* 48(2): 189–192.

Paret, M.L., E.G. Vallad, R.D. Averett et al. 2013b. Photocatalysis: Effect of light-activated nanoscale formulations of TiO_2 on *Xanthomonas perforans*, and control of bacterial spot of tomato. *Phytopathology* 103: 228–236.

Park, H-J., S.H. Kim, H.J. Kim et al. 2006. A new composition of nanosized silica-silver for control of various plant diseases. *Plant Pathol. J.* 22: 295–302.

Patlolla, A. K., A. Berry, L. May et al. 2012. Genotoxicity of Ag NPs in *Vicia faba*: A pilot study on the environmental monitoring of NPs. *Int. J. Environ. Res. Public Health* 9(5): 1649–1662. doi: 10.3390/ijerph9051649

Paulkumar, K., G. Gnanajobitha, M. Vanaja et al. 2014. Piper nigrum leaf and stem assisted green synthesis of silver nanoparticles and evaluation of its antibacterial activity against agricultural plant pathogens. *Sci. World J.* 2014: 829894. doi: 10.1155/2014/829894

Prakrashi, S., N. Jain, S. Dalai et al. 2014. In vivo genotoxicity assessment of titanium dioxide NPs by *Allium cepa* root tip assay at high exposure concentrations. *PLoS One* 9(2): e87789. doi: 10.1371/journal.pone.0087789

Prasad, R.G.S.V., D. Basavaraju, K.N. Rao et al. 2011. Nanostructured TiO_2 and TiO_2-Ag antimicrobial thin films. In *International Conference on Nanoscience, Technology and Societal Implications (NSTSI)*. (pp. 1–6). Proceedings of meeting held 8–10 December 2011, Bhubneshwar, India, p. 300. Edited by Majhi, Chand, and Pradhan. Published by IEEE.

Rajiv, P., S. Rajeshwari and R. Venkatesh. 2013. Bio-fabrication of zinc oxide nanoparticles using leaf extract of *Parthenium hysterophorous* L. and its size dependent antifungal activity against plant fungal pathogens. *Spectrochim. Acta Part A* 112: 384–387.

Raghupathi, K.R., R.T. Koodali and A.C. Manna. 2011. Size-dependent bacterial growth inhibition and mechanism of antibacterial activity of zinc oxide NPs. *Langmuir* 27(7): 4020–4028. doi: 10.1021/la104825u

Rajeshwari, A., S. Kavitha, S.A. Alex et al. 2015. Cytotoxicity of aluminum oxide NPs on *Allium cepa* root tip: Effects of oxidative stress generation and bio uptake. *Environ. Sci. Pollut. Res. Int.* 22: 11057–11066.

Raliya, R., P. Biswas, and J.C. Tarafdar. 2015. TiO_2 nanoparticle biosynthesis and its physiological effect on mung bean (*Vigna radiata* L.). *Biotechnol. Rep.* 5: 22–26. doi: 10.1016/j.btre.2014.10.009

Raliya, R. and J.C. Tarafdar. 2013. ZnO nanoparticle biosynthesis and its effect on phosphorous –mobilizing enzyme secretion and gum contents in cluster bean (*Cyamopsis tetragonoloba* L.). *Agric. Res.* 2: 48–57. doi: 10.1007/s40003-012-0049-z.

Raliya, R., J.C. Tarafdar, S.K. Singh et al. 2014. MgO nanoparticles biosynthesis and its effect on chlorophyll contents in the leaves of Clusterbean (*Cyamopsis tetragonoloba* L.). *Adv. Sci. Eng. Med.* 6: 538–545. doi: 10.1166/asem.2014.1540

Rao, K.J. and S. Paria. 2013. Use of sulphur nanoparticles as green pesticides on *Fusarium solani* and *Venturia inaequalis* phytopathogens. *RSC Adv.* 3(26): 10471–10478.

Ratwani, D., N. Khatri, S. Tyagi et al. 2018. Nanotechnology-based recent approaches for sensing and remediation of pesticides. *J. Environ. Manag.* 206: 749–762. doi: 10.1016/jenvman.2017.11.037

Rawat, M., P. Yadukrishnan and N. Kumar. 2018. Mechanisms of action of nanoparticles in living systems. In P. Sharma and A. Sharma, *Environmental Monitoring and Cleanup*, p. 427. Hershey, PA: IGI Global. doi: 10.4018/978-1-5225-3126-5 ch.014

Research and Markets. 2015. Global nanotechnology Market Outlook 2015–2020. Available at: http://www.prnewswire.com/news-release/global-nanotechnology-market-outlook-2015-2020---industry-will-grow-to-reach-us-758-billion-507155671.html

Rico, C. M., M. I. Morales, R. McCreary et al. 2013. Cerium oxide NPs modify the antioxidative stress enzyme activities and macromolecule composition in rice seedlings. *Environ. Sci. Technol.* 47(24): 14110–14118. doi: 10.1021/es4033887

Roy, A., S.S. Gauri, M. Bhattacharya et al. 2013. Antimicrobial activity of CaO NPs. *J Biomed. Nanotechnol.* 9(9): 1570–1578. doi: 10.1166/jbn.2013.1681

Sabir, S., M. Arshad and S.K. Chaudhari. 2014. Zinc oxide nanoparticles for revolutionizing agriculture: Synthesis and applications. *Sci. World J.* 2014: 925494. doi: 10.1155/2014/925494

Sagili, J.L., R.S. Roopa Bai, H. Sharanagouda et al. 2018. Effect of biosynthesized zinc oxide nanoparticles coating on quality parameters of fig (*Ficus carica* L.) fruit. *J Pharmacog. Phytochem.* 7(3): 10–14.

San, K.M. and J.G. Scott. 2016. The next generation of insecticides: DsRNA is stable as a foliar-applied insecticide. *Pest Manag. Sci.* 72: 801–809.

Sardella, D., R. Gatt and V. Valdramidis. 2017. Physiological effects and mode of action of ZnO nanoparticles against postharvest fungal contaminants. In *Paper Presented at the 109th Annual Meeting of American Phytopathological Society, San Antonio, TX.*

Sarlak, N., A. Taherifar and F. Salehi. 2014. Synthesis of nano pesticides by encapsulating pesticides nanoparticles using functional carbon nanotubes and application of new nano composite for plant disease treatment. *J. Agric. Food Chem.* 62(21): 4833–4838.

Shang, Y., M.K. Hasan, G.J. Ahammed et al. 2019. Application of nanotechnology in plant growth and crop protection: A review. *Molecules* 24(14): 2558. doi: 10.3390/molecules24142558

Sharma, N. 2014. Biologicals-green alternatives for plant disease management. In N. Sharma (Ed.), *Biological Controls for Preventing Food Deterioration: Strategies for Pre-and Postharvest Management* (pp. 1–25). Hoboken, NJ: John Wiley & Sons Ltd.

Sharon, M., A.K. Choudhary and R. Kumar. 2010. Nanotechnology in agricultural diseases and food safety. *J. Phtology* 2(4): 83–92.

Sharon, M. and M. Sharon. 2008. Carbon nanomaterials: Applications in Physico-Chemical and Bio-Systems. *Def. Sci. J.* 58(4): 5491–5516.

Shastry, K., H.B. Rashmi and N.H. Rao. 2010. Nanotechnology patents as R&D indicators for disease management strategies in agriculture. *J. Intellectual Property Rights* 15: 197–205.

Shen, C.X., Q.F. Zhang, J. Li et al. 2010. Induction of programmed cell death in Arabidopsis and rice by single-wall carbon nanotubes. *Am. J. Bot.* 97(10): 1602–1609. doi: 10.3732/ajb.1000073

Shoala, T. 2018. Positive impacts of nanoparticles in plant resistance against different stimuli. In K. Abd Elsalam and R. Prasad (Eds.), *Nanobiotechnology Applications in Plant Protection. Nanotechnology in Life Sciences.* Cham, Switzerland: Springer.

Singh, P., A. Garg, S. Pandit et al. 2018b. Antimicrobial effects of biogenic nanoparticles. *Nanomaterials* 8(12): 1009. doi: 10.3390/nano8121009

Singh, P., Y.J. Kim, H. Singh et al. 2015. Biosynthesis of anisotropic silver nanoparticles by *Bhargavaea indica* and their synergistic effect with antibiotics against pathogenic microorganisms. *J. Nanomater.* 2015: 10.

Singh, P., S. Pandit, J. Garnaes et al. 2018a. Green synthesis of gold and silver nanoparticles from *Cannabis sativa* (industrial hemp) and their capacity for biofilm inhibition. *Int. J. Nanomed.* 13: 3571–3591.

Sintubin, L., B. De Gusseme, P. Van der Meeren et al. 2011. The antibacterial activity of biogenic Ag and its mode of action. *Appl. Microbiol. Biotechnol.* 91(1): 153–162. doi: 10.1007/s00253-011-3225-3

Solgi, M., M. Kafi, T.S. Taghavi et al. 2009. Essential oils and silver nanoparticles (SNP) as novel agents to extend vase-life of gerbera (*Gerbera jasmesonii* cv. 'Dune') flowers. *Postharvest Biol. Technol.* 53(3): 155–158.
Strayer-Scherer, A.L., Y.Y. Liao, M. Young et al. 2018. Advanced copper composites against copper tolerant *Xanthomonas perforans* and tomato bacterial spot. *Phytopathology* 108: 196–205.
Tamjidi, F., M. Shahedi, J. Varshosaz et al. 2013. Nanostructured lipid carriers (NLC): A potential delivery system for bioactive food molecules. *Innov. Food Sci. Emerg. Technol.* 19: 29–43.
Thairu, M., I. Skidmore, R. Bansal et al. 2017. Efficacy of RNA interference knockdown using aerosolized short interfering RNAs bound to nanoparticles in three diverse aphid species. *Insect Mol. Boil.* 26: 356–368.
Tripathi, S., S.K. Sonkar and S. Sarkar. 2011. Growth stimulation of gram (*Cicer arientium*) plant by water soluble carbon nanotubes. *Nanoscale* 3(3): 1176–1181.
van Viet, P., H.N. Nguyen, T.M. Cao et al. 2016. Fusarium antifungal activities of copper nanoparticles synthesized by chemical reduction method. *Nanomaterials* 2016: e1957612, p. 7.
Vermeulen, S.J., P.K. Aggarwal, A. Ainslie et al. 2012. Options for support to agriculture and food security under climate change. *Environ. Sci. Policy* 15: 136–144. doi: 10.1016/j.envsci.2011.09.003
Wang, H.H., X.M. Kou, Z.G. Pei et al. 2011. Physiological effects of magnetite (Fe_3O_4) NPs on perennial ryegrass (*Lolium perenne* L.) and pumpkin (*Cucurbita mixta*) plants. *Nanotoxicology* 5(1): 30–42. doi: 10.3109/17435390.2010.489206
Wang, X., X. Liu, J. Chen et al. 2014. Evaluation and mechanism of antifungal effects of carbon nanomaterials in controlling plant fungal pathogen. *Carbon* 68: 798–806.
Wang, F.Y., X.Q. Liu, Z.Y. Shi et al. 2016b. Arbuscular mycorrhizae alleviate negative effects of zinc oxide nanoparticle and zinc accumulation in maize plants–A soil microcosm experiment. *Chemosphere* 147: 88–97. doi: 10.1016/j.chemosphere.2015.12.076
Wang, M., A. Weiberg, F.M. Lin et al. 2016a. Bidirectional cross-kingdom RNAi and fungal uptake of external RNAs confer plant protection. *Nat. Plants* 2: 16151.
Wani, A.H. and M.A. Shah. 2012. A unique and profound effect of MgO and ZnO nanoparticles on some plant pathogenic fungi. *J. Appl. Pharm. Sci.* 2(3): 40–44.
Worrall, E.A., A. Hamid, K.T. Mody et al. 2018. Nanotechnology for plant disease management. *Agronomy* 8: 285. doi: 10.3390/agronomy8 120285.
Yang, Z., J. Chen, R. Dou et al. 2015. Assessment of the phytotoxicity of metal oxide NPs on two crop plants, maize (*Zea mays* L.) and rice (*Oryza sativa* L.). *Int. J. Environ. Res. Public Health* 12(12): 15100–15109. doi: 10.3390/ijerph121214963
Yang, H., C. Liu, D. Yang et al. 2009. Comparative study of cytotoxicity, oxidative stress and genotoxicity induced by four typical nanomaterials: The role of particle size, shape and composition. *J. Appl. Toxicol.* 29(1): 69–78. doi: 10.1002/jat.1385
Young, M., A. Ozcan, M.E. Myers et al. 2017. Multimodal generally recognized as safe ZnO/nanocopper composite: A novel antimicrobial material for management of citrus phytopathogens. *J. Agric. Food Chem.* 66(24): 6604–6608. https://doi.org/10.1021/acs.jafc.7b02526
Zhang, L., Y. Jiang, Y. Ding et al. 2007. Investigation into the antibacterial behavior of suspensions of ZnO NPs (ZnO nanofluids). *J. Nanopart. Res.* 9(3): 479–489. doi: 10.1007/s11051-006-9150-1

14 Cold Atmospheric Pressure Plasma

A Novel Approach to Managing Fungal Spoilage and Mycotoxin Decontamination

Avantina Sharma Bhandari

CONTENTS

14.1 INTRODUCTION

The incessantly increasing global population requires substantial resources for food production. Food production needs to increase by 50 to 70% in the next 30 years to avoid global food insecurity. The danger of food insecurity is particularly serious for developing countries where more people suffer from hunger and this situation is expected to be aggravated in the future.

The challenges of safely and securely feeding people become intensified in a world facing other problems such as shrinking arable land, less and more expensive fossil fuels, increasingly limited supplies of water, social unrest, economic uncertainty and within a scenario of a rapidly changing climate. Moreover, the world food situation is aggravated by the fact that, in spite of all the available means of plant protection, the impact of plant diseases cannot be overestimated.

Agricultural crops can suffer colossal losses due to the ravages of various pests, including bacteria, fungi, viruses, insects, rodents and nematodes, destroying a major fraction of the yearly output of food commodities. The impact of fungal diseases and new variants of existing pathogens on agriculturally important crops is considered to be one of the main threats to worldwide food availability and safety. It was figured that diseases on our most important agricultural crops resulted in damages that were enough to feed 8.5% of the world's population. Owing to favorable and conducive climatic conditions and a congenial environment, stored food products can also be contaminated and severely destroyed by fungi, insects and rodents. Generally, tropical conditions such as high temperature, moisture, unseasonal rains during harvest, flash floods, poor harvesting practices and improper packing during transport and storage contribute to the proliferation of toxicogenic fungi. Recent studies have revealed a correlation between the increased presence of mycotoxins and global climate change (Miraglia et al., 2009). Parameters including elevated temperature, moisture level and plant stress-related response stimulate fungal growth and, consequently, the production of mycotoxins (Narasaiah et al., 2006; Mousa et al., 2013). Furthermore, climate change plays a significant role in the global economy, where food is transported over long distances from producer to consumer, and may be subject to different local climates, transport and prolonged storage times. All these factors may contribute to increased food contamination (Nellemann, 2009).

14.2 MYCOTOXINS: THE HIDDEN CULPRITS

Mycotoxins, the naturally occurring, ubiquitous secondary metabolites of several toxicogenic fungi, contaminate the whole food chain, from agricultural products such as fresh fruits, dried fruits and nuts to cereals and other grains that are ultimately consumed by living beings (Marin et al., 2013). They can harm human health through a wide range of toxic effects, including carcinogenicity, teratogenicity, hepatotoxicity, mutagenicity, neurotoxicity effects and the disruption of both immune and endocrine systems (da Rocha et al., 2014). The recent and rapid increase in mycotoxin food contamination represents one of the main concerns in the field of agriculture and food production (Stoev, 2013). Their production occurs under certain environmental conditions during and/or after plant colonization (Mattsson et al., 2007). These mycotoxins can be produced in the field during the growth of the fungus on the crop or later as a result of substandard handling or storage.

Presently, more than 400 structurally divergent mycotoxins have been identified, including aflatoxins (AFs), ochratoxins, fumonisins (FBs), patulin, zearalenone (ZEN), ergot alkaloids and trichothecenes, including deoxynivalenol (DON), HT-2 and T-2 toxin (Murphy et al., 2006). These mycotoxins are thermolabile and chemically stable and exposure to mycotoxins either in the short and/or long-term can lead to diverse toxic effects on a wide range of organisms (Bryla et al., 2016; Lawley et al., 2008). Often, these fungal toxins are not only harmful to vertebrates and invertebrates (mycotoxins) but also to plants (phytotoxins). These toxigenic plant pathogens that produce mycotoxins, secondary metabolites of unrelated chemical structures and biological properties have very broad toxic effects to humans and livestock, so, in addition to posing a threat for food security, these pathogens also pose a threat to food safety and the health of living beings (Marin et al., 2013). Economically, these natural contaminants hamper international trade and significantly affect the world economy due to rejection when mycotoxin concentrations exceed the maximum permissible levels.

14.3 LOSSES DUE TO MYCOTOXIN CONTAMINATION

Since 1999, the estimate of 25% mycotoxin incidence in food crops at the global level presented in the Food and Agriculture Organization (FAO) journal has been widely cited in many publications. Over the years, it has therefore become "THE" reference value (Eskola et al., 2019). Notably in the recent past, data on large surveys on mycotoxin occurrence across the world have been released suggesting higher mycotoxin prevalence than 25% stated by the FAO. Streit et al. (2013) reported that, overall, 72% of around 17,300 feed samples, gathered from different parts of the world and collected over the period of eight years, contained mycotoxins. Karlovsky et al. (2016) analyzed and reported that mycotoxin contamination in feeds could be up to 79% or even more in about 2000 samples collected from 52 countries. Biomin (2017) in a large survey comprising 72 countries across the globe also observed similar high frequencies of mycotoxins in feed. Although in these surveys, animal feed samples have mainly been studied, they point out that the prevalence of mycotoxin-contaminated food crops could be higher than the 25% estimation of the FAO.

14.4 CONSEQUENCES OF MYCOTOXIN CONSUMPTION

Health hazards from food can be caused by infectious agents and toxic compounds. Fungi and mycotoxins can affect the shelf life as well as the quality of foods as they can reduce nutritional quality and seed viability (Betina, 1984; Bhat et al., 2010). Living microorganisms ingested with food can cause infectious diseases, while toxic substances lead to acute poisoning or long-term negative impacts on the health of consumers. Chemical food contaminants mainly originate from the following sources: unintentional pollutants; intentionally added compounds at levels exceeding legal limits or in commodities for which they have not been approved; toxic plant metabolites; contaminants generated by processing; and toxic microbial metabolites (Klich et al., 2007).

Some of the carcinogenic mycotoxins, such as aflatoxins, zearalenone, deoxynivalenol (DON, vomitoxin) and fumonisins are toxic to humans and can also cause mutations (Park et al., 2007). Aflatoxins are notorious mycotoxins produced by the fungi *Aspergillus flavus* and *A. parasiticus*. Agricultural products such as corn, peanut and cottonseed can be susceptible to aflatoxin production, especially at high temperatures (25 to 32 °C) and humidity conditions (>85%). Vomitoxins or DON are produced by *Fusarium* species in grains such as wheat, corn and sorghum, and is a potential health hazard to humans.

To reduce the potential danger to human health, many countries adopted strict legislation to control the mycotoxin presence in food and feed. In the European Union, the presence of mycotoxins in food and feed is regulated by Regulation (EC) No 1881/2006, Directive 2002/32/EC, Recommendations 2006/576/EC and 2013/165/EU, and their amendments (in the references). The harmful impact of plant protection products on the environment and human and animal health has prompted the European Union (EU Directive 2009/128/EC) to encourage research on alternative and ecofriendly solutions. Although the production of mycotoxins by these toxicogenic plant pathogens is of economic importance, many research groups do not take them into account when studying non-chemical control strategies.

14.5 STRATEGIES TO MANAGE FOOD SPOILAGE AND MYCOTOXINS

Mycotoxin-contaminated food represents a significant and increasing threat to human health and an enormous burden for the global economy. Fungal spoilage is not only problematic for the fresh produce and meat industries, but also for those relying on fruit products (for example, minimally processed fruit yogurts containing berries) and cheese as ingredients. Fungal species such as *Fusarium*, *Aspergillus*, *Penicillium* and *Alternaria* produce chemically stable mycotoxins.

Several of the postharvest operations of grain and other food commodities can significantly reduce the spoilage of the product as well as the quantity of mycotoxins present in them; however, complete removal of mycotoxins or decontamination of these foods is impossible during these operations. Hence, additional processing of these commodities may be required for safe consumption. The commonly used methods employed to remove mycotoxins from food and feeds include physical, chemical, enzymatic and microbial decontamination methods (Karlovsky et al., 2016). The physical processing operations used are optical sorting, sieving, floatation, washing, dehulling, steeping, milling, thermal treatments, ultraviolet and gamma treatments. The physical separation process segregates inferior infected quality from bulk based on color and density difference, visual identification, shape and size. Unfortunately, these methods can be inaccurate, unreliable, laborious, time consuming and not suitable for inline measurement. Adopting several strategies such as the cultivation of resistant cultivars, the use of sound crop rotation schemes and the use of chemical control can manage plant diseases.

Current management strategies for controlling mycotoxin occurrence are mostly based on chemical, biological and physical approaches (Mattsson et al., 2007). Amid these, ultraviolet (UV) irradiation is one of the most frequently used non-thermal food processing methods for the decontamination of mycotoxins such as AFs, trichothecenes and ZEN.

Nevertheless, degradation efficiencies can vary widely due to differences in irradiation conditions, meaning the exposure times that are necessary to achieve a satisfactory level of decontamination are often too long to be relevant for large-scale food production. UV treatment has shown effectiveness in reducing fungi such as *Penicillium expansum* and mycotoxins; however, UV radiation cannot penetrate food materials (Murata et al., 2008; Syamaladevi et al., 2015, 2014). Gamma irradiation has been a promising intervention against mycotoxins and accepted by the food industry for the treatment of dry food. Gamma irradiation produces high-energy photons, which interact with microbial cells and induce DNA rupture (Udomkun et al., 2017).

The chemical treatments using acetic acid, citric acid, lactic acid under simulated cooking (Aiko and Mehta, 2016) and bases such as NH_3, Ca $(OH)_2$, Na_2 CO_3 (Park et al., 1988), oxidizing agents (ozone) and reducing agents ($NaHSO_3$) are also used to reduce mycotoxins (Temba et al., 2016). Enzymatic detoxification has also been investigated (for example, amylases, glucanases, proteases) to reduce mycotoxins in foods. The biological control of mycotoxin detoxification includes fermentation, prevention of ingestion of mycotoxin content of contaminated food by microorganisms of the gastrointestinal tract and introduction of active enzymes by genetic engineering in plant genes capable of detoxification (Halász et al., 2009). It is worthwhile noting that traditional mycotoxin degradation methods can take a long processing time for effective mycotoxin reduction, hence are not energy efficient and are expensive.

Although several methods have been proposed to degrade mycotoxins in foods, the food industry continues to seek rapid and effective technology. The combination of modified climatic conditions and a tendency for consumers to eat healthier and fresher foods makes it imperative that new, sustainable and more effective approaches in agriculture, processing, transportation and storage methods are developed. Therefore, there is an urgent requirement to search for alternative methods and strategies to manage fungal spoilage and decontamination of mycotoxins from food products. In line with this attitude, to combat the growing challenges of food safety and new consumer demands, several technologies have been (and are being) explored by the food industry (Misra et al., 2018a). Given the diverse nature of the food industry, it is continually seeking and willing to adopt new technologies for sustainability, safety, profitability, consumer trust and continued success (Pal et al., 2016).

Also, the growing "green and clean label" trend has resulted in a shift toward the use of natural anti-mycotic agents, which are generally expensive and increase the product cost.

Besides their high toxicity, mycotoxins are highly stable to physical, chemical or biological detoxification. New mycotoxin-decontamination technologies will play a significant role in all stages of the supply chain. Therefore, treatment with cold atmospheric plasma could be one approach to reduce the number of mycotoxins in different products. One thing has to be kept in mind that alongside managing mycotoxin decontamination, novel methods should also have to preserve the quality of food products, be environmentally benign and economically suitable. Considering the above-mentioned requirements, cold plasma technology offers a promising non-thermal mycotoxin-decontamination approach (Raviteja et al., 2019).

The terms cold plasma, cool plasma, atmospheric pressure plasma, cold atmospheric gas plasma and other comparable terms have been used. In other cases, the plasma is described by the generative technology, for example, dielectric barrier discharge, jet, uniform glow discharge plasma, gliding arc discharge and so on (Derek et al., 2018).

14.6 WHY USE PLASMA TECHNOLOGY?

Plasma is generally known as the fourth state of matter, next to solids, liquids and gases. The term "Plasma" was used for the first time by Irving Langmuir in 1928 to define this fourth state of matter, who discovered a partially or wholly ionized state of gas with plasma oscillations in the ionized gas (Misra et al., 2016a). Plasma is an ionized gas containing atoms or molecules in a metastable state with a roughly zero net electrical charge (Turner, 2016). Examples of natural plasma include the sun and aurora in the sky, whereas fluorescent lamps, plasma televisions and commercial ozonizers are man-made plasmas. The sun being at a very high temperature is an ideal example of thermal (hot) plasma. The aurora light (polar light) is an example of low temperature (cold) plasma. A plasma state is reached by increasing the energy level of a substance from a solid state through liquid and gaseous states of matter, ending in an ionized state of gas, which has unique physical and chemical properties (Ni et al., 2016). In electrically created plasmas, energy is delivered in the form of an electric field from an electrical power source; seed electrons produced by UV or background radiation are accelerated by the applied electric field leading to the excitation, dissociation or ionization of the background gas (Figure 14.1). Ionization, caused by the collision of an energetic electron with a neutral atom or molecule, results in the production of further electrons which are also accelerated in the electric field. These free and energetic electrons subsequently collide with other surrounding molecules and atoms present in the gas, resulting in an avalanche process. Through the simultaneous generation and interaction among electrons, neutrals, metastables and ions, a vast number of reactions occur, yielding a wide variety of reactive chemical species. Plasma is quasi-neutral ionized gas, primarily composed of photons, ions and free electrons, as well as atoms in their ground or excited states with a net neutral charge. The strong electric fields used in generating cold plasma accelerate electrons to energies capable of ionizing, electronically exciting, dissociating and heating the constituents of the background gas. In air, this results in over 50 distinct chemical species, including the ionic species of H^+, H^-, O^+, O^-, H_3O^+, OH^- and N_2^+; the excited and ground states of N_2, O, NO and OH; and longer-lived species such as O_3, NO_2 and N_2O; collectively, these are referred to as reactive oxygen and nitrogen species (RONS) (Sakiyama et al., 2012). In complex gas mixtures, such as humid air, a large number of reactive chemical species are created which take part in many hundreds of reactions. In addition, molecules or atoms in an excited state can emit photons with wavelengths in the UVC, UVB and UVA range. It has been reported that plasma

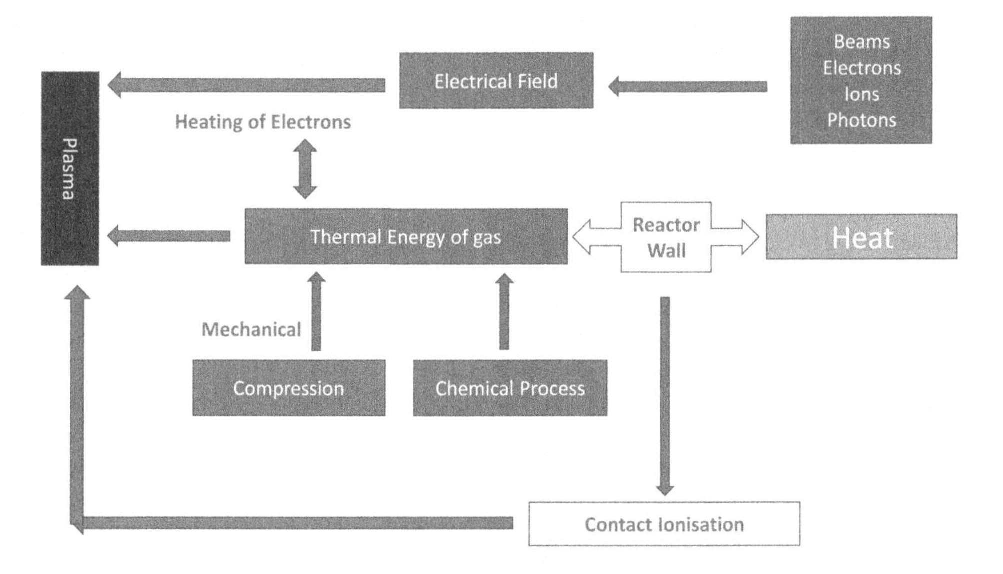

FIGURE 14.1 Synthesis of cold plasma.

could be obtained either in low temperature, non-equilibrium glow discharge or high temperature, equilibrium thermal plasma.

In the past, cold plasma was used for the sterilization of sensitive materials. The application of plasma technology as a surface-cleaning tool has been commercially adopted for the removal of disinfection chemicals applied to medical devices manufactured from heat-sensitive plastics. In the biomedical sector, plasma technology is used for the cold sterilization of instruments and prostheses, as well as many thermolabile materials. It is used for its particular advantages, including the moderate or negligible impact on substrate materials and use on non-toxic compounds. Based on the properties of plasma, it is used in various fields like textiles, electronics, life sciences, packaging and so on (Misra et al., 2016b), and now it has been extended to the food industry as a novel technology.

Conventionally, sterilization methods such as heat and chemical solutions are used for the surface disinfection of fruits, seeds, spices and so on, which are often time consuming and damaging or have toxic residues. Feichtinger et al. (2003) reported that cold plasma technology is preferred as an alternative source for surface sterilization and the disinfection process, as it can act on both vegetative cells and spores within shorter periods of time.

Van de Veen et al. (2015) reported that the effect of cold plasma on bacterial spores is more than conventional techniques like heat, chemicals and UV treatment, and plasma treatment can effectively inactivate a wide range of microorganisms.

Plasma can be produced under low pressure or even atmospheric pressure conditions. Typically, low-pressure plasma systems require a discharge generator, a gas source and an expensive vacuum system, consisting of pumps and a vacuum chamber. Such systems are widely used for applications in material processing. Nevertheless, they are not suitable for materials sensitive to low-pressure conditions including biological material (Liebermann and Lichtenberg, 2008). The use of atmospheric pressure plasma avoids the disadvantages of vacuum systems and enables the treatment of biological materials. Common examples of atmospheric pressure plasma systems include arc, corona and dielectric barrier discharges.

The most perspective discharges for the treatment of biological materials are those that are in thermal non-equilibrium, have a gas temperature that is close to room temperature and are typically referred to as cold plasmas. Recent developments in cold atmospheric pressure plasma (CAP) sources and the ability to tailor discharges to produce highly reactive species in high concentrations, but at temperatures close to room temperature, have paved the way for a wide number of biological applications. Such CAPs are also suitable for use in electronics, surface modifications in the polymer and textile industry, synthesis of nanoparticles and the degradation of pollutants (Thirumdas et al., 2014).

The new findings and developments in plasma science through the last decade reveal the great potential of CAP as an innovative technology in the field of biology, moving from the treatment of inanimate materials to living or cellular objects. Such applications include CAP treatments in medicine as well as in agriculture and the food industry.

CAP technology is, on the other hand, a newcomer to the field of agriculture and the food industry. The CAP treatment of food products refers to its high chemical reactivity, achieved through the reactive species generated, and consequently its ability to deactivate harmful agents such as pathogenic bacteria and toxic pollutants in short processing times and at low temperatures with an almost negligible impact on the treated food products.

One of the important challenges associated with cold plasma technology is ensuring high microbial inactivation while maintaining sensory qualities that ensure a fresh appearance. Perhaps even more importantly, plasma treatments did not significantly influence the organoleptic characteristics of the treated foods or their nutritional properties (Raviteja et al., 2019). Despite promising results, the use of CAP in the field of food spoilage and mycotoxin decontamination needs further exploration to uncover the CAP-related decontamination chemical processes.

14.7 HOW PLASMA WORKS

In the plasma state, the free electric charges, viz., ions and electrons, make plasma electrically conductive, internally interactive and strongly responsive to electromagnetic fields (Fridman, 2008). The chemical composition of plasma contains free radicals; highly reactive species and radiation are often generated in varying ranges from UV to visible. It is believed that the role of different constituents depends on the gas and operating pressure. The destruction of microbial DNA by UV irradiation, the volatilization of compounds from spores and the so-called "etching" of the spore surface by adsorption is because of reactive species like free radicals (Laroussi et al., 2012). Among all plasma species, many studies have highlighted the key role played by atomic oxygen (O), the hydroxyl radical (•OH), ozone (O_3), hydrogen peroxide (H_2O_2) and peroxynitrite in CAP-related decontamination effects, since they all possess a very high oxidative potential.

When the substrate gas has oxygen as one of its components, reactive oxygen species (ROS), such as ground state atomic oxygen [O(3P)], hydroxyl radicals (•OH), singlet oxygen molecules [O2(1_g)], superoxide anions (•O2$^-$) and ozone (O3) are responsible for effective inactivation of microorganisms (Hashizume et al., 2015). Among the ROS, ozone is a powerful oxidant, second only to the hydroxyl radical (Segat et al., 2014). The importance of reactive nitrogen species (RNS) and ultraviolet light, in addition to ROS, has also been reported (Lu et al., 2016; Moiseev et al., 2014).

Several research investigations have highlighted different modes of action that show a reduced growth of microorganisms via etching phenomenon, cell disruption by electroporation etc. The reaction mechanisms resulting in the formation of active plasma chemical species include electronic impact processes (vibration, excitation, dissociation, attachment and ionization), ion–ion neutralization, ion–molecule reactions, Penning ionization, quenching, three-body neutral recombination and neutral chemistry, as well as photoemission, photo-absorption and photo ionization (Misra et al., 2016c; Sahu et al., 2017).

In biological systems such as bacterial and fungal cells, the short-lived O and OH_ first react with cell walls and membranes and with all the compounds composing these two structures (lipids, proteins and polysaccharides). The lipids are the most sensitive to oxidation. The mechanism of OH_ reaction with lipids refers to its H-abstraction from the unsaturated carbon bonds of the fatty acids, ending in lipid peroxidation (Lukes et al., 2012).

O_3 is also a powerful oxidant; ozonation alone represents one of the most potent sanitizing and detoxifying approaches in the food industry and mycotoxin decontamination. O_3 has high reactivity, penetrability and spontaneous decomposition into non-toxic oxygen without forming harmful oxygen species. Compared with OH_, O_3 induced reaction kinetics are slower (Karaca et al., 2010). In addition, the antimicrobial activity of H_2O_2 is well explored. Generally, cytotoxicity caused by H_2O_2 begins with penetration into cells and then transformation to OH_ through Fenton's reaction causing intercellular damage (Laurita et al., 2015).

It has recently been found that peroxynitrite also plays an important role in oxidative stress and various neurodegenerative diseases, AIDS, arteriosclerosis and so on (Beckman et al., 1996). It oxidizes biomolecules directly or through H^+- or CO_2-catalyzed homolysis. As for direct reactivity, it has affinity on key parts in proteins such as thiols, iron–sulfur center and zinc fingers. The lifetime of peroxynitrite is relatively short, nonetheless, it can still cross membranes and reach deep within the cell, which allows it to interact with most of the important biomolecules (Naitali et al., 2010, 2012). Regarding the CAP decontamination of toxic compounds, OH_ as one of the strongest oxidative species initiates the toxic molecule oxidation, resulting in its degradation. However, other slower reaction pathways such as those caused by O_3 and H_2O_2 are shunted or even bypassed (Jiang et al., 2014).

Detailed studies to unravel the role of singlet oxygen species in inactivation of fungi have been carried out by Hashizume et al. (2013, 2014, 2015), Iseki et al. (2011) and Ishikawa et al. (2012). Through a comprehensive set of experiments, they reported that ROS, particularly O (3Pj), from an atmospheric pressure oxygen plasma potentially inactivates fungi. The action of atomic oxygen has been found to result in non-culturable, but viable fungal cells, where the spores of the resulting hyphae appear wrinkled.

Avramidis et al. (2010) applied atmospheric pressure DBD plasma treatment to *Ascochyta pinodella* and *Fusarium culmorum* and during light microscopy observations they found that after plasma treatment the cell walls and cell membranes structures were damaged, resulting in leakage of cytoplasm. These changes became prominent only after 60 s treatments and after 180 s the cells were flattened. Suhem et al. (2013) have reported results from a light microscopic study of radio-frequency plasma jet treated *Aspergillus flavus* cells. They observed that post-treatment, the conidiophores and vesicle were broken, which resulted in cell leakage and loss of viability. Hayashi et al. (2014) demonstrated a prominent role of atomic oxygen, O (1D) at 60 ppm for inactivation of *Aspergillus oryzae* and *Penicillium digitatum*. Lee et al. (2015) observed through scanning electron microscopy (SEM) that plasma treatment (using an atmospheric jet) resulted in considerable morphological alterations in fungal spores of *Cordyceps bassiana*, causing rupture, flattening and shrinkage, with surface wrinkling. SEM observations of Dasan et al. (2017) also revealed that the spores of *A. parasiticus* lose their integrity after plasma treatment for 30 s using a fluidized plasma bed and the cell contents disperse into clusters. Based on these observations, it may be hypothesized that plasma treatment results in cell wall destruction, making it permeable and thus allowing the leakage of intracellular components. This has been partially confirmed through circular dichroism (CD) and fluorescence studies of fungal spores where the spectral peak corresponding to cell wall protein was found to decay after plasma treatment. It has also been observed that plasma treatment results in destruction of the DNA in fungal spores, which has been confirmed from the decaying CD spectrum signature and loss of band intensity after gel electrophoresis (Lee et al., 2015). The extent of changes in fungal spores is due to the level of exposure to plasma species, as reported by Panngom et al. (2014), which showed considerable changes in plasma treated spore morphology of *Fusarium oxysporum* were not observed under electron microscopy, although the germination levels decreased and cells underwent apoptosis. Even when the active species density was not sufficiently high to cause spore structure destruction, the cells could undergo physiological changes because of apoptosis, causing an increase in accumulation of lipid bodies. This is likely due to the resultant Ar ions with relatively low antimicrobial activity (Panngom et al., 2014).

The fungal inactivation occurs through one or more of the following cellular effects:

- Inhibition of the cell membrane function for short plasma treatment times (ca. 30 s); subsequent treatment results in incomplete inactivation.
- At low doses, fungal cells could undergo apoptosis (Hashizume et al., 2015).
- Drastic morphological change of the cellular structures is not a necessity for fungal inactivation. However, intracellular nanostructural changes will be evident.
- Morphological changes, if they occur, can be observed as distinct changes in the cell membrane and an increase in its permeability (Cerioni et al., 2010; Dasan et al., 2016a; Kang et al., 2014).
- The ROS from plasma causes oxidation of intracellular organelles, particularly, in lipid phosphates which are oxidized by ROS to lipid peroxide through a chain reaction. In later stages, the oxidation of genomic DND and cellular proteins may also occur (Kang et al., 2014; Lu et al., 2014; Panngom et al., 2014).

So far, these studies have emphasized on the dominant role of ROS, while air plasma is a mixture of ROS and RNS. The roles and contributions of RNS and ultraviolet radiation in cold plasma remain less researched considering the difficulty in separating the effects of ROS and RNS. It has been suggested that fungal spores contain the protective pigment melanin in the cell wall layers, which confers resistance to external stresses, including UV (Eisenman and Casadevall, 2012).

Therefore, further research to understand fungal interactions with UV and RNS from plasma sources is desirable. The efficiency of microbial inactivation depends on the surface of the treating produce, plasma device, gas composition and mode of exposure. Produce like potatoes and strawberries take more time for complete destruction of microbes due to their grooves and uneven surfaces (Misra et al., 2019). The efficiency of microbial reduction was improved with an increase in the humidity of air.

These studies also demonstrated that sterilization efficacy increased with increased applied voltage and frequency of the plasma generating system. *A. flavus* cells were more sensitive to plasma compared with *A. parasiticus*, suggesting that the difference in sterilization effect was not directly related to plasma system efficacy, but it would be essential to perform a molecular and structural dependent study against the response of plasma species to cells during the treatment process.

14.8 MECHANISM OF MYCOTOXIN DEGRADATION

Being an emerging area, the degradation products of mycotoxins and the pathways during cold plasma treatment are sparse in the literature. The mycotoxin degradation pathways during cold plasma treatment are inevitably related to their molecular structure, the nature of the plasma chemistry and thus, the species interaction with toxin molecules (Pankaj et al., 2018b). Drawing an analogy from polymer science, one could argue that the presence of aromatic structures in polymers often slows down the degradation process during plasma treatments (Klarhöfer et al., 2010; ten Bosch et al., 2017). However, the mycotoxin degradation pathways during plasma treatment turn out to be remarkably different, being relatively small molecules. Wang et al. (2015) studied the chemistry of low-pressure plasma treated AFB1 and based on mass-spectrometry proposed degradation pathways. They anticipated the formation of an intermediate with C17H15O7, which is also a major degradation product of AFB1 after UV treatment. The cold plasma degradation of mycotoxins could be directly related to the free radicals (for example, O• and OH•) produced during the treatments. In a very recent study, Shi et al. (2017) also confirmed the same pathways when treating AFB1 using a high voltage DBD plasma source.

The reactive gas species that have been identified as primary contributors to aflatoxin degradation by humid air cold plasma include ozone, hydroxyl, and aldehyde radicals that form from the ionization of oxygen, water molecules and carbon dioxide precursors. The majority of the reactions are due to ozonolysis, involving sequential addition and cleavage reactions. The degradation pathways primarily involve sequential addition of a water molecule, hydrogen atom or aldehyde group to AFB1, or the epoxidation and oxidation reactions via action of the hydroperoxyl radical (HO2•). Several studies investigating the reaction of oxidative stressors with AFB1 have also suggested the breakdown of AFB1 at the C8 to C9 double bond of the dihydrofuran rings (Chen et al., 2014; Diao et al., 2013b).

It may be noted that the emission intensity of ultraviolet light during cold plasma treatment is much less than the UV intensity required for the effective degradation of aflatoxin (Laroussi and Leipold, 2004; Liu et al., 2009).

Wang et al. (2015) proposed reduced toxicity of the degradation products of AFB1 after cold plasma treatments, probably due to the loss of the double bond in the terminal furan ring since the furfuran moiety of AFB1 is important for its toxicity and carcinogenicity. The degradation of mycotoxins by changing their structure during plasma treatment could be related to the presence of UV photons, ozone or reactive ions and electrons. Significant contributions of reactive species other than ozone and UV photons to degrade mycotoxins or synergistic action of these species along with ozone and UV photons during cold plasma treatments could be possible, as the mycotoxin degradation efficacy of cold plasma technology is greater than that of ozone or UV treatments alone (Diao et al., 2013a; Luo et al., 2014; Mao et al., 2016). Further work in this direction is likely to generate better insights regarding the influence of each of the major reactive components.

14.9 FUNGAL INACTIVATION BY PLASMA IN FOODS

Some of the notable food sectors that could benefit from antifungal efficacy of cold plasma include fresh produce, food grains, nuts, spices, herbs, dried meat and fish industries (Mishra et al., 2016a). The range of microorganisms that have been shown to be inactivated using cold plasmas include Gram-positive as well as Gram-negative bacteria (Sysolyatina et al., 2014), biofilm forming bacteria (Puligundla and Mok, 2017), bacterial spores (Patil et al., 2014), yeast and fungi (Ishikawa et al., 2012), prions (Julák et al., 2011) and viruses (Puligundla an Mok, 2016). The differences in susceptibilities of fungi compared with bacteria in the action of cold plasma are likely associated with differences in cytology, morphology, the reproductive cycle and growth. The technology can be applied to different types of food products in both solid and liquid forms. In addition, the low energy consumption of such discharges and price-value inputs contribute to CAP being considered as an economically acceptable method.

14.9.1 Fruits and Vegetables

The fresh produce industry is frequently facing challenges of food-borne pathogen outbreaks. The pathogen inactivation effects of cold plasma potentially offer a viable treatment step for fresh produce to reduce the microbial load without adversely affecting nutritional and other key characteristics (Min et al., 2017, 2018; Misra et al., 2015; Niemira, 2012); they also ensure better food quality and an extended shelf life. Lacombe et al. (2015) studied the effects of a cold plasma jet in air on blueberries inoculated with yeasts and molds over treatment durations ranging between 15 and 120 s where the samples were placed at a distance of 7 cm from the plasma jet. It was observed that yeast and molds showed 0.8 to 1.6 log CFU/g order of reduction after 1 day and 1.5 to 2.0 log CFU/g reduction after 7 days. Plasma exposure longer than 60 s was reported to cause considerable reduction in firmness, anthocyanins and color.

The importance of operating conditions was evident in the studies carried out by Won et al. (2017), where a microwave powered nitrogen cold plasma treatment of mandarins at 0.7 kPa pressure and 1 L/min flow rate for 10 min was reported to inactivate *Penicillium italicum* and reduce the disease occurrence by 84%. The inhibition of *P. italicum* was found to depend on the gas, power input and treatment duration. The inhibition increased with an increase in plasma power and treatment time, whereas maximum inhibition was attained using N_2 at 900 W for 10 min as compared with helium and 4:1 N_2/O_2 mixtures. Schnabel et al. (2015) carried out a detailed study on the applicability of microwave plasma processed air for decontamination of the yeast *Candida albicans* in fresh produce. They observed complete inactivation of yeast by 6.2 $\log_{10}$ steps on apple peel and strawberry, whereas only 3 $\log_{10}$ steps in apple pulp, indicating that surface features also have an impact on the efficacy of plasma. A limited number of studies have also evaluated the evolution of spoilage microorganisms in plasma treated plant produce during storage. Min et al. (2017) reported the inactivation of yeasts and molds ranging from 0.9 to 1.7 log CFU/g in packaged lettuce exposed to a high voltage (34.8 kV, at 1.1 kHz frequency) dielectric barrier pin type discharge system after 5 min. This study also demonstrated that the post-packaged treatment storage of lettuce after 7 days did not have any effect on decontamination. Such time dependent effects point at the likelihood of fungal cell recovery post-plasma treatment and deserve consideration for practical applications.

The use of fungicides for controlling fungi in the field or after harvest is associated with certain drawbacks. However, fungicide application raises serious concerns about residues of these compounds in food products, particularly because many fungicides are suspected or potential oncogenes (Saharan et al., 2004). In order to be efficient, any fungicide must be completely lethal against the fungal species, otherwise, it could stimulate mycotoxin production *in vitro* (Jouany, 2007). Interestingly, it has been observed and suggested that cold plasma can effectively dissipate fungicide residues in strawberries and blueberries (Misra, 2015; Misra et al., 2014; Sarangapani et al., 2017).

14.9.2 Action of Plasma on Endogenous Enzymes

Fruits and vegetables, besides getting bio-deteriorated, are also spoiled due to enzymatic browning, which is considered as a secondary loss during postharvest handling and storage. The endogenous enzymes, particularly polyphenol oxidase and peroxidase, are the major causes for enzymatic browning as they oxidize phenols at the expense of H_2O_2, leading to "off" flavors (Surowsky et al., 2014). Different methods used to prevent the enzymatic browning are heating, blanching and commercial sterilization (Elez-Martinez et al., 2009). The other non-conventional techniques used in the inactivation of endogenous enzymes are pulsed electric fields (Zhong et al., 2007), irradiation (Zhang et al., 2006) and high pressure processing. Enzymes are inactivated through oxidation reactions mediated by free radicals and atomic oxygen. Thus, plasma can be applied to manage these biological compounds as there was also a decrease in enzymatic activity of trypsin (zero at 4 Jcm^{-2}) after the application of plasma. It was also seen that plasma was able to change the 3D structure of proteins in trypsin enzymes due to the cleavage of peptides bonds. It was found that the activity of polyphenol oxidase in cold plasma treated guava (*Psidium guajava*) pulp and the whole fruit was reduced by 70% and 10% in 300 s at 2 kv, respectively (Mishra et al., 2019). In a recent investigation, the kinetics of the inactivation of tomato peroxidase enzymes fitting in different kinetic modeling like first-order, Weibull and logistic models were made by Pankaj et al. (2013). They observed that there is a decrease in enzyme activity at different voltages using atmospheric air dielectric barrier discharge plasma.

14.9.3 Action of Plasma on Herbs and Spices

Dry food systems with low water activity are often susceptible to fungal contamination and their safety remains a challenge for the processors (Syamaladevi et al., 2016). Despite a higher resistance observed in dry food systems, plasma treatment has been found to be considerably effective in reducing molds as compared with other known methods. Common examples of such food classes include herbs and spices, which, when contaminated, can cause rapid spoilage of the foods to which they are applied. A further challenge with the decontamination of herbs and spices is their susceptibility to loss of volatiles on exposure to heat (Hertwig et al., 2015).

The ability of plasma treatment to reduce the yeast and mold counts on pepper seeds and crushed oregano was first reported by Hertwig et al. (2015). It was found that 5 min plasma treatment completely inactivated yeast and mold on black pepper seeds, while 60 min treatment of crushed oregano reduced the population of microorganisms by 1.8 log_{10} CFU/g. Kim et al. (2013) investigated the effect of treatment time, power and gas composition on the decontamination of red pepper inoculated with *Aspergillus flavus* using a microwave powered cold plasma system. The highest inactivation of *A. flavus* of 2.5 log was observed using N_2 plasma at operating power of 900 W after 20 min, whereas for other gases such as He, N_2+O_2 and He-O_2 mixture, the inactivation levels were 2.0, 0.3 and 0.3 log, respectively.

Kim et al. (2017) explored the effects of microwave cold plasma treatment on *Aspergillus brasiliensis* inoculated into onion powder. They observed that plasma treatment following vacuum drying of the powder exhibited greater fungal reduction (1.5 log_{10} spores/cm^2) compared with that after hot air drying (0.7 log_{10} spores/cm^2). This difference was attributed to the surface cracks in hot air-dried samples versus a smooth surface in vacuum dried samples. While mold growth in tea is a rare incidence, tea leaves served as a model dry herb in research studies. Amini and Ghorannevi (2016) reported that the decontamination effect of argon cold plasma jet in black and green tea against mold and yeast was dependent on treatment time and initial microbial concentration. They observed a complete inactivation of the molds and yeasts after 7 min of treatment, starting from initial populations of 3.30 and 3.0 log_{10} CFU/g, respectively.

Overall, the success of cold plasma for herbs/spices is established, as far as inactivation of molds is concerned. However, no studies have attempted to look at the volatile profile of plasma treated herbs/spices. This is crucial information for attracting industries and further studies in this direction are desired.

14.9.4 Food Grains and Nuts

Owing to their extensive use as human foods and livestock feeds, the microbiology and safety of grains, seeds, nuts and their products deserve high importance. Fungal attack in cereal grains could be due to field fungi, which attack grains at high moistures, or storage fungi that attack stored grains at relatively low moistures (Los et al., 2018b). Typical examples of field fungi include species of *Alternaria*, *Cladosporium* and *Fusarium*, whereas storage fungi include *Eurotium*, *Aspergillus* and *Penicillium*.

Aspergillus and *Penicillium* are ubiquitous invaders of nuts and can produce aflatoxins. Basaran et al. (2008) reported the sterilizing effect of low-pressure cold plasma against *Aspergillus parasiticus*, where the cells were reduced by 1 $\log_{10}$ CFU/g after 5 min treatment and by an additional 1 $\log_{10}$ CFU/g after 10 min treatment. A 5 $\log_{10}$ CFU/g decrease for *A. parasiticus* populations on hazelnut, peanut and pistachio surfaces was recorded after 20 min while a 3-log reduction of *Aspergillus* spp. and *Penicillium* spp. was recorded at the same time.

Dasan et al. (2016b) studied the sterilization effect of fluidized bed plasma treatment on *Aspergillus flavus* and *Aspergillus parasiticus* inoculated on maize grains. A maximum reduction of 5.48 and 5.20 $\log_{10}$ CFU/g in *A. flavus and A. parasiticus*, respectively, after 5 min plasma treatment was observed. In a more recent study, an ambient air corona discharge plasma jet operating at 20 kV, 58 kHz at flow rate of 2.5 m/s was tested on rapeseed surface. About 1.8 and 2.0 $\log_{10}$ CFU/g reduction in yeast and mold cells respectively was observed after 3 min of plasma exposure (Puligundla et al., 2017).

Pignata et al. (2014) reported a 2 $\log_{10}$ reduction in the fungal population of *Aspergillus brasiliensis* on pistachios after 1 min of low-pressure plasma treatment in argon/oxygen (10:1 v/v) using a DC discharge (600 W, 15.56 MHz RF).

Amini and Ghoranneviss (2016) reported complete destruction of *A. flavus* inoculated on fresh and dried walnut surfaces after 11 min and 10 min of argon plasma jet exposure. Moreover, argon plasma jet exposed walnut showed insignificant residue after 15 and 30 days storage period at 4 °C, and there were no changes in phenolic content and antioxidant activity of treated and control walnut after treatment and 30 days of storage. In another study, Dasan et al. (2016a) reported a reduction in *A. flavus* and *A. parasiticus* cells on hazelnuts with 4.50 and 4.19 $\log_{10}$ CFU/g reduction in *A. flavus* and *A. parasiticus* population, respectively, after 5 min plasma treatment in dry air. Moreover, argon and oxygen CAP proved to be efficient against *A. brasiliensis* contaminating pistachios (Pingata et al., 2014). Nearly 70% decontamination of AFB1 from dehulled hazelnuts was observed when CAP treatment with different mixtures of oxygen and nitrogen were performed (Siciliano et al., 2016). It is worthwhile noting that although external contamination can easily be decontaminated using gas plasma, whether the active species can penetrate the internal sites of colonization in large storages of grains and nuts remains unknown.

14.10 EFFECT ON PACKING MATERIAL

Food packaging materials are responsible for the protection of food materials from the outside environment during handling, transportation and distribution. Applications of nanotechnology in packaging have become widespread to improve the barrier properties of packing material and this can be achieved by cold plasma processing (Pahwa and Kumar, 2018; Modic et al., 2017). Plasma processing is well known to change or make surface modifications in packing materials. Plasma deposition of heat-sensitive materials such as vitamins, antioxidants and antimicrobials into packaging material may be sought as potential alternatives in the emerging field of antimicrobial and active packaging. It serves purposes for surface treatments such as cleaning, coating, printing, painting and adhesive bonding (Pankaj et al., 2014). Cold plasma is used in the decontamination of packing material externally where the chance of shadow effects is negligible as plasma flows all around the surface (Sakila et al., 2012).

One of the most interesting applications of CAP is the in-package treatment of food. The background microflora containing fungal species was reduced by a 2 log reduction in strawberries when treated with CAP generated between the electrode gap and inside a sealed package. Misra et al. (2014) reported the sterilization efficacy of in-package treatment by indirect cold plasma generated in atmospheric pressure. They observed that a 5 min treatment at 60 kV resulted in a 3.3 log cycle reduction of naturally occurring yeasts and molds on strawberry surface after 24 hr of in-pack storage. However, unlike Lacombe et al. (2015), they did not find significant differences in the respiration rate, color and firmness in plasma treated strawberries.

Low temperature gas plasma sterilization allows for fast and safe sterilization of packaging materials such as plastic bottles, lids and biofilms (Modic et al., 2017) without adversely affecting the properties of the material or leaving any residues. This technology can be used for sterilization of heat-sensitive packing materials like polythene ethylene and polycarbonate, as the temperature is low.

Surfaces of polymers, particularly for edible packaging films, should be more hydrophobic with lower surface energies (Vesel and Mozetic, 2017). Using plasma as the transport mechanism and the catalyst, one material can be deposited (in a very thin layer) onto the surface of another material, thereby transferring some of its qualities. It was found in studies that the oxygen transport properties of the SiOx coating on polyethylene terephthalate (PET), low and high-density polyethylene (LDPE, HDPE) and polypropylene (PP) films were less than for normal material for diffusivity, obtained using cold plasma technology, in comparison with experimental data with computer models.

Pankaj et al. (2014) studied the surface topography of zein film using atomic force microscopy. The roughness of plasma treated zein film increased with a root mean square of 100 nm from 20 nm in untreated films, this is due to surface etching that occurred during plasma treatment.

14.11 MYCOTOXINS DEGRADATION BY PLASMA

In essence, the food supply chain urgently needs a mycotoxin decontamination method that is low-cost, highly effective and can be applied to contaminated food to minimize waste and enhance safety for the consumer.

Recently, exciting preliminary results have shown that CAP has considerable potential to reduce food contamination and improve food safety (Hojnik et al., 2017). The concept of treating food with non-thermal plasma for the purposes of microbial inactivation has been under consideration for well over a decade (Pankaj et al., 2018a), and it has been shown that the RONS produced within plasma are highly effective antimicrobial agents that are also capable of degrading a wide variety of toxic compounds, including mycotoxins (Schmidt et al., 2018).

In an early study, Park et al. (2007) employed a microwave powered atmospheric pressure cold argon plasma treatment to evaluate the effects on mycotoxins. They found that mycotoxins such as aflatoxin B1 (AFB1), deoxynivalenol (DON, vomitoxin) and nivalenol (NIV) were degraded completely within 5 s of treatment. In the subsequent year, Basaran et al. (2008) investigated the possibility of dissipating mycotoxin from nut surfaces using a low-pressure inductively coupled plasma source. They found that 20 min of air plasma allowed the concentration of a mixture of aflatoxins (B1, B2, G1 and G2) to decrease by up to 50%. Some years later, Ouf et al. (2015) reported a complete degradation of fumonisin B2 and ochratoxin A produced by *Aspergillus niger* inoculated onto date palm slices by double jet argon plasma treatment at 3.5 L/min. This treatment also eliminated the *A. niger* spores inoculated onto date palm fruit slices. The study not only suggested the potential of cold plasma treatment in degrading mycotoxins in food but also for reducing the capability of harmful fungi in mycotoxin production in food and feed. This capability of cold plasma may be related to its significant influence on genes regulating the biosynthesis of mycotoxins in fungi. Recently, aflatoxin B1 was shown to degrade up to 88% when treated using 300 W RF plasma for 10 min and the degradation products of aflatoxin B1 were noted to be less toxic (Wang et al., 2015). Cold plasma treatment of aflatoxin-inoculated hazelnut resulted in a more than 70% reduction in the concentration of aflatoxins after 12 min (Siciliano et al., 2016). Aflatoxins B1 and G1 were more sensitive to plasma treatments compared with aflatoxins B2 and G2, respectively. It has also been suggested that cold plasma can be scaled up and included in the hazelnut processing line after dehulling and before roasting operations.

Considering this, CAP technology also has a high potential as a decontamination tool for both mycotoxin-producing fungi and the mycotoxins they produce. Different set-ups in both low and atmospheric pressure conditions have been used on mycotoxins such as aflatoxin B1, deoxynivalenol and nivalenol (AFB1, DON and NIV, respectively), resulting in high decontamination rates in only a matter of seconds. Treatments of the mycotoxin-producing fungi, for example, one of the main mycotoxin producers *Aspergillus* spp. and *Penicillium* spp., with plasma, demonstrated very promising results as well.

In a recent investigation (Misra et al., 2014), it was reported that the decrease in total mesophilic count was 12–85% and yeast and mold count by 44–95% in cold plasma treated strawberries. In raw milk treated with low temperature plasma for 3 min, destruction of *Escherichia coli* was to the tune of 54% (Gurol et al., 2012). Application of air plasma on nuts (peanuts, hazelnuts and pistachio) led to a 50% reduction in total aflatoxins and 20% reduction using SF6 (Basaran et al., 2008; Dasan et al., 2016b). Treating with plasma for 15 min could achieve 2.72, 1.76 and 0.94 log reductions of *Salmonella typhimurium* on lettuce, strawberry and potato, respectively.

The use of an atmospheric pressure fluidized bed plasma system with air and nitrogen as a feed gas for the inactivation of *A. flavus* and *A. parasiticus* contaminated maize resulted in a 5.48 log reduction (Ziuizina et al., 2014). Furthermore, the production of fumonisin B2 and ochratoxin A (FB2 and OTA, respectively) was inhibited after the exposure of date fruits infested with *A. niger* to an argon CAP source.

In a recent study, Devi et al. (2017) investigated the effect of cold plasma treatment on the growth of *A. flavus* and *A. parasiticus* and production of aflatoxins in groundnut inoculated with fungal species. Application of air plasma treatment at 60 W reduced the *A. flavus* and *A. parasiticus* growth in groundnut by 97.9% and 99.3%, respectively. Additionally, a reduction in aflatoxins B1 production by 70% and 90% was reported for plasma exposure at 40 W for 15 min and 60 W for 12 min, respectively.

Ten Bosch et al. (2017) used atmospheric DBD cold plasma to degrade deoxynivalenol, zearalenone, enniatins, fumonisin B1 and T2 toxin produced by *Fusarium* spp., sterigmatocystin produced by *Aspergillus* spp. and AAL toxin produced by *Alternaria alternata*. The temperatures of the gas and substrate (which was a cover glass) during cold plasma treatments were lower than 60 °C. They observed almost complete degradation of mycotoxins in 60 s but the inactivation efficacy of cold plasma was dependent on the type of mycotoxins and the matrix. For instance, when extracts of rice cultures of fungal strains producing these mycotoxins containing approximately 100 μg/mL of each toxin were exposed to plasma, the mycotoxin degradation was smaller compared with mycotoxins in solution. This is most likely because the reactive species in plasma could be scavenged by the different components in the matrix.

No stable residues of mycotoxins were observed with HPLC-MS after cold plasma treatment, which was suggested to be due to the conversion of mycotoxins to volatile compounds (ten Bosch et al., 2017). In a recent study, Shi et al. (2017) studied the degradation of aflatoxin in corn kernels using high voltage DBD plasma operating in air and modified atmospheres. They observed a rapid degradation of the toxins in a high oxygen atmosphere and at elevated humidity levels. This observation can be attributed to the favorable plasma chemistry involving the production of greater amounts of hydroxyl radicals and ozone with high humidity and oxygen, as was experimentally demonstrated in previous studies (Moiseev et al., 2014; Patil et al., 2014).

Reports pertinent to mycotoxin breakdown using plasma have focused on several aspects—the type of mycotoxin, the substrate, the plasma source and the process parameters.

14.12 LIMITATIONS AND TOXICOLOGY OF PLASMA TREATMENT

Cold plasma is a novel, non-thermal technology that has shown good potential for food decontamination. However, there are some reports about the limitations of plasma processing such as a reduction in color, a decrease in the firmness of fruits, an increase in acidity and an increase in the oxidation of lipids. Another main disadvantage of this technology is that it is not possible to use it in the inactivation of endogenous enzymes, which are present intact in the whole fruits because the plasma effect is a surface phenomenon.

Mycotoxins produced by fungi in cereals and feed materials can be confined to surfaces, which can be effectively degraded using plasma treatments without inducing significant changes in nutritional components. However, the effect of cold plasma on mycotoxins in flours and resulting quality changes has not received the attention it deserves. In addition, it is recommended that researchers should explore the suitability of cold plasma treatment as an additional decontamination step along with conventional approaches to understand the synergistic or additive potential to reduce fungi and mycotoxins.

Plants produce stress-induced secondary metabolites in their epidermal layer when they are exposed to adverse conditions like UV radiations to protect their cells. These secondary metabolites may induce pathogenesis-related proteins, some of which have high allergenicity.

The allergic activity of proteins will either remain stable or can be lowered, but it is rare that an increase is observed. Kasera et al. (2012) investigated the allergenicity of legume proteins using the combined effects of cooking and irradiation. It was reported that there was a reduction of IgE binding to both soluble and insoluble proteins that resulted in attenuating allergenicity of legume proteins. Lectins treated with gamma irradiations showed changes in the hydrophobic surface of proteins, resulting in protein misfolding and aggregation to reduce or eliminate allergenicity (Antonio et al., 2011). These reports concluded that exposure with gamma irradiation can reduce allergenicity of proteins in lectins.

Until now, no investigation has been carried out on the formation of any toxic compounds after the application of cold plasma in food products. Any technology used for food processing should not affect allergenicity of food constituents. However, there is presently no such data available on the allergenicity of plasma treated food products. No studies have been conducted on the formation of toxic compounds in plasma treated foods.

The effectiveness of cold plasma treatment depends on multiple intrinsic and extrinsic parameters (Smet et al., 2016, 2018), including surface characteristics, type of food and feed materials, nature and structure of mycotoxins, type of fungi and their attachment to surfaces, ability of antimicrobial species to diffuse and spread on surfaces, their lifetime during and after treatments, time of treatment, cost effectiveness and so on. A better understanding of the gas phase chemistry of air plasma is required as it depends on the characteristics of the surrounding air and atmospheric conditions such as relative humidity and temperature.

However, most of the research is largely focused on microbial inactivation studies, with limited emphasis on food quality. Cold plasma processing has been shown to affect the quality attributes of the food products during treatment as well as in storage. It presents a research opportunity to further explore the effects of cold plasma on the physico-chemical and sensory properties of the food products at the molecular level. The differences in the reported studies demonstrated the need for mechanistic studies to understand the interaction of plasma reactive species with food components. Optimization studies are also required to avoid the negative impacts on quality, such as accelerated lipid oxidation, loss of vitamins and sensory characteristics. The precise understanding of the mechanisms and control over the quality attributes will be required for cold plasma technology to realize its full potential on the commercial scale (Pankaj et al., 2018b).

14.13 FUTURE PERSPECTIVES

Cold plasma is one of the most promising and unique preservation treatments. It is effective at ambient temperatures, responsible for microbial destruction and surface modification of the substrate and deactivates mycotoxins in/on food. Therefore, it has minimum thermal effects on nutritional and sensory quality parameters of food with no chemical residues compared with conventional preservatives techniques that have some detrimental effects on nutritional quality. It has been convincingly demonstrated at the experimental level that this technique can efficiently destroy fungi present on the surface of food and decontaminate the mycotoxins that these organisms secrete.

Cold plasma could be a potential alternative to reduce fungi on food materials including grains, spices, fruits, vegetables, meat and so on (Schmidt et al., 2018). Compared with several conventional and non-thermal approaches, cold plasma treatments act rapidly against molds, require low energy input and have a relatively milder impact on quality. Several findings also indicate that cold plasma could very well serve as a rapid, effective and economically viable technology in degrading mycotoxins as compared with UV, heat or chemical treatments (Misra et al., 2016a, 2011; Surowsky et al., 2014). Plasma sterilization provides high efficacy, preservation and does not introduce toxicity to the medium. The most important part is to select (choosing) particular gases, which already possess germicidal properties so that the efficiency of plasma sterilization can be increased.

In favor over many of the traditional food decontamination methods, plasma-based decontamination methods are generally lower-cost and ecologically benign. The total energy and costs associated with plasma treatments are less than or comparable with conventional mycotoxin reduction technologies used in the food and feed industry (Hojnick et al., 2019).

Most importantly, plasma-based mycotoxin decontamination of food has been demonstrated significantly more efficient in both the mycotoxin degradation level and the speed of decontamination in comparison with conventional decontamination methods, as presented in the case of one of the most toxic mycotoxins, AFB1.

Cold plasma is used efficiently for the sterilization and modification of packaging polymers but there is a huge application in food processing. Plasma can be used for starch modification as an additive and as a filler component in packing materials

Within the context of non-thermal technologies, cold plasma is of contemporary interest to researchers in food science and physics, as well as the food industry.

The amount of energy consumption and stability depends on the type of discharges used for treatment. Based on these parameters, the application of plasma should be optimized for maximum efficiency at a low cost of operation. Many scientists successfully applied plasma on foods (solids and liquids) for the microbial inactivation but they did not explain its effects on the nutritional qualities and toxicology of treated foods. Therefore, it is necessary that the application of plasma on foods should be recognized as generally recognized as safe (GRAS) after intense study and research (*in vitro* and *in vivo*) in this field.

However, further investigations into this subject are required for its practical usage. Overall, cold plasma technology also offers opportunities for multiple points of application within the food chain, while ensuring microbiological as well as chemical safety. Future research activities are likely to yield more information for the development of validation protocols for antimycotic or mycotoxin reduction action in food and feed, paving the path for potential upscaling of the technology for industrial adoption.

Cold plasma technology has the potential to reduce fungi and degrade mycotoxins in/on food and feed materials effectively as it could be a more sustainable method, requiring smaller energy input and investment. However, scaling up of cold plasma technology to reduce pathogenic fungi and mycotoxins in/on food and feed materials will need several aspects addressed. Cold plasma treatments should be able to handle bulk quantities of food and feed materials. Therefore, studies exploring the feasibility of batch or continuous plasma systems to handle large quantities of food and feed materials are required.

Although cold plasma technology is not yet used commercially on a large scale, the equipment should be readily scalable. However, research efforts must be taken to evaluate the expenditure for the treatment of large quantities of food commodities at industry level and also the quality, safety and wholesomeness of the food commodities.

Before commercialization of cold plasma technology can be realized, the molecular mechanisms and kinetics of plasma-based mycotoxin decontamination should be better characterized in order to become standardized. In the future, we hope that plasma processing becomes a common process for food industries.

To support the use of plasma technology in minimizing fungal deterioration and the decontamination of mycotoxins, additional experimental work is needed to address the following:

- Designing plasma-forming systems for efficient mycotoxin decontamination of various types and sizes of food products.
- Drawing correlations between the composition of the plasma and the structure of the mycotoxin degradation products. As toxicities of the mycotoxin degradation products can be experimentally determined, the mycotoxin decontamination efficiency can be defined as well.
- Drawing firm correlations between different plasma operating parameters and the specific reactive chemical species formed.
- Examining the effects of different plasma treatments on the quality of food products, for example, on their nutritional value and organoleptic qualities.
- Testing if hybrid plasma-conventional systems for mycotoxin decontamination of food products can be even more effective.

REFERENCES

Aiko, V. and A. Mehta. 2016. Prevalence of toxigenic fungi in common medicinal herbs and spices in India. *Biotechnology* 6(2): 159.

Amini, M. and M. Ghoranneviss. 2016. Effects of cold plasma treatment on antioxidants activity, phenolic contents and shelf life of fresh and dried walnut (*Juglans regia* L.) cultivars during storage. *LWT - Food Sci. Technol.* 73: 178–184.

Antonio, F.M., R.M. Vaz, P.B. Costa et al. 2011. Gamma irradiation as an alternative treatment to abolish allergenicity of lectins in food. *Food Chem.* 124(4): 1289–1295.

Avramidis, G., B. Stüwe, R. Wascher et al. 2010. Fungicidal effects of an atmospheric pressure gas discharge and degradation mechanisms. *Surf. Coat. Technol.* 205: S405–S408.

Basaran, P., N. Basaran and L. Oksuz. 2008. Elimination of Aspergillus parasiticus from nut surface with low-pressure cold plasma (LPCP) treatment. *Food Microbiol.* 25: 626–632.

Beckman, J.S. and W.H. Koppenol. 1996. Nitric oxide, superoxide, and peroxynitrite: The good, the bad, and ugly. *Am. J. Physiol. Cell Physiol.* 271: C1424–C1437.

Betina, V. 1984. *Mycotoxins: Production, Isolation, Separation and Purification*. Amsterdam, The Netherlands: Elsevier.

Bhat, R., R.V. Rai and A.A. Karim. 2010. Mycotoxins in food and feed: Present status and future concerns. *Com. Rev. Food Sci. Food Saf.* 9: 57–81.

Biomin. 2017. Biomin World mycotoxin survey 2017. Annual Report No. 14. Accessed May 11, 2018. https://www.biomin.net/en/blogposts/2017-biomin-mycotoxin-survey-results/

Bryla, M., A. Was'kiewicz, G. Podolska et al. 2016. Occurrence of 26 Mycotoxins in the grain of cereals cultivated in Poland. *Toxin* 8: 160. doi: 10.10.3390/toxins8060160

Cerioni, L., S. I. Volentini, F. E. Prado et al. 2010. Cellular damage induced by a sequential oxidative treatment on *Penicillium digitatum*. *J. Appl. Microbiol.* 109(4): 1441–1449.

Chen, R., F. Ma, P.W. Li et al. 2014. Effect of ozone on aflatoxins detoxification and nutritional quality of peanuts. *Food Chem.* 146: 284–288.

da Rocha, M.E.B., F.D.C.O. Freire, F.E.F. Maia et al. 2014. Mycotoxins and their effects on human and animal health. *Food Control* 36: 159–165.

Dasan, B.G., I.H. Boyaci and M. Mutlu. 2016a. Inactivation of aflatoxigenic fungi (*Aspergillus* spp.) on granular food model, maize, in an atmospheric pressure fluidized bed plasma system. *Food Cont.* 70: 1–8.

Dasan, B.G., I.H. Boyaci and M. Mutlu. 2017. Nonthermal plasma treatment of *Aspergillus* spp. spores on hazelnuts in an atmospheric pressure fluidized bed plasma system: Impact of process parameters and surveillance of the residual viability of spores. *J. Food Eng.* 196: 139–149.

Dasan, B.G., M. Mutlu and I.H. Boyaci. 2016b. Decontamination of *Aspergillus flavus* and *Aspergillus parasiticus* spores on hazelnuts via atmospheric pressure fluidized bed plasma reactor. *Int. J. Food Microbiol.* 216: 50–59.

Devi, Y., R. Thirumdas, C. Sarangapani et al. 2017. Influence of cold plasma on fungal growth and aflatoxins production on groundnuts. *Food Control* 77: 187–191.

Diao, E., H. Hou, B. Chen et al. 2013a. Ozonolysis efficiency and safety evaluation of aflatoxin B1 in peanuts. *Food Chem. Toxicol.* 55: 519–525.

Diao, E., H. Hou and H. Dong. 2013b. Ozonolysis mechanism and influencing factors of aflatoxin B1: A review. *Trends Food Sci. Technol.* 33(1): 21–26.

Durek, J., O. Schlüter, A. Roscher et al. 2018. Inhibition or stimulation of ochratoxin A synthesis on inoculated barley triggered by diffuse coplanar surface barrier discharge plasma. *Front. Microbiol.* 9: 2782. ldoi: 10.3389/fmicb.2018.02782

Eisenman, H.C. and A. Casadevall. 2012. Synthesis and assembly of fungal melanin. *Appl. Microbiol. Biotechnol.* 93(3): 931–940.

Elez-Martinez, P., R. Solvia-Fortuny and O. Martin-Belloso. 2009. Impact of high intensity pulsated electric fields on bioactive compounds in Mediterranean plant-based foods. *Nat. Prod. Commun.* 4(5): 741–746.

Eskola, M., G. Kos, C.T. Elliott et al. 2019. Worldwide contamination of food-crops with mycotoxins: Validity of the widely cited 'FAO estimate' of 25%. *Crit. Rev. Food Sci. Nutr.* doi: 10.1080/10408398.2019.1658570

European Commission. 2006a. Regulation (EC) no 1881/2006 of 19 December 2006 setting maximum levels for certain contaminants in foodstuffs. *Off. J. Eur. Union* 364: 5–24.

European Commission. 2006b. Recommendation 2006/576/ec of 17 August 2006 on the presence of deoxynivalenol, zearalenone, ochratoxin A, T-2 and HT-2 and fumonisins in products intended for animal feeding. *Off. J. Eur. Union* 229: 7–9.

European Commission. 2013. Recommendation 2013/165/EU of 27 March 2013 on the presence of T-2 and HT-2 toxin in cereals and cereal products. *Off. J. Eur. Union* 91: 12–15.

Feichtinger, J., A. Schultz, M. Walker et al. 2003. Sterilisation with low-pressure microwave plasmas. *Surf. Coat. Technol.* 174–175: 564–569. doi: 10.1016/S0257-8972(03)00404-3

Fridman, A. 2008. *Plasma Chemistry*. New York, NY: Cambridge University Press.

Gurol, C., F.Y. Ekinci, N. Aslan et al. 2012. Low temperature Plasma for decontamination of *E. coli* in milk. *Int. J. Food Microbiol.* 15: 157. doi: 10.1016/j.ijfoodmicro.2012.02.016

Halász, A., R. Lásztity, T. Abonyi et al. 2009. Decontamination of mycotoxin-containing food and feed by biodegradation. *Food Rev. Inter.* 25(4): 284–298.

Hashizume, H., T. Ohta, J. Fengdong et al. 2013. Inactivation effects of neutral reactive-oxygen species on *Penicillium digitatum* spores using non-equilibrium atmospheric-pressure oxygen radical source. *Appl. Phys. Lett.* 103(15): 153708.

Hashizume, H., T. Ohta, K. Takeda et al. 2014. Oxidation mechanism of *Penicillium digitatum* spores through neutral oxygen radicals. *Jap. J. Appl. Phys.* 53(1): 010209.

Hashizume, H., T. Ohta, K. Takeda et al. 2015. Quantitative clarification of inactivation mechanism of *Penicillium digitatum* spores treated with neutral oxygen radicals. *Jap. J. Appl. Phys.* 54(1S): 01AG05.

Hayashi, N., Y. Yagyu, A. Yonesu et al. 2014. Sterilization characteristics of the surfaces of agricultural products using active oxygen species generated by atmospheric plasma and UV light. *Jap. J. Appl. Phys.* 53(5S1): 05FR03.

Hertwig, C., K. Reineke, J. Ehlbeck et al. 2015. Impact of remote plasma treatment on natural microbial load and quality parameters of selected herbs and spices. *J. Food Eng.* 167: 12–17.

Hojnik, N., U. Cvelbar, G. Tavcar-Kalcher et al. 2017. Mycotoxin decontamination of food: Cold atmospheric pressure plasma versus "Classic" decontamination. *Toxins* 9(5): 151. doi: 10.3390/toxins9050151

Hojnik, N., M. Modic, G. Tavcar-Kalcher et al. 2019. Mycotoxin decontamination efficacy of Atmospheric Pressure Air Plasma. *Toxins* 11(4): 219. doi: 10.3390/toxins11040219

Iseki, S., H. Hashizume, F. Jia et al. 2011. Inactivation of *Penicillium digitatum* spores by a high-density ground-state atomic oxygen-radical source employing an atmospheric-pressure plasma. *Appl. Phys. Express* 4(11): 116201.

Ishikawa, K., H. Mizuno, H. Tanaka et al. 2012. Real-time in situ electron spin resonance measurements on fungal spores of *Penicillium digitatum* during exposure of oxygen plasmas. *Appl. Phys. Lett.* 101: 013704.

Jiang, B., J. Zheng, S. Qiu et al. 2014. Review on electrical discharge plasma technology for wastewater remediation. *Chem. Eng. J.* 236: 348–368.

Jouany, J. P. 2007. Methods for preventing, decontaminating and minimizing the toxicity of mycotoxins in feeds. *Anim. Feed Sci. Technol.* 137(3–4): 342–362.

Julák, J., O. Janoušková, V. Scholtz et al. 2011. Inactivation of prions using electrical DC discharges at atmospheric pressure and ambient temperature. *Plasma Processes Polym.* 8(4): 316–323.

Kang, M. H., Y.J. Hong, P. Attri et al. 2014. Analysis of the antimicrobial effects of nonthermal plasma on fungal spores in ionic solutions. *Free Radical Biol. Med.* 72: 191–199.

Karaca, H., Y.S. Velioglu and S. Nas. 2010. Mycotoxins: Contamination of dried fruits and degradation by ozone. *Toxin Rev.* 29: 51–59.

Karlovsky, P., M. Suman, F. Berthiller et al. 2016. Impact of food processing and detoxification treatments on mycotoxin contamination. *Mycotoxin Res.* 32(4): 179–205.

Kasera, R., A.B. Singh, R. Kumar et al. 2012. Effect of thermal processing and γ-irradiation on allergenicity of legume proteins. *Food Chem. Toxicol.* 50: 3456–3461.

Kim, J. E., D.U. Lee and S.C. Min. 2013. Microbial decontamination of red pepper powder by cold plasma. *Food Microbiol.* 38: 128–136.

Kim, J. E., Y.J. Oh, M.Y. Won et al. 2017. Microbial decontamination of onion powder using microwave-powered cold plasma treatments. *Food Microbiol.* 62: 112–123.

Klarhöfer, L., W. Viöl and W. Maus-Friedrichs. 2010. Electron spectroscopy on plasma treated lignin and cellulose. *Holzforschung* 64(3): 331–336.

Klich, M. A. 2007. Environmental and developmental factors influencing aflatoxin production by *Aspergillus flavus* and *Aspergillus parasiticus*. *Mycoscience* 48: 71–80.

Lacombe, A., B.A. Niemira, J.B. Gurtler et al. 2015. Atmospheric cold plasma inactivation of aerobic microorganisms on blueberries and effects on quality attributes. *Food Microbiol.* 46: 479–484.

Laroussi, M., M. Kong and G. Morfill. 2012. *Plasma Medicine: Applications of Low Temperature Gas Plasmas in Medicine and Biology*. Cambridge, UK: Cambridge University Press.

Laroussi, M. and F. Leipold. 2004. Evaluation of the roles of reactive species, heat, and UV radiation in the inactivation of bacterial cells by air plasmas at atmospheric pressure. *Int. J. Mass Spectrom.* 233(1–3): 81–86.

Laurita, R., D. Barbieri, M. Gherardi et al. 2015. Chemical analysis of reactive species and antimicrobial activity of water treated by nanosecond pulsed DBD air plasma. *Clin. Plasma Med.* 3: 53–61.

Lawley, L., L. Curtis and J. Davis. 2008. *The Food Safety Hazard Guidebook*. Cambridge, UK: RSC Publishing.

Lee, G. J., G.B. Sim, E.H. Choi et al. 2015. Optical and structural properties of plasma-treated *Cordyceps bassiana* spores as studied by circular dichroism, absorption, and fluorescence spectroscopy. *J. Appl. Phys.* 117(2): 023303.

Liebermann, M.A. and A.J. Lichtenberg. 2008. *Principles of Plasma Discharge and Materials Processing*. (2nd ed.). Hoboken, NJ: Wiley-Interscience.

Liu, R., Q. Jin, G. Tao et al. 2009. LC–MS and UPLC–quadrupole time-of-flight MS for identification of photodegradation products of aflatoxin B1. *Chromatographia* 71(1–2): 107–112.

Los, A., D. Ziuzina and P. Bourke. 2018b. Current and future technologies for microbiological decontamination of cereal grains. *J. Food Sci.* 83(6): 1484–1493.

Lu, X., G.V. Naidis, M. Laroussi et al. 2016. Reactive species in non-equilibrium atmospheric-pressure plasmas: Generation, transport, and biological effects. *Phys. Rep.* 630: 1–84.

Lu, Q., D. Liu, Y. Song et al. 2014. Inactivation of thetomato pathogen Cladosporium fulvumby an atmospheric-pressure cold plasma jet. *Plasma Processes Polym.* 11(11): 1028–1036.

Lukes, P., J.L. Brisset and B.R. Locke. 2012. Biological effects of electrical discharge plasma in water and in gas–liquid environments. In V.I. Parvulescue, M. Magureanu and P. Lukes (Eds.), *Plasma Chemistry and Catalysis in Gases and Liquids* (pp. 309–352). Weinheim, Germany: Wiley-VCH Verlag GmbH & Co. KGaA.

Luo, X., R. Wang, L. Wang et al. 2014. Detoxification of aflatoxin in corn flour by ozone. *J. Sci. Food Agric.* 94(11): 2253–2258.

Mao, J., B. He, L. Zhang et al. 2016. A structure identification and toxicity assessment of the degradation products of aflatoxin B1 in peanut oil under UV irradiation. *Toxins* 8(11): 332.

Marin, S., A.J. Ramos, G. Cano-Sancho et al. 2013. Mycotoxins: Occurrence, toxicology, and exposure assessment. *Food Chem. Toxicol.* 60: 218–237.

Mattsson, J.L. 2007. Mixtures in the real world: The importance of plant self-defense toxicants, mycotoxins, and the human diet. *Toxicol. Appl. Pharmacol.* 223: 125–132.

Min, S.C., S.H. Roh, B.A. Niemira et al. 2017. In-package inhibition of *E. coli* O157:H7 on bulk Romaine lettuce using cold plasma. *Food Microbiol.* 65: 1–6. doi: 10.1016/j.fm.2017.01.010

Min, S. C., S.H. Roh, B.A. Niemira et al. 2018. In-package atmospheric cold plasma treatment of bulk grape tomatoes for microbiological safety and preservation. *Food Res. Int.* 108: 378–386.

Miraglia, M., H.J.P. Marvin, G.A. Kleter et al. 2009. Climate change and food safety: An emerging issue with special focus on Europe. *Food Chem. Toxicol.* 47: 1009–1021.

Mishra, R., S. Bhatia, R. Pal et al. 2016a. Cold plasma: Emerging as the new standard in food safety. *Res. Inv. Int. J. Eng. Sci.* 6(2): 15–20.

Misra, N.N., L. Han, B. Tiwari et al. 2014. Nonthermal plasma technology for decontamination of foods. In I.S. Boziaris (Ed.), *Novel Food Preservation and Microbial Assessment Techniques* (pp. 155–183). Boca Raton, FL: CRC Press.

Misra, N.N., K.M. Keener, P. Bourke et al. 2015. Generation of in-package cold plasma and efficacy assessment using methylene blue. *Plasma Chem. Plasma Process.* 35(6): 1043–1056.

Misra, N.N., S.K. Pankaj, A. Segat et al. 2016c. Cold plasma interactions with enzymes in foods and model systems. *Trends Food Sci. Technol.* 55: 39–47.

Misra, N.N., O. Schlüter and P.J. Cullen. 2016b. *Cold Plasma in Food and Agriculture*. Cambridge, MA: Academic Press.

Misra, N.N., B.K. Tiwari, K.S.M.S. Raghavarao et al. 2011. Nonthermal plasma inactivation of food-borne pathogens. *Food Eng. Rev.* 3: 159–170.

Misra, N.N., B. Yadav, M.S. Roopesh et al. 2019. Cold plasma for effective fungal and mycotoxin control in foods: Mechanisms, inactivation effects, and applications. *Compr. Rev. Food Sci. Food Saf.* 18: 106–120. doi: 10.1111/1541-4337-12398

Modic, M., N.P. McLeod, J.M. Sutton et al. 2017. Cold atmospheric pressure plasma elimination of clinically important single-and mixed-species biofilms. *Int. J. Antimicrob. Agents* 49: 375–378.

Moiseev, T., N.N. Misra, S. Patil et al. 2014. Post-discharge gas composition of a large-gap DBD in humid air by UV-Vis absorption spectroscopy. *Plasma Sources Sci. Technol.* 23(6): 065033.

Mousa, W., F.M. Ghazali, S. Jinap et al. 2013. Modeling growth rate and assessing aflatoxins production by *Aspergillus flavus* as a function of water activity and temperature on polished and brown rice. *J. Food Sci.* 78: M56–M63.

Murata, H., M. Mitsumatsu and N. Shimada. 2008. Reduction of feed-contaminating mycotoxins by ultraviolet irradiation: An in vitro study. *Food Addit. Contam., Part A* 25(9): 1107–1110.

Murphy, P.A., S. Hendrich, C. Landgren et al. 2006. Food mycotoxins: An update. *J. Food Sci.* 71: 51–65.

Naitali, M., J.M. Herry, E. Hnatiuc et al. 2012. Kinetics and bacterial inactivation induced by peroxynitrite in electric discharges in air. *Plasma Chem. Plasma Process* 32: 675–692.

Naitali, M., G. Kamgang-Youbi, J.M. Herry et al. 2010. Combined effects of long-living chemical species during microbial inactivation using atmospheric plasma-treated water. *Appl. Environ. Microbiol.* 76: 7662–7664.

Narasaiah, K.V., R. Sashidhar and C. Subramanyam. 2006. Biochemical analysis of oxidative stress in the production of aflatoxin and its precursor intermediates. *Mycopathologia* 162: 179–189.

Nellemann, C. 2009. *The Environmental Food Crisis: The Environment's Role in Averting Future Food Crises: A Unep Rapid Response Assessment*. Nairobi, Kenya: UNEP/Earthprint.

Ni, Y., M.J. Lynch, M. Modic et al. 2016. A solar powered handheld plasma source for microbial decontamination applications. *J. Phys. D Appl. Phys.* 49: 355203.

Niemira, B. A. 2012. Cold plasma decontamination of foods. *Annu. Rev. Food Sci. Technol.* 3(1): 125–142.

Ouf, S.A., A.H. Basher and A.A. Mohamed. 2015. Inhibitory effect of double atmospheric pressure argon cold plasma on spores and mycotoxin production of *Aspergillus niger* contaminating date palm fruits. *J. Sci. Food Agric.* 95(15): 3204–3210.

Pahwa, A. and H. Kumar. 2018. Influence of cold plasma technology on food packaging material: A review. *Int. J. Chem. Studies* 6(2): 594–603.

Pal, P., P. Kaur, N. Singh et al. 2016. Effect of nonthermal plasma on physico-chemical, amino acid composition, pasting and protein characteristics of short and long grain rice flour. *Food Res. Int.* 81: 50–57.

Pankaj, S. K., H. Shi and K.M. Keener. 2018a. A review of novel physical and chemical decontamination technologies for aflatoxin in food. *Trends Food Sci. Technol.* 71: 73–83.

Pankaj, S.K., C. Bueno-Ferrer, N.N. Misra et al. 2014a. Application of cold plasma technology in food packaging. *Trends Food Sci. Technol.* 35: 5–17.

Pankaj, S.K., C. Bueno-Ferrer, N.N. Misra et al. 2014b. Zein film: Effects of dielectric barrier discharge atmospheric cold plasma. *J. Appl. Polym. Sci.* 131(18): 40801–40806.

Pankaj, S.K., N.N. Misra and P.J. Cullen. 2013. Kinetics of tomato peroxidase inactivation by atmospheric cold plasma based on dielectric barrier discharge. *Innov. Food Sci. Emerg. Technol.* 19: 153–157. doi: 10.1016/j.ifset.2013.03.001

Pankaj, S.K., Z. Wan and K.M. Keener. 2018b. Effect of cold Plasma on food quality: A review. *Foods* 7: 4. doi: 103390/foods7010004

Panngom, K., S.H. Lee, D.H. Park et al. 2014. Non-thermal plasma treatment diminishes fungal viability and up-regulates resistance genes in a plant host. *PLoS One* 9(6): e99300.

Park, D., L. Lee, R. Price et al. 1988. Review of the decontamination of aflatoxins by ammoniation: Current status and regulation. *J. Assoc. Off. Anal. Chem.* 71(4): 685–703.

Park, B.J., K. Takatori, Y. Sugita-Konishi et al. 2007. Degradation of mycotoxins using microwave-induced argon plasma at atmospheric pressure. *Surf. Coat. Technol.* 201(9–11): 5733–5737.

Patil, S., T. Moiseev, N.N. Misra et al. 2014. Influence of high voltage atmospheric cold plasma process parameters and role of relative humidity on inactivation of *Bacillus atrophaeus* spores inside a sealed package. *J. Hosp. Infect.* 88(3): 162–169.

Pignata, C., D. d'Angelo, D. Basso et al. 2014. Low-temperature, low-pressure gas plasma application on *Aspergillus brasiliensis, Escherichia coli* and pistachios. *J. Appl. Microbiol.* 116(5): 1137–1148.

Puligundla, P., J.W. Kim and C. Mok. 2017. Effect of corona discharge plasma jet treatment on decontamination and sprouting of rapeseed (*Brassica napus* L.) seeds. *Food Control* 71: 376–382.

Puligundla, P. and C. Mok. 2016. Non-thermal plasmas (NTPs) for inactivation of viruses in abiotic environment. *Res. J. Biotechnol.* 11(6): 91–96.

Puligundla, P. and C. Mok. 2017. Potential applications of nonthermal plasmas against biofilm-associated micro-organisms in vitro. *J. Appl. Microbiol.* 122(5): 1134–1148.

Raviteja, T., S.K. Dayam and J. Yashwanth. 2019. A study on cold plasma for food preservation. *J. Sci. Res. Rep.* 23(4): 1–14. doi: 10.9743/JSRR/2019/v23i430126

Saharan, M.S., J. Kumar, A.K. Sharma et al. 2004. Fusarium head blight (FHB) or head scab of wheat—A review. *Proc. Ind. Nat. Sci. Acad., Part B, Biol. Sci.* 70: 255–268.

Sahu, B. B., J. G. Han and H. Kersten. 2017. Shaping thin film growth and microstructure pathways via plasma and deposition energy: A detailed theoretical, computational and experimental analysis. *Phys. Chem. Chem. Phys.* 19(7): 5591–5610.

Sakila Banu M., P. Sasikala, A. Dhanapal et al. 2012. Cold plasma as a novel food processing technology. *Int. J. Emerg. Trends Eng. Dev.* 4(2): 803–818.

Sakiyama, Y., D.B. Graves, H.W. Chang et al. 2012. Plasma chemistry model of surface microdischarge in humid air and dynamics of reactive neutral species. *J. Phys. D Appl. Phys.* 45: 425201.

Sarangapani, C., G. O'Toole, P.J. Cullen et al. 2017. Atmospheric cold plasma dissipation efficiency of agrochemicals on blueberries. *Innov. Food Sci. Emerg. Technol.* 44: 235–241.

Schmidt, M., E. Zannini, K.M. Lynch et al. 2018. Novel approaches for chemical and microbiological shelf life extension of cereal crops. *Crit. Rev. Food Sci. Nutr.* 59(21): 3395–3419.

Schnabel, U., R. Niquet, O.Schlüter et al. 2015. Decontamination and sensory properties of microbiologically contaminated fresh fruits and vegetables by microwave plasma processed air (PPA). *J. Food Process. Preserv.* 39(6): 653–662.

Segat, A., N.N. Misra, A. Fabbro et al. 2014. Effects of ozone processing on chemical, structural and functional properties of whey protein isolate. *Food Res. Intern.* 66: 365–372.

Shi, H., B. Cooper, R.L. Stroshine et al. 2017. Structures of degradation products and degradation pathways of aflatoxin B1 by high-voltage atmospheric cold plasma (HVACP) treatment. *J. Agric. Food Chem.* 65(30): 6222–6230.

Shi, H., K. Ileleji, R.L. Stroshine et al. 2017. Reduction of aflatoxin in corn by high voltage atmospheric cold plasma. *Food Bioprocess Technol.* 10(6): 1042–1052.

Siciliano, I., D. Spadaro, A. Prelle et al. 2016. Use of cold atmospheric plasma to detoxify hazelnuts from aflatoxins. *Toxins (Basel)* 8(5): 125.

Smet, C., M. Baka, L. Steen et al. 2018. Combined effect of cold atmospheric plasma, intrinsic and extrinsic factors on the microbial behavior in/on (food) model systems during storage. *Innov. Food Sci. Emerg. Technol.* 48: 296.

Smet, C., E. Noriega, F. Rosier et al. 2016. Influence of food intrinsic factors on the inactivation efficacy of cold atmospheric plasma: Impact of osmotic stress, suboptimal pH and food structure. *Innov. Food Sci. Emerg. Technol.* 38, 393–406.

Stoev, S.D. 2013. Food safety and increasing hazard of mycotoxin occurrence in foods and feeds. *Crit. Rev. Food Sci. Nutr.* 53: 887–901.

Streit, E., K. Naehrer, I. Rodrigues et al. 2013. Mycotoxin occurrence in feed and feed raw materials worldwide: Long term analysis with special focus on Europe and Asia. *J. Sci. Food Agric.* 93(12): 2892–2899.

Suhem, K., N. Matan, M. Nisoa et al. 2013. Inhibition of *Aspergillus flavus* on agar media and brown rice cereal bars using cold atmospheric plasma treatment. *Int. J. Food Microbiol.* 161(2): 107–111.

Surowsky, B., O. Schlüter and D. Knorr. 2014. Interactions of non-thermal atmospheric pressure plasma with solid and liquid food systems: A review. *Food Eng. Rev.* 7(2): 82–108.

Syamaladevi, R.M., A. Adhikari, S.L. Lupien et al. 2015. Ultraviolet-C light inactivation of *Penicillium expansum* on fruit surfaces. *Food Control* 50: 297–303.
Syamaladevi, R.M., S.L. Lupien, K. Bhunia et al. 2014. UV-C light inactivation kinetics of *Penicillium expansum* on pear surfaces: Influence on physicochemical and sensory quality during storage. *Postharvest Biol. Technol.* 87: 27–32.
Sysolyatina, E., A. Mukhachev, M. Yurova et al. 2014. Role of the charged particles in bacteria inactivation by plasma of a positive and negative corona in ambient air. *Plasma Process. Polym.* 11(4): 315–334.
Syamaladevi, R.M., J. Tang and Q. Zhong. 2016. Water diffusion from a bacterial cell in low-moisture foods. *J. Food Sci.* 81(9): R2129–R2134.
Temba, B.A., Y. Sultanbawa, D.J. Kriticos et al. 2016. Tools for defusing a major global food and feed safety risk: Non biological postharvest procedures to decontaminate mycotoxins in foods and feeds. *J. Agric. Food Chem.* 64(47): 8959–8972.
ten Bosch, L., K. Pfohl, G. Avramidis et al. 2017. Plasma-based degradation of mycotoxins produced by *Fusarium, Aspergillus* and *Alternaria* species. *Toxins* 9: 97.
Thirumdas, R., C. Sarangpani and U.S. Annapure. 2014. Cold plasma: A novel non-thermal technology for food processing. *Food Biophys.* 10: 1–14. doi: 10.1007/s11483-014-9382-z
Turner, M. 2016. Physics of cold plasma. In N.N. Misra et al. (Eds.), *Cold Plasma in Food and Agriculture* (pp.17–51). San Diego, CA: Academic Press.
Udomkun, P., A. N. Wiredu, M. Nagle et al. 2017. Innovative technologies to manage aflatoxins in foods and feeds and the profitability of application—A review. *Food Control* 76: 127–138.
van de Veen, H.B., H. Xie, E. Esveld et al. 2015. Inactivation of chemical and heat- resistant spores of *Bacillus* and *Geobacillus* by nitrogen cold atmospheric plasma evokes distinct changes in morphology and integrity of spores. *Food Microbiol.* 45: 26–33.
Vesel, A. and M. Mozetic. 2017. New developments in surface functionalization of polymers using controlled plasma treatments. *J. Phys. D, Appl. Phys.* 50(29): 293001.
Wang, S.-Q., G.-Q. Huang, Y.-P. Li et al. 2015. Degradation of aflatoxin B1 by low-temperature radio frequency plasma and degradation product elucidation. *Eur. Food Res. Technol.* 241(1): 103–113.
Won, M.Y., S.J. Lee and S.C. Min. 2017. Mandarin preservation by microwave-powered cold plasma treatment. *Innov. Food Sci. Emerg. Technol.* 39: 25–32.
Zhang, L., Z. Lu, F. Lu et al. 2006. Effect of γ irradiation on quality- maintaining of fresh cut lettuce. *Food Control* 17: 225–228.
Zhong, K., J. Wu, Z. Wang et al. 2007. Inactivation kinetics and secondary structural changes of PEF-treated pOD and PPO. *Food Chem.* 100: 115–123.
Ziuizina, D., S. Patil, P.J. Cullen et al. 2014. Atmospheric cold plasma inactivation of *Escherichia coli, Salmonella entrica serovar Typhimurium and Listeria monocytogenes* inoculated on fresh produce. *Food Microbiol.* 42: 109–116. doi: 10.1016/j.fm.2014.02.007

Index

D

E

H

I

J

K

L

M

Q

R

S

T

Printed in the United States
By Bookmasters